AIRCRAFT SYSTEMS

FOR PILOTS

DALE DE REMER, PhD

JEPPESEN.
Sanderson Training Products

JS312686B

Table of Contents

Introduction

This is not a book of new knowledge. It is, rather, an arrangement of existing knowledge from many sources into a concise presentation of what pilots should know about basic aircraft systems, based on my experiences over thirty-five years of flying and fourteen years of teaching Aircraft Systems to university students.

This text includes a brief study of the fundamentals of physical matter (from which airplanes are made) and mechanics (how airplane parts act and react) and sufficient study of each type of system which, when understood, will allow the professional pilot to stay abreast of the learning which must occur as the pilot advances into management of more and more complex aircraft.

How To Use This Textbook

Bold words have been chosen to call the reader's attention to their importance. When you read a bold word, be alert because the word will be defined or described within the sentence OR it has been so defined earlier OR shortly will be defined.

Bold Words:

1. Are about to be defined or described—be alert for this.

2. Are important to the pilot's knowledge base—learn them!

3. Are very important to the meaning of the sentence. If they are not defined, they are considered to be common knowledge. If you don't know the word and can't find it defined nearby in the text, seek its meaning in the classroom or from other books.

4. If you run across a word you don't know that isn't bold, stop and consult your dictionary. It is the only way you will be able to fully understand the sentence you have just read and besides, this is how we all build our vocabulary.

Insofar as possible, publishing standards (abbreviations, etc.) follow those of the publications manual of the APA as it is a standard that is widely accepted by aviation programs in higher education.

This book is designed to permit the student to self-learn. To utilize your instructor most efficiently, learn the assignment BEFORE coming to class. Use the study questions, as one method, to see how well you have understood what you have studied. Make notes or questions on what you don't understand so you can get further explanation and clarification from your instructor. This way, the instructor won't need to use valuable class time to teach you what you can learn on your own. In the class time saved, your instructor can take you "beyond the book." By this means, even more can be learned in the time available. Remember—it's YOUR class time!

For the pilot who is not involved in a formal study program, welcome to another opportunity to have fun—learning more about airplanes! I hope you enjoy the adventure.

Acknowledgements

The author wishes to thank the administration, faculty and staff of the Center for Aerospace Sciences for providing the environment and encouragement during the long process of researching and writing this book. Special thanks go to Mr. Greg Wagner, Assistant Professor of Aviation, for his assistance in proofing many of the chapters; Dr. Duane Cole, Professor of Physics, UND, for his assistance with Chapter One; David Blumkin for his critique of Chapter Nine; Mr. Mike Miller, IA for his proofreading efforts; and Mr. Jeff Boerboon for his contribution about preflight inspection in Chapter Sixteen. Also, many thanks to Dale Hurst and the staff at IAP, Inc., as well as Becky See, for the skill and professionalism that made completion of this book a very pleasant experience.

Chapter I
Physics

General Characteristics Of Matter

Physics is the term applied to that area of knowledge regarding the basic and fundamental nature of matter and energy. It does not attempt to determine why matter and energy behave as they do in their relation to physical phenomena, but rather how they behave.

The persons who fly, maintain and repair aircraft should have a knowledge of basic physics in order to be able to understand the interactions of matter and energy.

This may be a review for those who have a background in physics. I suggest you read it anyway. You may find the aircraft applications interesting!

Matter

Although matter is the most basic of all things related to the field of physics and the material world, it is the hardest to define. Since it cannot be rigidly defined, this chapter will point out those characteristics which are easily recognizable.

Matter itself cannot be destroyed, but it can be changed from one state into another state by chemical or physical means. Matter is often considered in terms of the energy it contains, absorbs, or gives off. Under certain controlled conditions, it can be made to aid man in the process of flight.

Matter is any substance that occupies space and has mass. There are **four states** of matter: (1) **Solids**, (2) **liquids**, (3) **gases**, and (4) **plasma**. **Solids** have a definite volume and a definite shape; **liquids** have a definite volume, but they take the shape of the containing vessel; gases have neither a definite volume nor a definite shape. Gases not only take the shape of the containing vessel, but they expand and fill the vessel, no matter what its volume. **Plasma** is made up of very hot, ionized gases. The gases are so hot that thermal collisions dissociate all of the atoms into positive ions and electrons. Most of the matter in the universe is plasma. The Sun and all the stars are giant balls of plasma. About 99% of the total mass of the universe is in this plasma state.

Water is a good example of matter changing from one state to another. At high temperature it is in the gaseous state known as steam. At moderate temperatures it is a liquid, and at low temperatures it becomes ice, a solid state. In this example, the temperature is the dominant factor in determining the state that the substance assumes. **Pressure** is another important factor that will effect changes in the state of matter. At pressures lower than atmospheric, water will boil and thus change into steam at temperatures lower than 212° F (100° C). For example, the vapor pressure of water at 98.6° F (37° C) is equal to atmospheric pressure at about 63,000 feet. This means that blood will boil at that pressure altitude! Pressure is a critical factor in changing some gases to liquids or solids. Normally, when pressure and chilling are both applied to a gas, it assumes a liquid state. Liquid air, which is a mixture of oxygen and nitrogen, is produced in this manner.

All matter has certain characteristics or general properties. These properties are defined elementally and broadly at this point, and more specifically in applications throughout the text. Among these properties and relationships are:

a. **Volume**—meaning to occupy space; having some measurements such as length, width, and height. It may be measured in cubic inches, cubic centimeters, liters, or the like.

b. **Inertia** is the characteristic of matter that resists change in motion (velocity and direction). Newton's first law is sometimes called the law of inertia: "A body at rest will remain at rest and a body in motion will continue in motion with a constant speed along a straight line path (constant velocity) unless acted upon by some net force".

c. **Mass** is a measure of the inertia of a body, and therefore is a measure of the quantity of matter associated with the body. Units for measuring mass are usually considered fundamental units for a measurement system. The gram and kilogram are units for measuring mass in the metric system of measurement and the corresponding English unit for measuring mass is the less familiar unit called the **slug.**

	METRIC SYSTEM	ENGLISH SYSTEM	EQUIVALENTS
LENGTH (DISTANCE)	**METER** 1 CENTIMETER = 10 MILLIMETERS 1 DECIMETER = 10 CENTIMETERS 1 METER = 100 CENTIMETERS 1 KILOMETER = 1000 METERS	**FOOT** 1 FOOT = 12 INCHES 1 YARD = 3 FEET 1 STATUTE MILE = 5,280 FEET 1 NAUTICAL MILE = 6,080.27 FEET	1 INCH = 2.54 CENTIMETERS 1 FOOT = 30.5 CENTIMETERS 1 METER = 39.37 INCHES 1 KILOMETER = 0.62 MILE (ST.)
WEIGHT (MASS)	**GRAM** 1 GRAM = 1000 MILLIGRAMS 1 KILOGRAM = 1000 GRAMS	**POUND** 1 POUND = 16 OUNCES 1 TON = 2,000 POUNDS	1 POUND = 453.6 GRAMS 1 KILOGRAM = 2.2 POUNDS
VOLUME	**LITER** 1 LITER = 1000 MILLILITERS 1 MILLILITER = 1 CUBIC CENTIMETER	**GALLON** 1 GALLON = 4 QUARTS 1 QUART = 2 PINTS	1 LITER = .26417 GALLONS (U.S.) 1 LITER = .21998 GALLONS (BR.)
TIME	**SECOND** SAME AS FOR ENGLISH SYSTEM	**SECOND** 1 SECOND = $\dfrac{1}{86,400}$ of average solar day.	TIME SAME FOR BOTH SYSTEMS

Figure 1-1. Comparison of metric and English systems of measurement.

d. **Gravitation**, sometimes called **mass attraction**, is a **force** that results from the characteristic of particles of matter that causes attraction or pull on other particles of matter. This mutual attraction of the mass of particles can be described in terms of gravitational forces with Newton's law of universal gravitation. Gravitational forces, like all other forces (pulls and pushes), are measured with the English unit of the **pound** or the metric unit of the **Newton**.

e. **Weight** is the name commonly used for the gravitational force of attraction betweeen the Earth and a body (mass) near the Earth. Weight is a force and is described with force units such as the pound or Newton. Since the weight of a body (gravitational force acting on the mass of the body) is proportional to the mass of the body, these different physical quantities, weight and mass, are sometimes confused.

f. **Density** is a quantity which is useful when describing matter, especially when in the liquid or gaseous state. Depending upon the application, density can be defined as either **weight density** (weight per unit volume) or as **mass density** (mass per unit volume).

Systems Of Measurement

The two most commonly used systems of measurement are the English system, which is still in general use in the United States, and the metric system, used in most European countries and now adopted by the Armed Forces of the United States.

The metric system is normally used in all scientific applications.

The metric system is sometimes called the **cgs** system because it uses as basic measuring units, the centimeter (c) to measure length, the gram (g) to measure mass, and the second (s) to measure time. The metric system is also referred to as the **mks** system (meter, kilogram, second).

The English system uses different units for the measurement of mass and length. The pound is the unit of weight; the foot and inch are used to measure length. The second is used to measure time as in the metric system.

The units of one system can be converted to units in the other system by using a conversion factor or by referring to a chart similar to that shown in figure 1-1. In this figure the English and the metric systems are compared; in addition, a column of equivalents is included which can be used to convert units from one system to the other.

Fluids

Because both liquids and gases flow freely, they are called fluids, from the Latin word "fluidus," meaning to flow. A **fluid** is defined as a substance which changes its shape easily and takes the shape of its container. This applies to both liquids and gases. The characteristics of liquids and gases may be grouped under similarities and differences.

Similar characteristics are as follows:

1. Each has no definite shape but conforms to the shape of the container.

2. Both readily transmit pressures.

Differential characteristics are as follows:

1. Gases fill their containers completely, but liquids may not.

2. Gases are lighter than equal volumes of liquids.

3. Gases are highly compressible, but liquids are essentially not compressible.

These differences are described in the appropriate areas of the following discussion concerning the properties and characteristics of fluids at rest. Also included are some of the factors which affect fluids in different situations.

Machines

General

Ordinarily, a machine is thought of as a complex device, such as an internal-combustion engine or a typewriter. These are machines, but so is a hammer, a screwdriver, or a wheel. A **machine** is any device with which **work** may be accomplished. Machines are used to transform energy, as in the case of a generator transforming mechanical energy into electrical energy. Machines are used to transfer energy from one place to another, as in the examples of the connecting rods, crankshaft, and reduction gears transferring energy from an aircraft's cylinder to its propeller.

A main purpose of machines is to multiply force; for example, a system of pulleys may be used to lift a heavy load. The pulley system enables the load to be raised by exerting a force which is smaller than the weight of the load.

Machines are also used to multiply speed. A good example is the bicycle, by which speed can be gained by exerting a greater force.

Finally, machines can be used to change the direction of a force. An example of this use is the flag hoist. A downward force on one side of the rope exerts an upward force on the other side, raising the flag toward the top of the pole.

There are only six simple machines. They are the lever, the pulley, the wheel and axle, the inclined plane, the screw, and the gear. However, physicists recognize only two basic principles in machines; namely, the lever and the inclined plane. The wheel and axle, the block and tackle, and gears may be considered levers. The wedge and the screw use the principle of the inclined plane.

An understanding of the principles of simple machines provides a necessary foundation for the study of compound machines, which are combinations of two or more simple machines.

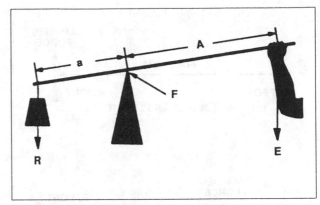

Figure 1-2. A simple lever.

The Lever

The simplest machine, and perhaps the most familiar one, is the lever. A seesaw is a familiar example of a lever in which one weight balances the other.

There are three basic parts in all levers; namely, the fulcrum "F," a force or effort "E," and a resistance "R." Shown in figure 1-2 are the pivotal point "F" (fulcrum); the effort "E," which is applied at a distance "A" from the fulcrum; and a resistance "R," which acts at a distance "a" from the fulcrum. Distances "A" and "a" are the lever arms.

Classes of Levers

The three classes of levers are illustrated in figure 1-3. The location of the fulcrum (the fixed or pivot point) with relation to the resistance (or weight) and the effort determines the lever class.

First-Class Levers

In the first-class lever (A of figure 1-3), the fulcrum is located between the effort and the resistance. As mentioned earlier, the seesaw is a good example of the first-class lever. The amount of weight and the distance from the fulcrum can be varied to suit the need.

Second-Class Levers

The second-class lever (B of figure 1-3) has the fulcrum at one end; the effort is applied at the other end. The resistance is somewhere between these points. The wheelbarrow is a good example of a second-class lever.

Both first- and second-class levers are commonly used to help in overcoming big resistances with a relatively small effort.

Third-Class Levers

There are occasions when it is desirable to speed up the movement of the resistance even though a large amount of effort must be used. Levers that

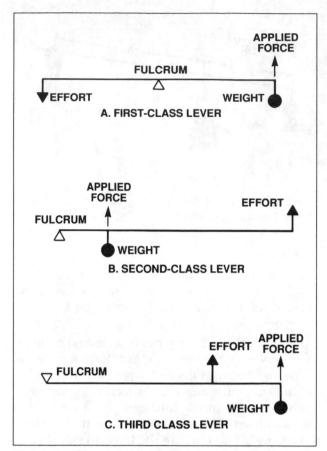

Figure 1-3. Three classes of levers.

help accomplish this are third-class levers. As shown in C of figure 1-3, the fulcrum is at one end of the lever and the weight or resistance to be overcome is at the other end, with the effort applied at some point between. Third-class levers are easily recognized because the effort is applied between the fulcrum and the resistance.

This relationship can be stated in general terms: The length of the effort arm is the same number of times greater than the length of the resistance arm as the resistance to be overcome is greater than the effort that must be applied.

The mathematical equation for this relationship is:

$$\frac{A}{a} = \frac{R}{E}$$

where:

A = Length of effort arm.

a = Length of resistance arm.

R = Resistance weight or force.

E = Effort force.

Remember that all distances must be in the same units, and all forces must also be in the same units.

Mechanical Advantage Of Levers

Levers may provide mechanical advantages, since they can be applied in such manner that they can magnify an applied force. This is true of first- and second-class levers. The third-class lever provides what is called a fractional disadvantage, i.e., one in which a greater force is required than the force of the load lifted.

Mechanical advantage machines are used throughout the aircraft to help the pilot or a motor or hydraulic or pneumatic system to accomplish a task where work is involved.

Work, Energy, And Power

Work

The study of machines, both simple and complex, is in one sense a study of the energy of mechanical work. This is true because all **machines** transfer input energy, or the work done on the machine to output energy, or the work done by the machine.

Work is done when a resistance is overcome by a force acting through a measurable distance. Two factors are involved: (1) **Force** and (2) movement through a **distance**. As an example, suppose a small aircraft is stuck in the snow. Two men push against it for a period of time, but the aircraft does not move. According to the technical definition, no **work** was done in pushing against the aircraft. By definition, work is accomplished only when an object is displaced some distance against a resistive force.

In equation form, this relationship is,

Work = Force (F) × distance (d).

The physicist defines **work** as "work is force times displacement. Work done by a force acting on a body is equal to the magnitude of the force multiplied by the distance through which the force acts."

In the metric system, the unit of work is the *joule*, where one joule is the amount of work done by a force of one newton when it acts through a distance of one meter. That is,

1 joule = 1 newton-m

Hence we can write the definition in the form

$$W \text{ (joules)} = F \text{ (newtons)} \times d \text{ (meters)}$$

If we push a box for 8 m across a floor with a force of 100 newtons, the work we perform is

$$W = Fd = 100 \text{ newtons} \times 8 \text{ m} = 800 \text{ joules}$$

How much work is done in raising a 500-kg (kilogram) elevator cab from the ground floor of a building to its tenth floor, 30 m (meters) higher? We note that the force needed is equal to the weight of the cab.

In the metric system, mass rather than weight is normally specified. To find the weight in **newtons** (the metric unit of force) of something whose mass in kilograms is known, the weight or gravitational force $F = mg$ is used with $g = 9.8 \text{ m/sec}^2$.

$$F \text{ (newtons)} = m \text{ (kilograms)} \times g \text{ (9.8 m/sec}^2)$$

and

$$W \text{ (joules)} = m \text{ (kilograms)} \times g \text{ (9.8 m/sec}^2) \times d \text{ (meters)}$$

$$W = Fd = mgd = 500 \text{ kg} \times 9.8 \text{ m/sec}^2 \times 30 \text{m}$$
$$W = 147,000 \text{ joules}$$
$$W = 1.47 \times 10^5 \text{ joules}$$

Force Parallel To Displacement

If force is expressed in pounds and distances in feet, work will be expressed in foot-pounds (ft-lbs).

Example:
How much work is accomplished in lifting a 40-pound weight to a vertical height of 25 feet?

$$W = Fd$$
$$W = 40 \text{ lb} \times 25 \text{ ft}$$
$$W = 1,000 \text{ ft-lb}$$

Example:
How much work is accomplished in pushing a small aircraft into a hangar a distance of 115 feet if a force of 75 pounds is required to keep it moving?

$$W = Fd$$
$$W = 75 \text{ lb} \times 115 \text{ ft}$$
$$W = 8,625 \text{ ft-lb}$$

Force Not Parallel To Displacement

In the equation above, F is assumed to be in the same direction as d. If it is not, for example the

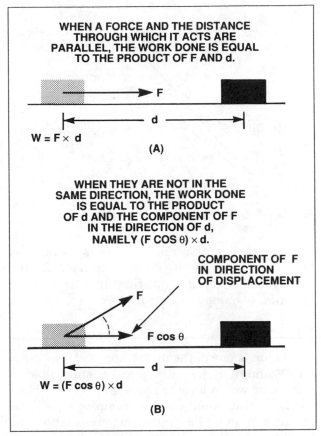

Figure 1-4. Direction of work.

case of a body pulling a wagon with a rope not parallel to the ground, we must use F for the component of the applied force that acts in the direction of the motion, figure 1-4(B).

The component of a force in the direction of a displacement d is:

$$F \cos \theta$$

where θ is the angle between F and d. Hence the most general equation for work is

$$\textbf{W = Fd cos } \boldsymbol{\theta}$$

When F and d are parallel, $\theta = 0°$ and $\cos \theta = 1$, so that Fd cos θ reduces to just Fd. When F and d are perpendicular, $\theta = 90°$ and $\cos \theta = 0$, so that no work is done. A force that is perpendicular to the motion of an object can do no work upon it. Thus gravity, which results in a downward force on everything near the earth, does no work on objects moving horizontally along the earth's surface However, if we drop an object, as it falls to the ground work is definitely done upon it.

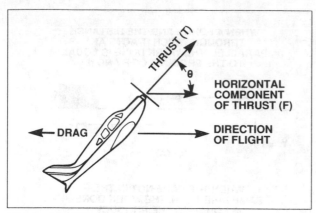

Figure 1-5. Direction of work in level flight.

In the case of an aircraft in level slow-flight at 60 KTS true airspeed, F must be used for the component of the force (thrust) that acts in the direction of motion. See figure 1-5.

Friction

Friction is one of the most important aspects of life. Without friction it would be impossible to walk. One would have to shove oneself from place to place, and would have to bump against some obstacle to stop at a destination. Yet friction is a liability as well as an asset, and requires consideration when dealing with any moving mechanism.

In experiments relating to friction, measurement of the applied forces reveals that there are three kinds of friction. One force is required to start a body moving, while another is required to keep the body moving at constant speed. Also, after a body is once in motion, a definitely larger force is required to keep it sliding than to keep it rolling.

Thus, the three kinds of friction may be classified as: (1) Starting (static) friction, (2) sliding friction, and (3) rolling friction.

Static Friction

When an attempt is made to slide a heavy object along a surface, the object must first be broken loose or started. Once in motion, it slides more easily. The "breaking loose" force is proportional to the weight of the body. The force necessary to start the body moving slowly is designated F, and F is the force pressing the body against the surface (usually its weight). Since the nature of the surfaces rubbing against each other is important,

they must be considered. The nature of the surfaces is indicated by the coefficient of starting friction which is designated by the letter "k." This coefficient can be established for various materials and is often published in tabular form. Thus, when the load (weight of the object) is known, starting friction can be calculated by using the equation,

$$F = kF'$$

For example, if the coefficient of static friction of a smooth iron block on a smooth, horizontal surface is 0.3, the force required to start a 10-pound block would be 3 pounds; a 40-pound block, 12 pounds.

Starting friction for objects equipped with wheels and roller bearings is much smaller than that for sliding objects. Nevertheless, a locomotive would have difficulty getting a long train of cars in motion all at one time. Therefore, the couples between the cars are purposely made to have a few inches of play. When the engineer is about to start the train, he backs the engine until all the cars are pushed together. Then, with a quick start forward the first car is set in motion. This technique is employed to overcome the static friction of each wheel (as well as the inertia of each car). It would be impossible for the engine to start all of the cars at the same instant, for static friction, which is the resistance of being set in motion, would be greater than the force exerted by the engine. Once the cars are in motion, however, static friction is greatly reduced and a smaller force is required to keep the train in motion than was required to start it.

Dynamic Or Sliding Friction

Sliding friction, sometimes called **kinetic friction** or **dynamic friction**, is the resistance to motion offered by an object sliding over a surface. It pertains to friction produced after the object has once been set into motion, and is always less than starting friction. The amount of sliding resistance is dependent on the nature of the surface of the object, the surface over which it slides, and the normal force between the object and the surface. This resistive force may be computed by the formula,

$$F = \mu N$$

where: "F" is the resistive force due to friction expressed in pounds; "N" is the force exerted on or by the object perpendicular (normal) to the surface over which it slides; and "μ" (**mu**) is the coefficient of dynamic friction. (On a horizontal surface, N is equal to the weight of the object in pounds.) The area of the sliding object exposed to the sliding surface has no effect on the results. A block of wood, for example, will not slide any easier on one of the broad sides than it will on a narrow side, (assuming all sides have the same smoothness). Therefore, area does not enter into the equation above.

The coefficient of static friction will always be significantly higher than the coefficient of dynamic friction if other factors don't change. What application does this knowledge have for the pilot? Consider this: During landing deceleration, will there be more braking (stopping) ability with the brakes locked or with maximum braking but wheels turning? That's why anti-skid brake systems have been developed!

Rolling Friction

Resistance to motion is greatly reduced if an object is mounted on wheels or rollers. The force of friction for objects mounted on wheels or rollers is called rolling friction. This force may be computed by the same equation used in computing sliding friction, but the values of μ will be much smaller. For example, μ for rubber tires on concrete or macadam is about .02. The value of μ for roller bearings is very small, usually ranging from .001 to .003 and is often disregarded.

Example:
An aircraft with a gross weight of 79,600 lb. is towed over a level concrete ramp. What force must be exerted by the towing vehicle to keep the airplane rolling after once set in motion?

$$F = \mu N$$

$$F = .02 \times 79,600 = 1,592 \text{ lb}$$

Power

Power is a badly abused term. In speaking of power-driven equipment, people often confuse the term "power" with the ability to move heavy loads. This is not the meaning of power. A sewing machine motor is powerful enough to rotate an aircraft engine propeller providing it is connected to the crankshaft through a suitable mechanism. It could not rotate the propeller at 2,000 RPM, however, for it is not powerful enough to move a large load at a high speed. **Power**, thus, means rate

of doing work. It is measured in terms of work accomplished per unit of time. In equation form, it reads:

$$\text{Power} = \frac{\text{Work}}{\text{Time}}$$

$$\text{Power} = \frac{\text{Force} \times \text{Distance}}{\text{time}}$$

or,

$$P = \frac{Fd}{t}$$

If force is expressed in pounds, distance in feet, and time in seconds, then power is given in ft-lbs/sec (foot-pounds per second). Time may also be given in minutes. If time in minutes is used in this equation, then power will be expressed in ft-lbs/min.

$$\text{Power} = \frac{\text{pounds} \times \text{feet}}{\text{seconds}} = \text{ft–lbs/sec}$$

or,

$$\text{Power} = \frac{\text{pounds} \times \text{feet}}{\text{minutes}} = \text{ft–lbs/min}$$

Example:
An aircraft engine weighing 3,500 pounds was hoisted a vertical height of 7 feet in order to install it on an aircraft. The hoist was hand-powered and required 3 minutes of cranking to raise the engine. How much power was developed by the man cranking the hoist? (Neglect friction in the hoist.)

$$\text{Power} = \frac{Fd}{t}$$

$$P = \frac{3,500 \text{ pounds} \times 7 \text{ feet}}{3 \text{ minutes}}$$

$$P = 8,167 \text{ ft–lbs/min.}$$

Power is often expressed in units of horsepower. One **horsepower** is equal to 550 ft-lbs/sec or 33,000 ft-lbs/min.

$$1 \text{ HP} = 550 \text{ ft-lbs/sec}$$

Example:

In the hoist example above, calculate the horsepower developed by the man.

$$P = 8{,}167 \frac{\text{ft–lbs}}{\text{min}} \times \left(\frac{1 \text{ HP}}{33{,}000 \frac{\text{ft–lbs}}{\text{min}}} \right)$$

$$P = 0.247 \text{ HP}$$

Power is rate of doing work:

$$P = \frac{W}{t}$$

In the metric system the unit of power is the *watt*, where

$$1 \text{ watt} = 1 \text{ joule/sec}$$

The **watt** is the metric unit of power, thus a motor with a power output of 5,000 watts is capable of doing 5,000 joules of work per second.

A *kilowatt* (kw) is equal to 1,000 watts. Hence the above motor has a power output of 5 kw. For conversion, one HP is equal to 746 watts.

$$1 \text{ hp} = 746 \text{ watts}$$

How much time does the elevator cab weighing 500 kg need to ascend 30 meters if it is being lifted by a 5 kw motor? We rewrite $P = W/t$ in the form

$$t = \frac{W}{P}$$

and then substitute $W = 1.47 \times 10^5$ joules and $P = 5 \times 10^3$ watts to find that

$$t = \frac{W}{P} = \frac{1.47 \times 10^5 \text{ joules}}{5 \times 10^3 \text{ watts}} = 29.4 \text{ sec}$$

Energy

In many cases when work is done on an object, something is given to the object which it retains and which later enables it to do work. When a weight is lifted to a certain height such as in the case of a trip-hammer, or when a clock spring is wound, the object acquires, through having work done on it, the ability to do work itself. In storage batteries and gasoline, energy is stored which can be used later to do work. Energy stored in coal or food can be used to do work. Thus, **energy** can be defined as the ability to do work. The same units, ft-lbs or Joules, are used to describe both work and energy.

In general, a change in energy is equal to the work done; the loss in energy of a body may be measured by the work it does, or the gain in energy of a body may be measured by the amount of work done on it. Energy which bodies possess is classified into two categories: (1) potential and (2) kinetic.

Potential energy may be classified into three groups: (1) that due to position, (2) that due to distortion of an elastic body, and (3) that which produces work through chemical action. Water in an elevated reservoir, and the lifted weight of a pile-driver are examples of the first group; a stretched rubber band or compressed spring are examples of the second group; and energy in coal, food, and batteries are examples of the third group.

Bodies in motion required work to put them in motion. Thus, they possess energy of motion. Energy due to motion is known as **kinetic energy**. A moving vehicle, a rotating flywheel, and a hammer in motion are examples of kinetic energy.

Energy is expressed in the same units as those used to express work. The quantity of **potential energy** possessed by an elevated weight may be computed by the equation,

$$\text{Potential Energy} = \text{Weight} \times \text{Height}$$

If weight is given in pounds and height in feet, the final unit of energy will be ft-lbs (foot-pounds).

Example: An aircraft with a gross weight of 110,000 pounds is flying at an altitude of 15,000 feet above the surface of the earth. How much potential energy does the airplane possess with respect to the earth?

$$\text{Potential Energy} = \text{Weight} \times \text{Height}$$

$$PE = 110{,}000 \times 15{,}000$$

$$PE = 1{,}650{,}000{,}000 \text{ ft–lbs}$$

Forms Of Energy

The most common **forms of energy** are heat (thermal), mechanical, electrical, and chemical. The various forms of energy can be changed, or transformed, into another form in many different ways. For example, in the case of mechanical energy, the energy of work done against friction is always converted into heat energy, and the mechanical energy that turns an electric generator develops electrical energy at the output of the generator.

Force And Motion Of Bodies

General

The study of the relationship between the motion of bodies or objects and the forces acting on them is often called the study of **force and motion** or **dynamics.** In a more specific sense, the relationship between velocity, acceleration, and distance is known as **kinematics.**

Uniform Motion

Motion may be defined as a continuing change of position or place, or as the process in which a body undergoes displacement. When an object is at different points in space at different times, that object is said to be in motion, and if the distance the object moves remains the same for a given period of time, the motion may be described as uniform. Thus, an object in uniform motion always has a constant speed.

Speed And Velocity

In everyday usage, speed and velocity often mean the same thing. In physics they have definite and distinct meanings. Speed refers to how fast an object is moving, or how far the object will travel in a specific time. The speed of an object tells nothing about the direction an object is moving. For example, if the information is supplied that an airplane leaves New York City and travels 8 hours at a speed of 150 MPH, this information tells nothing about the direction in which the airplane is moving. At the end of 8 hours, it might be in Kansas City, or if it traveled in a circular route, it could be back in New York City.

Velocity is that quantity in physics which denotes both the speed of an object and the direction in which the object moves. Velocity can be defined as the rate of motion in a particular direction, which is a vector quantity.

Acceleration

Acceleration is defined by the physicist as the rate of change of velocity. If the velocity of an object is increased from 20 MPH to 30 MPH, the object has been accelerated. If the increase in velocity is 10 MPH in 5 seconds, the rate of change in velocity is 10 MPH in 5 seconds, or $\frac{2\ MPH}{sec}$. Expressed as an equation,

$$a = \frac{\Delta V}{\Delta t}$$

$$a = \frac{V_f - V_i}{\Delta t}$$

where:

  a = acceleration.
  V_f = the final velocity (30 MPH).
  V_i = the initial velocity (20 MPH)
  Δt = change in time or the elapsed time.
  ΔV = change in velocity.

The example used can be expressed as follows:

$$A = \frac{30\ MPH - 20\ MPH}{5\ sec}$$

$$A = \frac{2\ MPH}{sec}$$

If the object accelerated to 22 MPH in the first second, 24 MPH in the next second, and 26 MPH in the third second, the change in velocity each second is 2 MPH. The acceleration is said to be constant, and the motion is described as uniformly accelerated motion. Since **velocity** denotes both speed and direction, a change in direction of a mass is considered to be an acceleration. An aircraft executing a standard rate turn (3 degrees per second) is considered to be in uniformly accelerated motion.

If a body has a velocity of 3 MPH at the end of the first second of its motion, 5 MPH at the end of the next second, and 8 MPH at the end of the third second, its motion is described as acceleration, but it is variable accelerated motion.

Newton's Laws Of Motion

When a magician snatches a tablecloth from a table and leaves a full setting of dishes undisturbed, he is not displaying a mystic art; he is demonstrating the principle of **inertia**.

Inertia is responsible for the discomfort felt when a car is brought to a sudden halt in the parking area and the passengers are thrown forward in their seats, and inertia provides the feeling of being pushed back into the seat felt by occupants of a rapidly accelerating aircraft during takeoff. Inertia is a property of matter. This property of matter is described by **Newton's first law of motion,** which states:

Objects at rest tend to remain at rest; objects in motion tend to remain in motion at the same speed and in the same direction.

Bodies in motion have the property called **momentum**. A body that has great momentum has a strong tendency to remain in motion and is therefore hard to stop. For example, a train moving at even low velocity is difficult to stop because of its large mass. **Newton's second law** applies to this property. It states:

When forces act upon a body, the momentum of the body can be changed. The rate of change of momentum of the body is proportional to the vector sum of the forces (**net force**) applied to the body. The momentum (p) of a body is defined as the product of its mass times its velocity.

$$\text{Momentum} = \text{mass} \times \text{velocity or,}$$

$$p = mV$$

Now if a force is applied, the momentum changes at a rate equal to the vector sum of the forces applied or the new force (ΣF):

$$\Sigma F = \text{rate of change of momentum}$$

$$\Sigma F = \frac{M_f - M_i}{\Delta t}$$

Substituting mV for M:

$$\Sigma F = \frac{m_f V_f - m_i V_i}{\Delta t}$$

Since the mass does not usually change, $m_f = m_i = m$. Then

$$\Sigma F = \frac{mV_f - mV_i}{\Delta t}$$

$$\Sigma F = m\frac{V_f - V_i}{\Delta t}$$

From the previous section the second term is recognized as acceleration. Then the second law becomes:

$$F = ma$$

If the only force acting upon a body near the Earth is the gravitational force (its weight), then the body will experience an acceleration which is called the acceleration due to gravity, usually designated as **"g,"** and equal in English units to 32 ft/SEC^2, directed toward the center of the Earth. The weight **"W"** of a body is the commonly used name for the force due to gravity acting upon that body, and if this is the only force acting upon the body of mass, **m**, that body accelerates toward the center of the Earth at 32 feet per second per second. If these ideas are used along with Newton's second law:

$$F = ma \quad \text{becomes} \quad W = mg, \text{ which can be}$$
rearranged as

$$m = \frac{W}{g}$$

If the mass m of a body is known, then the weight W of that body near the Earth is given by:

$$W = mg$$

If the weight W of a body near the surface of the Earth is known, then the mass m of that body is given by:

$$m = \frac{W}{g}$$

Distance between the bodies being attracted to each other affects the strength of the attracting force (mass attraction). For example, an airplane that weighs 3000 pounds at sea level will weigh about ¾ pound less at 10,000' ASL, but its mass remains the same.

The following examples illustrate the use of this formula.

Example:
A train weighs 32,000 lbs. and is traveling at 10 ft/sec. What force is required to bring it to rest in 10 seconds?

$$F = \frac{W}{g} \ (a)$$

$$= \frac{W}{g} \frac{(V_f - V_i)}{t}$$

$$= \frac{32,000}{32} \frac{(0 - 10)}{10}$$

$$= \frac{32,000 \times (-10)}{32 \times 10}$$

$$= -1,000 \text{ lbs}$$

The negative sign means that the force must be applied against the train's motion.

Example:

An aircraft weighs 6,400 pounds. How much force is needed to give it an acceleration of 6 ft/sec^2 (neglecting drag)?

$$F = \frac{W}{g} (a)$$

$$= \frac{6400 \times 6}{32} = 1,200 \text{ lb}$$

Newton's third law of motion is often called the law of action and reaction. It states that for every action there is an equal and opposite reaction. This means that if a force is applied to an object, the object will supply a resistive force exactly equally to and in the opposite direction of the force applied. It is easy to see how this might apply to objects at rest. For example, as a man stands on the floor, the floor exerts a force against his feet exactly equal to his weight. But this law is also applicable when a force is applied to an object in motion.

When an aircraft propeller pushes a stream of air backward with a force of 500 pounds, the air pushes the blades forward with a force of 500 pounds. This forward force causes the aircraft to move forward. In like manner, the discharge of exhaust gases from the tailpipe of a turbine engine is the action which causes the aircraft to move forward.

The three laws of motion which have been discussed here are closely related. In many cases, all three laws may be operating on a body at the same time.

In order to keep straight in one's mind the relationships between Newton's three laws of motion, it is well to remember that Newton's third law involves interactions between two or more bodies. The action and reaction forces act on the different bodies which are also producing forces which are interacting with the bodies. Thus, when considering the resulting effect on the bodies, the **net force**, or vector sum of all the forces must be considered. Newton's second law describes the effect of all the forces which act on **one** body (mass). Newton's first law describes the action of a body when there is no net force acting upon it.

Vibration

The term **vibration** indicates periodic, continuing motion, usually of a solid object, but may also apply to a liquid (wave) or a gas (sound). The motion may be in the form of a **pulse** (a motion

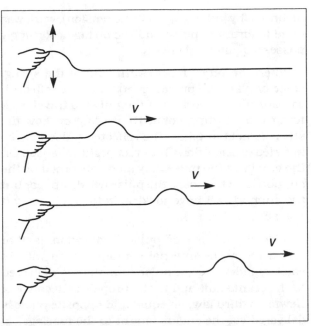

Figure 1-6. A pulse moves along a stretched string with a constant velocity v.

that comes along once-in-a-while) or waveforms, where any part of the vibrating mass is continuously moving.

If we give one end of a stretched string a quick shake, a kink or **pulse** travels down the string at some velocity v (figure 1-6). If the string is uniform and completely flexible, the pulse keeps the same shape and velocity. The velocity of the pulse depends on the properties of the string, its density and flexibility, and on how tightly it is stretched rather than on the shape of the pulse or how it is produced.

The energy content of a moving pulse is partly kinetic and partly potential. As the pulse travels, its forward part is moving upward and its rear part is moving downward (figure 1-7). Because the string has mass, there is kinetic energy associated with these up-and-down motions. Because the stretched string had to be stretched further to

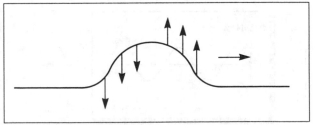

Figure 1-7. The forward part of a traveling pulse is moving upward and the rear part is moving downward.

deform it (by pulling against the tension), work was done to form the pulse and the string accordingly possesses potential energy.

When the pulse reaches the end of the string, some or all of the pulse's energy may be reflected, causing all or a portion of the pulse to travel back toward its starting point. Depending on how the string is held in place, the reflected pulse may be inverted (upside-down) or erect (right-side-up). Or, the energy of the pulse may all be absorbed by the support, in which case the pulse will disappear but the support will have to "deal with" the energy it has absorbed.

A good example of pulsing vibration is that produced by the internal combustion engine. As each cylinder fires, a pulse of motion is delivered to the crankshaft and to the propeller. Because of Newton's third law, an equal and opposite pulse is delivered via the engine mount to the fuselage.

In a **periodic wave**, one pulse follows another in regular succession, so a certain **waveform** (the shape of the individual waves) is repeated at regular intervals. Periodic waves usually have **sinusoidal** waveforms (the same appearance as a graph of *sin x (or cos x)* versus the angle *x* (figure 1-8).

Sinusoidal waveforms are common in matter because the particles of matter undergo simple harmonic motion when displaced by the passage of a wave. Because the link between two particles of matter is not totally rigid, the movement of one particle causes the next particle to move a moment later (like a row of toppling dominoes).

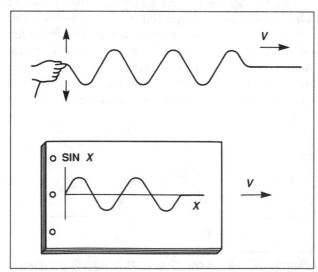

Figure 1-8. Most periodic waves have sinusoidal waveforms.

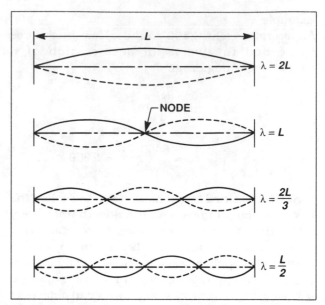

Figure 1-9. Standing waves in a stretched string.

Some related terms are used to describe periodic waves: **wave velocity** describes the distance each wave moves per second, **wavelength** (symbolized by the Greek letter *lambda*) is the distance between adjacent crests or troughs, **frequency** is the number of waves that pass a given point per second (cycles per second, or Hertz, *Hz*), **period** is the time it takes for one complete cycle to pass a given point (the reciprocal of the frequency), **amplitude** is the maximum displacement of the particle from its normal position. The amplitude of a wave in a stretched string is the height of the crests above or the troughs below the original line of the string.

Standing Waves

When a string whose ends are fixed in place is plucked, the string will vibrate in one or more loops (figure 1-9). These **standing waves** may be thought of as being the result of waves that travel down the string in both directions, are reflected at the ends, proceed across the string to the opposite ends and are again reflected, and so on.

Pilots should attempt to visualize how the types of vibrations being described here can travel through the various parts of the aircraft.

Principle Of Superposition

Standing waves in a mass interact with each other according to the **principle of superposition**: *When two or more waves of the same nature travel past a point at the same time, the displacement at that point is the sum of the instantaneous displacements of the individual waves.*

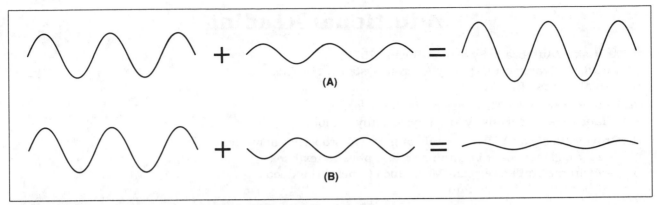

Figure 1-10(A). Constructive interference. (B) Destructive interference.

The principle of superposition holds for all types of waves. What it means is that every wave train proceeds independently of any other that may also be present but the effect of waves present at any point is the additive result of all waves present. For example, should two waves with the same wavelength come together in such a way that crest aligns with crest and trough aligns with trough, the resulting composite wave will have an amplitude greater (the sum of the two amplitudes) than that of either of the original waves. The waves are said to interfere constructively with each other. But if the waves come together so that one wave's crest aligns with the other's trough, the effect is cancelling and the amplitude will be decreased (to zero, if the amplitudes of the individual waves are equal). In this condition, the waves interfere destructively with each other. See figure 1-10.

Resonance

Consider for a moment a string where nodes occur at each end of the string, as in figure 1-9. The lowest possible frequency of oscillation corresponds to the longest wavelength, $\lambda = 2L$, as depicted by the top waveform in figure 1-9. Higher frequencies correspond to the shorter wavelengths. The lowest frequency is called the **fundamental frequency**. The higher frequencies are 2, 3, 4, etc. times higher (always whole numbers) and are called **overtones**.

The fundamental frequency and overtones of a mass are its **natural frequencies of vibration**. Pulsing a mass will cause it to vibrate at one or more of these frequencies. Eventually, internal friction in the mass will cause the various vibrations to die out.

However, the vibration can be caused to continue by applying to the mass a periodic force whose frequency is exactly the same as that of one of the mass's natural frequencies. If the force applied at the correct frequency is greater than that needed to overcome internal friction, the amplitude of the standing wave will increase until the mass ruptures (a column of soldiers can destroy a flimsy bridge by marching across it in step with one of the bridge's natural frequencies although the bridge is capable of holding up the static load of the soldier's weight). This phenomenon is called **resonance** and is based upon the concept of superposition of waves.

When periodic impulses are given to a mass at frequencies other than its natural ones, hardly any response occurs. The pilot must keep in mind that each structural part of the aircraft has its own natural frequencies of oscillation, at which a small amount of excitation can cause large amplitudes of oscillation. It is possible that an aircraft structure (as large as the fuselage itself or as small as the hat-rack structure) will have a resonant frequency the same as certain vibrations from the engine, propeller or even the air in the intake duct. For this reason, aircraft undergo extensive flight and vibration testing to minimize the chances of resonance excitation. For example, a propeller should never be used on an aircraft without first determining that it has been tested and certified for use with that particular engine and airframe combination.

Additional Reading

1. Anti-skid/Autobrake Systems—Boeing 757
   Wild, T.; Transport Category Aircraft Systems; IAP, Inc.
   1990, pages 167-170.

2. Unit conversions—There are many sources, such as:
   Handbook of Chemistry and Physics, any edition.

3. Gravity effects and further study of physical properties of matter.
   – Any high school or beginning college physics textbook.
   – Airframe and Powerplant Mechanics General Handbook,
     Chapter 7; IAP, Inc., Publ.

Study Questions And Problems

1. Neglecting drag, how much power is needed to cause a 2500 pound airplane to climb at 1000 ft/min? Express your answer in ft-lbs/min, watts and HP.

2. Three minutes after takeoff, how much potential energy will the aircraft of Question #1 possess (neglect remaining fuel in the tanks)?

3. What kind of energy does the aircraft of Question #2 possess?

4. What are the states of matter? Define or describe each.

5. What distinguishes a liquid from a gas and what is a fluid?

6. Show the value of pressure of the standard day atmosphere at sea level in five different units of measure.

7. Show the value of temperature of the standard day atmosphere at sea level in four different units of measure.

8. What kind of a lever is your forearm?

9. What kind of a lever is the aircraft control yoke, with respect to aileron movement?

10. What kind of a lever is a hydraulic wobble pump handle?

11. If you fell from an airplane at a high altitude where the air was so thin that there was negligible drag, how fast would you be falling after 5 seconds, in ft/sec, MPH, KTS and Km/hr?

12. What are the differences between inertia and momentum?

Chapter II
Aircraft Engine Types And Construction

The Heat Engine

For an aircraft to remain in level unaccelerated flight, a thrust must be provided that is equal to and opposite in direction to the aircraft drag. This thrust, or propulsive force, is provided by a suitable type of **heat engine.**

All **heat engines** have in common the ability to convert fuel (chemical energy) into heat energy, then into mechanical energy, by the flow of some fluid mass through the engine. In all cases, the heat energy is released at a point in the cycle where the pressure is high, relative to atmospheric.

These engines are customarily divided into groups or types depending upon:

(1) The working fluid used in the engine cycle.

(2) The means by which the mechanical energy is transmitted into a propulsive force.

(3) The method of compressing the engine working fluid.

The types of engines are illustrated in figure 2-1.

The propulsive force is obtained by the displacement of a working fluid (not necessarily the same fluid used within the engine) in a direction opposite to that in which the airplane is propelled. This is an application of Newton's third law of motion. Air is the principal fluid used for propulsion in every type of powerplant except the rocket, in which only the by-products of combustion are accelerated and displaced.

The propellers of aircraft powered by reciprocating or turboprop engines accelerate a large mass of air through a small velocity change. The fluid (air) used for the propulsive force is a different quantity than that used within the engine to produce the mechanical energy. Turbojets, ramjets, and pulse-jets accelerate a smaller quantity of air through a large velocity change. They use the same working fluid for propulsive force that is used within the engine. A rocket carries its own oxidizer rather than using ambient air for combustion. It discharges the gaseous by-products of combustion through the exhaust nozzle at an extremely high velocity.

Engines are further characterized by the means of compressing the working fluid before the addition of heat. The basic methods of compression are:

(1) The turbine-driven compressor (turbine engine).

(2) The positive displacement, piston-type compressor (reciprocating engine).

(3) Ram compression due to forward flight speed (ramjet).

(4) Pressure rise due to combustion (pulse-jet and rocket).

A more specific description of the major engine types used in commercial aviation is given later in this chapter.

ENGINE TYPE	MAJOR MEANS OF COMPRESSION	ENGINE WORKING FLUID	PROPULSIVE WORKING FLUID
TURBOJET	TURBINE-DRIVEN COMPRESSOR	FUEL/AIR MIXTURE	SAME AS ENGINE WORKING FLUID
TURBOPROP	TURBINE-DRIVEN COMPRESSOR	FUEL/AIR MIXTURE	AMBIENT AIR
RAMJET	RAM COMPRESSION DUE TO HIGH FLIGHT SPEED	FUEL/AIR MIXTURE	SAME AS ENGINE WORKING FLUID.
PULSE-JET	COMPRESSION DUE TO COMBUSTION	FUEL/AIR MIXTURE	SAME AS ENGINE WORKING FLUID
RECIPROCATING	RECIPROCATING ACTION OF PISTONS	FUEL/AIR MIXTURE	AMBIENT AIR
ROCKET	COMPRESSION DUE TO COMBUSTION	OXIDIZER/FUEL MIXTURE	SAME AS ENGINE WORKING FLUID

Figure 2-1. Types of engines.

Comparison Of Aircraft Powerplants

In addition to the differences in the methods employed by the various types of powerplants for producing thrust, there are differences in their suitability for different types of aircraft. The following discussion points out some of the important characteristics which determine their suitability.

General Requirements

All engines must meet certain general requirements of **efficiency, economy,** and **reliability.** Besides being economical in fuel consumption, an aircraft engine must be economical (the cost of original procurement and the cost of maintenance) and it must meet exacting requirements of efficiency and low weight per horsepower ratio. It must be capable of sustained high-power output with no sacrifice in reliability; it must also have the durability to operate for long periods of time between overhauls. It needs to be as compact as possible, yet have easy accessibility for maintenance. It is required to be as vibration free as possible and be able to cover a wide range of power output at various speeds and altitudes.

These requirements dictate the use of ignition systems that will deliver the firing impulse to the spark plugs or igniter plugs at the proper time in all kinds of weather and under other adverse conditions. Fuel-metering devices are needed that will deliver fuel in the correct proportion to the air ingested by the engine regardless of the attitude, altitude, or type of weather in which the engine is operated. The engine needs a type of oil system that delivers oil under the proper pressure to all of the operating parts of the engine when it is running. Also, it must have a system of damping units to damp out the vibrations of the engine when it is operating.

Power and Weight

The useful output of all aircraft powerplants is the force **thrust,** the force which propels the aircraft. Since the reciprocating engine is rated in BHP (brake horsepower) and the gas turbine engine is rated in pounds of thrust, no direct comparison can be made. However, since the reciprocating engine/propeller combination receives its thrust from the propeller, a comparison can be made by converting the horsepower developed by the reciprocating engine to thrust.

If desired, the thrust of a gas turbine engine can be converted into THP (thrust horsepower). But it is necessary to consider the speed of the aircraft. This conversion can be accomplished by using the formula:

$$THP = \frac{\text{thrust} \times \text{aircraft speed (MPH)}}{375 \text{ mile–pounds per hour}}$$

The value 375 mile-pounds per hour is derived from the basic horsepower formula as follows:

$$1HP = 33,000 \text{ ft–lb per minute}$$

$$33,000 \text{ ft–lb/min} \times 60 \text{ min}/1 \text{ hr} = 1,980,000 \text{ ft–lb per hour}$$

$$\frac{1,980,000 \text{ ft–lb/hr}}{5,280 \text{ ft/mi}} = 375 \text{ mile–pounds per hour}$$

One horsepower equals 33,000 ft-lb per minute or 375 mile-pounds per hour. Under static conditions, thrust is figured as equivalent to approximately 2.6 pounds per hour.

If a gas turbine is producing 4,000 pounds of thrust and the aircraft in which the engine is installed is traveling at 500 MPH, the THP will be:

$$\frac{4000 \times 500}{375} = 5,333.33 \text{ THP}$$

It is necessary to calculate the horsepower for each speed of an aircraft, since the horsepower varies with speed. Therefore, it is not practical to try to rate or compare the output of a turbine engine on a horsepower basis.

The aircraft engine operates at a relatively high percentage of its maximum power output throughout its service life. The aircraft engine is at full power output whenever a takeoff is made. It may hold this power for a period of time up to the limits set by the manufacturer. The engine is seldom held at a maximum power for more than 2 minutes, and usually not that long. Within a few seconds after lift-off, the power is reduced to a power that is used for climbing and that can be maintained for longer periods of time. After the aircraft has climbed to cruising altitude, the power of the engine(s) is further reduced to a cruise power which can be maintained for the duration of the flight.

If the weight of an engine per brake horsepower (called the **specific weight** of the engine) is decreased, the useful load that an aircraft can carry and/or the performance of the aircraft ob-

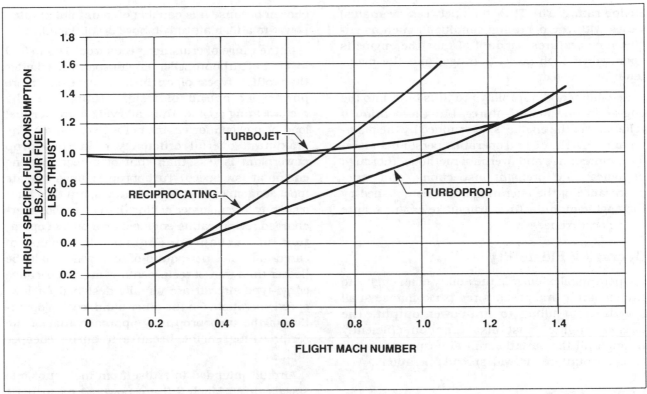

Figure 2-2. Comparison of fuel consumption for three types of engines at rated power at sea level.

viously are increased. Every excess pound of weight carried by an aircraft engine reduces its performance. Tremendous gains in reducing the weight of the aircraft engine through improvement in design and metallurgy have resulted in reciprocating engines now producing approximately 1 HP for each pound of weight.

Fuel Economy

The basic parameter for describing the fuel economy of aircraft engines is usually **specific fuel consumption.** Specific fuel consumption for turbojets and ramjets is the fuel flow (lbs/hr) divided by thrust (lbs), and for reciprocating engines the fuel flow (lbs/hr) divided by brake horsepower. These are called **"thrust specific fuel consumption"** and **"brake specific fuel consumption,"** respectively. Equivalent specific fuel consumption is used for the turboprop engine and is the fuel flow in pounds per hour divided by a turboprop's equivalent shaft horsepower. Comparisons of efficiency can be made between the various engines on a specific fuel consumption basis.

At low speed, the reciprocating and turbopropeller engines have better economy than the turbojet engines. However, at high speed, because of losses in propeller efficiency, the reciprocating

or turbopropeller engine's efficiency becomes less than that of the turbojet. Figure 2-2 shows a comparison of average thrust specific fuel consumption of three types of engines at rated power at sea level.

Durability And Reliability

Durability and **reliability** are usually considered identical factors since it is difficult to mention one without including the other. An aircraft engine is reliable when it can perform at the specified ratings in widely varying flight attitudes and in extreme weather conditions. Standards of powerplant reliability are agreed upon by the FAA, the engine manufacturer, and the airframe manufacturer. The engine manufacturer ensures the reliability of his product by design, research, and testing. Close control of manufacturing and assembly procedures is maintained, and each engine is tested before it leaves the factory.

Durability is the amount of engine life obtained while maintaining the desired reliability. The fact that an engine has successfully completed its type or proof test indicates that it can be operated in a normal manner over a long period before requiring overhaul. However, no definite time interval between overhauls is specified or implied in the

17

engine rating. The TBO (time between overhauls) varies with the operating conditions such as engine temperatures, amount of time the engine is operated at high-power settings, and the maintenance received.

Reliability and durability are thus built into the engine by the manufacturer, but the continued reliability of the engine is determined by the maintenance, overhaul, and operating personnel. Careful maintenance and overhaul methods, thorough periodical and preflight inspections, and strict observance of the operating limits established by the engine manufacturer will make engine failure a rare occurrence.

Operating Flexibility

Operating flexibility is the ability of an engine to run smoothly and give desired performance at all speeds from idling to full-power output. The aircraft engine must also function efficiently through all the variations in atmospheric conditions encountered in widespread operations.

Compactness

To effect proper streamlining and balancing of an aircraft, the shape and size of the engine must be as compact as possible. In single-engine aircraft, the shape and size of the engine also affect the view of the pilot, making a smaller engine better from this standpoint, in addition to reducing the drag created by a large frontal area.

Weight limitations, naturally, are closely related to the compactness requirement. The more elongated and spread out an engine is, the more difficult it becomes to keep the specific weight within the allowable limits.

Powerplant Selection

Engine specific weight and specific fuel consumption were discussed in the previous paragraphs, but for certain design requirements, the final powerplant selection may be based on factors other than those which can be discussed from an analytical point of view. For that reason, a general discussion of powerplant selection is included here.

For aircraft whose cruising speeds will not exceed 250 MPH the reciprocating engine is the usual choice. When economy is required in the low-speed range, the conventional reciprocating engine is chosen because of its excellent efficiency. When high-altitude performance is required, the turbosupercharged reciprocating engine may be

chosen because it is capable of maintaining rated power to a high altitude (above 30,000 feet).

In the range of cruising speeds from 180 to 350 MPH the turbopropeller engine performs better than other types of engines. It develops more power per pound of weight than does the reciprocating engine, thus allowing a greater fuel load or payload for engines of a given power. The maximum overall efficiency of a turboprop powerplant is less than that of a reciprocating engine at low speed. Turboprop engines operate most economically at high altitudes, but they have a slightly lower service ceiling than do turbosupercharged reciprocating engines. Economy of operation of turboprop engines, in terms of cargo-ton-miles per pound of fuel, will usually be poorer than that of reciprocating engines because cargo-type aircraft are usually designed for low-speed operation. On the other hand, cost of operation of the turboprop may approach that of the reciprocating engine because it burns cheaper fuel.

Aircraft intended to cruise from high subsonic speeds up to Mach 2.0 are powered by turbojet engines. Like the turboprop, the turbojet operates most efficiently at high altitudes. High-speed, turbo-jet-propelled aircraft fuel economy, in terms of miles per pound of fuel, is poorer than that attained at low speeds with reciprocating engines.

However, reciprocating engines are more complex in operation than other engines. Correct operation of reciprocating engines requires about twice the instrumentation required by turbojets or turboprops, and it requires several more controls. A change in power setting on some reciprocating engine installations may require the adjustment of five controls, but a change in power on a turbojet requires only a change in throttle setting. Furthermore, there are a greater number of critical temperatures and pressures to be watched on reciprocating engine installations than on turbojet or turboprop installations.

Types Of Reciprocating Engines

Many types of reciprocating engines have been designed. However, manufacturers have developed some designs that are used more commonly than others and are therefore recognized as conventional. Reciprocating engines may be classified according to cylinder arrangement with respect to the crankshaft (in-line, V-type, radial, and opposed) or according to the method of cooling (liquid cooled or air cooled). Actually, all engines are cooled by transferring excess heat to the surrounding air. In air-cooled engines, this heat transfer is direct from the cylinders to the air. In liquid-cooled engines, the heat is transferred from the cylinders to the coolant, which is then sent through tubing and cooled within a radiator placed in the airstream. The radiator must be large enough to cool the liquid efficiently. Heat is transferred to air more slowly than it is to a liquid. Therefore, it is necessary to provide thin metal fins on the cylinders of an air-cooled engine in order to have increased surface for sufficient heat transfer. Most aircraft engines are air cooled.

In-line Engines

An in-line engine generally has an even number of cylinders, although some three- and five-cylinder engines have been constructed. This engine may be either liquid cooled or air cooled and has only one crankshaft, which is located either above or below the cylinders. If the engine is designed to operate with the cylinders below the crankshaft, it is called an inverted engine.

The in-line engine has a small frontal area and is better adapted to streamlining. When mounted with the cylinders in an inverted position, it offers the added advantages of a shorter landing gear and greater pilot visibility. The in-line engine has a higher weight-to-horsepower ratio than most other engines. With increase in engine size, the air cooled, in-line type offers additional handicaps to proper cooling; therefore, this type of engine is, to a large degree, confined to low- and medium-horsepower engines used in light aircraft.

Opposed Or O-type Engines

The opposed-type engine, shown in figure 2-3, has two banks of cylinders opposite each other with a crankshaft in the center. The pistons of both cylinder banks are connected to the single crankshaft. Although the engine can be either liquid cooled or air cooled, the air-cooled version is used predominantly in aviation. It can be mounted with the cylinders in either a vertical or horizontal position, although horizontal provides better visibility and eliminates the problem of fluid lock on bottom cylinders.

The opposed-type engine has a low weight-to-horsepower ratio, and its narrow silhouette makes it ideal for horizontal installation on the aircraft wings. Another advantage is its comparative freedom from vibration.

V-type Engines

In the V-type engines, the cylinders are arranged in two in-line banks generally set 30-60° apart. Engines of this type have an even number of cylinders and are liquid cooled or air cooled. The engines are designated by a V, followed by a dash and the piston displacement in cubic inches, for example, V-1710.

Radial Engines

The radial engine consists of a row, or rows, of cylinders arranged radially about a central crankcase (see figure 2-4). This type of engine has

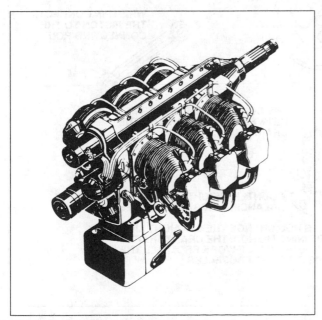

Figure 2-3. Opposed engine.

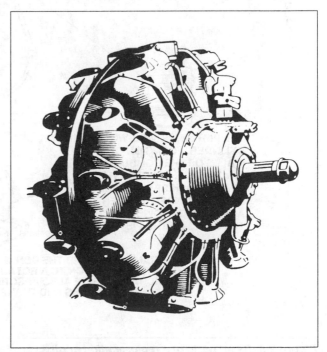

Figure 2-4. Radial engine.

proven to be very rugged and dependable. The number of cylinders composing a row may be either three, five, seven, or nine. Some radial engines have two rows of seven or nine cylinders arranged radially about the crankcase. One type has four rows of cylinders with seven cylinders in each row.

The power output from the different sizes of radial engines varies from 100 to 3,800 horsepower.

Reciprocating Engine Design And Construction

The basic parts of a reciprocating engine are the **crankcase, cylinders, pistons, connecting rods, valves, valve-operating mechanism,** and **crankshaft.** In the **head** of each cylinder are the **valves** and **spark plugs.** One of the valves is in a passage leading from the induction system; the other is in a passage leading to the exhaust system. Inside each cylinder is a movable **piston**

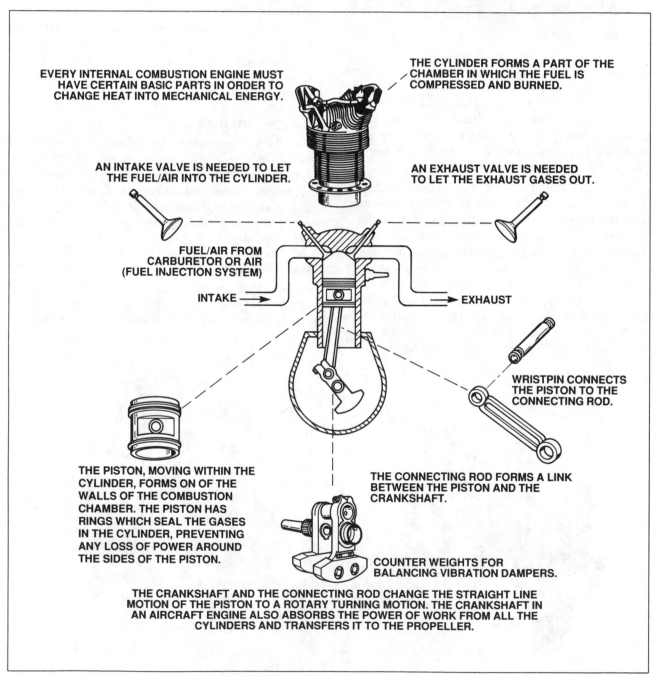

Figure 2-5. Basic parts of a reciprocating engine.

connected to a **crankshaft** by a **connecting rod.** Figure 2-5 illustrates the basic parts of a reciprocating engine.

Crankcases

The foundation of an engine is the **crankcase.** It contains the bearings in which the crankshaft revolves. Besides supporting itself, the crankcase must provide a tight enclosure for the lubricating oil and must support various external and internal mechanisms of the engine. It also provides support for attachment of the cylinder assemblies, and the powerplant to the aircraft. It must be sufficiently rigid and strong to prevent misalignment of the crankshaft and its bearings. Cast or forged aluminum alloy is generally used for crankcase construction because it is light and strong. Forged steel crankcases are used on some of the high-power output engines.

The crankcase is subjected to many variations of vibrational and other forces. Since the cylinders are fastened to the crankcase, the tremendous expansion forces tend to pull the cylinder off the crankcase. The unbalanced centrifugal and inertia forces of the crankshaft acting through the main bearing subject the crankcase to bending moments which change continuously in direction and magnitude. The crankcase must have sufficient stiffness to withstand these bending moments without objectional deflections. If the engine is equipped with a propeller reduction gear, the front or drive end will be subjected to additional forces.

In addition to the thrust forces developed by the propeller under high-power output, there are severe centrifugal and gyroscopic forces applied to the crankcase due to sudden changes in the direction of flight, such as those occurring during maneuvers of the airplane. Gyroscopic forces are, of course, particularly severe when a heavy propeller is installed.

Radial Engines

The engine shown in figure 2-6 is a single-row, nine-cylinder **radial engine** of relatively simple construction, having a one-piece nose and a two-section main crankcase.

The larger twin-row engines are of slightly more complex construction than the single-row engines. For example, the crankcase of the Wright R-3350 engine is composed of the crankcase front section,

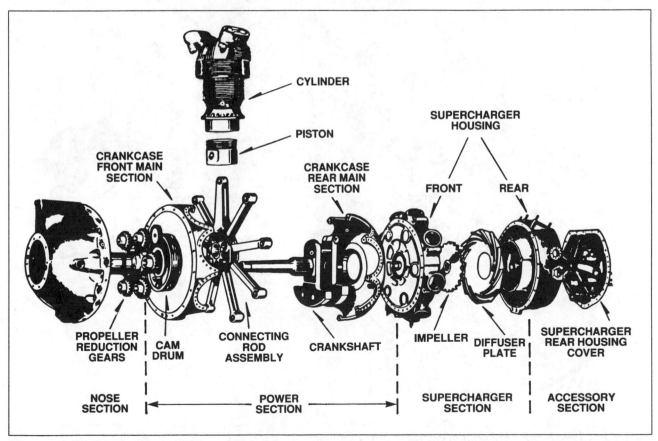

Figure 2-6. Engine sections, radial engines.

four crankcase main sections (the front main, the front center, the rear center, and the rear main sections), the rear cam and tappet housing, the supercharger front housing, the supercharger rear housing, and the supercharger rear housing cover. Pratt and Whitney engines of comparable size incorporate the same basic sections, although the construction and the nomenclature differ considerably.

Opposed And In-line Engines

The **crankcases** used on engines having opposed or in-line cylinder arrangements vary in form for the different types of engines, but in general they are approximately cylindrical. One or more sides are surfaced to serve as a base to which the cylinders are attached by means of capscrews, bolts, or studs. These accurately machined surfaces are frequently referred to as **cylinder pads.**

The **crankshaft** is carried in a position parallel to the longitudinal axis of the crankcase and is generally supported by a **main bearing** between each throw. The crankshaft main bearings must be supported rigidly in the crankcase. This usually is accomplished by means of transverse **webs** in the crankcase, one for each main bearing. The webs form an integral part of the structure and, in addition to supporting the main bearings, add to the strength of the entire case.

The crankcase is divided into two sections in a longitudinal plane. This division may be in the plane of the crankshaft so that one-half of the main bearing (and sometimes crankshaft bearings) are carried in one section of the case and the other half in the opposite section. (See figure 2-7.) Another method is to divide the case in such a manner that the main bearings are secured to only one section of the case on which the cylinders are attached,

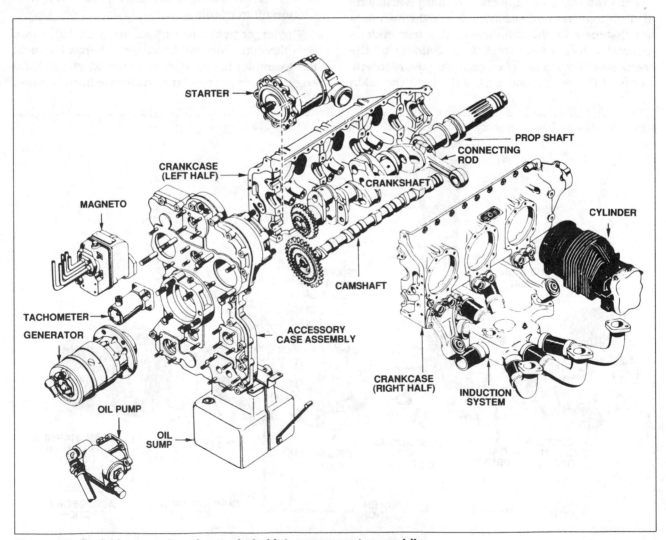

Figure 2-7. Typical opposed engine exploded into component assemblies.

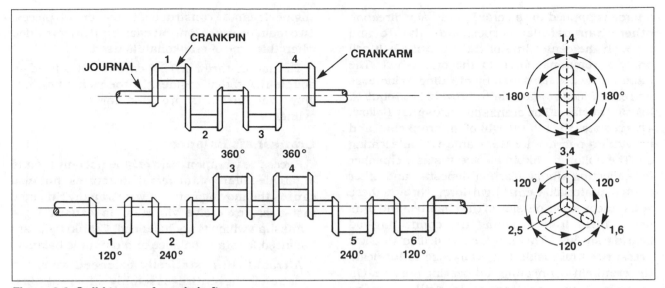

Figure 2-8. Solid types of crankshafts.

thereby providing means of removing a section of the crankcase for inspection without disturbing the bearing adjustment.

Crankshafts

The **crankshaft** is the backbone of the reciprocating engine. It is subjected to most of the forces developed by the engine. Its main purpose is to transform the reciprocating motion of the piston and connecting rod into rotary motion for rotation of the propeller. The **crankshaft** as the name implies, is a shaft composed of one or more **cranks** located at specified points along its length. The **cranks,** or **throws** are formed by forging offsets into a shaft before it is machined. Since crankshafts must be very strong, they generally are forged from a very strong alloy, such as chromium-nickel-molybdenum steel.

A crankshaft may be of single-piece or multipiece construction. Figure 2-8 shows two representative types of solid crankshafts used in aircraft engines. The four-throw construction may be used either on four-cylinder horizontal opposed or four-cylinder in-line engines.

The six-throw shaft is used on six-cylinder in-line engines, 12-cylinder V-type engines, and six-cylinder opposed engines.

Crankshafts of radial engines may be the single-throw, two-throw, or four-throw type, depending on whether the engine is the single-row, twin-row, or four-row type. A single-row radial engine crankshaft is shown in figure 2-9.

No matter how many throws it may have, each **crankshaft** has three main parts—a **journal, crankpin,** and **crankcheek.** Counterweights and

dampers, although not a true part of a crankshaft, are usually attached to it to reduce engine vibration.

The **journal** is supported by, and rotates in, a main bearing. It serves as the center of rotation of the crankshaft. It is surface-hardened to reduce wear.

The **crankpin** is the section to which the connecting rod is attached. It is off-center from the main journals and is often called the **throw.** Two crankcheeks and a crankpin make a throw. When

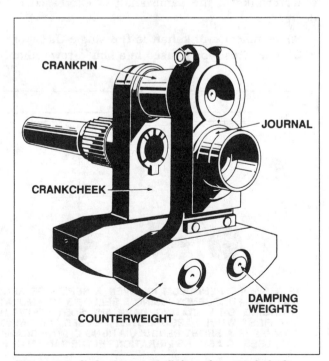

Figure 2-9. A single-row radial engine crankshaft.

a force is applied to the crankpin in any direction other than parallel or perpendicular to and through the center line of the crankshaft, it will apply a rotational force to the crankshaft. The outer surface is hardened by nitriding to increase its resistance to wear and to provide the required bearing surface. The crankshaft is usually hollow. This reduces the total weight of the crankshaft and provides a passage for the transfer of lubricating oil. The hollow crankpin also serves as a chamber for collecting sludge, carbon deposits, and other foreign material. Centrifugal force throws these substances to the outside of the chamber and thus keeps them from reaching the connecting-rod bearing surface. The crankpin is drilled to allow pressurized oil inside the crankcase to lubricate the crankpin-connecting rod bearing surface. On some engines, a passage is drilled in the crankcheek to allow oil from the hollow crankshaft to be sprayed on the cylinder walls.

The **crankcheek** connects the **crankpin** to the **main journal.** In some designs, the cheek extends beyond the journal and carries a counterweight to balance the crankshaft. The crankcheek must be of sturdy construction to obtain the required rigidity between the crankpin and the journal.

In all cases, the type of crankshaft and the number of crankpins must correspond with the cylinder arrangement of the engine. The position of the cranks on the crankshaft in relation to the other cranks of the same shaft is expressed in degrees.

The simplest crankshaft is the single-throw or 360° type. This type is used in a single-row radial engine. It can be constructed in one or two pieces. Two main bearings (one on each end) are provided when this type of crankshaft is used.

The double-throw or 180° crankshaft is used on double-row radial engines. In the radial-type engine, one throw is provided for each row of cylinders.

Crankshaft Balance

Excessive vibration in an engine not only results in fatigue failure of the metal structures, but also causes the moving parts to wear rapidly. In some instances, excessive vibration is caused by a crankshaft which is not balanced. Crankshafts are balanced for static balance and dynamic balance.

A crankshaft is **statically balanced** when the weight of the entire assembly of crankpins, crankcheeks, and counterweights is balanced around the axis of rotation. When testing the crankshaft for static balance, it is placed on two knife edges. If the shaft tends to turn toward any one position during the test, it is out of static balance.

A crankshaft is **dynamically balanced** when all the forces created by crankshaft rotation and power impulses are balanced within themselves so that little or no vibration is produced when the engine is operating. To reduce vibration to a minimum during engine operation, **dynamic dampers** are incorporated on the crankshaft. A dynamic damper is merely a pendulum which is so fastened to the crankshaft that it is free to move in a small arc. It is incorporated in the counterweight assembly. Some crankshafts incorporate two or

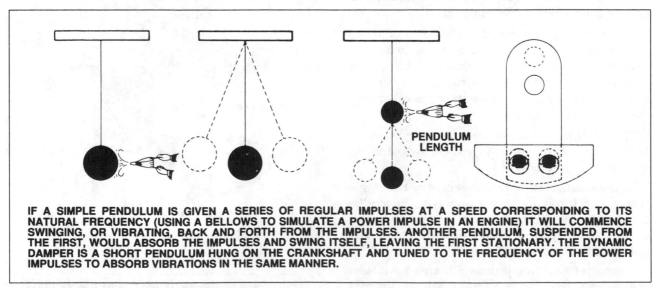

IF A SIMPLE PENDULUM IS GIVEN A SERIES OF REGULAR IMPULSES AT A SPEED CORRESPONDING TO ITS NATURAL FREQUENCY (USING A BELLOWS TO SIMULATE A POWER IMPULSE IN AN ENGINE) IT WILL COMMENCE SWINGING, OR VIBRATING, BACK AND FORTH FROM THE IMPULSES. ANOTHER PENDULUM, SUSPENDED FROM THE FIRST, WOULD ABSORB THE IMPULSES AND SWING ITSELF, LEAVING THE FIRST STATIONARY. THE DYNAMIC DAMPER IS A SHORT PENDULUM HUNG ON THE CRANKSHAFT AND TUNED TO THE FREQUENCY OF THE POWER IMPULSES TO ABSORB VIBRATIONS IN THE SAME MANNER.

Figure 2-10. Principles of a dynamic damper.

more of these assemblies, each being attached to a different crankcheek. The distance the pendulum moves and its vibrating frequency correspond to the frequency of the power impulses of the engine. When the vibration frequency of the crankshaft occurs, the pendulum oscillates out of time with the crankshaft vibration, thus reducing vibration to a minimum. See figure 2-10.

Dynamic Dampers

The construction of the dynamic damper used in one engine consists of a movable slotted-steel counterweight attached to the crankcheek. Two spool-shaped steel pins extend into the slot and pass through oversized holes in the counterweight and crankcheek. The difference in the diameter between the pins and the holes provides a pendulum effect. An analogy of the functioning of a dynamic damper is shown in figure 2-10.

Connecting Rods

The **connecting rod** is the link which transmits forces between the piston and the crankshaft. Connecting rods must be strong enough to remain rigid under load and yet be light enough to reduce the inertia forces which are produced when the rod and piston stop, change direction, and start again at the end of each stroke.

There are three types of connecting-rod assemblies: (1) the **plain-type** connecting rod, (2) the **fork-and-blade** connecting rod, and (3) the **master-and-articulated-rod** assembly. (See figure 2-11.)

Master-And-Articulated Rod Assembly

The master-and-articulated rod assembly is commonly used in radial engines. In a radial engine the piston in one cylinder in each row is connected to the crankshaft by a master rod. All other pistons in the row are connected to the master rod by an articulated rod. In an 18-cylinder engine which has two rows of cylinders, there are two master rods and 16 articulated rods. The articulated rods are constructed of forged steel alloy in either the I- or H-shape, denoting the cross-sectional shape. Bronze bushings are pressed into the bores in each end of the articu-

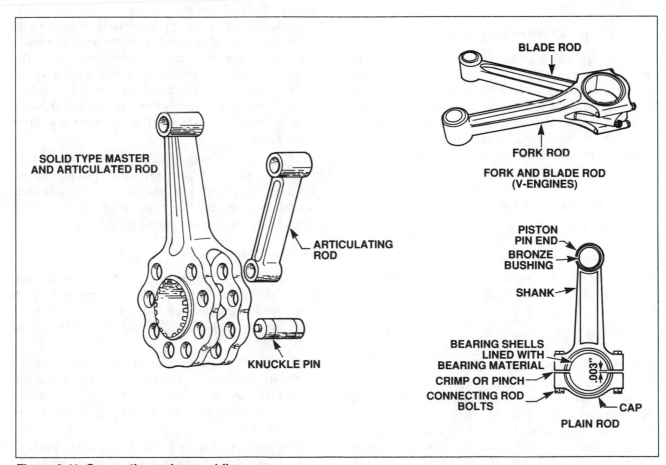

Figure 2-11. Connecting rod assemblies.

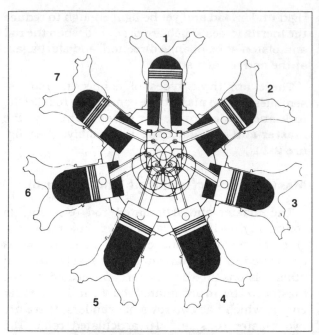

Figure 2-12. Elliptical travel path of knuckle pins in an articulated rod assembly (radial engine).

lated rod to provide knuckle-pin and piston-pin bearings.

Plain-Type Connecting Rods

Plain-type connecting rods are used in in-line and opposed engines. The end of the rod attached to the crankpin is fitted with a cap and a two-piece bearing. The bearing cap is held on the end of the rod by bolts or studs. To maintain proper fit and balance, connecting rods should always be replaced in the same cylinder and in the same relative position.

Fork-And-Blade Rod Assembly

The fork-and-blade rod assembly is used primarily in V-type engines. The forked rod is split at the crankpin end to allow space for the blade rod to fit between the prongs. A single two-piece bearing is used on the crankshaft end of the rod.

Pistons

The **piston** of a reciprocating engine is a cylindrical member which moves back and forth within a steel cylinder. The piston acts as a moving wall within the combustion chamber. As the piston moves down in the cylinder, it draws in the fuel/air mixture. As it moves upward, it compresses the charge, ignition occurs, and the expanding gases force the piston downward. This force is transmitted to the crankshaft through the connecting rod. On the return upward stroke, the piston forces the exhaust gases from the cylinder.

Piston Construction

The majority of aircraft engine **pistons** are machined from aluminum alloy forgings. Grooves are machined in the outside surface of the piston to receive the piston rings, and cooling fins are provided on the inside of the piston for greater heat transfer to the engine oil.

Pistons may be either the trunk type or the slipper type; both are shown in figure 2-13. Slipper-type pistons are not used in modern, high-powered engines because they do not provide adequate strength or wear resistance. The top face of the piston, or head, may be either flat, convex, or concave. Recesses may be machined in the piston head to prevent interference with the valves.

As many as six grooves may be machined around the piston to accommodate the compression rings and oil rings. (See figure 2-13.) The compression rings are installed in the three uppermost grooves; the oil control rings are installed immediately above the piston pin. The piston is usually drilled at the oil control ring grooves to allow surplus oil scraped from the cylinder walls by the oil control rings to pass back into the crankcase. An oil scraper ring is installed at the base of the piston wall or skirt to prevent excessive oil consumption. The portions of the piston walls that lie between each pair of ring grooves are called the ring lands.

In addition to acting as a guide for the piston head, the piston skirt incorporates the piston-pin bosses. The piston-pin bosses are of heavy construction to enable the heavy load on the piston head to be transferred to the piston pin.

Piston Pin

The **piston pin,** sometimes called the **wrist pin,** joins the piston to the connecting rod. It is machined in the form of a tube from a nickel steel alloy forging, casehardened and ground to precise dimensions.

The piston pin used in modern aircraft engines is the full-floating type, so called because the pin is free to rotate in both the piston and in the connecting rod piston-pin bearing, resulting in more even wear.

The piston pin must be held in place to prevent the pin ends from scoring the cylinder walls. In earlier engines, spring coils were installed in

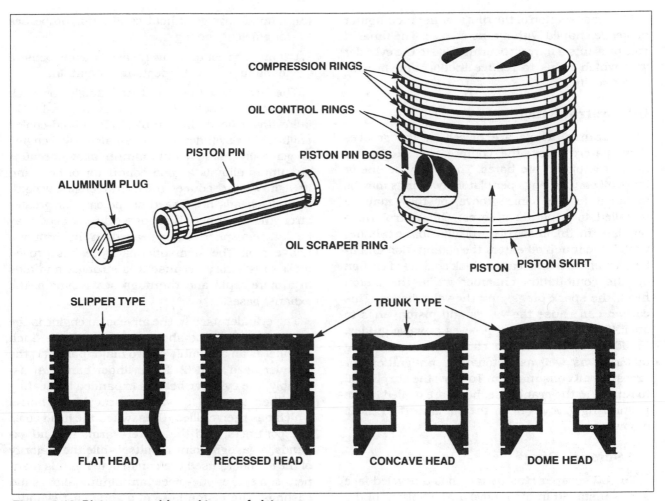

Figure 2-13. Piston assembly and types of pistons.

grooves in the piston-pin bores at either end of the pin. The current practice is to install a plug of relatively soft aluminum in the pin ends to provide a good bearing surface against the cylinder wall.

Piston Rings

The **piston rings** prevent leakage of gas pressure from the combustion chamber and reduce to a minimum the seepage of oil into the combustion chamber. The rings fit into the piston grooves but spring out to press against the cylinder walls; when properly lubricated, the rings form an effective gas seal.

Piston Ring Construction

Most piston rings are made of high-grade cast iron. After the rings are made, they are ground to the cross section desired. They are then split so that they can be slipped over the outside of the piston and into the ring grooves which are machined in the piston wall. Since their purpose is to seal the clearance between the piston and the cylinder wall, they must fit the cylinder wall snugly enough to provide a gastight fit; they must exert equal pressure at all points on the cylinder wall; and they must make a gastight fit against the sides of the ring grooves.

Gray cast iron is most often used in making piston rings. However, many other materials have been tried. In some engines, chrome-plated mild steel piston rings are used in the top compression ring groove because these rings can better withstand the high temperatures present at this point.

Compression Ring

The purpose of the **compression rings** is to prevent the escape of gas past the piston during engine operation. They are placed in the ring grooves immediately below the piston head. The number of compression rings used on each piston is determined by the type of engine and its design, although most aircraft engines use two compression rings plus one or more oil control rings.

The cross section of the ring is either rectangular or wedge shaped with a tapered face. The tapered face presents a narrow bearing edge to the cylinder wall which helps to reduce friction and provide better sealing.

Oil Control Rings

Oil control rings are placed in the grooves immediately below the compression rings and above the piston pin bores. There may be one or more oil control rings per piston; two rings may be installed in the same groove, or they may be installed in separate grooves. Oil control rings regulate the thickness of the oil film on the cylinder wall. If too much oil enters the combustion chamber, it will burn and leave a thick coating of carbon on the combustion chamber walls, the piston head, the spark plugs, and the valve heads. This carbon can cause the valves and piston rings to stick if it enters the ring grooves or valve guides. In addition, the carbon can cause spark plug misfiring as well as detonation, preignition, or excessive oil consumption. To allow the surplus oil to return to the crankcase, holes are drilled in the piston ring grooves or in the lands next to these grooves.

Oil Scraper Ring

The **oil scraper ring** usually has a beveled face and is installed in the groove at the bottom of the piston skirt. The ring is installed with the scraping edge away from the piston head or in the reverse position, depending upon cylinder position and the engine series. In the reverse position, the scraper ring retains the surplus oil above the ring on the upward piston stroke, and this oil is returned to the crankcase by the oil control rings on the downward stroke.

Cylinders

The portion of the engine in which the power is developed is called the **cylinder.** The cylinder provides a combustion chamber where the burning and expansion of gases take place, and it houses the piston and the connecting rod.

There are four major factors that need to be considered in the design and construction of the cylinder assembly. These are:

(1) It must be strong enough to withstand the internal pressures developed during engine operation.

(2) It must be constructed of a lightweight metal to keep down engine weight.

(3) It must have good heat-conducting properties for efficient cooling.

(4) It must be comparatively easy and inexpensive to manufacture, inspect, and maintain.

The head is either produced singly for each cylinder in air-cooled engines, or is cast "in-block" (all cylinder heads in one block) for liquid-cooled engines. The cylinder head of an air-cooled engine is generally made of aluminum alloy, because aluminum alloy is a good conductor of heat and its light weight reduces the overall engine weight. Cylinder heads are forged or die-cast for greater strength. The inner shape of a cylinder head may be flat, semispherical, or peaked, in the form of a house roof. The semispherical type has proved most satisfactory because it is stronger and aids in a more rapid and thorough scavenging of the exhaust gases.

The cylinder used in the air-cooled engine is the overhead valve type shown in figure 2-14. Each cylinder is an assembly of two major parts: (1) the cylinder head, and (2) the cylinder barrel. At assembly, the cylinder head is expanded by heating and then screwed down on the cylinder barrel which has been chilled; thus, when the head cools and contracts, and the barrel warms up and expands, a gastight joint results. While the majority of the cylinders used are constructed in this manner, some are one-piece aluminum alloy sand castings. The piston bore of a sand cast cylinder is fitted with a steel liner which extends the full length of the cylinder barrel section and projects

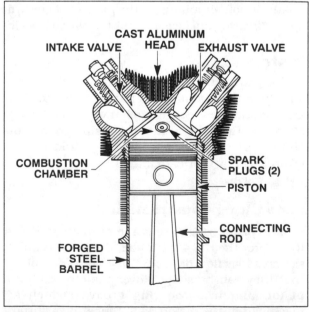

Figure 2-14. Cutaway view of the cylinder assembly.

below the cylinder flange of the casting. This liner is easily removed, and a new one can be installed in the field.

Cylinder Heads

The purpose of the cylinder head is to provide a place for combustion of the fuel/air mixture and to give the cylinder more heat conductivity for adequate cooling. The fuel/air mixture is ignited by the spark in the combustion chamber and commences burning as the piston travels toward top dead center on the compression stroke. The ignited charge is rapidly expanding at this time, and pressure is increasing so that as the piston travels through the top dead center position, it is driven downward on the power stroke. The intake and exhaust valve ports are located in the cylinder head along with the spark plugs and the intake and exhaust valve actuating mechanisms.

After casting, the spark plug bushings, valve guides, rocker arm bushings, and valve seats are installed in the cylinder head. Spark plug openings may be fitted with bronze or steel bushings that are shrunk and screwed into the openings. Stainless steel Heli-Coil® spark plug inserts are used in many engines currently manufactured. Bronze or steel valve guides are usually shrunk or screwed into drilled openings in the cylinder head to provide guides for the valve stems. These are generally located at an angle to the center line of the cylinder. The valve seats are circular rings of hardened metal which protect the relatively soft metal of the cylinder head from the hammering action of the valves and from the exhaust gases.

The cylinder heads of air-cooled engines are subjected to extreme temperatures; it is therefore necessary to provide adequate fin area, and to use metals which conduct heat rapidly. Cylinder heads of air-cooled engines are usually cast or forged singly. Aluminum alloy is used in the construction for a number of reasons. It is well adapted for casting or for the machining of deep, closely spaced fins, and it is more resistant than most metals to the corrosive attack of tetraethyl lead in gasoline. The greatest improvement in air cooling has resulted from reducing the thickness of the fins and increasing their depth. In this way the fin area has been increased from approximately 1,200 square inches to more than 7,500 square inches per cylinder in modern engines. Cooling fins taper from 0.090 inch at the base to 0.060 inch at the tip end. Because of the difference in temperature in the various sections of the cylinder head, it is necessary to provide more cooling-fin area on some sections than on others. The exhaust valve region is the hottest part of the internal surface; therefore, more fin area is provided around the outside of the cylinder in this section.

Cylinder Barrels

In general, the cylinder barrel in which the piston operates must be made of a high-strength material, usually steel. It must be as light as possible, yet have the proper characteristics for operating under high temperatures. It must be made of a good bearing material and have high tensile strength.

The cylinder barrel is made of a steel alloy forging with the inner surface hardened to resist wear of the piston and the piston rings which bear against it. This hardening is usually done by exposing the steel to ammonia or cyanide gas while the steel is very hot. The steel soaks up nitrogen from the gas which forms iron nitrides on the exposed surface. As a result of this process, the metal is said to be **nitrided.**

Cylinder walls of used cylinders that are worn out-of-round can be ground out to make them round again and re-nitrided or **chrome plated.** The chrome is plated onto the cylinder wall using a current reversal process that leaves small channels in the surface of the chrome where oil "hides" to improve lubrication of the very hard, durable chromed cylinder wall. Chromed cylinders can usually be recognized by the orange colored paint mark on the outside of the cylinder. More recent processes called *cermachrome* and *ceramisteel* provide other methods of hardening cylinder walls to make them more durable.

In some instances the barrel will have threads on the outside surface at one end so that it can be screwed into the cylinder head. Some air-cooled cylinder barrels have replaceable aluminum cooling fins attached to them, while others have the cooling fins machined as an integral part of the barrel.

Cylinder Numbering

Occasionally it is necessary to refer to the left or right side of the engine or to a particular cylinder. Therefore, it is necessary to know the engine directions and how cylinders of an engine are numbered.

The propeller shaft end of the engine is always the **front** end, and the accessory end is the **rear** end, regardless of how the engine is mounted in an aircraft. When referring to the right side or left side of an engine, always assume you are viewing

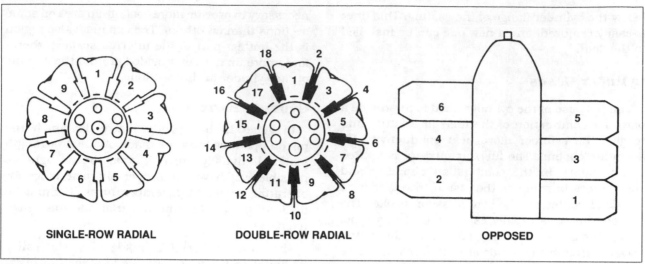

Figure 2-15. Numbering of engine cylinders, viewed from behind (radial) or above (opposed).

it from the rear or accessory end (pilot's view). As seen from this position, crankshaft rotation is referred to as either clockwise or counterclockwise.

Radial engine cylinders are numbered clockwise as viewed from the accessory end. In-line and V-type engine cylinders are usually numbered from the rear. In V-engines, the cylinder banks are known as the right bank and the left bank, as viewed from the accessory end.

The numbering of engine cylinders is shown in figure 2-15. Note that the cylinder numbering of the opposed engine shown begins with the right rear as No. 1, and the left rear as No. 2. The one forward of No. 1 is No. 3; the one forward of No. 2, is No. 4, and so on. The numbering of opposed engine cylinders is by no means standard. Some manufacturers number their cylinders from the rear and others from the front of the engine. Always refer to the appropriate engine manual to determine the correct numbering system used by the manufacturer.

Single-row radial engine cylinders are numbered clockwise when viewed from the rear end. Cylinder No. 1 is the top cylinder. In double-row engines, the same system is used, in that the No. 1 cylinder is the top one in the rear row. No. 2 cylinder is the first one clockwise from No. 1, but No. 2 is in the front row. No. 3 cylinder is the next one clockwise to No. 2, but is in the rear row. Thus, all odd-numbered cylinders are in the rear row, and all even-numbered cylinders are in the front row.

Firing Order

The firing order of an engine is the sequence in which the power event occurs in the different cylinders. The firing order is designed to provide for balance and to eliminate vibration to the greatest extent possible. In radial engines the firing order must follow a special pattern, since the firing impulses must follow the motion of the crankthrow during its rotation. In in-line engines the firing orders may vary somewhat, yet most orders are arranged so that the firing of cylinders is evenly distributed along the crankshaft. Six-cylinder in-line engines generally have a firing order of 1-5-3-6-2-4. Cylinder firing order in opposed engines can usually be listed in pairs of cylinders, as each pair fires across the center main bearing. The firing order of six-cylinder opposed engines is 1-4-5-2-3-6. The firing order of one model four-cylinder opposed engine is 1-4-2-3, but on another model it is 1-3-2-4.

Valves

The fuel/air mixture enters the cylinders through the intake valve ports, and burned gases are expelled through the exhaust valve ports. The head of each valve opens and closes these cylinder ports. The valves used in aircraft engines are the conventional poppet type. The valves are also typed by their shape and are called either mushroom or tulip because of their resemblance to the shape of these plants. Figure 2-16 illustrates various shapes and types of these valves.

Valve Construction

The **valves** in the cylinders of an aircraft engine are subjected to high temperatures, corrosion, and operating stresses; thus, the metal alloy in the valves must be able to resist all these factors.

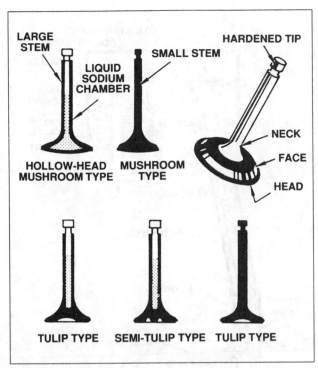

Figure 2-16. Valve types.

Because **intake valves** operate at lower temperatures than **exhaust valves,** they can be made of chrome-nickel steel. Exhaust valves are usually made of nichrome, silchrome, or cobalt-chromium steel.

The valve head has a ground face which forms a seal against the ground valve seat in the cylinder head when the valve is closed. The face of the valve is usually ground to an angle of either 30° or 45°. In some engines, the intake-valve face is ground to an angle of 30°, and the exhaust valve face is ground to a 45° angle.

Valve faces are often made more durable by the application of a material called stellite. About 1/16 inch of this alloy is welded to the valve face and ground to the correct angle. Stellite is resistant to high-temperature corrosion and also withstands the shock and wear associated with valve operation. Some engine manufacturers use a nichrome facing on the valves. This serves the same purpose as the stellite material.

The valve stem acts as a pilot for the valve head and rides in the valve guide installed in the cylinder head for this purpose. The valve stem is surface-hardened to resist wear. The neck is the part that forms the junction between the head and the stem. The tip of the valve is hardened to withstand the hammering of the valve rocker arm as it opens the valve. A machined groove on the stem near the tip receives the split-ring stem keys.

These stem keys form a lock ring to hold the valve spring retaining washer in place.

Some intake and exhaust valve stems are hollow and partially filled with metallic sodium. This material is used because it is an excellent heat conductor. The sodium will melt at approximately 208 °F, and the reciprocating motion of the valve circulates the liquid sodium and enables it to carry away heat from the valve head to the valve stem, where it is dissipated through the valve guide to the cylinder head and the cooling fins. Thus, the operating temperature of the valve may be reduced as much as 300° to 400 °F. Under no circumstances should a sodium-filled valve be cut open or subjected to treatment which may cause it to rupture. Exposure of the sodium in these valves to the outside air will result in fire or explosion with possible personal injury.

The most commonly used intake valves have solid stems, and the head is either flat or tulip shaped. Intake valves for low-power engines are usually flat headed.

In some engines, the intake valve may be the tulip type and have a smaller stem than the exhaust valve, or it may be similar to the exhaust valve but have a solid stem and head. Although these valves are similar, they are not interchangeable since the faces of the valves are constructed of different material. The intake valve will usually have a flat milled on the tip to identify it.

Valve-Operating Mechanism

For a reciprocating engine to operate properly, each valve must open at the proper time, stay open for the required length of time, and close at the proper time. Intake valves are opened just before the piston reaches top dead center, and exhaust valves remain open after top dead center. At a particular instant, therefore, both valves are open at the same time (end of the exhaust stroke and beginning of the intake stroke). This valve-overlap permits better volumetric efficiency and lowers the cylinder operating temperature. This timing of the valves is controlled by the valve-operating mechanism.

The **valve lift** (distance that the valve is lifted off its seat) and the **valve duration** (length of time the valve is held open) are both determined by the shape of the **cam lobes.**

Typical cam lobes are illustrated in figure 2-17. The portion of the lobe that gently starts the valve-operating mechanism moving is called a ramp, or step. The ramp is machined on each side of the cam lobe to permit the rocker arm to be eased

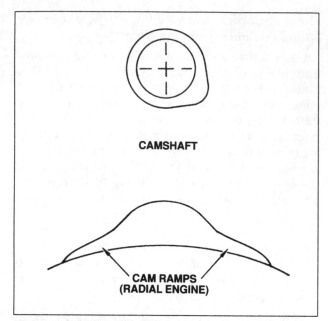

Figure 2-17. Typical cam lobes.

into contact with the valve tip and thus reduce the shock load which would otherwise occur.

The valve-operating mechanism consists of a cam ring or camshaft equipped with lobes, which work against a cam roller or a cam follower. (See figures 2-18 and 2-19.) The cam follower, in turn, pushes a push rod and ball socket, which, in turn, actuates a rocker arm which opens the valve. Springs, which slip over the stem of the valves and which are held in place by the valve-spring retaining washer and stem key, close each valve and push the valve mechanism in the opposite direction when the cam roller or follower rolls along a low section of the cam ring.

Camshaft

The valve mechanism of an opposed engine is operated by a camshaft. The camshaft is driven by a gear that mates with another gear attached to the crankshaft (see figure 2-20). The camshaft always rotates at one-half the crankshaft speed. As the camshaft revolves, the lobes cause the tappet assembly to rise in the tappet guide, transmitting the force through the push rod and rocker arm to open the valve.

Tappet Assembly

The tappet assembly consists of:

(1) A cylindrical tappet, which slides in and out in a tappet guide installed in one of the crankcase sections around the cam ring.

(2) A cam follower or tappet roller, which follows the contour of the cam ring and lobes.

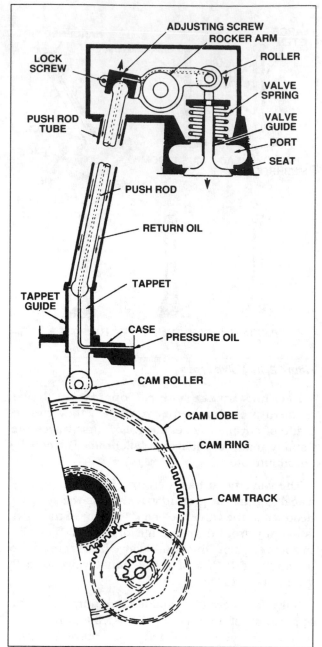

Figure 2-18. Valve-operating mechanism (radial engine).

(3) A tappet ball socket or push rod socket.

(4) A tappet spring.

The function of the tappet assembly is to convert the rotational movement of the cam lobe into reciprocating motion and to transmit this motion to the push rod, rocker arm, and then to the valve tip, opening the valve at the proper time. The purpose of the tappet spring is to take up the clearance between the rocker arm and the valve tip to reduce the shock load when the valve is opened. A hole is drilled through the tappet to

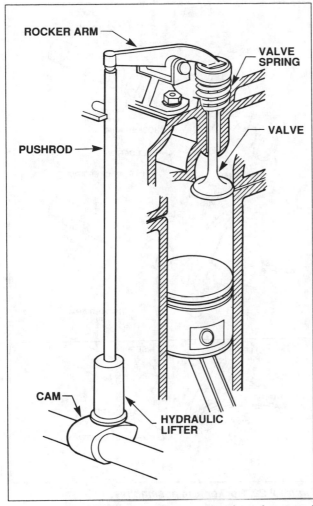

Figure 2-19. Valve-operating mechanism (opposed engine).

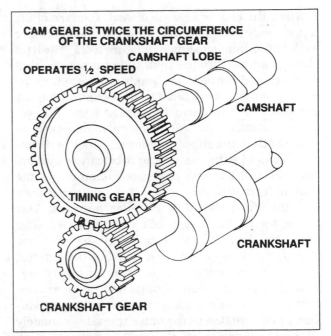

Figure 2-20. Cam drive mechanism opposed-type aircraft engine.

allow engine oil to flow to the hollow push rods to lubricate the rocker assemblies.

Hydraulic Valve Tappets

Some aircraft engines incorporate hydraulic tappets which automatically keep the valve clearance at zero, eliminating the necessity for any valve clearance adjustment mechanism. A typical hydraulic tappet (zero-lash valve lifter) is shown in figure 2-21.

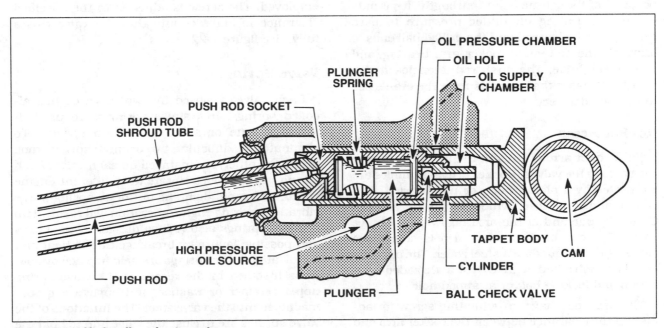

Figure 2-21. Hydraulic valve tappets.

When the engine valve is closed, the face of the tappet body (cam follower) is on the base circle or back of the cam, as shown in figure 2-21. The light plunger spring lifts the hydraulic plunger so that its outer end contacts the push rod socket, exerting a light pressure against it, thus eliminating any clearance in the valve linkage. As the plunger moves outward, the ball check valve moves off its seat. Oil from the supply chamber, which is directly connected with the engine lubrication system, flows in and fills the pressure chamber. As the camshaft rotates, the cam pushes the tappet body and the hydraulic lifter cylinder outward. This action forces the ball check valve onto its seat; thus, the body of oil trapped in the pressure chamber acts as a cushion. During the interval when the engine valve is off its seat, a predetermined leakage occurs between plunger and cylinder bore which compensates for any expansion or contraction in the valve train. Immediately after the engine valve closes, the amount of oil required to fill the pressure chamber flows in from the supply chamber, preparing for another cycle of operation.

Push Rod

The **push rod,** tubular in form, transmits the force from the valve tappet to the rocker arm. A hardened-steel ball is pressed over or into each end of the tube. One ball end fits into the socket of the rocker arm. In some instances the balls are on the tappet and rocker arm, and the sockets are on the push rod. The tubular form is employed because of its lightness and strength. It permits engine lubricating oil under pressure to pass through the hollow rod and the drilled ball ends to lubricate the ball ends, rocker-arm bearing, and valve-stem guide. The push rod is enclosed in a tubular housing that extends from the crankcase to the cylinder head.

Rocker Arms

The **rocker arms** transmit the lifting force from the cams to the valves. Rocker arm assemblies are supported by a plain, roller, or ball bearing, or a combination of these, which serves as a pivot. Generally one end of the arm bears on the valve stem. One end of the rocker arm is sometimes slotted to accommodate a steel roller. The opposite end is constructed with either a threaded split clamp and locking bolt or a tapped hole.

The arm may have an adjusting screw for adjusting the clearance between the rocker arm and the valve stem if hydraulic tappets are not

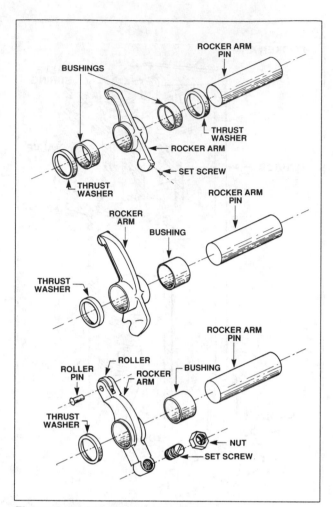

Figure 2-22. Rocker arm assemblies.

employed. The screw is adjusted to the specified clearance to make certain that the valve closes fully. See figure 2-22.

Valve Springs

Each valve is closed by two or three helical-coiled springs. If a single spring were used, it would vibrate or surge at certain speeds. To eliminate this difficulty, two or more springs (one inside the other) are installed on each valve. Each spring will therefore vibrate at a different engine speed, and rapid damping out of all spring-surge vibrations during engine operation will result. Two or more springs also reduce danger of weakness and possible failure by breakage due to heat and metal fatigue. The springs are held in place by split locks installed in the recess of the valve spring upper retainer or washer, and engage a groove machined into the valve stem. The functions of the valve springs are to close the valve and to hold the valve securely on the valve seat.

Bearings

A **bearing** is any surface which supports, or is supported by, another surface. A good bearing must be composed of material that is strong enough to withstand the pressure imposed on it and should permit the other surface to move with a minimum of friction and wear. The parts must be held in position within very close tolerances to

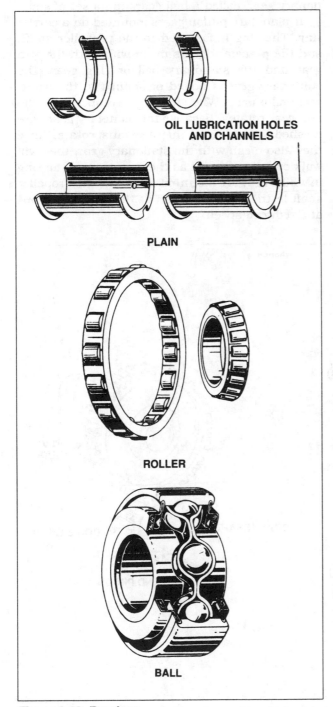

OIL LUBRICATION HOLES AND CHANNELS

PLAIN

ROLLER

BALL

Figure 2-23. Bearings.

provide efficient and quiet operation, and yet allow freedom of motion. To accomplish this, and at the same time reduce friction of moving parts so that power loss is not excessive, lubricated bearings of many types are used. Bearings are required to take radial loads, thrust loads, or a combination of the two. There are two ways in which bearing surfaces move in relation to each other. One is by the sliding movement of one metal against the other, and the second is for one surface to roll over the other. The three different types of bearings in general use are plain, roller, and ball (see figure 2-23).

Plain Bearings

Plain bearings are generally used for the crankshaft, cam ring, camshaft, connecting rods, and the accessory drive shaft bearings. Such bearings are usually subjected to radial loads only, although some have been designed to take thrust loads.

Plain bearings are usually made of nonferrous (having no iron) metals, such as silver, bronze, aluminum, and various alloys of copper, tin, or lead. Master rod or crankpin bearings in some engines are thin shells of steel, plated with silver on both the inside and the outside surfaces and with lead-tin plated over the silver on the inside surface only. Smaller bearings, such as those used to support various shafts in the accessory section, are called bushings. Porous Oilite bushings are widely used in this instance. They are impregnated with oil so that the heat of friction brings the oil to the bearing surface during engine operation.

Ball Bearings

A ball bearing assembly consists of grooved inner and outer races, one or more sets of balls, and, in bearings designed for disassembly, a bearing retainer. They are used for supercharger impeller shaft bearings and rocker arm bearings in some engines. Special deep-groove ball bearings are used in aircraft engines to transmit propeller thrust to the engine nose section.

Roller Bearings

Roller bearings are made in many types and shapes, but the two types generally used in the aircraft engine are the straight roller and the tapered roller bearings. Straight roller bearings are used where the bearing is subjected to radial loads only. In tapered roller bearings, the inner- and outer-race bearing surfaces are cone shaped. Such bearings will withstand both radial and thrust loads. Straight roller bearings are used in high-power aircraft engines for the crankshaft main bearings. They are also used in other applications where radial loads are high.

Propeller Reduction Gearing

The increased brake horsepower delivered by a high-horsepower engine may be the result of increased crankshaft RPM. It is therefore necessary to provide reduction gears to limit the propeller rotation speed to a value at which efficient operation of the propeller is obtained. Whenever the speed of the blade tips approaches the speed of sound, the efficiency of the propeller decreases rapidly.

Since reduction gearing must withstand extremely high stresses, the gears are machined from steel forgings. Many types of reduction gearing systems are in use. The three types (figure 2-24) most commonly used are:

(1) Spur planetary

(2) Bevel planetary

(3) Spur and pinion

The planetary reduction gear systems are used with radial and opposed engines, and the spur and pinion system is used with in-line and V-type engines. Two of these types, the spur planetary and the bevel planetary, are discussed here.

The spur planetary reduction gearing consists of a large driving gear or sun gear splined (and sometimes shrunk) to the crankshaft, a large stationary gear, called a bell gear, and a set of small spur planetary pinion gears mounted on a carrier ring. The ring is fastened to the propeller shaft, and the planetary gears mesh with both the sun gear and the stationary bell or ring gear. The stationary gear is bolted or splined to the front-section housing. When the engine is operating, the sun gear rotates. Because the planetary gears are meshed with this ring, they also must rotate. Since they also mesh with the stationary gear, they will walk or roll around it as they rotate, and the ring in which they are mounted will rotate the propeller shaft in the same direction as the crankshaft but at a reduced speed.

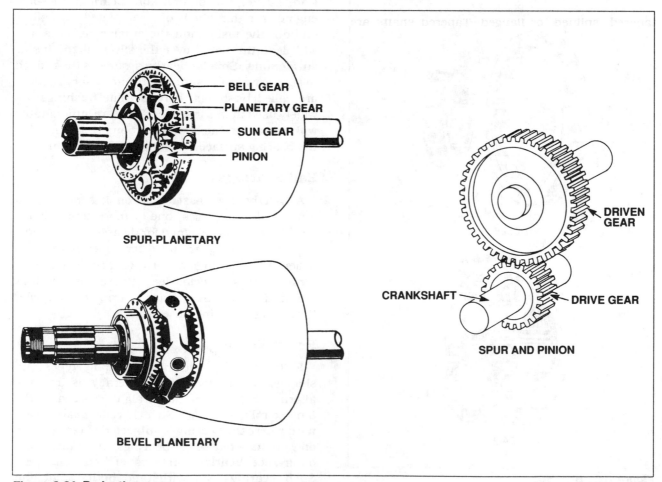

Figure 2-24. Reduction gears.

In some engines, the bell gear is mounted on the propeller shaft, and the planetary pinion gear cage is held stationary. The sun gear is splined to the crankshaft and thus acts as a driving gear. In such an arrangement, the propeller travels at a reduced speed, but in opposite direction to the crankshaft.

In the bevel planetary reduction gearing system, the driving gear is machined with beveled external teeth and is attached to the crankshaft. A set of mating bevel pinion gears is mounted in a cage attached to the end of the propeller shaft. The pinion gears are driven by the drive gear and walk around the stationary gear, which is bolted or splined to the front-section housing. The thrust of the bevel pinion gears is absorbed by a thrust ball bearing of special design. The drive and the fixed gears are generally supported by heavy-duty ball bearings. This type of planetary reduction assembly is more compact than the other one described and can therefore be used where a smaller propeller gear step-down is desired.

Propeller Shafts

Propeller shafts may be three major types; tapered, splined, or flanged. Tapered shafts are identified by taper numbers. Splined and flanged shafts are identified by SAE numbers.

The propeller shaft of most low-power output engines is forged as part of the crankshaft. It is tapered and a milled slot is provided so that the propeller hub can be keyed to the shaft. The keyway and key index of the propeller are in relation to the #1 cylinder top dead center. The end of the shaft is threaded to receive the propeller retaining nut. Tapered propeller shafts are common on older and in-line engines.

The propeller shaft of a high-output engine generally is splined. It is threaded on one end for a propeller hub nut. The thrust bearing, which absorbs propeller thrust, is located around the shaft and transmits the thrust to the nose-section housing. The shaft is threaded for attaching the thrust-bearing retaining nut. On the portion protruding from the housing (between the two sets of threads), splines are located to receive the splined propeller hub. The shaft is generally machined from a steel-alloy forging throughout its length. The propeller shaft may be connected by reduction gearing to the engine crankshaft, but in smaller engines the propeller shaft is simply an extension of the engine crankshaft. To turn the propeller shaft, the engine crankshaft must revolve.

Flanged propeller shafts are used on medium or low powered reciprocating and turbojet engines. One end of the shaft is flanged with drilled holes to accept the propeller mounting bolts. The installation may be a short shaft with internal threading to accept the distributor valve to be used with a controllable propeller. The flanged propeller shaft is a normal installation on most approved reciprocating engines.

Study Questions And Problems

1. What is the principal fluid used for propulsion by most heat engines?

2. Define Thrust.

3. Derive the THP to thrust conversion factor when airspeed is in knots rather than MPH. (Your answer should be in units of nautical mile-pounds per hour.)

4. What is THP? (Name and define or describe it.)

5. How much thrust is being used by an aircraft that is using 100 THP to go 100 MPH?

6. How much more or less thrust than the airplane of Question #5 is another aircraft using that is using 100 THP, just like the airplane of Question #5, but is going 90 KTS?

7. Trace the energy that is used to open the intake or exhaust valves from its source (the fuel burning above the piston), naming each engine part in proper sequence that transfers energy (force) to open the valve.

8. What are the advantages of a hydraulic lifter?

9. Name all the parts in a typical cylinder assembly; describe the function of each and what material it is made of.

10. What is the purpose of the process called nitriding?

11. What is the main function of a piston?

12. Describe the purpose and operation of a dynamic damper.

Chapter III
Reciprocating Engine Theory Of Operation

Reciprocating Gasoline Engine Operating Principles

A study of this section will help in understanding the basic operating principles of reciprocating engines. The principles which govern the relationship between the pressure, volume, and temperature of gases is primary to understanding engine operation.

An internal-combustion engine is a device for converting chemical energy of fuel into heat energy, then into mechanical energy. Gasoline is vaporized and mixed with air, forced or drawn into a cylinder, compressed by a piston, and then ignited by an electric spark. The conversion of the resultant heat energy into mechanical energy and then into work is accomplished within the cylinder. Figure 3-1 illustrates the various engine components necessary to accomplish this conversion and also presents the principal terms associated with engine operation.

The **operating cycle** of an internal combustion reciprocating engine includes the series of events required to induct, compress, ignite, burn, expand the fuel/air charge in the cylinder, and to exhaust the by-products of the combustion process.

When the compressed mixture of fuel and air is ignited, the resultant gases of combustion expand very rapidly and force the piston to move away from the cylinder head. This downward motion of the piston, acting on the crankshaft through the connecting rod, is converted to a circular or rotary motion by the crankshaft.

A valve in the top or head of the cylinder opens to allow the burned gases to escape, and the momentum of the crankshaft and the propeller forces the piston back up in the cylinder where it is ready for the next event in the cycle. The intake valve in the cylinder head then opens to let in a fresh charge of the fuel/air mixture.

The valve allowing for the escape of the burning exhaust gases is called the **exhaust valve**, and the valve which lets in the fresh charge of the fuel/air mixture is called the **intake valve**. These valves are opened and closed mechanically at the proper times by the valve-operating mechanism.

The **bore** of a cylinder is its inside diameter. The **stroke** is the distance the piston moves from one end of the cylinder to the other, specifically, from **top dead center** (TDC) to **bottom dead center** (BDC), or vice versa (see figure 3-1).

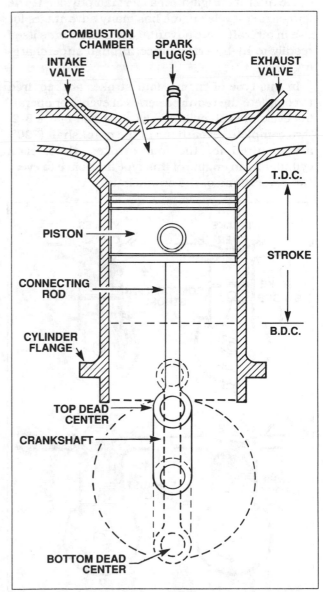

Figure 3-1. Components and terminology of engine operation.

39

Operating Cycles

There are two operating cycles in general use: (1) the two-stroke cycle, and (2) the four-stroke cycle. At the present time, except for a few very light aircraft, the two-stroke-cycle engine is fast disappearing from the aviation scene. As the name implies, two-stroke-cycle engines require only one upstroke and one downstroke of the piston to complete the required series of events in the cylinder. Thus the engine completes the operating cycle in one revolution of the crankshaft.

Four-Stroke Cycle

Most aircraft reciprocating engines operate on the four-stroke cycle, sometimes called the **Otto cycle** after its originator, a German physicist. The four-stroke-cycle engine has many advantages for use in aircraft. One advantage is that it lends itself readily to high performance through supercharging.

In this type of engine, four strokes are required to complete the required series of events or operating cycle of each cylinder, as shown in figure 3-2. Two complete revolutions of the crankshaft (720°) are required for the four strokes; thus, each cylinder in an engine of this type fires once in every two revolutions of the crankshaft.

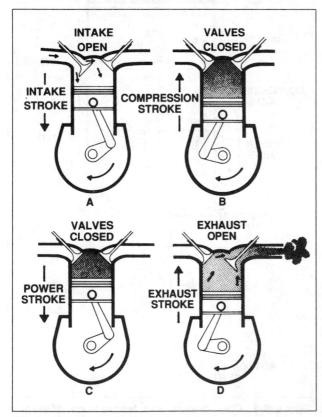

Figure 3-2. Four-stroke cycle.

In the following discussion of the four-stroke-cycle engine operation, it should be realized that the **timing** of the ignition and the valve events will vary considerably in different engines. Many factors influence the timing of a specific engine, and it is most important that the engine manufacturer's recommendations in this respect be followed in maintenance and overhaul. The timing of the valve and ignition events is always specified in degrees of crankshaft travel.

In the following paragraphs, the timing of each event is specified in terms of degrees of crankshaft travel on the stroke during which the event occurs. It should be remembered that a certain amount of crankshaft travel is required to open a valve fully; therefore, the specified timing represents the start of opening rather than the full-open position of the valve.

Intake Stroke

During the intake stroke, the piston is pulled downward in the cylinder by the rotation of the crankshaft. This reduces the pressure in the cylinder and causes air under atmospheric pressure to flow through the carburetor, which meters the correct amount of fuel. The fuel/air mixture passes through the **intake manifold pipes** and **intake valves** into the cylinders. The quantity or weight of the fuel/air charge depends upon the degree of throttle opening.

The intake valve is opened considerably before the piston reaches top dead center on the exhaust stroke, in order to induce a greater quantity of the fuel/air charge into the cylinder and thus increase the horsepower. The distance the valve may be opened before top dead center, however, is limited by several factors, such as the possibility that hot gases remaining in the cylinder from the previous cycle may flash back into the intake manifold, burning the fuel/air mixture in the induction system (backfire).

In all high-power aircraft engines, both the intake and the exhaust valves are off the valve seats at top dead center at the start of the intake stroke. As mentioned above, the intake valve opens before top dead center on the exhaust stroke (valve lead), and the closing of the exhaust valve is delayed considerably after the piston has passed top dead center and has started the intake stroke (valve lag). This timing is called **valve overlap** and is designed to aid in cooling the cylinder internally by circulating the cool incoming fuel/air mixture, to increase the amount of the fuel/air mixture induced into the cylinder, and to aid in scavenging the by-products of combustion.

The intake valve is timed to close about 50° to 75° past bottom dead center on the compression stroke depending upon the specific engine, to allow the momentum of the incoming gases to charge the cylinder more completely. Because of the comparatively large volume of the cylinder above the piston when the piston is near bottom dead center, the slight upward travel of the piston during this time does not have a great effect on the incoming flow of gases. This late timing can be carried too far because the gases may be forced back through the intake valve and defeat the purpose of the late closing.

Compression Stroke

After the intake valve is closed, the continued upward travel of the piston compresses the fuel/air mixture to obtain the desired burning and expansion characteristics.

The charge is fired by means of an electric spark as the piston approaches top dead center. The time of ignition will vary from 20° to 35° before top dead center, depending upon the requirements of the specific engine, to ensure complete combustion of the charge by the time the piston is slightly past the top dead center position.

Many factors affect ignition timing, and the engine manufacturer has expended considerable time in research and testing to determine the best setting. All engines incorporate devices for adjusting the ignition timing, and it is most important that the ignition system be timed according to the engine manufacturer's recommendations.

Power Stroke

As the piston moves through the top dead center position at the end of the compression stroke and starts down on the power stroke, it is pushed downward by the rapid expansion of the burning gases within the cylinder head with a force that can be greater than 15 tons (30,000 psi) at maximum power output of the engine. The temperature of these burning gases may be between 3,000° and 4,000° F.

As the piston is forced downward during the power stroke by the pressure of the burning gases exerted upon it, the downward movement of the connecting rod is changed to rotary motion by the crankshaft. Then the rotary movement is transmitted to the propeller shaft to drive the propeller. As the burning gases expand, and the downward movement of the piston causes increased volume of the chamber, the temperature drops to within

safe limits before the exhaust gases flow out through the exhaust port.

Valve Timing

The timing of the exhaust valve opening is determined by, among other considerations, the desirability of using as much of the expansive force as possible and of scavenging the cylinder as completely and rapidly as possible. The valve is opened considerably before bottom dead center on the power stroke (on some engines at 50° to 75° before BDC) while there is still some pressure in the cylinder. This timing is used so that the pressure can start moving gases out of the exhaust port as soon as possible. This process frees the cylinder of waste heat after the desired expansion has been obtained and avoids overheating the cylinder and the piston. Thorough scavenging of exhaust gases is very important, because any exhaust products remaining in the cylinder will dilute the incoming fuel/air charge at the start of the next cycle.

Exhaust Stroke

As the piston travels through bottom dead center at the completion of the power stroke, and starts upward on the exhaust stroke, it will begin to push the burned exhaust gases out the exhaust port. Near the end of the exhaust stroke, the speed of the exhaust gases leaving the cylinder creates a low pressure in the cylinder. This low or reduced pressure speeds the flow of the fresh fuel/air charge into the cylinder as the intake valve is beginning to open. The intake valve opening is timed to occur at 8° to 55° before top dead center on the exhaust stroke on various engines.

Engine Power And Efficiency

All aircraft engines are rated according to their ability to do work and produce power. This section presents an explanation of work and power and how they are calculated. Also discussed are the various efficiencies that govern the power output of a reciprocating engine.

Work

The physicist defines work as "Work is force times distance. Work done by a force acting on a body is equal to the magnitude of the force multiplied by the distance through which the force acts."

Work (W) = Force (F) × Distance (D).

Work is measured by several standards, the most common English system unit is called the

foot-pound. If a 1-pound mass is raised 1 foot, 1 foot-pound (ft-lb) of work has been performed. The greater the mass and the greater the distance, the greater the work.

Horsepower

The common English system unit of mechanical power is the horsepower (HP). Late in the 18th century, James Watt, the inventor of the steam engine, found that an English workhorse could work at the rate of 550 ft-lb per second, or 33,000 ft-lb per minute, for a reasonable length of time. To calculate the HP rating of an engine, divide the power developed in ft-lb per minute by 33,000, or the power in ft-lb per second by 550.

$$HP = \frac{\text{ft-lb per min}}{33,000}$$

or

$$HP = \frac{\text{ft-lb per sec}}{550}$$

Work is performed not only when a force is applied for lifting; force may be applied in any direction. If a 100-lb weight is dragged along the ground, a force is still being applied to perform work, although the direction of the resulting motion is approximately horizontal. The amount of this force would depend upon the roughness of the ground.

If the weight were attached to a spring scale graduated in pounds, then dragged by pulling on the scale handle, the amount of force required could be measured. Assume that the force required is 90 lbs, and the 100-lb weight is dragged 660 ft in 2 min. The amount of work performed in the 2 min will be 59,400 ft-lb or 29,700 ft-lb per min. Since 1 HP is 33,000 ft-lb per min, the HP expended in this case will be 29,700 divided by 33,000 or 0.9 HP.

If a 4,000 pound airplane was climbing 1,000 ft/min at a true airspeed of 100 knots while experiencing 600 pounds of total drag, the power required would be:

Power required to overcome drag:

$$\frac{100 \cancel{\text{NM}}}{\cancel{\text{hr}}} \times \frac{6080 \text{ ft}}{\cancel{\text{NM}}} \times \frac{1 \cancel{\text{hr}}}{60 \text{ min}} \times 600 \text{ lbs}$$

$$= 6.08 \times 10^6 \ \frac{\text{ft-lbs}}{\text{min}} \div 33,000 = 184 \text{ HP}$$

Power required to climb:

$$\frac{1000 \text{ ft}}{\text{min}} \times 4000 \text{ lbs}$$

$$= 4 \times 10^6 \ \frac{\text{ft-lbs}}{\text{min}} \div 33,000 = 121 \text{ HP}$$

Total power required:

$$184 \text{ HP} + 121 \text{ HP} = 305 \text{ HP}$$

Piston Displacement

When other factors remain equal, the greater the piston displacement the greater the maximum horsepower an engine will be capable of developing. When a piston moves from bottom dead center to top dead center, it displaces a specific volume. The volume displaced by the piston is known as **piston displacement** and is expressed in cubic inches for most American-made engines and cubic centimeters for others.

The piston displacement of one cylinder may be obtained by multiplying the area of the cross section of the cylinder by the total distance the piston moves in the cylinder in one stroke. For multi-cylinder engines this product is multiplied by the number of cylinders to get the total piston displacement of the engine.

Since the volume (V) of a geometric cylinder equals the area (A) of the base multiplied by the height (H), it is expressed mathematically as:

$$V = A \times H$$

The area of the base (A) is the area of the piston top, and height (H) is equal to the length of the piston stroke.

Example:

Compute the piston displacement of the PWA 14 cylinder engine having a cylinder with a 5.5-inch diameter and a 5.5-inch stroke. Formulas required are:

$$R = \frac{D}{2}$$

$$A = \pi R^2$$

$$V = A \times H$$

Total V = V × N (number of cylinders)

Substitute values into these formulas and complete the calculation.

$$R = \frac{D}{2} \quad R = 5.5/2 = 2.75$$

$$A = \pi R^2 \quad A = 3.1416 \,(2.75 \times 2.75)$$

$$A = 3.1416 \times 7.5625 = 23.7584 \text{ sq in}$$

$$V = A \times H \quad V = 23.7584 \times 5.5 \quad V = 130.6712$$

Total $V = V \times N$ Total $V = 130.6712 \times 14$ Total

$$V = 1829.3968 \text{ cu in}$$

Rounded off to the next whole number, total piston displacement equals 1830 cu in.

Compression Ratio

All internal-combustion engines must compress the fuel/air mixture to receive a reasonable amount of work from each power stroke. The fuel/air charge in the cylinder can be compared to a coil spring, in that the more it is compressed the more work it is potentially capable of doing, because combustion is a chemical reaction (see Chapter 6), and chemical reactions occur more quickly and completely if the reacting materials are closer together (high pressure) and more physically active (higher temperature).

The **compression ratio** of an engine (see figure 3-3) is a comparison of the volume of space in a cylinder when the piston is at the bottom of the stroke to the volume of space when the piston is at the top of the stroke. This comparison is expressed as a ratio, hence the term "compression ratio." Compression ratio is a controlling factor in the maximum horsepower developed by an engine, but it is limited by present-day fuel grades and the high engine speeds and manifold pressures required for takeoff. For example, if there are 140 cu in. of space in the cylinder when the piston is at the bottom and there are 20 cu in. of space when the piston is at the top of the stroke, the compression ratio would be 140 to 20. If this ratio is expressed in fraction form, it would be 140/20, or 7 to 1, usually represented as 7:1.

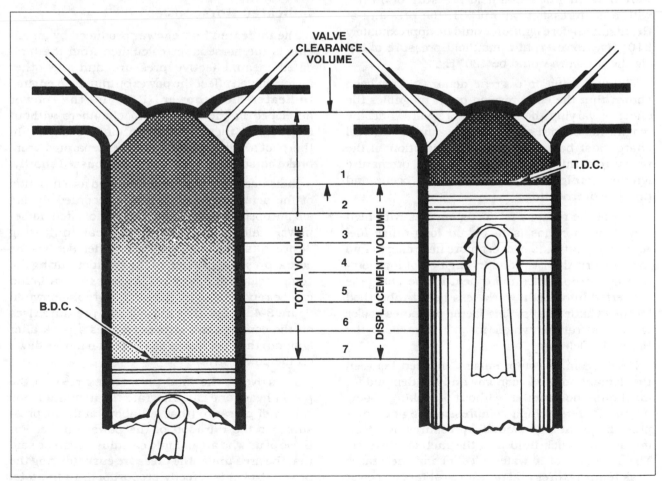

Figure 3-3. Compression ratio.

Generally it can be said that, within limits of proper engine operation, the higher the engine compression ratio, the higher will be engine efficiency (and horsepower output).

Manifold Absolute Pressure (MAP)

To grasp more thoroughly the limitation placed on compression ratios, manifold pressure and its effect on compression pressures should be understood. **Manifold pressure** is the average **absolute pressure** of the air or fuel/air charge in the intake manifold and is measured in units of inches of mercury (Hg). Manifold pressure is dependent mostly on ambient air pressure, engine speed, throttle setting and supercharging.

Compression ratio and manifold pressure determine the pressure in the cylinder in that portion of the operating cycle when both valves are closed. The pressure of the charge before compression is determined by manifold pressure, while the pressure at the height of compression (just prior to ignition) is determined by manifold pressure times the compression ratio. For example, if an engine were operating at a manifold pressure of 30″ Hg with a compression ratio of 7:1, the pressure at the instant before ignition would be approximately 210″ Hg. However, at a manifold pressure of 60″ Hg the pressure would be 420″ Hg.

Without going into great detail, it has been shown that the compression event magnifies the effect of varying the manifold pressure, and the magnitude of both affects the pressure of the fuel charge just before the instant of ignition. If the pressure at this time becomes too high, premature ignition (**preignition**) or **knock** will occur and produce overheating.

One of the reasons for using engines with high compression ratios is to obtain long-range fuel economy, that is, to convert more heat energy into useful work than is done in engines of low compression ratio. Since more heat of the charge is converted into useful work, less heat is absorbed by the cylinder walls. This factor promotes cooler engine operation, which in turn increases the thermal efficiency.

Here, again, a compromise is needed between the demand for fuel economy and the demand for maximum horsepower without knocking. Some manufacturers of high-compression engines suppress knock at high manifold pressures by injecting an antiknock fluid into the fuel/air mixture. The injection of a water/alcohol mix decreases peak temperatures and pressures at takeoff power settings but increases average pressures and, therefore, increases power output for short periods, such as at takeoff and during emergencies, when power is critical.

Absolute And Gauge Pressure

The word **absolute** in the term "manifold absolute pressure (MAP) identifies the pressure measurement as one that is based on a comparison of the pressure in the manifold with pressure at absolute zero (no pressure at all). The other method of measuring pressure, called **gauge** pressure, compares the pressure being measured against ambient pressure. As an example, the pressure in an aircraft's tire measured at sea level is 32 psig. Tire gauges measure "gauge" pressure, so the pressure in the tire is 32 psi higher than atmospheric pressure which is 14.7 psi (standard day sea level). The absolute pressure of the tire would be 32 + 14.7 = 46.7 psia. If that tire's pressure is measured at the same tire temperature but at progressively higher altitudes, the psig would increase but the psia would remain the same.

Indicated Horsepower

The **indicated horsepower** produced by an engine is the horsepower calculated from the indicated mean effective pressure and the other factors which affect the power output of an engine. **Indicated horsepower** (IHP) is the power developed in the combustion chambers without reference to friction losses within the engine. In the pilot's mind, IHP should be separated from brake horsepower, which will be discussed shortly.

Indicated horsepower is calculated as a function of the actual cylinder pressure recorded during engine operation. To facilitate the indicated horsepower calculations, a mechanical indicating device, attached to the engine cylinder, scribes the actual pressure existing in the cylinder during the complete operating cycle. This pressure variation can be represented by the kind of graph shown in figure 3-4. Notice that the cylinder pressure rises on the compression stroke, reaches a peak after ignition, then decreases as the piston moves down on the power stroke.

To arrive at the total pressure exerted on the piston head, one can integrate the instantaneous values of pressure (that are above ambient pressure) for the duration of the power stroke. For those of us who are not into calculus, it can be said that the area under the pressure curve (during the power stroke) is directly proportional to the total force applied to the piston head. From this must

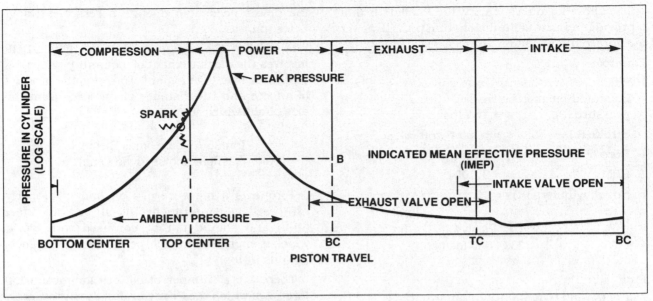

Figure 3-4. Cylinder pressure during power cycle.

be subtracted the area below the ambient pressure line because ambient pressure is pushing up on the bottom of the piston since ambient pressure exists in the crankcase.

Since the cylinder pressure varies during the operating cycle, an average pressure, line AB, is computed. This average pressure, if applied steadily during the time of the power stroke, would do the same amount of work as the varying pressure during the same period.

The area under the average pressure line equals the area under the actual pressure curve minus the area under the ambient pressure curve. The decrease of ambient pressure as the aircraft flies at higher altitudes is one of the reasons why an engine will produce more power at altitude if all else (including MAP) remains the same. Other reasons for this phenomena include lower exhaust back pressure at altitude (which provides better removal of exhaust gases resulting in more fuel/air mixture entering for the next power stroke) and usually colder temperatures at altitude (which increases the number of oxygen molecules that can be packed into each cylinder).

This average pressure is known as **indicated mean effective pressure** and is included in the indicated horsepower calculation with other engine specifications. If the characteristics and the indicated mean effective pressure of an engine are known, it is possible to calculate the indicated horsepower rating.

The indicated horsepower for a four-stroke-cycle engine can be calculated from the following for-

mula, in which the letter symbols in the numerator are arranged to spell the word "plank" to assist in memorizing the formula:

$$IHP = \frac{PLANK}{33,000}$$

Where:

P = Indicated mean effective pressure in psi

L = Length of the stroke in ft or in fractions of a foot

A = Area of the piston head or cross-sectional area of the cylinder, in sq in.

N = Number of power strokes per minute; $\frac{RPM}{2}$

K = Number of cylinders

In the formula above, the area of the piston times the indicated mean effective pressure gives the force acting on the piston in pounds. This force multiplied by the length of the stroke in feet gives the work performed in one power stroke, which, multiplied by the number of power strokes per minute, gives the number of ft-lb per minute of work produced by one cylinder. Multiplying this result by the number of cylinders in the engine gives the amount of work performed, in ft-lb, by the engine. Since HP is defined as work done at the rate of 33,000 ft-lb per min., the total number of ft-lb of work performed by the engine is divided by 33,000 to find the indicated horsepower.

The above formula's main value to pilots is that it provides a sense of the factors that provide power output from a piston engine.

Example:
Given:

Indicated mean effective pressure (P)	= 165 lbs / sq in
Stroke (L)	= 6 in or .5 ft
Bore	= 5.5 in
RPM	= 3,000
No. of cylinders (K)	= 12

$$IHP = \frac{PLANK}{33,000 \text{ ft–lbs/min}}$$

Find IHP:

A is found by using the equation

$$A = 1/4\pi D^2$$

$$A = 1/4 \times 3.1416 \times 5.5 \times 5.5$$

$$= 23.76 \text{ sq in}$$

N is found by multiplying the RPM by 1/2:

$$N = 1/2 \times 3,000$$

$$= 1,500 \text{ RPM}$$

Now, substituting in the formula:

$$\text{Indicated HP} = \frac{165 \times .5 \times 23.76 \times 1,500 \times 12}{33,000 \text{ ft–lbs/min}}$$

$$= 1069.20 \text{ HP}$$

Brake Horsepower

The **indicated horsepower** calculation discussed in the preceding paragraph is the theoretical power of a frictionless engine. The total horsepower lost in overcoming friction must be subtracted from the indicated horsepower to arrive at the actual horsepower delivered to the propeller. The power delivered to the propeller for useful work is known as brake horsepower (BHP). The difference between indicated and brake horsepower is known as friction horsepower, which is the horsepower required to overcome mechanical losses such as the pumping action of the pistons, the pistons and the friction of all moving parts and the power needed to operate accessories (magnetos, pumps, etc.).

In practice, the measurement of an engine's BHP involves the measurement of a quantity known as torque, or twisting moment. Torque is the product of a force and the distance of the force from the axis about which it acts, or

Torque = Force times Distance
(at right angles to the force).

Torque is a measure of load and is properly expressed in pound-inches (lb-in) or pound-feet (lb-ft) and should not be confused with work, which is expressed in inch-pounds (in-lbs) or foot-pounds (ft-lbs).

There are a number of devices for measuring torque, of which the **Prony brake, dynamometer**, and **torquemeter** are examples. Typical of these devices is the Prony brake (figure 3-5), which measures the usable power output of an engine on a test stand. It consists essentially of a hinged collar, or brake, which can be clamped to a drum splined to the propeller shaft. The collar and drum form a friction brake which can be adjusted. An arm of a known length is rigidly attached to or is a part of the hinged collar and terminates at a point which bears on a set of scales. As the propeller shaft rotates, it tends to carry the hinged collar of the brake with it and is prevented from doing so only by the arm that bears on the scale. The scale reads the force necessary to arrest the motion of the arm. If the resulting force registered on the scale is multiplied by the length of the arm, the resulting product is the torque exerted by the rotating shaft. Example: If the scale registers 200 lbs and the length of the arm is 3.18 ft, the torque exerted by the shaft is:

$$200 \text{ lb} \times 3.18 \text{ ft} = 636 \text{ lb-ft}$$

Once the torque is known, the work done per revolution of the propeller shaft can be computed without difficulty by the equation:

Work per revolution = $2\pi \times$ torque

If work per revolution is multiplied by the RPM, the result is work per minute, or power. If the work is expressed in ft-lbs per min, this quantity is divided by 33,000; the result is the brake horsepower of the shaft. In other words:

Power = Work per revolution × RPM

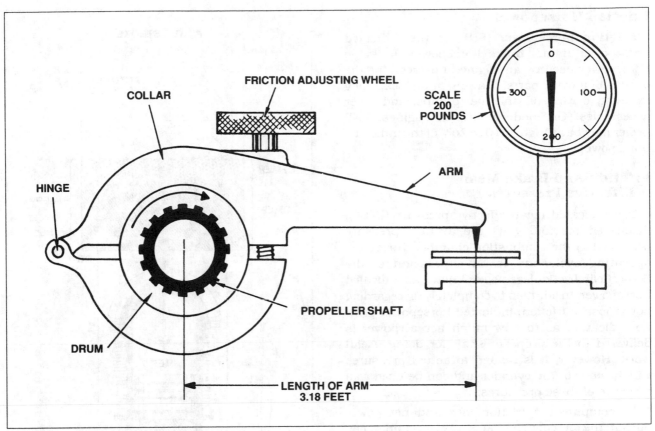

Figure 3-5. Typical Prony brake.

$$BHP = \frac{\text{Work per revolution} \times \text{RPM}}{33{,}000}$$

or

$$BHP = \frac{\begin{array}{c}2\pi \times \text{force on the scales (lbs)} \\ \times \text{ length of arm (ft)} \times \text{RPM}\end{array}}{33{,}000}$$

Example:
Given:

Force on scales = 200 lbs

Length of arm = 3.18 ft

RPM = 3,000

π = 3.1416

Find BHP:

Substituting in equation:

$$BHP = \frac{6.2832 \times 200 \times 3.18 \times 3{,}000}{33{,}000}$$

$$= 363.2 \text{ BHP}$$

As long as the friction between the brake collar and propeller shaft drum is great enough to impose an appreciable load on the engine, but is not great enough to stop the engine, it is not necessary to know the amount of friction between the collar and drum to compute the BHP. If there were no load imposed, there would be no torque to measure, and the engine would "run away." If the imposed load is so great that the engine stalls, there may be considerable torque to measure, but there will be no RPM. In either case it is impossible to measure the BHP of the engine. However, if a reasonable amount of friction exists between the brake drum and the collar and the load is then increased, the tendency of the propeller shaft to carry the collar and arm about with it becomes greater, thus imposing a greater force upon the scales. As long as the torque increase is proportional to the RPM decrease, the horsepower delivered at the shaft remains unchanged. This can be seen from the equation in which 2π and 33,000 are constants and torque and RPM are variables. If the change in RPM is inversely proportional to the change in torque, their product will remain unchanged. Therefore, BHP remains unchanged. This is important. It shows that horsepower is the function of both torque and RPM, and can be changed by changing either torque or RPM, or both.

Friction Horsepower

Friction horsepower (FHP) is the indicated horsepower minus brake horsepower. It is the horsepower used by an engine in overcoming the friction of moving parts, drawing in fuel, expelling exhaust, driving oil and fuel pumps, and other accessories. On modern aircraft engines, FHP power may be as high as 10 to 20% of the indicated horsepower.

Friction And Brake Mean Effective Pressures

The indicated mean effective pressure (IMEP), discussed previously, is the average pressure produced in the combustion chamber during the operating cycle and is an expression of the theoretical, frictionless power known as indicated horsepower. In addition to completely disregarding power lost to friction, indicated horsepower gives no indication as to how much actual power is delivered to the propeller shaft for doing useful work. However, it is related to actual pressures which occur in the cylinder and can be used as a measure of these pressures.

To compute the friction loss and net power output, the indicated horsepower of a cylinder may be thought of as two separate powers, each producing a different effect. The first power overcomes internal friction, and the horsepower thus consumed is known as friction horsepower. The second power, known as brake horsepower, produces useful work at the propeller. Logically, therefore, that portion of IMEP that produces brake horsepower is called **brake mean effective pressure** (BMEP). The remaining pressure used to overcome internal friction is called friction mean effective pressure (FMEP). This is illustrated in figure 3-6. IMEP is a useful expression of total cylinder power output, but is not a real physical quantity; likewise, FMEP and BMEP are theoretical but useful expressions of friction losses and net power output.

Although BMEP and FMEP have no real existence in the cylinder, they provide a convenient means of representing pressure limits, or rating engine performance throughout its entire operating range. This is true since there is a relationship between IMEP, BMEP, and FMEP.

One of the basic limitations placed on engine operation is the pressure developed in the cylinder during combustion. In the discussion of compression ratios and indicated mean effective pressure, it was found that, within limits, the increased pressure resulted in increased power. It was also

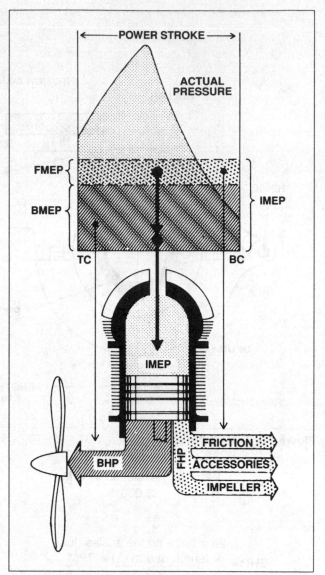

Figure 3-6. Powers and pressures.

noted that if the cylinder pressure was not controlled within close limits, it would impose dangerous internal loads that might result in engine failure. It is therefore important to have a means of determining these cylinder pressures as a protective measure and for efficient application of power.

If the BHP is known, the BMEP can be computed by means of the following equation:

$$BMEP = \frac{BHP \times 33,000}{LANK}$$

Example:
Given:

$$BHP = 1,000$$
$$Stroke = 6 \text{ in.}$$
$$Bore = 5.5 \text{ in.}$$
$$RPM = 3,000$$
$$\text{No. of cyls.} = 12$$

48

Find BMEP:

Find length of stroke (in feet):

$$L = 0.5$$

Find area of cylinder bore:

$$A = 1/4\pi D^2$$

$$= 0.7854 \times 5.5 \times 5.5$$

$$= 23.76 \text{ sq in.}$$

Find No. of power strokes per min:

$$N = 1/2 \times RPM$$

$$= 1/2 \times 3{,}000$$

$$= 1{,}500$$

Then substituting in the equation:

$$BMEP = \frac{1{,}000 \times 33{,}000}{.5 \times 23.76 \times 1{,}500 \times 12}$$

$$= 154.32 \text{ lbs per sq in.}$$

Thrust Horsepower

Thrust horsepower (THP) can be considered as the result of the engine and the propeller working together. If a propeller could be designed to be 100% efficient, the thrust and brake horsepower would be the same. However, the efficiency of the propeller varies with the engine speed, attitude, altitude, temperature, and airspeed, thus the ratio of thrust horsepower and brake horsepower delivered to the propeller shaft will never be equal. For example, if an engine develops 1,000 BHP and it is used with a propeller having 85 percent efficiency, the thrust horsepower of that engine-propeller combination is 85 percent of 1,000 or 850 THP. Of the four types of horsepower discussed, it is the **thrust horsepower** that determines the performance of the engine-propeller combination.

Remember:

$$THP = BHP \times \text{Propeller Efficiency}$$

For details on propeller efficiency, see Chapter 5.

Efficiencies

Thermal Efficiency

Any study of engines and power involves consideration of heat as the source of power. The heat produced by the burning of gasoline in the cylinders causes a rapid expansion of the gases in the cylinder, and this, in turn, moves the pistons and creates mechanical energy.

It has long been known that mechanical work can be converted into heat and that a given amount of heat contains the energy equivalent of a certain amount of mechanical work. Heat and work are theoretically interchangeable and bear a fixed relation to each other. Heat can therefore be measured in work units (for example, ft-lbs) as well as in heat units. The **British thermal unit** (BTU) of heat is the quantity of heat required to raise the temperature of 1 lb of water 1 °F. It is equivalent to 778 ft-lbs of mechanical work. A pound of petroleum fuel, when burned with enough air to consume it completely, produces about 20,000 BTU, the equivalent of 15,560,000 ft-lbs of mechanical work. These quantities express the heat energy of the fuel in heat and work units, respectively.

The ratio of useful work done by an engine to the heat energy of the fuel it uses, expressed in work or heat units, is called the **thermal efficiency** of the engine. If two similar engines use equal amounts of fuel, obviously the engine which converts into work the greater part of the energy in the fuel (higher thermal efficiency) will deliver the greater amount of power. Furthermore, the engine which has the higher thermal efficiency will have less waste heat to dispose of to the valves, cylinders, pistons, and cooling system of the engine. A high thermal efficiency also means a low specific fuel consumption and, therefore, less fuel for a flight of a given distance at a given power. Thus, the practical importance of a high thermal efficiency is threefold, and it constitutes one of the most desirable features in the performance of an aircraft engine.

Of the total heat produced, 25 to 30% is utilized for power output; 15 to 20% is removed by cooling (heat radiated from cylinder head fins and oil cooler); 5 to 10% is utilized in overcoming friction of moving parts, most of which produces heat that is carried away by the lubricating oil; and 40 to 45% is lost through the exhaust. Anything which increases the heat content that goes into mechanical work on the piston, which reduces the friction and pumping losses, or which reduces the quan-

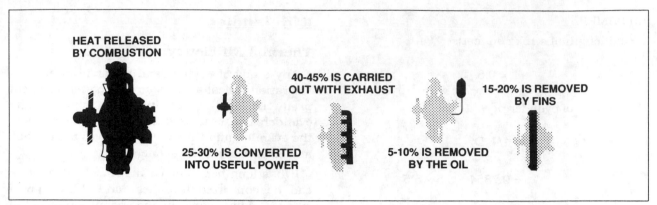

Figure 3-7. Thermal distribution in an engine.

tity of unburned fuel or the heat lost to the engine parts, increases the thermal efficiency. See figure 3-7.

The portion of the total heat of combustion which is turned into mechanical work depends to a great extent upon the **compression ratio.** Other things being equal, the higher the compression ratio, the larger is the proportion of the heat energy of combustion turned into useful work at the crankshaft. On the other hand, increasing the compression ratio increases the cylinder head temperature and the need for higher octane fuel.

Mechanical Efficiency

Mechanical efficiency is the ratio that shows how much of the power developed by the expanding gases in the cylinder is actually delivered to the output shaft. It is a comparison between the BHP and the IHP. It can be expressed by the formula:

$$\text{Mechanical efficiency} = \frac{\text{BHP}}{\text{IHP}}$$

The factor that has the greatest effect on mechanical efficiency is the friction within the engine itself. The friction between moving parts in an engine remains relatively constant throughout an engine's speed range. Therefore, the mechanical efficiency of an engine will be highest when the engine is running at the RPM at which maximum BHP is developed. Mechanical efficiency of the average aircraft reciprocating engine approaches 90%.

Volumetric Efficiency

Volumetric efficiency, another engine efficiency, is a ratio expressed in terms of percentages. It is a comparison of the volume of fuel/air charge (corrected for temperature and pressure) inducted into the cylinders to the total piston displacement

of the engine. Various factors cause departure from a 100% volumetric efficiency.

The pistons of an unsupercharged engine displace the same volume each time they sweep the cylinders from top dead center to bottom dead center. The amount of charge that fills this volume on the intake stroke depends on the existing pressure and temperature of the surrounding atmosphere and how completely the burned gases have been removed from the cylinder during the previous exhaust stroke. Therefore, to find the volumetric efficiency of an engine, standards for atmospheric pressure and temperature had to be established. The U.S. standard atmosphere was established in 1958. It provides the necessary pressure and temperature values to calculate volumetric efficiency.

The standard sea-level temperature is 59 °F or 15 °C. At this temperature the pressure of one atmosphere is 14.69 lbs/sq in., and this pressure will support a column of mercury 29.92 in. high. These standard sea-level conditions determine a standard density, and if the engine draws in a volume of charge of this density exactly equal to its piston displacement, it is said to be operating at 100% volumetric efficiency. An engine drawing in less volume than this has a volumetric efficiency lower than 100%. An engine equipped with a high-speed internal or external blower may have a volumetric efficiency greater than 100%. The equation for volumetric efficiency is:

$$\text{Volumetric efficiency} = \frac{\text{Volume of charge (corrected for temperature and pressure)}}{\text{Piston displacement}}$$

Many factors decrease volumetric efficiency; some of these are:

(1) Part-throttle operation.

(2) Long intake pipes of small diameter.

(3) Sharp bends in the induction system.

(4) Carburetor air temperature too high.

(5) Cylinder-head temperature too high.

(6) Incomplete scavenging of exhaust.

(7) Improper valve timing.

(8) High density altitude.

Study Questions And Problems

1. What three energy forms does the reciprocating engine utilize or convert?

2. Describe the operating cycle of the typical aircraft reciprocating engine.

3. What percentage of each engine revolution is one cylinder producing power?

4. How much horsepower is being used to climb a 3000 lb airplane at 1200 fpm at 120 KTS if its total drag is 350 pounds? What kind of horsepower is this?

5. How much thrust is the airplane of problem #4 using?

6. How much power and thrust is the airplane of problem #4 using if its climb speed is reduced to 100 KTS and drag is reduced to 320 pounds but ROC remains the same at 1200 fpm?

7. What would gauge pressure be in an intake manifold where the MAP is 22" Hg if the airplane is flying at sea level? AT 5000 ft ASL?

8. What is BHP? How is it measured?

9. On a short-field takeoff, what can the pilot do to reduce FHP and thus increase BHP and THP?

Chapter IV
Engine Lubrication And Cooling

Principles Of Engine Lubrication

The primary purpose of a lubricant is to reduce friction between moving parts. Because liquid lubricants (oils) can be circulated readily, they are used universally in aircraft engines.

In theory, fluid lubrication is based on the actual separation of the surfaces so that no metal-to-metal contact occurs. As long as the oil film remains unbroken, metallic friction is replaced by the internal fluid friction of the lubricant. Under ideal conditions, friction and wear are held to a minimum.

In addition to **reducing friction,** the oil film acts as a **cushion** between metal parts. This cushioning effect is particularly important for such parts as reciprocating engine crankshaft and connecting rods, which are subject to shock-loading.

As oil circulates through the engine, it **absorbs heat** from the parts. Pistons and cylinder walls in reciprocating engines are especially dependent on the oil for **cooling.** The oil also aids in forming a **seal** between the piston and the cylinder wall to prevent leakage of the gases from the combustion chamber. Oils also **reduce abrasive wear** by picking up foreign particles and carrying them to a filter, where they are removed, thus **cleansing** the engine.

Oil serves other purposes by changing propeller pitch (Chapter 5), moving the turbocharger waste gate (Chapter 8), heating fuel, and protecting metal surfaces from oxidation and corrosion.

Requirements & Characteristics Of Reciprocating Engine Lubricants

While there are several important properties which a satisfactory reciprocating engine oil must possess, it's **viscosity** is most important in engine operation. The **resistance** of an oil to **flow** is known as its **viscosity.** An oil which flows slowly is viscous or has a high viscosity. If it flows freely, it has a low viscosity. Unfortunately, the viscosity of oil is affected by temperature. It is not uncommon for some grades of oil to become practically solid in cold weather. This increases drag and makes circulation almost impossible. Other oils may become so thin at high temperature that the oil film is broken, resulting in rapid wear of the moving parts. The oil selected for aircraft engine lubrication must be light enough to circulate freely, yet heavy enough to provide the proper oil film at engine operating temperatures. Since lubricants vary in properties and since no one oil is satisfactory for all engines and all operating conditions, it is extremely important that only the recommended grade be used.

The following characteristics of lubricating oils measure their grade and suitability:

(1) **Flash point** and **fire point** are determined by laboratory tests that show the temperature at which a liquid will begin to give off ignitable vapors (flash) and the temperature at which there are sufficient vapors to support a flame (fire). These points are established for engine oils to determine that they can withstand the high temperatures encountered in an engine.

(2) **Cloud point** and **pour point** also help to indicate suitability. The cloud point of an oil is the temperature at which its wax content, normally held in solution, begins to solidify and separate into tiny crystals, causing the oil to appear cloudy or hazy. The pour point of an oil is the lowest temperature at which it will flow or can be poured.

(3) **Specific gravity** is a comparison of the weight of the substance to the weight of an equal volume of distilled water at a specified temperature. As an example, water weighs approximately 8 lbs to the gallon; an oil with a specific gravity of 0.9 would weigh 7.2 lbs to the gallon.

Generally, commercial aviation oils are classified numerically, such as 80, 100, 140, etc., which are an approximation of their viscosity as measured by a testing instrument called the Saybolt Universal Viscosimeter. In this instrument a tube holds a specific quantity of the oil to be tested. The oil is brought to an exact temperature by a liquid bath surrounding the tube. The time in seconds required for exactly 60 cubic centimeters of oil to flow through an accurately calibrated orifice is recorded as a measure of the oil's viscosity.

If actual Saybolt values were used to designate the viscosity of oil, there probably would be several hundred grades of oil. To simplify the selection of oils, they often are classified under an SAE (Society

COMMERCIAL AVIATION NO.	COMMERCIAL SAE NO.	ARMY & NAVY SPECIFICATION NO.
65	30	1065
80	40	1080
100	50	1100
120	60	1120
140	70	

Figure 4-1. Grade designations for aviation oils.

of Automotive Engineers) system, which divides all oils into seven groups (SAE 10 to 70) according to viscosity at either 130° or 210° F.

SAE ratings are purely arbitrary and bear no direct relationship to the Saybolt or other ratings. The letter "W" occasionally is included in the SAE number giving a designation such as SAE 20W. This letter "W" indicates that the oil, in addition to meeting the viscosity requirements at the testing temperature specifications, is a satisfactory oil for winter use in cold climates.

Although the SAE scale has eliminated some confusion in the designation of lubricating oils, it must not be assumed that this specification covers all the important viscosity requirements. An SAE number indicates only the viscosity (grade) or relative viscosity; it does not indicate quality or other essential characteristics. It is well known that there are good oils and inferior oils that have the same viscosities at a given temperature and, therefore, are subject to classification in the same grade. The SAE letters on an oil container are not an endorsement or recommendation of the oil by the Society of Automotive Engineers.

Although each grade of oil is rated by an SAE number, depending on its specific use, it may be rated with a commercial aviation grade number or an Army and Navy specification number. The correlation between these grade-numbering systems is shown in figure 4-1.

Lubricating Oil Types

Since many internal engine parts are steel, presence of an oil film **protects against corrosion**.

Straight mineral oil is often used in reciprocating engines after overhaul or when new to facilitate wear-in or seating of piston rings. This oil's main limitation is that it oxides when exposed to temperatures higher than normal, forming carbonaceous deposits which combine with partly burned fuel, moisture and lead compounds to form **sludge**.

Ashless dispersant (AD) oil usually replaces mineral oil after the piston rings have seated, as evidenced by decreased oil consumption. This oil has less tendency to oxidize to carbon, nor does it form ash deposits. Additives tend to cause particle dispersion so particles don't clump together, forming sludges, but repel each other, staying in suspension until trapped in the oil filter. AD oils have better lubricating properties than straight mineral oils.

Compatibility of oils within their basic categories is better than popular opinion might indicate. Although it is good practice to use the brand and grade of oil already in the engine when adding oil, if necessary in order to have proper operating oil levels, brands of the above oil types can be mixed.

Synthetic oils and many different types of **additives** are marketed with a wide variety of claims made for benefits of use. Not many are endorsed by the engine manufacturers. For the user, decisions about the use of these products are made difficult by the lack of operating history.

Internal Lubrication Of Reciprocating Engines

The lubricating oil is distributed to the various moving parts of a typical internal-combustion engine by one of the three following methods: (1) pressure, (2) splash, or (3) a combination of pressure and splash.

Combination Splash-And-Pressure Lubrication

In a typical **pressure-lubrication** system (figure 4-2), a mechanical pump supplies oil under pressure to the bearings throughout the engine. The oil flows into the inlet or suction side of the oil pump through a line connected to the tank or sump at a point higher than the bottom of the oil sump. This prevents sediment which falls into the sump from being drawn into the pump.

The rotation of the pump, which is driven by the engine, causes the oil to pass around the outside of the gears in the manner illustrated in figure 4-3. This develops a pressure in the crankshaft oiling system (drilled passage-holes). The variation in the speed of the pump from idling to full-throttle and the fluctuation of oil viscosity because of temperature changes are compensated for by the tension on the **relief valve** spring (figure 4-3). The pump is designed to create a greater pressure than probably will ever be required to compensate for wear of the bearings or thinning out of oil. The parts oiled by pressure throw a lubricating spray onto the cylinder and piston assemblies. After lubricating the various

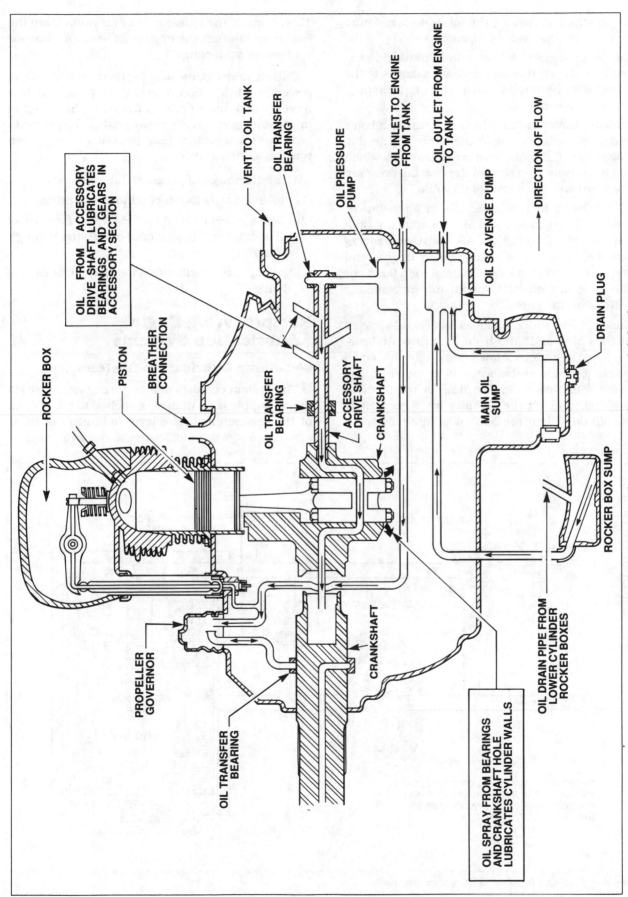

Figure 4-2. Schematic showing pressure dry-sump lubrication system.

ROCKER BOX

PISTON

BREATHER CONNECTION

OIL FROM ACCESSORY DRIVE SHAFT LUBRICATES BEARINGS AND GEARS IN ACCESSORY SECTION

VENT TO OIL TANK

OIL TRANSFER BEARING

OIL PRESSURE PUMP

OIL INLET TO ENGINE FROM TANK

OIL OUTLET FROM ENGINE TO TANK

DIRECTION OF FLOW

OIL TRANSFER BEARING

ACCESSORY DRIVE SHAFT

CRANKSHAFT

OIL SCAVENGE PUMP

DRAIN PLUG

MAIN OIL SUMP

ROCKER BOX SUMP

PROPELLER GOVERNOR

OIL TRANSFER BEARING

CRANKSHAFT

OIL DRAIN PIPE FROM LOWER CYLINDER ROCKER BOXES

OIL SPRAY FROM BEARINGS AND CRANKSHAFT HOLE LUBRICATES CYLINDER WALLS

units on which it sprays, the oil drains back into the sump and the cycle is repeated.

The pump forces the oil into a manifold that distributes the oil through drilled passages to the crankshaft bearings and other bearings throughout the engine.

Oil flows from the main bearings through holes drilled in the crankshaft to the lower connecting rod bearings. Each of these holes through which the oil is fed is located so that the bearing pressure at the point will be as low as possible.

Oil reaches a hollow camshaft (in an in-line or opposed engine), or a camplate or camdrum (in a radial engine), through an **oil transfer bearing** (figure 4-2) which is a connection with the end bearing, or the main oil manifold; oil then flows out to the various camshaft, camdrum, or camplate bearings and the cams.

The engine cylinder surfaces receive oil sprayed from the crankshaft and from the crankpin bearings. Since oil seeps slowly through the small crankpin clearances before it is sprayed on the cylinder walls, considerable time is required for enough oil to reach the cylinder walls, especially on a cold day when the oil flow is more sluggish.

This is one of the chief reasons for preheating the engine or diluting the engine oil with gasoline for cold weather starting.

Splash lubrication may be used in addition to pressure lubrication on aircraft engines, but it is never used by itself; hence, aircraft-engine lubrication systems are always either the pressure type or the combination pressure-and-splash type, usually the latter.

The advantages of pressure lubrication are:
(1) Positive introduction of oil to the bearings.
(2) Cooling effect caused by the large quantities of oil which can be (pumped) circulated through a bearing.
(3) Satisfactory lubrication in various attitudes of flight.

Reciprocating Engine Lubrication Systems

Wet-Sump Lubrication Systems

The system consists of a sump or pan, in which the oil supply is contained, attached to the bottom of the engine case. The level (quantity) of oil is

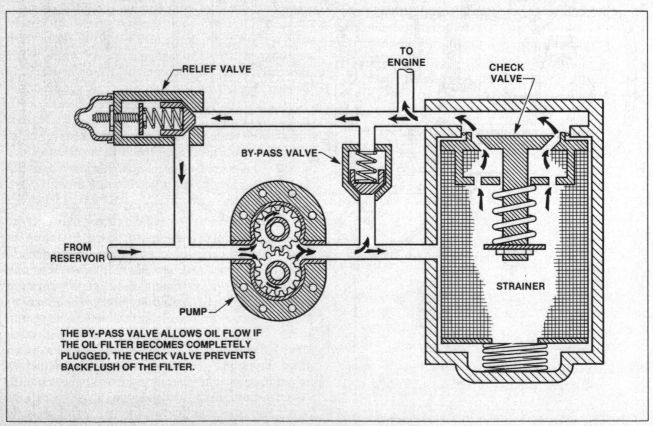

Figure 4-3. Engine oil pump and associated units.

indicated or measured by a vertical rod that protrudes into the oil from an elevated hole on top of the crankcase. In the bottom of the sump (oil pan) is a screen strainer having a suitable mesh or series of openings to strain undesirable particles from the oil and yet pass sufficient quantity to the inlet or (suction) side of the oil pressure pump.

The main disadvantages of the wet-sump system are:

(1) The oil supply is limited by the sump (oil pan) capacity.

(2) Provisions for cooling the oil are more difficult to arrange because the system is a self-contained unit.

(3) Oil temperatures are likely to be higher on large engines because the oil supply is so close to the engine and is continuously subjected to high operating temperatures.

(4) The system is not readily adaptable to inverted flying since the entire oil supply will flood the engine and oil supply to the pressure oil pump will be interrupted.

Advantages of the wet-sump system include:

(1) Because the wet-sump is attached to the bottom of the case, it is integral and complete, requiring no external parts and fittings to complicate installation and maintenance. Therefore, the system is simpler, lighter weight and less costly.

(2) No second (scavenge) pump is required. Chance of oil pump failure is reduced.

(3) This system can be operated in much colder ambient temperatures without concern that the oil will congeal in the external lines.

For these and other reasons, wet-sump systems are the most common in todays general aviation fleet.

Dry-Sump Systems

Some reciprocating aircraft engines have pressure **dry-sump** lubrication systems. The oil supply in this type of system is carried in a separate tank. A pressure pump circulates the oil through the engine in the same manner as with the wet-sump system; scavenger pumps then return the oil to the tank as quickly as it accumulates in the engine sumps.

Radial engine design is such that the oil can't be carried inside the engine itself. Other types of engines capable of sustained inverted flight share in this problem. The dry sump system caters to this problem by gathering engine oil to and dispensing it from an external tank.

Although the arrangement of the oil systems in different aircraft varies widely and the units of which they are composed differ in construction details, the functions of all such systems are the same.

The principal units in a typical reciprocating engine **dry-sump oil system** include an oil supply tank, engine oil pressure and scavenger pumps, an oil cooler, an oil control valve, an actuator for an oil-cooler air-exit control, the necessary tubing, and quantity, pressure, and temperature indicators.

Lubrication System Operation Maintenance Practices

The following lubrication system practices are typical of those performed on small, single-engine aircraft. The oil system and components described are those used to lubricate a 225 HP, six-cylinder, horizontally opposed, air-cooled engine.

The oil system is the dry-sump type, using a pressure lubrication system sustained by engine-driven, positive-displacement, gear-type pump. The system (figure 4-4) consists of an oil cooler (radiator), a 3-gal. (U.S.) oil tank, oil pressure pump and scavenge pump, and the necessary inter-connecting oil lines. Oil from the oil tank is pumped to the engine, where it circulates under pressure, then collects in the cooler, and is returned to the oil tank. A thermostat in the cooler controls oil temperature by allowing part of the oil to flow through the cooler and part to flow directly into the oil supply tank. This arrangement allows hot engine oil, with a temperature still below 65°C (150° F), to mix with the cold uncirculated oil in the tank. This raises the complete engine oil supply to operating temperature in a shorter period of time.

The oil tank, constructed of welded aluminum, is serviced through a filler neck located on the tank and equipped with a spring-loaded locking cap. Inside the tank a weighted, flexible rubber oil hose is mounted so that it is repositioned automatically to ensure oil pickup during inverted maneuvers. A dipstick guard is welded inside the tank for the protection of the flexible oil hose assembly. During normal flight, the oil tank is vented to the engine crankcase by a flexible line at the top of the tank. However, during inverted flight the normal vent is covered or submerged below the oil level within the tank. Therefore, a secondary vent and check-valve arrangement is incorporated in the tank for inverted operation. During an inversion, when air in

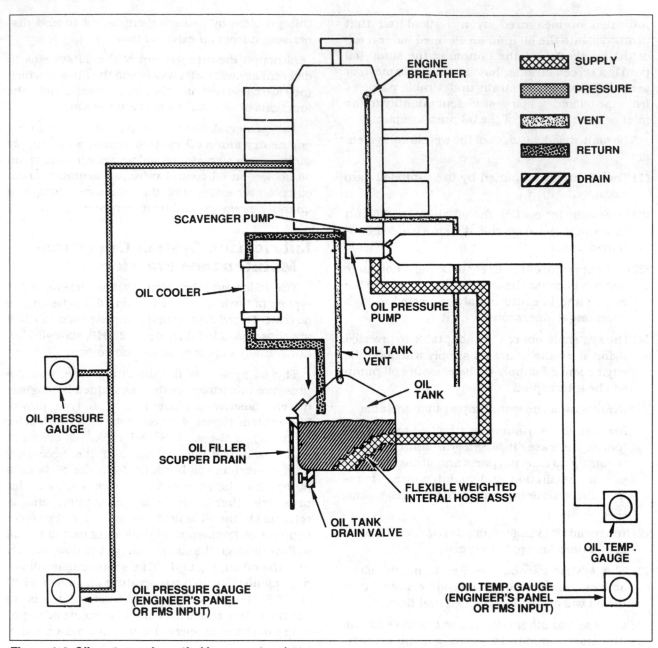

Figure 4-4. Oil system schematic (dry sump type).

the oil tank reaches a certain pressure, the check valve in the secondary vent line will unseat and allow air to escape from the tank. This assures an uninterrupted flow of oil to the engine.

Engine Cooling Systems

An internal-combustion engine is a heat machine that converts chemical energy in the fuel into mechanical energy at the crankshaft. It does not do this without some loss of energy, however, and even the most efficient aircraft engines may waste 60 to 70% of the original energy in the fuel. Unless most of this waste heat is rapidly removed,

the cylinders may become hot enough to cause complete engine failure.

Excessive heat is undesirable in any internal-combustion engine for three principal reasons:

(1) It affects the behavior of the combustion of the fuel/air charge.

(2) It weakens and shortens the life of engine parts. If the temperature inside the engine cylinder is too great, the fuel/air mixture will be preheated, and combustion will occur before the desired time. Since premature combustion causes detonation, knocking, and other undesirable condi-

tions, there must be a way to eliminate heat before it causes damage.

(3) It impairs lubrication. As temperature increases, oil viscosity decreases (oil becomes thinner), oil is not pumped as easily, oil pressure decreases and the lubricating qualities of the oil film decrease.

One gallon of aviation gasoline has enough heat value to boil 75 gallons of water; thus, it is easy to see that an engine releases a tremendous amount of heat. About one-fourth of the heat released is changed into useful power. The remainder of the heat must be dissipated so that it will not be destructive to the engine. In a typical aircraft powerplant, half of the heat goes out with the exhaust, and the other is absorbed by the engine. Circulating oil picks up part of this soaked-in heat and transfers it to the airstream through the oil cooler. The engine cooling system takes care of the rest.

Cooling is a matter of transferring the excess heat from the cylinders to the air, but there is more

to such a job than just placing the cylinders in the airstream.

A cylinder is roughly the size of a gallon jug. Its outer surface, however, is increased by the use of cooling fins so that it presents a barrel-sized exterior to the cooling air. Such an arrangement increases the heat transfer by radiation. If too much of the cooling fin area is broken off, the cylinder cannot cool properly and a hotspot will develop. Therefore, cylinders are normally replaced when a specified number of square inches of fins are missing.

Cowling and **baffles** are designed to force air over the **cylinder cooling fins** (figure 4-5). The baffles direct the air close around the cylinders and prevent it from forming hot pools of stagnant air while the main streams rush by unused. Blast tubes are built into the baffles to direct jets of cooling air onto the bottom spark plug elbows of each cylinder to prevent overheating of ignition leads.

An engine can have an operating temperature that is too low. For the same reasons that an

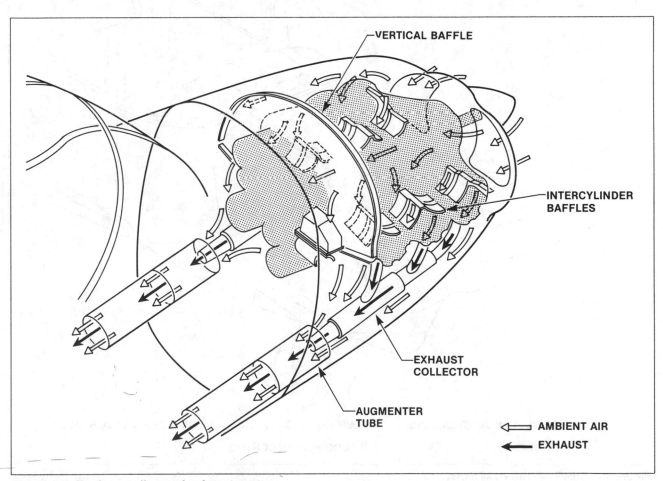

Figure 4-5. Engine cooling and exhaust system.

engine is warmed up before takeoff, it is kept warm during flight. Fuel evaporation and distribution and oil circulation depend on an engine being kept warm. The automobile engine depends on a thermostatic valve in the water system to keep the engine in its most efficient temperature range. The aircraft engine, too, has temperature controls. These controls regulate air circulation over the engine. Unless some controls are provided, the engine will overheat on takeoff and get too cold in high-speed and low-power letdowns.

The most common means of controlling cooling is the use of **cowl flaps,** as illustrated in figure 4-6. These flaps are opened and closed by electric-motor-driven jackscrews, by hydraulic actuators, or manually in some light aircraft. When extended for increased cooling, the cowl flaps produce drag and sacrifice streamlining for the added cooling. On takeoff, the cowl flaps are opened only enough to keep the engine below the red-line temperature. Heating above the normal range may be allowed so that drag will be as low as possible. During ground operations, the cowl flaps should be opened wide since drag does not matter and cooling airflow is decreased greatly due to very low forward speed.

Some aircraft use **augmenters** (figure 4-5) to provide additional cooling airflow, especially for low speed, high power conditions (climb). The exhaust collectors feed exhaust gas into the augmenter tubes. The high velocity exhaust gas mixes with air that has passed over the engine and heats it to form

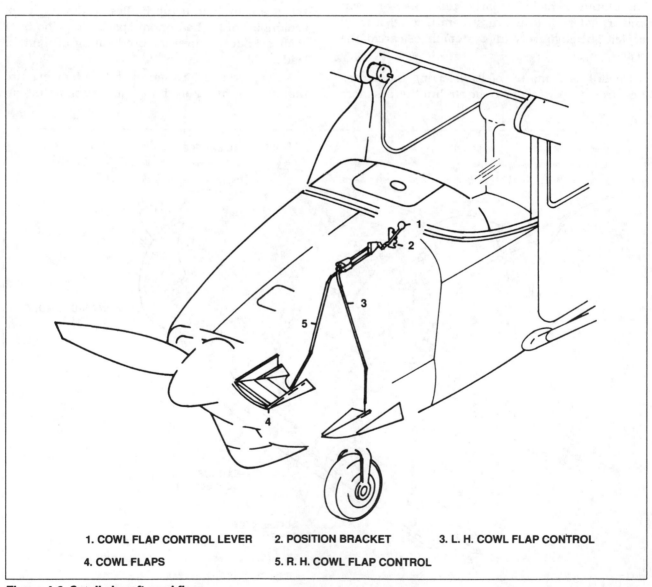

1. COWL FLAP CONTROL LEVER   2. POSITION BRACKET   3. L. H. COWL FLAP CONTROL

4. COWL FLAPS   5. R. H. COWL FLAP CONTROL

Figure 4-6. Small aircraft cowl flaps.

a high-temperature, low-pressure, jetlike exhaust. This low-pressure area in the augmenters draws additional cooling air over the engine.

Augmenters use exhaust gas velocity to cause an airflow over the engine so that cooling is not entirely dependent on the prop wash. Vanes installed in the augmenters control the volume of air. These vanes usually are left in the trail position to permit maximum flow. They can be closed to increase the heat for cabin or anti-icing use or to prevent the engine from cooling too much during descent from altitude. In addition to augmenters, some aircraft have residual heat doors or nacelle flaps that are used mainly to let the retained heat escape after engine shutdown. The nacelle flaps can be opened for more cooling than that provided by the augmenters.

On most light aircraft with horizontally opposed engines (figure 4-5), the engine is pressure-cooled by air taken in through two openings in the nose cowling, one on each side of the propeller spinner. A pressure chamber is sealed off on the top side of the engine with baffles properly directing the flow of cooling air to all parts of the engine. Warm air is drawn from the lower part of the engine compartment by the venturi action of the exhaust gases through the augmenter tubes (if installed) or by the low pressure caused by open cowl flaps. Augmenter tubes eliminate the use of controllable cowl flaps and assures adequate engine cooling at all operating speeds. They are especially effective during low-speed, high power climbs.

Many light aircraft use only one or two engine cowl flaps to control engine temperature. As shown in figure 4-6, two cowl flaps, operated by a single control in the cabin, are located at the lower aft end of the engine nacelle. Cutouts in the flaps permit extension of engine exhaust stacks through the nacelle. The flaps are operated by a manual control in the cockpit to control the flow of air directed by baffles around the cylinders and other engine components.

The engine cooling system of most reciprocating engines usually consists of the engine **cowling, cylinder baffles, cylinder fins,** and some type of **cowl flaps.** In addition to these major units, there is also some type of temperature-indicating system (cylinder head temperature).

The **cowling** performs two functions:

(1) It streamlines the bulky engine to reduce drag.

(2) It forms an envelope around the engine which forces air to pass around and between the cylinders.

The **cylinder baffles** are metal shields, designed and arranged to direct the flow of air evenly around all cylinders. This even distribution of air aids in preventing one or more cylinders from being excessively hotter than the rest.

The **cylinder fins** radiate heat from the cylinder walls and heads. As the air passes over the fins, it absorbs this heat, carries it away from the cylinder, and is exhausted overboard through the cowl flaps.

The controllable cowl flaps provide a means of decreasing or increasing the exit area at the rear of the engine cowling. Closing the cowl flaps decreases the exit area, which effectively decreases the amount of air that can circulate over the cylinder fins. The decreased airflow cannot carry away as much heat; therefore, there is a tendency for the engine temperature to increase. Opening the cowl flaps makes the exit area larger. The flow of cooling air over the cylinders increases, absorbing more heat, and the tendency is then for the engine temperature to decrease. Good inspection and maintenance in the care of the engine cooling system will aid in overall efficient and economical engine operation.

Cylinder Head Temperature Indicating System

This system usually consists of an indicator, electrical wiring, and a **thermocouple.** The wiring is between the instrument and the nacelle firewall. At the firewall, one end of the thermocouple leads connect to the electrical wiring, and the other end of the thermocouple leads connect to the cylinder.

The **thermocouple** consists of two dissimilar metals, generally constantan and iron, connected by wiring to an indicating system. If the temperature of the junction is different from the temperature where the dissimilar metals are connected to wires, a voltage is produced. This voltage sends a current through wires to the indicator, a current-measuring instrument graduated in degrees.

The thermocouple end that connects to the cylinder is either the bayonet or gasket type. The bayonet type is threaded into the cylinder head. The gasket type fits under the spark plug and replaces the normal spark plug gasket. See also "cylinder head temperature" in Chapters 10 and 17.

When installing a thermocouple lead, it must not be cut off because it is too long, but coiled and tied-up. The thermocouple is designed to produce a given amount of resistance. If the length of the lead is reduced, an incorrect temperature reading will result. The bayonet or gasket of the thermocouple is inserted or installed on the hottest cylinder of the engine, as recommended by the manufacturer.

When the thermocouple is installed and the wiring connected to the instrument, the indicated reading is the cylinder temperature. Prior to operating the engine, provided it is at ambient temperature, the cylinder head temperature indicator will indicate the free outside air temperature; that is one test for determining that the instrument is working correctly. A check to see that cylinder head temperature is near ambient temperature (cool engine) should be a part of the pilot's pre-start procedure. Most cylinder head gauges work without the master switch on because thermocouple's generate their own current.

Pilot Responsibility For Engine Temperature Control

Most aircraft engine installations are engineered for proper operating temperatures under normal operating conditions. However, not all operations are conducted under normal conditions of airspeed, power settings and ambient temperatures. Therefore, the pilot must have the final responsibility for keeping engine operating temperatures within normal limits.

Pilots have several "tools" at their disposal with which to control engine temperature, including:

(1) **Cowl flaps.** Open cowl flaps increase cooling airflow over the engine and through the oil cooler (but hot and cold airflow into the cabin is reduced in most single engine installations due to reduced air pressure inside and in front of the cowling when cowl flaps are open).

(2) **Power setting.** Increased power increases the amount of heat which must be removed by engine cooling.

(3) **Airspeed.** Increased airspeed increases airflow over the engine and through the oil cooler. Remember, in most aircraft, climbing at speeds greater than Vy improves engine cooling, visibility over the nose, cabin airflow and passenger comfort!

(4) **Fuel mixture.** A rich mixture cools the engine internally as heat is absorbed to vaporize the

Figure 4-7. *View into cylinder exhaust port showing cracked valve guide housing caused by thermal shock.*

additional fuel. The excess fuel doesn't produce heat because it isn't burned within the engine.

Thermal Shock

Thermal shock occurs when engine parts that are operating at high temperatures are quickly cooled. Some parts (cooling fins, exhaust ports, etc.) are cooled much more rapidly than others. The metal there contracts (shrinks in size) faster than the warmer, surrounding metal. Great internal stress occurs in solids under these conditions, causing cracks in the cylinder head casting which renders the part unairworthy (see figure 4-7).

A quick way to damage most engines is to allow the engine to cool rapidly (idle power, high airspeed, cool or cold air) after high power operation. Example: practicing a power failure just after takeoff and full power climb. With many engines, thermal shock can occur during minimum power letdown after cruise power use.

So, it is important that the pilot, when operating air cooled engines, manage power, mixture, airspeed and cowl flaps with the engine's welfare in mind, especially during training and descents.

Additional Reading

Powerplant Section Textbook; EA-ITP-P; Chapter 7, Engine Cooling Systems; Chapter 8, Engine Lubrication Systems; IAP, Inc., Publ.; 1983

Study Questions And Problems

1. List all of the useful functions of lubricating oil within the engine.

2. What is viscosity?

3. Why is viscosity important when selecting a lubricant?

4. What is the relationship between aviation grade numbers and SAE numbers for lubricating oil?

5. Under what conditions would a dry sump system be advantageous, over a wet sump?

6. What are the advantages of a wet sump system, compared to the dry sump?

7. What type of pump is the engine oil pump? (at least two descriptive terms are appropriate here, to fully describe this pump).

8. What is the purpose of the bypass valve and the relief valve in figure 4-3?

9. How are the cylinder walls lubricated in a typical opposed aircraft engine?

10. How does the cowling contribute to engine cooling?

11. What is the function of an augmenter tube?

12. How does opening cowl flaps affect engine cooling? Cabin cooling?

Chapter V
Propellers And Governors

General

The propeller, the unit which must absorb the power output of the engine, has passed through many stages of development. Great increases in power output have resulted in the development of four- and six-bladed propellers of large diameters. However, there is a limit to the RPM at which these large propellers can be turned.

As an outgrowth of the problems involved in operating large propellers, a variable-pitch, constant-speed propeller system was developed. This system makes it necessary to vary the engine RPM only slightly during various flight conditions and therefore increases flying efficiency. Roughly, such a system consists of a flyweight-equipped governor unit, which controls the pitch angle of the blades so that the engine speed remains constant. The governor can be regulated by controls in the cockpit so that any desired blade angle setting and engine operating speed can be obtained. A low-pitch, high-RPM setting, for example, can be utilized for takeoff; then after the aircraft is airborne, a higher pitch and lower RPM setting can be used. More on governors later in this chapter.

Propeller Principles

The aircraft **propeller** consists of two or more blades and a central **hub** to which the blades are attached. Each **blade** of an aircraft propeller is essentially a rotating wing. As a result of their construction, the propeller blades produce forces that create thrust to pull or push the airplane through the air.

The power needed to rotate the propeller blades is furnished by the engine. The propeller is mounted on a shaft, which may be an extension of the crankshaft on low-horsepower engines; on high-horsepower engines, it is mounted on a propeller shaft which is geared to the engine crankshaft. In either case, the engine rotates the airfoils of the blades through the air at high speeds, and the propeller transforms the rotary motion (power) of the engine into thrust.

The engine supplies **brake horsepower** through a rotating shaft, and the propeller converts it into **thrust horsepower.** In this conversion, some power is wasted. For maximum efficiency, the propeller must be designed to keep this waste as small as possible. Since the efficiency of any machine is the ratio of the useful power output to the power input, propeller efficiency is the ratio of thrust horsepower to brake horsepower. The usual symbol for propeller efficiency is the Greek letter η (eta). Propeller efficiency varies from 50% to 87%, depending on how much the propeller "slips."

$$THP = BHP \times \text{Propeller Efficiency}$$

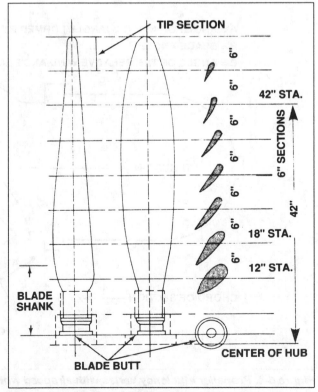

Figure 5-2. Propeller blade design.

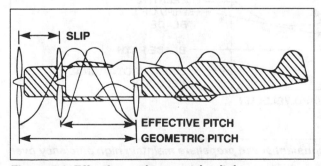

Figure 5-1. Effective and geometric pitch.

Propeller **slip** is the difference between the **geometric pitch** of the propeller and its **effective pitch** (see figure 5-1). Geometric pitch is the distance a propeller should advance in one revolution; effective pitch is the distance it actually advances. Thus, geometric or theoretical pitch is based on no slippage, but actual, or effective pitch, recognizes propeller slippage in the air.

The typical propeller blade can be described as a twisted airfoil of irregular planform. Two views of a propeller blade are shown in figure 5-2. For purposes of analysis, a blade can be divided into segments, which are located by station numbers in inches from the center of the blade hub. The cross sections of each 6-in. blade segment are shown as airfoils in the right-hand side of figure 5-2. Also identified in figure 5-2 are the blade shank and the blade butt. The blade shank is the thick, rounded portion of the propeller blade near the hub, which is designed to give strength to the blade. The blade butt, also called the blade base or root, is the end of the blade which fits in the propeller hub. The blade tip is that part of the propeller blade farthest from the hub, generally defined as the last 6 in. of the blade.

A cross section of a typical propeller blade is shown in figure 5-3. This **section** or blade element

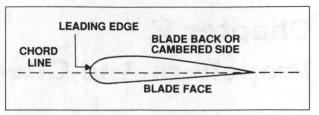

Figure 5-3. Cross section of a propeller blade.

is an airfoil comparable to a cross section of an aircraft wing. The blade **back** is the cambered or curved side of the blade, similar to the upper surface of an aircraft wing. The blade **face** is the flat side of the propeller blade ("facing" the pilot, if the propeller is up front in the **tractor** position). The **chord** line is an imaginary line drawn through the blade from the leading edge to the trailing edge. The **leading edge** is the thick edge of the blade that meets the air as the propeller rotates.

Blade angle, usually measured in degrees, is the angle between the chord line of the blade and the plane of rotation (figure 5-4). The chord of the propeller blade is determined in about the same manner as the chord of an airfoil. In fact, a propeller blade can be considered as being made up of an infinite number of thin blade elements, each of which is a miniature airfoil section whose

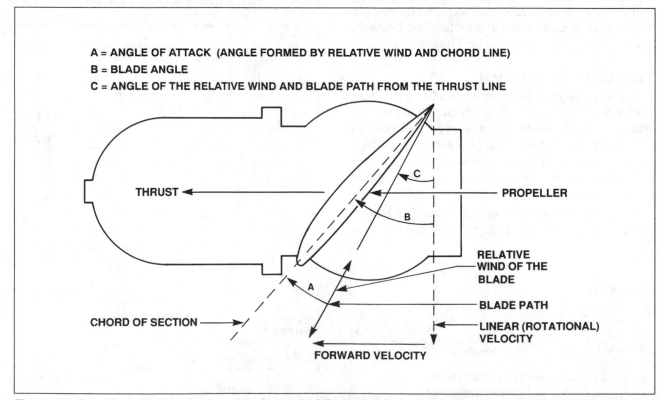

Figure 5-4. Propeller efficiency varies with airspeed while constant speed propellers maintain high efficiency over a wide range of airspeeds.

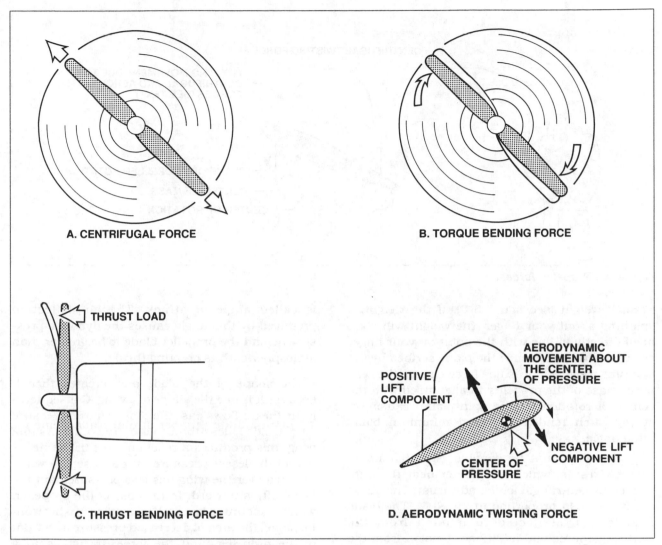

A. CENTRIFUGAL FORCE

B. TORQUE BENDING FORCE

THRUST LOAD

C. THRUST BENDING FORCE

AERODYNAMIC MOVEMENT ABOUT THE CENTER OF PRESSURE

POSITIVE LIFT COMPONENT

CENTER OF PRESSURE

NEGATIVE LIFT COMPONENT

D. AERODYNAMIC TWISTING FORCE

Figure 5-5. Forces acting on a rotating propeller.

chord is the width of the propeller blade at that section. Because most propellers have a flat blade face, **pitch** is easily measured by finding the angle between a line drawn along the face of the propeller blade and a line scribed by the plane of rotation. Pitch is not the same as blade angle, but, because pitch is largely determined by blade angle, the two terms are often used interchangeably. An increase or decrease in one is usually associated with an increase or decrease in the other.

Forces Acting On The Propeller

A rotating propeller is acted upon by **centrifugal**, **twisting**, and **bending** forces. The principal forces acting on a rotating propeller are illustrated in figure 5-5.

Centrifugal force (A of figure 5-5) is a physical force that tends to throw the rotating propeller blades away from the hub. **Torque bending force**

(B of figure 5-5), in the form of air resistance, tends to bend the propeller blades opposite to the direction of rotation. **Thrust bending force** (C of figure 5-5) is the thrust load that tends to bend propeller blades forward as the aircraft is pulled through the air. **Aerodynamic twisting force** (D of figure 5-5) creates a rotational force (twisting moment) about the center of pressure, causing the blade to tend to pitch to a lower blade angle (streamline).

At high angles of attack this twisting moment is reduced as the center of lift moves forward. At very high blade angles of attack the blades may exhibit a weak tendency to pitch toward a greater blade angle.

Centrifugal twisting force also twists the blade to flat pitch (unless the blade is counterweighted so its center of mass is behind the center of rotation). This is a strong force at normal propeller speeds. Imagine a string tied to your finger and to

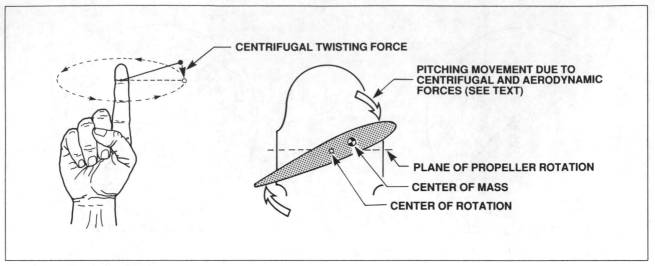

CENTRIFUGAL TWISTING FORCE

PITCHING MOVEMENT DUE TO CENTRIFUGAL AND AERODYNAMIC FORCES (SEE TEXT)

PLANE OF PROPELLER ROTATION

CENTER OF MASS

CENTER OF ROTATION

Figure 5-6. Propeller forces.

a small weight (see figure 5-6). If the weight is spinning about your finger, the weight will align itself directly in line with the point on your finger where it is tied (reference the plane scribed by the spinning weight). This same force causes the center of mass of the propeller to align itself with the center of rotation on the spin-plane, causing a strong pitch tendency toward minimum blade pitch angle.

A propeller must be capable of withstanding severe stresses, which are greater near the hub, caused by centrifugal force and thrust. The stresses increase in proportion to the RPM. The blade face is also subjected to tension from the centrifugal force and additional tension from the bending. For these reasons, nicks or scratches on the blade may cause very serious consequences.

A propeller must also be rigid enough to prevent **fluttering**, a type of vibration in which the ends of the blade twist back and forth at high frequency around an axis perpendicular to the engine crankshaft. Fluttering is accompanied by a distinctive noise often mistaken for exhaust noise. The constant vibration tends to weaken the blade and eventually causes failure.

Propeller Operation

To understand the action of a propeller, consider first its motion, which is both rotational and forward. Thus, as shown by the vectors of propeller forces in figure 5-4, a section of a propeller blade moves downward and forward. As far as the forces are concerned, the result is the same as if the blade were stationary and the air were coming at it from a direction opposite its path. The angle at which this air (relative wind) strikes the propeller blade

is called angle of attack. The air deflection produced by this angle causes the dynamic pressure behind the propeller blade to be greater than atmospheric, thus creating thrust.

The shape of the blade also creates thrust, because it is like the shape of a wing. Consequently, as the air flows past the propeller, the pressure on one side is less than that on the other. As in a wing, this produces a reaction force in the direction of the lesser pressure. In the case of a wing, the area over the wing has less pressure, and the force (lift) is upward. In the case of the propeller, which is mounted in a vertical instead of a horizontal plane, the area of decreased pressure is in front of the propeller, and the force (thrust) is in a forward direction. Aerodynamically, thrust is the result of the propeller shape and the angle of attack of the blade.

Another way to consider thrust is in terms of the mass of air handled. In these terms, **thrust** is equal to the mass of air handled times the slipstream velocity minus the velocity of the airplane. Thus, the power expended in producing thrust depends on the mass of air moved per second. On the average, thrust constitutes approximately 80% of the torque (total horsepower) absorbed by the propeller. The other 20% is lost in friction and slippage. For any speed of rotation, the horsepower absorbed by the propeller balances the horsepower delivered by the engine. For any single revolution of the propeller, the amount of air handled depends on the blade angle, which determines how big a "bite" of air the propeller takes. Thus, the blade angle is an excellent means of adjusting the load on the propeller to control the engine RPM.

The blade angle is also an excellent method of adjusting the angle of attack of the propeller. On constant-speed propellers, the blade angle must be adjusted to provide the desired engine and airplane speeds. Lift versus drag curves, which are drawn for propellers as well as wings, indicate that the most efficient angle of attack is a small one varying from 2° to 4° positive. The actual blade angle necessary to maintain this small angle of attack varies with the forward speed of the airplane. Just as airspeed of a wing varies with its angle of attack, RPM of the propeller varies with propeller blade angle of attack.

Imagine an aircraft trimmed for level, cruise flight. Without changing anything else, the pilot trims the aircraft slightly nose high. The wing angle of attack is increased so the aircraft slows. With forward velocity decreased, propeller angle of attack increases (see figure 5-4), thus loading the engine. As a result, RPM decreases. This decrease is sensed by the governor which causes propeller blade angle to decrease which decreases propeller angle of attack. When propeller angle of attack becomes the same as it was before the aircraft slowed, RPM will also have returned to the same value as before the aircraft slowed.

An explanation of how the governor senses RPM change and causes blade angle to change is to be found later in this chapter, in the discussion of governor principles.

Fixed-pitch and ground-adjustable propellers are designed for best efficiency at one rotation and forward speed. In other words, they are designed to provide maximum efficiency for a given airplane and engine combination at a given airspeed. A propeller may be used that provides the maximum propeller efficiency for takeoff, climb, cruising, or high speeds. Any change in these conditions results in lowering the efficiency of both the propeller and the engine. See figure 5-7.

A constant-speed propeller, however, keeps the blade angle adjusted for maximum efficiency for most conditions encountered in flight. During take-off, when maximum power and thrust are required, the constant-speed propeller is at a low propeller blade angle or pitch. The low blade angle keeps the angle of attack small. This light load allows the engine to turn at high RPM and to convert the maximum amount of fuel into heat energy in a given time (maximum engine power). Actually, it should be said that, for maximum power and thrust, the propeller pitch is set at a pitch that will give the maximum possible blade angle of attack that will allow the engine to develop maximum RPM. The high RPM and maximum possible blade angle and angle of attack creates maximum thrust; for, although the mass of air handled per revolution is smaller, the number of revolutions per minute are many, the slipstream velocity is high, and, with the low airplane speed, the thrust is maximum.

After lift-off, as the speed of the airplane increases, angle of attack decreases so the constant-speed propeller must change to a higher blade angle (or pitch). The higher blade angle keeps the angle of attack constant. This maintains constant engine RPM until takeoff power is reduced.

For climb after takeoff, the power output of the engine is reduced to climb power by decreasing the manifold pressure (partially closing the throttle) and increasing the blade angle to lower the RPM. Thus, horsepower absorbed by the propeller is reduced to match the reduced power of the engine.

At cruising altitude, when the airplane is in level flight and less power is required than is used in takeoff or climb, engine power is again reduced by lowering the manifold pressure and increasing the blade angle to decrease the RPM. Again, this reduces engine power; for, although the mass of air handled per revolution is greater, it is more than offset by a decrease in slipstream velocity and an increase in airspeed.

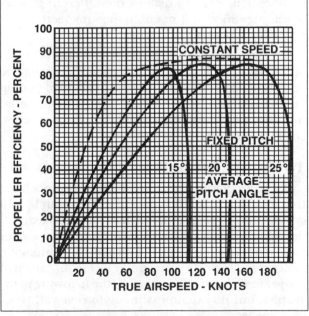

Figure 5-7. Fixed pitch propeller efficiency varies with air speed while constant speed propellers maintain high efficiency over a wide range of airspeeds.

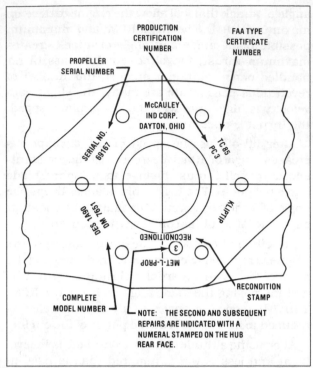

Figure 5-8. Typical hub stamping on a fixed-pitch propeller.

Propeller Regulations

FAR 23 (14CFR Part 23), Airworthiness Standards for Normal, Utility and Acrobatic Aircraft, and FAR 25, Airworthiness Standards for Transport Category Aircraft, specify the requirements for propellers and their control systems for aircraft certification. FAR 43 defines the different classes of maintenance and maintenance inspections for propeller systems.

Aircraft propellers are certified under FAR 35, and FAR 45 specifies what information must be permanently affixed to each propeller, including manufacturer's name, model designation, serial number, type certificate number and production certificate number. See figure 5-8.

Types Of Propellers

There are various types or classes of propellers, the simplest of which are the fixed-pitch and ground-adjustable propellers. The complexity of propeller systems increases from these simpler forms to controllable-pitch and complex automatic systems. Various characteristics of several propeller types are discussed in the following paragraphs, but no attempt is made to cover all types of propellers.

Fixed-Pitch Propeller

As the name implies, a **fixed-pitch propeller** has the blade pitch, or blade angle, built into the propeller. The blade angle generally is not changed after the propeller is built. Usually, this type of propeller is one piece and is constructed of wood or aluminum alloy.

Fixed-pitch propellers are designed for best efficiency at one rotational and forward speed. They are designed to fit one set of conditions of both airspeed and engine speed and any change in these conditions reduces the efficiency of both the propeller and the engine. See the efficiency curves of figure 5-7.

From figure 5-7 it can be seen that, since THP is a function of engine power (BHP) and propeller efficiency, the fixed pitch propeller of 15 degrees will produce more THP during the takeoff run but is limited at high speeds because angle of attack becomes small. This propeller would be called a **"climb prop"** and should be found installed on aircraft that must operate from short, obstacled strips with heavy loads where takeoff performance is more important than cruise performance.

The 20 degree pitch prop might be considered an **"all purpose prop"** and the one with 25 degrees of pitch would be termed a **"cruise prop"** for use on an aircraft that flies from long runways and spends most of its flight time in cross country flight.

When the efficiency curves for many propellers of a wide range of pitch are plotted, their peak efficiencies become the plotting points for the **constant speed propeller,** which changes pitch to provide the most efficient angle of attack for all speeds within the speed range of the engine-airframe combination. This makes the added weight, cost and complexity of the constant speed propeller worthwhile for aircraft with higher power and a greater range of operating speeds.

Ground-Adjustable Propeller

The **ground-adjustable propeller** operates as a fixed-pitch propeller. The pitch or blade angle can be changed only when the propeller is not turning. This is done by loosening the clamping mechanism which holds the blades in place. After the clamping mechanism has been tightened, the pitch of the blades cannot be changed in flight to meet variable flight requirements. In the air, it is a fixed-pitch propeller.

Controllable-Pitch Propeller

The **controllable-pitch propeller** permits a change of blade pitch, or angle, while the propeller is rotating. This permits the propeller to assume a blade angle that will give the best performance for particular flight conditions. The number of pitch positions may be limited, as with a **two-position controllable propeller**; or the pitch may be adjusted to any angle between the minimum and maximum pitch settings of a given propeller.

The use of controllable-pitch propellers also makes it possible to attain the desired engine RPM for a particular flight condition. As an airfoil is moved through the air, it produces two forces, lift and drag. Increasing propeller blade angle increases the angle of attack and produces more lift and drag; this action increases the horsepower required to turn the propeller at a given RPM. Since the engine is still producing the same horsepower, the engine and propeller slow down. If the blade angle is decreased, the propeller speeds up. Thus, the engine RPM can be controlled by increasing or decreasing the blade angle.

Constant-Speed Propeller

The use of propeller governors to increase or decrease propeller pitch converts the controllable-pitch propeller into a **constant speed prop.**

The governors used to control the hydraulic propeller pitch-changing mechanisms are geared to the engine crankshaft and, thus, are sensitive to changes in RPM. The governor directs the pressurized oil for operation of the propeller hydraulic pitch-changing mechanisms. When RPM increases above the value for which a governor is set, the governor causes the propeller pitch-changing mechanism to turn the blades to a higher angle. This angle increases the load on the engine, and RPM decreases. When RPM decreases below the value for which a governor is set, the governor causes the pitch-changing mechanism to turn the blades to a lower angle; the load on the engine is decreases, and RPM increases. Thus, a propeller governor tends to keep engine RPM constant by keeping the blade angle of attack constant.

Most pitch-changing mechanisms are operated by oil pressure (hydraulically) and use some type of piston-and-cylinder arrangement. The piston may move in the cylinder, or the cylinder may move over a stationary piston. The linear motion of the piston is converted by several different types of mechanical linkage into the rotary motion necessary to change the blade angle. The mechanical connection may be through gears, the pitch-changing mechanism turning a drive gear or power gear that meshes with a gear attached to the butt of each blade.

In most cases the oil for operating these various types of hydraulic pitch-changing mechanisms comes directly from the engine lubricating system. When the engine lubricating system is used, the engine oil pressure is usually boosted by a pump that is integral with the governor to operate the propeller.

Additional refinements, such as pitch reversal and feathering features, are included in some propellers to improve still further their operational characteristics.

Reverse-Pitch Propellers

A **reverse-pitch propeller** is a controllable propeller in which the blade angles can be changed to a negative value during operation. The purpose of the reversible pitch feature is to produce a high negative thrust at low speed by using engine power. It is used principally as an aerodynamic brake to reduce ground roll after landing.

Feathering Propellers

A **feathering propeller** is a controllable propeller having a mechanism to change the pitch to an angle so that forward aircraft motion produces a minimum windmilling effect on a "power-off" propeller. Feathering propellers are used on multi-engine aircraft to reduce propeller drag to a minimum under engine failure conditions.

Tractor Propeller

Tractor propellers are those mounted on the up-stream end of a drive shaft in front of the supporting structure. Most aircraft are equipped with this type of propeller. A major advantage of the tractor propeller is that lower stresses are induced in the propeller as it rotates in relatively undisturbed air.

Pusher Propellers

Pusher propellers are those mounted on the downstream end of a drive shaft behind the supporting structure. Pusher propellers are constructed as fixed- or variable-pitch propellers. Seaplanes and amphibious aircraft have used a greater percentage of pusher propellers than other kinds of aircraft.

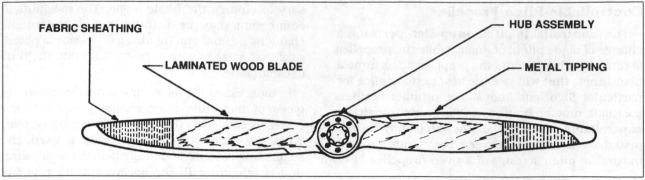

FABRIC SHEATHING

LAMINATED WOOD BLADE

HUB ASSEMBLY

METAL TIPPING

Figure 5-9. Fixed-pitch wooden propeller assembly.

On land planes, where propeller-to-ground clearance usually is less than propeller-to-water clearance of watercraft, pusher propellers are subject to more damage than tractor propellers. Rocks, gravel, and small objects, dislodged by the wheels, quite often may be thrown or drawn into a pusher propeller. Similarly, seaplanes with pusher propellers are apt to encounter propeller damage from water spray thrown up by the hull during landing or takeoff from water. Consequently, the pusher propeller quite often is mounted above and behind the wings to prevent such damage. On some aircraft, pusher propellers have proven to be more efficient. For example, the Cessna 337 has a higher single engine service ceiling with the rear engine, and Voyager feathered its front propeller for most of its record-setting, around-the-world-flight.

Fixed-Pitch Wooden Propellers

The construction of a fixed-pitch wooden propeller (figure 5-9) is such that its blade pitch cannot be changed after manufacture. The choice of the blade angle is decided by the normal use of the propeller on an aircraft during level flight, when the engine will perform at maximum efficiency.

The impossibility of changing the blade pitch on the fixed-pitch propeller restricts its use to small aircraft with low-horsepower engines, in which maximum engine efficiency during all flight conditions is of lesser importance than in larger aircraft. The wooden fixed-pitch propeller, because of its light weight, rigidity, economy of production, simplicity of construction, and ease of replacement, is well suited for such small aircraft.

A wooden propeller is not constructed from a solid block, but is built up of a number of separate layers of carefully selected and well-seasoned hardwoods. Many woods, such as mahogany, cherry, black walnut, and oak, are used to some

extent, but birch is the most widely used. Five to nine separate layers are used, each about ¾-inch thick. The several layers are glued together (laminated) with a waterproof, resinous glue and allowed to set. The "blank" is then roughed to the approximate shape and size of the finished product.

The roughed-out propeller is then allowed to dry for approximately a week to permit the moisture content of the layers to become equalized. This additional period of seasoning prevents warping and cracking that might occur if the blank were immediately carved. Following this period, the propeller is carefully constructed. Templates and bench protractors are used to obtain the proper contour and blade angle at all stations.

After the propeller blades are finished, a fabric covering is cemented to the outer 12 or 15 inches of each finished blade. A metal tipping is fastened to most of the leading edge and tip of each blade to protect the propeller from damage caused by flying particles in the air during landing, taxiing, or takeoff.

Metal tipping may be of terneplate, Monel metal, or brass. Stainless steel has been used to some extent. It is secured to the leading edge of the blade by countersunk wood screws and rivets. The heads of the screws are soldered to the tipping to prevent loosening, and the solder is filed to make a smooth surface. Since moisture condenses on the tipping between the metal and the wood, the tipping is provided with small holes near the blade tip to allow this moisture to drain away or be thrown out by centrifugal force. It is important that these drainholes be kept open at all times.

Since wood is subject to swelling, shrinking, and warping because of changes of moisture content, a protective coating is applied to the finished propeller to prevent a rapid change of moisture content.

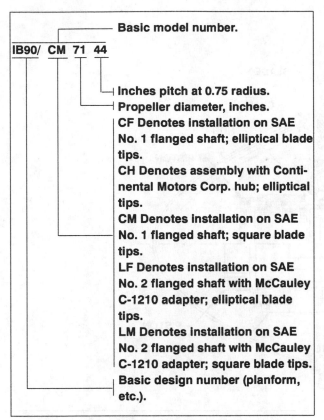

Basic model number.

IB90/ CM 71 44

Inches pitch at 0.75 radius.

Propeller diameter, inches.

CF Denotes installation on SAE No. 1 flanged shaft; elliptical blade tips.

CH Denotes assembly with Continental Motors Corp. hub; elliptical tips.

CM Denotes installation on SAE No. 1 flanged shaft; square blade tips.

LF Denotes installation on SAE No. 2 flanged shaft with McCauley C-1210 adapter; elliptical blade tips.

LM Denotes installation on SAE No. 2 flanged shaft with McCauley C-1210 adapter; square blade tips.

Basic design number (planform, etc.).

Figure 5-10. Complete propeller model number.

Metal Fixed-Pitch Propellers

Metal fixed-pitch propellers are similar in general appearance to a wooden propeller, except that the sections are usually thinner. The metal

fixed-pitch propeller is widely used on many models of light aircraft.

Many of the earliest metal propellers were manufactured in one piece of forged Duralumin. Compared to wooden propellers, they were lighter in weight because of elimination of blade-clamping devices; they offered a lower maintenance cost because they were made in one piece; they provided more efficient cooling because of the effective pitch nearer the hub; and because there was no joint between the blades and the hub, the propeller pitch could be changed, within limits, by twisting the blade slightly.

Propellers of this type are now manufactured of one-piece anodized aluminum alloy. They are identified by stamping the propeller hub with the serial number, model number, Federal Aviation Administration (FAA) type certificate number, production certificate number, and the number of times the propeller has been reconditioned. The complete model number of the propeller is a combination of the basic model number and suffix numbers to indicate the propeller diameter and pitch. An explanation of a complete model number, using the McCauley 1B90/CM propeller, is provided in figure 5-10.

Constant-Speed Propellers

Constant speed propellers for light aircraft are similar in operation. For **single-engine** installations, oil pressure from the governor usually

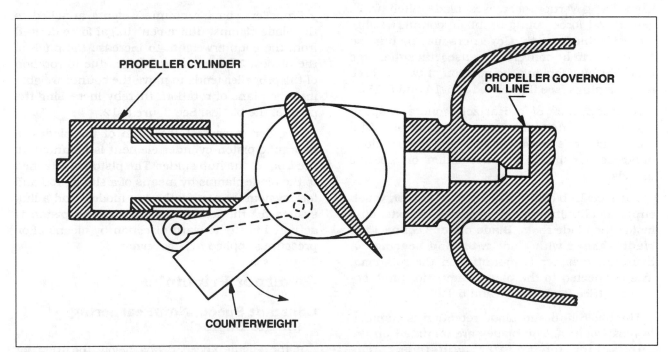

Figure 5-11. Pitch change mechanism for a multi-engine, counterweight propeller.

73

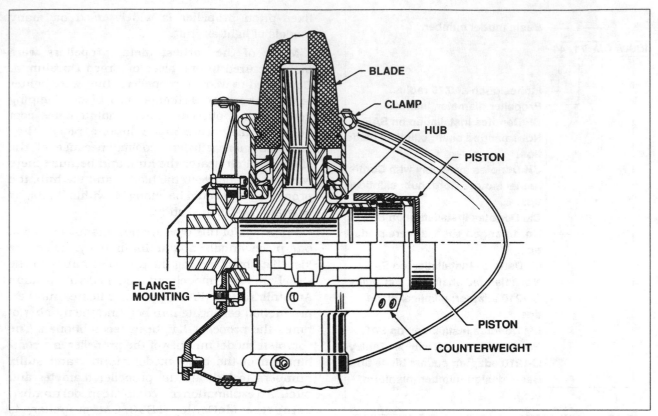

Figure 5-12. Constant Speed Prop.

increases blade pitch. Centrifugal and aerodynamic pitching moments decrease blade pitch. For **multi-engine aircraft,** it is necessary that the blades increase pitch to the feather position if engine oil pressure is lost, so oil pressure from the governor decreases blade pitch while centrifugal force acting on blade counterweights increases blade pitch. Governors may be built so that they can be converted for use with either type propeller by changing the position of two or three by-pass plugs (see figures 5-13, 5-14 and 5-15).

A description of a Hartzell constant speed propeller is used here as an example. The manufacturer's specifications and instructions must be consulted for information on specific models.

The steel hub consists of a central **spider,** which supports aluminum blades with a tube extending inside the blade roots. **Blade clamps** connect the **blade shanks** with blade **retention bearings.** A hydraulic cylinder is mounted on the rotational axis connected to the blade clamps for pitch actuation. (See figures 5-11 and 5-12.)

The basic hub and blade retention is common to most models. The blades are mounted on the hub spider for angular adjustment. The centrifugal force of the blades, amounting to as much as 25 tons, is transmitted to the hub spider through blade clamps and then through ball bearings. The propeller thrust and engine torque is transmitted from the blades to the hub spider through a bushing inside the blade shank.

Propellers, having counterweights attached to the blade clamps, utilize centrifugal force derived from the counterweights to increase the pitch of the blades. The centrifugal force, due to rotation of the propeller tends to move the counterweights into the plane of rotation, thereby increasing the pitch of the blades. See figure 5-12.

In order to control the pitch of the blades, a hydraulic piston-cylinder element is mounted on the front of the hub spider. The piston is attached to the blade clamps by means of a sliding rod and fork system for non-feathering models and a link system for the feathering models. The piston is actuated in the forward direction by means of oil pressure supplied by a governor.

Governor Principles

Constant Speed, Non-Feathering (Single Engine)

If the engine speed drops below the RPM for which the governor is set (see figure 5-15), the

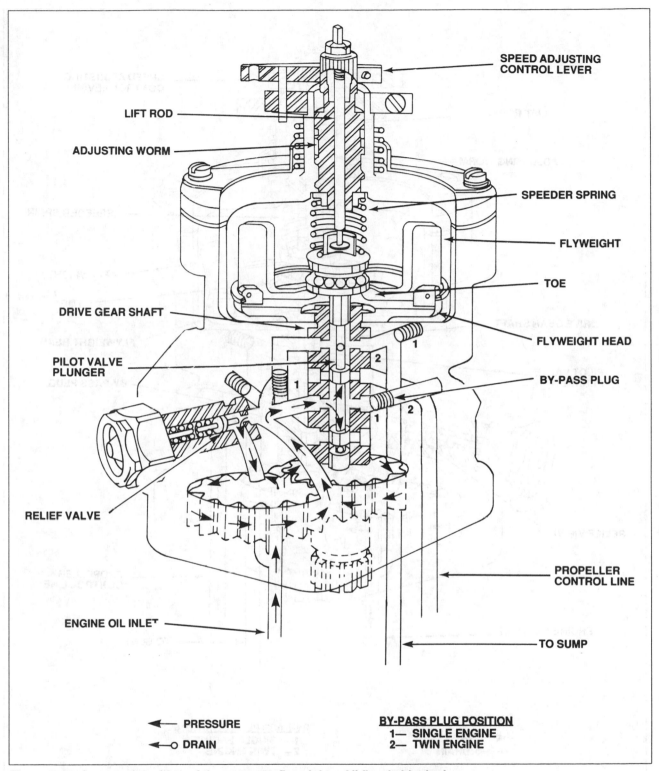

SPEED ADJUSTING CONTROL LEVER

LIFT ROD

ADJUSTING WORM

SPEEDER SPRING

FLYWEIGHT

TOE

DRIVE GEAR SHAFT

FLYWEIGHT HEAD

PILOT VALVE PLUNGER

BY-PASS PLUG

1

2

1

1

2

RELIEF VALVE

PROPELLER CONTROL LINE

ENGINE OIL INLET

TO SUMP

◄— **PRESSURE**

◄—o **DRAIN**

BY-PASS PLUG POSITION
1— SINGLE ENGINE
2— TWIN ENGINE

Figure 5-13. On-speed position of the governor flyweights. All flow is blocked.

rotational force on the engine driven governor flyweights becomes less. This allows the speeder spring to move the flyweights inward and the pilot valve downward. With the pilot valve in the downward position, oil from the gear boost pump is blocked and a passageway is opened to allow oil to flow from the propeller hub to the engine oil sump. Centrifugal and aerodynamic pitching moments are able to decrease blade pitch, forcing oil out of the hub. This in turn, decreases the blade

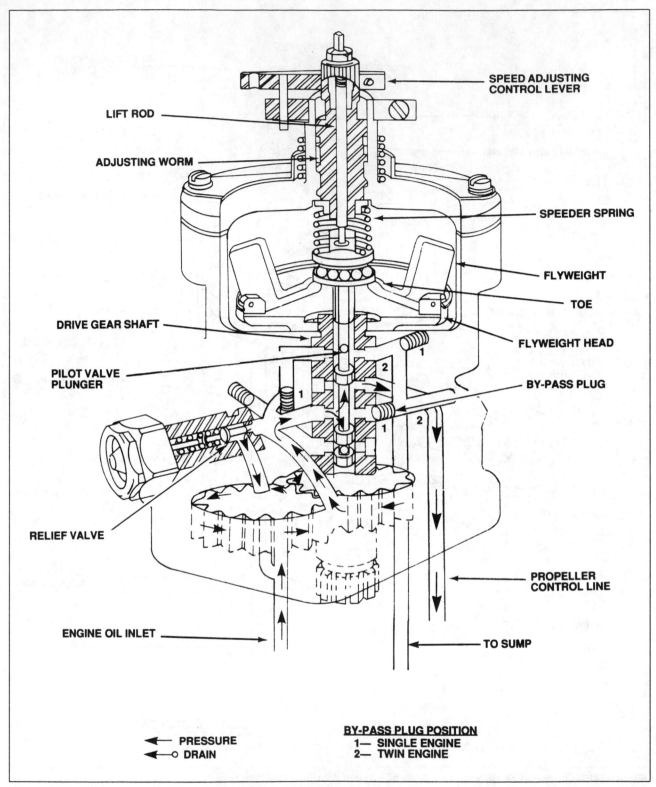

SPEED ADJUSTING
CONTROL LEVER

LIFT ROD

ADJUSTING WORM

SPEEDER SPRING

FLYWEIGHT

TOE

DRIVE GEAR SHAFT

FLYWEIGHT HEAD

PILOT VALVE
PLUNGER

BY-PASS PLUG

RELIEF VALVE

PROPELLER
CONTROL LINE

ENGINE OIL INLET

TO SUMP

← PRESSURE
←o DRAIN

BY-PASS PLUG POSITION
1— SINGLE ENGINE
2— TWIN ENGINE

Figure 5-14. Over-speed position of the governor flyweights.

angle and permits the engine to return to the on-speed setting.

If the engine speed increases above the RPM for which the governor is set, the flyweights move against the force of the speeder spring and raise the pilot valve. This permits high pressure oil to flow to the propeller hub, increasing blade pitch and angle of attack, decreasing RPM (figure 5-14).

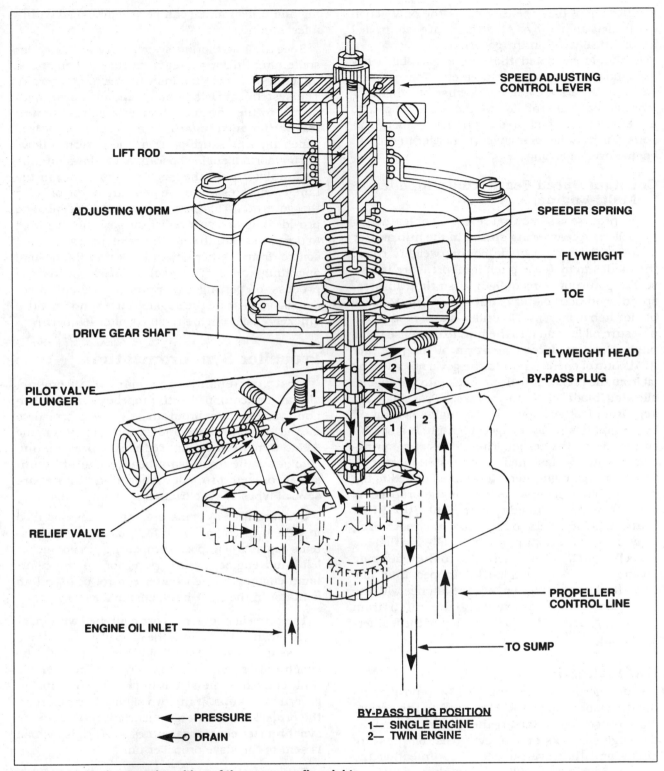

SPEED ADJUSTING CONTROL LEVER

LIFT ROD

ADJUSTING WORM

SPEEDER SPRING

FLYWEIGHT

TOE

DRIVE GEAR SHAFT

FLYWEIGHT HEAD

BY-PASS PLUG

PILOT VALVE PLUNGER

RELIEF VALVE

PROPELLER CONTROL LINE

ENGINE OIL INLET

TO SUMP

⟵ PRESSURE

⟵○ DRAIN

BY-PASS PLUG POSITION
1— SINGLE ENGINE
2— TWIN ENGINE

Figure 5-15. Under-speed position of the governor flyweights.

When the engine is exactly at the RPM set by the pilot, the centrifugal reaction of the flyweights balances the force of the speeder spring, positioning the pilot valve so that oil is neither supplied to nor drained from the propeller (figure 5-13). With this condition, propeller blade angle does not change until the next pilot-valve movement. The propeller functions as a fixed-pitch propeller. Note that the pilot controls the RPM setting by varying the amount of compression in the speeder spring.

Positioning of the speeder rack is the only action controlled manually. All others are controlled automatically within the governor.

It should be noted that there may be a few exceptions to almost any statement. For instance, some propellers are adapted for use on single-engine aerobatic aircraft so that the propeller goes to increased pitch if oil pressure is lost (which happens during some maneuvers) in order to prevent engine over-speeding.

Constant-Speed Feathering Propeller (Multi-Engine)

The process is somewhat reversed for feathering propellers. A **feathering spring** in the hub and/or centrifugal force on blade counterweights cause the blades to increase pitch if oil pressure is lost, so the governor is reconfigured so that, in over-speed condition, oil flows out of the hub to allow blade pitch to increase. In under-speed mode, high pressure oil is routed to the hub to decrease blade pitch. **Feather detent position** of the propeller RPM control opens a port in the governor allowing oil from the propeller to drain back into the engine, allowing blade pitch to increase. Propellers are kept from feathering when the engine is shut off by automatically disengaged high-pitch stops incorporated in the design. These consist of spring-loaded latches fastened to the stationary hub which engage high-pitch stop-plates bolted to the movable blade clamps. As long as the propeller is in rotation at speeds over 600-800 RPM, centrifugal force acts to disengage the **feathering latches** from the high-pitch stop-plates so that the propeller pitch may be increased to the feathering position. When the engine is stopped from idle RPM, the latch springs engage the latches with the high-pitch stops, preventing the pitch from increasing further due to the action of the feathering spring.

Unfeathering

Unfeathering is accomplished by repositioning the governor propeller RPM control to the normal flight range and restarting the engine. As soon as the engine cranks over a few turns, oil pressure builds and the governor starts to unfeather the blades. Soon, wind-milling takes place, which increases engine RPM and speeds up the process of unfeathering. In order to facilitate cranking of the engine, feathering blade angle is set at 80 to 85 degrees at the ¾ point on the blade, allowing the air to assist the engine starter. In general, restarting and unfeathering can be accomplished within a few seconds.

Special unfeathering systems are available for some aircraft, for which restarting the engine is difficult, or done often (as with training aircraft). The system consists of an oil **accumulator**, connected to the governor through a valve. The **unfeathering accumulator** is a pressure tank into which high pressure oil from the governor flows during normal engine operation. Oil flows into the accumulator until the gas (nitrogen or air) in the tank is compressed to the pressure of the oil from the governor. This stored energy is available to provide high pressure oil for the governor to direct to the propeller hub when the pilot moves the RPM control from feather detent back into the normal operating range. The **accumulator** speeds the process of getting the propeller unfeathered, decreases wear on the engine starter and provides quicker, easier engine re-starts after feathering.

Propeller Synchronization

Most four-engine, and many twin-engine, aircraft are equipped with propeller synchronization systems. Synchronization systems provide a means of controlling and synchronizing engine RPM. Synchronization reduces vibration and eliminates the unpleasant beat produced by unsynchronized propeller operation. There are several types of synchronizer systems in use.

Synchronizer systems are sometimes installed in light twin-engine aircraft. Typically, such systems consists of a special propeller governor on the left-hand engine, a slave governor on the right-hand engine, a synchronizer control unit and an actuator in the right-hand engine nacelle.

The propeller governors are equipped with magnetic pickups that count the propeller revolutions and send a signal to the synchronizer unit. The synchronizer, which is usually a transistorized unit, compares the signal from the two propeller governor pickups. If the two signals are different, the propellers are out of synchronization, and the synchronizer control generates a DC pulse which is sent to the slave propeller unit.

The control signal is sent to an actuator, which consists of two rotary solenoids mounted to operate on a common shaft. A signal to increase the RPM of the slave propeller is sent to one of the solenoids, which rotates the shaft clockwise. A signal to decrease RPM is sent to the other

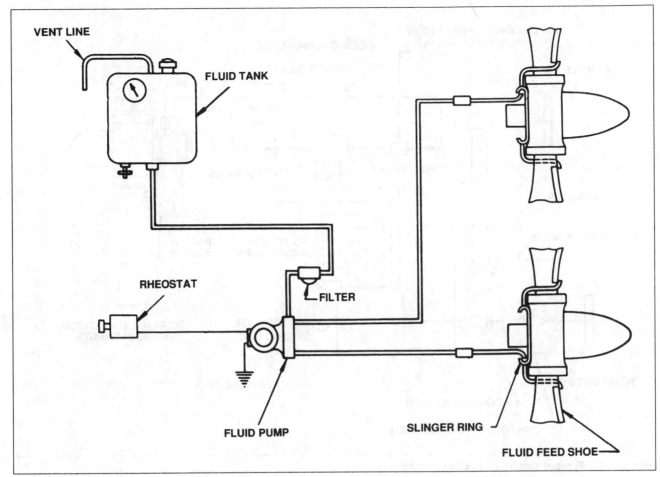

Figure 5-16. Typical propeller fluid anti-icing system.

solenoid, which moves the shaft in the opposite direction.

Each pulse signal rotates the shaft a fixed amount. This distance is called a "step." Attached to the shaft is a flexible cable, which is connected on its other end to a trimming unit. The vernier action of the trimming unit regulates the governor arm.

Propeller Ice Control Systems

Ice formation on a propeller blade, in effect, produces a distorted blade airfoil section which causes a loss in propeller efficiency and thrust. Generally, ice collects unsymmetrically on a propeller blade and produces propeller unbalance and destructive vibration.

Anti-Icing (Fluid) Systems

A typical fluid system (figure 5-16) includes a tank to hold a supply of anti-icing fluid. This fluid is forced to each propeller by a pump. The control system permits variation in the pumping rate so that the quantity of fluid delivered to a propeller can be varied, depending on the severity of icing. Fluid is transferred from a stationary nozzle on the engine nose case into a circular U-shaped channel (slinger ring) mounted on the rear of the propeller assembly. The fluid under pressure of centrifugal force is transferred through nozzles to each blade shank.

Because airflow around a blade shank tends to disperse anti-icing fluids to areas on which ice does not collect in large quantities, feed shoes, or **boots,** are installed on the blade leading edge. These feed shoes are narrow strips of rubber, extending from the blade shank to a blade station that is approximately 25% of the propeller radius. The feed shoes are molded with several parallel open channels in which fluid will flow from the blade shank toward the blade tip by centrifugal force. The fluid flows laterally from the channels, over the leading edge of the blade.

Isopropyl alcohol is used in some anti-icing systems because of its availability and low cost. Phosphate compounds are comparable to

79

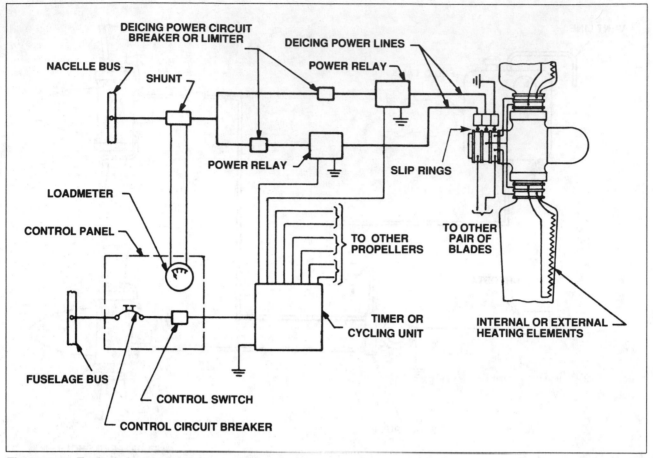

Figure 5-17. Typical electrical deicing system.

isopropyl alcohol in anti-icing performance and have the advantage of reduced flammability. However, phosphate compounds are comparatively expensive and, consequently, are not widely used.

Anti-icing systems are designed to prevent ice formation. They are not capable of removing ice once it has formed. Therefore, anti-icing systems must be operated continuously during icing conditions (and preferably before) ice accumulation starts.

Electrical Deicing Systems

An electrical propeller icing control system (figure 5-17) consists basically of an electrical energy source, a resistance heating element, system controls, and necessary wiring. The heating elements are mounted internally or externally on the propeller spinner and blades. Electrical power from the aircraft system is transferred to the propeller hub through electrical leads, which terminate in **slip rings** and brushes. Flexible connectors are used to transfer power from the hub to the blade elements.

Icing control is accomplished by converting electrical energy to heat energy in the heating element. Balanced ice removal from all blades must be obtained as nearly as possible if excessive vibration is to be avoided. To obtain balanced ice removal, variation of heating current in the blade elements is controlled so that similar heating effects are obtained in opposite blades.

Electrical **deicing** systems are usually designed for intermittent application of power to the heating elements to remove ice after formation but before excessive accumulation. Proper control of heating intervals aids in preventing runback, since heat is applied just long enough to melt the ice face in contact with the blade.

If heat supplied to an icing surface is more than that required to melt just the inner ice face, but insufficient to evaporate all the water formed, water will run back over the unheated surface and freeze. Runback of this nature causes ice formation on uncontrolled icing areas of the blade or surface.

Cycling timers are used to energize the heating element circuits for periods of 15 to 30 seconds, with a complete cycle time of 2 minutes. A cycling timer is an electric motor driven contactor which controls power contactors in separate sections of the circuit.

Controls for propeller electrical deicing systems include on-off switches, ammeters or loadmeters to indicate current in the circuits, and protective devices, such as current limiters or circuit breakers. The ammeters or loadmeters permit monitoring of individual circuit currents and reflect operation of the timer.

To prevent element overheating, the propeller deicing system is generally used only when the propellers are rotating, and for short periods of time during ground runup.

Preflight Inspection Of Propellers

I have often said that, if an enemy was shooting at me and I only had time to check two items on a preflight inspection, one of the two items I would check would be the propeller. Since the propeller operates under very high stress, chances for a catastrophic failure, although remote, do exist. NTSB data shows that propeller failure and other problems are involved in only about 1% of general aviation accidents and about .6% of accidents with fatalities. But, when they do happen, they are often catastrophic.

Therefore, propeller care is very important. Any rock-strike, nick or other irregularity, usually found on the leading edge or the face of the blade, should be examined and treated by a mechanic. For propellers with variable pitch blades, security (tightness) of the blades can be checked by holding the tip of the blade between thumb and one forefinger and attempting to move the blade fore and aft, up and down. If the blade moves from other than normal bending, consult a mechanic. The prop spinner should also be checked for security. Look for streaks of grease or oil on the propeller blade face and on the leading edge of the cowling. In flight, specks of oil on the windshield may indicate that weakening propeller seals need attention (or simply that someone was sloppy when putting that last can of oil into the engine).

Propeller Ground Handling

If you must re-position a propeller blade, it is best to move it in the opposite direction from normal rotation to keep the magnetos from firing the spark plugs. Always assume that the "gun is loaded". That is, that the ignition switch is "on" or

the **magneto P-leads** are discontinuous which results in the magnetos being able to spark the plugs even though the ignition switches are off. The only times the propeller should be moved by hand in normal rotation is when propping (hand-starting) or clearing or priming the engine. In these cases, proper techniques must be used, which assume that the engine will start if the propeller is moved.

There are some good reasons to "park" the two-bladed propeller in the 10 o'clock/4 o'clock position. In this position, the propeller is in proper position for hand starting, good walk-around clearance is provided, birds find the propeller tip a less satisfactory place to perch (especially if the propeller is square-tipped) so there is less adornment of the propeller and airframe by corrosive bird droppings and the propeller tip vertical height is decreased so that it won't catch a low overhead door while being hangared. Also, in that position, intake and exhaust valve openings are minimized on many engines.

If the aircraft is parked outside in regions of high rainfall, acid rain, exposure to salt air or exposed to repetitious freeze-thaw cycles, it may be valid to "park" the two-bladed propeller horizontally to prevent water from running down the blade and remaining for long periods in the blade-clamp/hub area.

Propeller Noise Patterns

Noise is a major public relations problem for aviators. Every pilot should keep in mind the geometry of the aircraft's noise signature and attempt to be a good neighbor with respect to noise. Propellers make a lot of noise — much more noise than does engine exhaust. Propeller blade noise is transmitted directionally. Noise is much greater off the tips of the propeller. Visualize extending the propeller disk to the ground. That is where noise will be greatest.

Noise can be reduced by decreasing propeller RPM, decreasing engine power and by increasing distance to the noise sensitive area.

Operation of Propeller Controls

RPM control of constant speed propellers is accomplished by the governor. Control of the governor is done by the laws of physics with some input from the pilot. It is important for the pilot to remember always that control of the propeller by the governor is not instantaneous. A little time is required for the governor to sense RPM, compare with desired value of RPM and send oil to make the proper blade angle change.

If the pilot makes rapid changes of the throttle, propeller RPM control or mixture control, it is possible for the engine to momentarily over-speed. Engine **over-speed** is a serious matter. It should not be permitted to occur at any time. Usually, **over-speed** is caused by rapid movement of the controls rather than by governor fault. Therefore, the pilot must always be smooth and gentle when moving engine controls.

Propeller Overhaul

Fixed pitch propellers don't go through the overhaul process but manufacturers recommend periodic reconditioning. One manufacturer recommends reconditioning after 1000 hours of use while some suggest that fixed pitch propellers be reconditioned on an "as needed" basis.

Reconditioning of metal propellers starts with paint being stripped from the propeller by bead-blasting or other stripping process. The blades are then warmed in a water bath. The expansion that takes place while warming amplifies the size of cracks in the metal so the dye penetrant that is applied next will readily show even very fine cracks. Cracks are not often found but if they are, the propeller (or blade) goes on the scrap pile. Length, chord line length and thickness at various stations is carefully checked. If dimensions become too small (due to wear, filing, reconditioning, etc.) the propeller is worn out and joins its cracked counterparts on the scrap pile. The next step in the process involves grinding 10 to 12 thousandths of an inch of the outside surface from the propeller to relieve surface stresses and fatigue lines and to remove corrosion pits.

Stresses that a propeller blade endures accumulate as surface fatigue. When a propeller is loaded, the surface is under compression or tension. A two-blade propeller turning at 2700 RPM has a tension load on it of nearly 20 tons. Picture the propeller blade with a locomotive hanging outward from each tip! Blade tension loading increases with the square of the tip speed or RPM, so an over-speed of 10% (270 RPM) adds another locomotive to the propeller! Manufacturers suggest reconditioning or overhaul after significant over-speeds.

Constant speed propellers have manufacturer recommended **TBO's** (time between overhaul) just like aircraft engines. A typical TBO might be 1500 hours or 5 years, whichever comes first. For the aircraft that operates only under the rules of Part 91, the TBO is only a recommendation. The owner/operator must decide whether to overhaul or defer. There are pros and cons that make the decision difficult which the operator must evaluate, after gathering facts from discussions with qualified mechanics and possibly the propeller maintenance facility.

Overhaul entails more than does reconditioning because the constant speed propeller is a more complex device. The propeller is first disassembled. Some parts (gaskets, seals, ball bearings and races) are discarded. Everything else is carefully inspected for cracks and wear. Aluminum parts go through dye penetrant testing for cracks. Steel parts are magnafluxed and dimensions checked. The propeller is reassembled and lubricated, then painted and statically balanced.

Additional Reading

1. Aircraft Propellers and Controls; Delp, F.; IAP, Inc., Publ; ISBN 0-89100-097-6; 1989.
2. Aircraft Powerplants; Bent & McKinley; McGraw-Hill, Publ; ISBN 0-0-004792-8; 1978; Chapters 17,18.
3. Airframe & Powerplant Mechanics Powerplant Handbook; IAP, Inc., Publ.; 1976; Chapter 7.
4. The Advanced Pilot's Flight Manual; Kershner, W.; Iowa State University Press; 1985; Chapter 12.
5. 1990 Pilot's Yearbook; Belvoir Publications (Aviation Consumer Magazine); Prop Overhaul; pp. 65-70.

Study Questions And Problems

1. Reference figure 5-2, what is the linear velocity of station 36 if the propeller RPM is 2400?* Express your answer in ft/sec, KTS and Mach number.

2. What does the blade angle of station 36 of the propeller in Question #1 need to be to produce an angle of attack of 4 degrees if TAS is 120 KTS?*

3. What does the blade angle of station 36 need to be to produce an angle of attack of 4 degrees if TAS is zero (before brake release on takeoff)?* (Assume rearward average airflow of 35 MPH through the propeller that is turning at 2400 RPM.)

4. Is the lowest blade angle near the shank or the tip? Why? Was it built that way? (several reasons)*

5. With a fixed RPM, does propeller blade angle of attack increase or decrease as airspeed increases?

6. What forces tend to twist the blade to decrease propeller blade angle?

7. What forces tend to twist the blade to increase propeller blade angle?

8. What is the tip speed of an 88″ diameter propeller turning at 2700 RPM? Express the answer in ft/sec, KTS and mach number (sea level, standard day).*

9. What are the advantages of leaving the propeller in the 10/4 o'clock position?

10. Why doesn't a fixed pitch propeller reach redline RPM during full throttle run-up on the ground?*

11. What is the difference between static and dynamic balancing?

12. What are the differences between a two position and a ground adjustable propeller?

13. What component(s) in the governor oppose the force of the speeder spring?

14. What evidence can the pilot expect to see, indicating that the oil supply from the engine to the governor has stopped?**

15. Presuming that all components are adjusted and operating correctly, what will cause momentary over-speeding of the propeller during takeoff? During flight?

16. What forces are used to feather a propeller?

17. Is the governor in figure 5-14 and 5-15 set up for a single engine or a multi-engine propeller?

18. Make a copy of figures 5-14 and 5-15. Reconfigure each governor for the type of propeller not chosen for your answer to question 17 (above). Show the correct location of the by-pass plugs and the path of oil flow.

*These questions are not answered directly in the text of the chapter. The reader will have to use what has been learned from the chapter and basic background knowledge to answer these questions.

**The reader may have to seek additional ideas from other books or from discussions with a knowledgeable flight instructor, mechanic or classroom instructor in order to be sure the reader's deductions have properly answered the question.

Chapter VI
Fuels And Fuel Systems

Fuel: The Energy Source For The Combustion Process

Source Of Energy

The internal combustion engine used in our modern aircraft is a form of **heat engine**; that is, it is a device which changes heat energy into mechanical energy. The heat energy used by internal combustion engines comes from the original source of energy, the sun. Solar energy was radiated to the earth where it was changed into chemical energy in vegetation and plant life. Some of this energy was stored in the bodies of animals that ate the vegetation. Millenniums ago, those plants and animals were buried in the earth. Heat, pressure and lots of time turned that organic material into fossil fuel, the petroleum products used in heat engines. Petroleum is an organic material made up of many different types of large molecules, each of which consists of many hydrogen and carbon atoms, of the family known as **hydrocarbons**.

Chemistry Of Combustion

In order to release heat energy from the hydrocarbons, a chemical reaction must take place. And for this to occur, the hydrocarbon fuel must be brought into contact with a source of oxygen and the temperature of the fuel and oxygen must be raised to its kindling point. When this happens, the oxygen will combine with the fuel, and oxidation, or burning, occurs. This process is called **combustion**.

Each molecule of aviation gasoline, our most widely used fuel for aircraft reciprocating engines, is made up of a variety of hydrocarbon molecules of different sizes and shapes. A typical or average sized molecule is one made up of eight atoms of carbon and eighteen atoms of hydrogen. With the chemical name **octane**, this molecule may be written as the chemical formula C_8H_{18}. In order to release all of the energy in the fuel, it must be completely burned; that is, all of the fuel must combine with oxygen.

Two atoms of oxygen exist in the atmosphere as one molecule of oxygen gas, and for complete burning, two molecules of octane and 25 molecules of oxygen gas combine to produce sixteen molecules of carbon dioxide, eighteen molecules of water and a lot of heat. Looking at this as a chemical equation, it can be seen that the following has taken place:

$$2\ C_8H_{18} + 25\ O_2 \rightarrow 16\ CO_2 + 18\ H_2O + HEAT$$

After ignition, a large increase in pressure in the cylinder (a closed container) is due to the large increase in temperature of the gasses in the cylinder. A much smaller contribution to increased pressure comes from the increase in the number of molecules (27 to 34) that occurs during combustion. Note, however, that there is no change in the number of atoms of carbon and hydrogen before and after combustion (matter cannot be created or destroyed).

Air is a physical mixture made up of several gases, principally nitrogen and oxygen. Since nitrogen is an inert gas, it does not enter into the chemical reaction. About fifteen pounds of air is needed to unite with one pound of gasoline to completely combine all of the gasoline and oxygen. If there is more air than is needed, oxygen will be left over after the burning is completed, and if there is more fuel in the mixture than is required for the amount of oxygen, free carbon will be left; this usually shows up as black smoke or soot.

Gasoline engine exhaust contains other products of combustion. Carbon monoxide (CO) is produced in small amounts because the burning process is not totally efficient. Since CO is very poisonous to humans and other animals, great care must be taken to keep exhaust fumes out of the aircraft cabin.

Other byproducts of combustion exist in the exhaust as well, including small quantities of oil, soot, lead oxides, and oxides of fuel additives that were in the gasoline.

Fuel:Air Ratio

The mixture ratio of fifteen pounds of air to one pound of gasoline is known as a **stoichiometric** mixture, which is a chemically correct mixture in which all of the chemical elements are used and none are left over. A 15:1 air:fuel ratio may also be expressed as a fuel:air ratio of 0.067 (1/15 = 0.067).

Combustion will occur with a mixture as rich as 8:1 (0.125) or as lean as 18:1 (0.055), but the maximum amount of heat energy is released with the stoichiometric mixture of 0.067. In a lean mixture there is less fuel and therefore less heat energy. But if the mixture is overly rich, there is not enough oxygen so some of the fuel will not be burned, and so a smaller amount of heat energy will be released.

It would seem that since the most heat energy is released from the fuel with a fuel-air mixture of 0.067, this ratio would be used to produce the most power. This is not actually the case, however. The design of the engine induction system and the valve timing requires a mixture that is slightly richer than chemically perfect in order to produce the maximum power. Maximum power is normally considered to be produced with a mixture of approximately 0.083 or 12:1.

Aircraft engines are built as light as possible and because so much power is produced in such a lightweight structure, they are highly susceptible to damage from too much heat. Cylinder head temperature and oil temperature are used to indicate the operating temperatures of air-cooled aircraft engines, but reaction time of these instruments are too slow to provide much information about the amount of heat being released from the fuel. So, in recent years EGT systems have been developed to measure the temperature of the exhaust gas to indicate the efficiency of the combustion inside the cylinder. A thermocouple probe is inserted into an exhaust stack where it can measure the temperature of the flow of exhaust gases as they come out of the cylinder.

Exhaust Gas Temperature

There is a direct relationship between the temperature of the exhaust gas and the mixture ratio being burned. In figure 6-1, it is apparent that, as the mixture ratio is leaned, the exhaust gas temperature rises until peak temperature is reached, and then it drops off. This same relationship exists regardless of the amount of power the engine is developing. The actual temperature depends upon the amount of fuel being burned, but the peak temperature will always be reached with the same fuel:air ratio, and so peak EGT can be used as a reference for adjusting the fuel:air mixture ratio.

Figure 6-2 shows a typical exhaust gas temperature gauge. Some have a pilot-adjustable reference pointer that most pilots set (at cruise-power) to peak EGT or peak minus 25 to 50 degrees (cruise

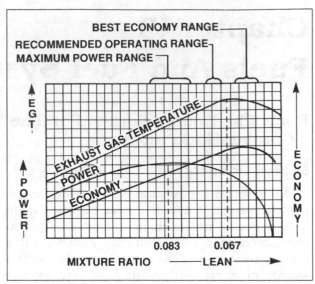

Figure 6-1. Relationship between fuel-air mixture ratio, engine power, engine economy, and exhaust gas temperature.

EGT setting). Some indicators have the actual temperature shown on the scale, but others have an asterisk or a red line about ⅘ the way up the scale. Many EGT indicators begin their indication at 1,200 ° F and go up to 1,700 °F, and if they are not calibrated with numbers, they are adjusted so the pointer is opposite the asterisk when the probe is sensing 1,600 °F.

Figure 6-2. Exhaust gas temperature gauge for a reciprocating engine.

To properly adjust the mixture ratio, the airplane is trimmed up for cruise flight and the power is adjusted with the propeller pitch control and the throttle to get the power required for cruise flight (see Chapter 7). The mixture is then leaned until the EGT of the hottest cylinder peaks. Then, the mixture control is moved back toward rich until the EGT drops about 25 °F. This procedure is typical, but it may be different for different airplanes. The recommendations of the aircraft manufacturer should be followed for that particular model.

The EGT probe is a bimetallic strip consisting of two strips of dissimilar metals bonded together. As the temperature increases, a small but increasing electrical current is generated by the strips and detected by the gauge. The probe is located in the exhaust stream of what has been determined to be the leanest cylinder at cruise power settings so that no cylinder will have to operate in a condition that is leaner than optimum.

However, it has been found that the leanest cylinder (the hottest cylinder) may change with changes of throttle setting, RPM, mixture or even cowl flaps. Therefore, EGT gauges that monitor all cylinders (a probe in each cylinder's exhaust) are sometimes used. The multi-cylinder EGT may display the temperature of one cylinder at a time with a pilot-operated switch or auto-scanning switch or temperatures of all cylinders may be displayed simultaneously.

Thermal Efficiency

Aviation gasoline has a nominal heat energy content of 20,000 British Thermal Units per pound, and one BTU of heat energy will produce 778 foot-pounds of work and is defined as the amount of heat energy required to raise the temperature of one pound of water one degree F.

When an airplane engine burns 12 gallons of aviation gasoline per hour, enough heat energy is released from the fuel to produce 566 horsepower. This is calculated by knowing that aviation gasoline has a nominal weight of six pounds per gallon, and so when the engine burns 12 gallons of fuel per hour it will release $12 \times 6 \times 20,000$ or 1,440,000 BTUs of heat energy per hour. This amount of energy will do $1,440,000 \times 778$ or 1,120,300,000 foot-pounds of work per hour. Since one horsepower is equal to 33,000 foot-pounds of work done in one minute, this amount of fuel will release enough heat to produce 565.8 horsepower.

However, an aircraft engine burning 12 gallons of aviation gasoline per hour does not produce nearly this much power. A typical reciprocating engine burning fuel at this rate will produce only about 135 BHP.

The brake thermal efficiency of the engine (the ratio of the amount of brake horsepower produced by the engine to the amount of horsepower in the fuel used to produce the power) is 135/566 or 23.83%. This is about typical for an aircraft reciprocating engine and is one reason much research remains to be done to make aircraft reciprocating engines fuel efficient. As seen in the next section, present day automotive engines are considerably more efficient than today's aircraft engines.

Specific Fuel Consumption

Thermal efficiency is seldom used to rate or to compare the performance of aircraft engines. Instead, a measure called specific fuel consumption is used. This is the number of pounds of fuel burned per hour for each horsepower developed. The engine used in the example for thermal efficiency has a brake specific fuel consumption of 0.53 pounds of fuel per brake horsepower per hour. This was found by using the formula:

$$\frac{\text{Pounds of fuel burned per hour}}{\text{Brake horsepower produced}} = \frac{72}{135} = 0.53$$

Some modern automobile engine installations in aircraft have produced BSFC values 50% better than commonly found in today's air-cooled aircraft engines.

Production Of Power

As discussed in the previous chapter, the horsepower produced by an aircraft engine may be computed by using the formula:

$$\text{Horsepower} = \frac{PLANK}{33,000}$$

P = Brake mean effective pressure. This is the average pressure inside the cylinder during the power stroke.

L = Length of the stroke in feet.

A = Area of the piston head in square inches.

N = Number of power strokes in one minute. This is the RPM of the engine divided by two, since only every other stroke is a power stroke.

K = Number of cylinders in the engine.

When the amount of pressure inside the cylinder, expressed in pounds per square inch, is multiplied by the area of the piston head, the number of pounds of force that is pushing down on the piston is found. If this force is multiplied by the length of the stroke, which is the distance the piston moves during each power stroke, the result is the number of foot-pounds of work done on each power stroke. The number of foot-pounds per stroke multiplied by the total number of power strokes per minute gives the number of foot-pounds of work done each minute. Dividing by the constant 33,000 produces the number of horse-power the engine is developing.

The pilot has no control over the area of the piston, length of the stroke or the number of cylinders, but there is control over the pressure inside the cylinder and the number of power strokes per minute. The pilot controls cylinder pressure with the throttle, while monitoring manifold absolute pressure (MAP). RPM is controlled with the governor control, while monitoring the tachometer. Thus, as further discussed in chapter 7, MAP and RPM are indicators of engine power for the pilot.

Reciprocating Engine Fuels

Requirements For Aviation Fuel

The specifications for aviation fuels have been established by the petroleum industry and accepted and approved by the Federal Aviation Administration. Every certificated aircraft has in its type certificate data sheets a list of the fuel that is approved for its use. The use of improper fuel may cause engine failure, or it can at least reduce the power output of the engine to below that required for the aircraft. Improper fuel use will probably void the aircraft's airworthiness and its insurance.

When a refinery produces a fuel for aviation use, it must consider two basic factors: the chemical and the physical characteristics of the fuel. Chemically, the fuel must have a high heat energy content, and it must be free from any constituents that will form acids or gums. It must have a high boiling point and a low freezing point. Its vapor pressure must be low enough that it will readily ignite from the spark plug, yet it must not be so low that it is hazardous for normal handling. Physically, the fuel must be free from contaminants, it must be easy to filter, and it must be pumpable at very low temperatures.

Aviation gasoline is the most widely used fuel for reciprocating aircraft engines, but constitutes only a very small fraction of refinery output. A look at world refinery output indicates that automotive gasoline is about 50%, diesel fuel is about 20%, fuel oil and jet fuel are each about 10%. Avgas, mostly 100LL, is less than 0.25% of production. This fact alone is reason enough for the considerable study that is being made of alternate fuels for aircraft.

Alternate Fuels

One of the major contenders as a substitute for aviation gasoline is automobile gasoline which, while high in price, is not nearly as costly as aviation gasoline and is considerably more plentiful. Several studies have been made, and are being made, regarding the suitability of automobile gasoline for aviation use. There seem to be some good arguments for adopting it within limits, of course. But, because of the liability involved and because of the lack of control the FAA may have over the production and uniformity of automotive gasoline, both the engine manufacturers and the FAA have been slow to approve the use of automotive gasoline as a legal substitute for aviation gasoline in certificated aircraft.

Presently, the EAA and Peterson Aviation of Minden, Nebraska offer STC's which permit the use of "autogas" or "mogas" in engines that were originally designed to use 80 octane avgas. Because of the cost savings, use of autogas has become quite popular. Some pilots have calculated that the savings realized by using autogas instead of avgas in 1000 hours of engine operation will pay for their next engine overhaul. No changes or adjustments to the engine are required in order to use autogas and it may be used interchangeably with avgas in aircraft that are properly STC'd.

Composition Of Fuels

Aviation gasoline is a hydrocarbon fuel refined from crude oil. The crude petroleum is distilled, and the fractions, as they boil off, are condensed to form the various petroleum products. Gasoline produced in this manner is called straight-run gasoline, and it makes up the largest amount of the gasoline used in aircraft. Some of the heavier fractions which are unsuitable for use as aviation gasoline are further treated by a process known as cracking. Here the hydrocarbon is heated under pressure with a catalyst to break it down into products having high volatility which are suitable for use in gasoline. Thus, a higher percentage of a barrel of crude oil can be made into gasoline.

All gasolines are blends of different hydrocarbons and additives, including the following:

The paraffin series is the most stable series of hydrocarbons. They are clean burning and have a high heat energy content for their weight, but because they are so light, their heat energy content for a unit volume is low. The paraffin series all have very low boiling points.

The cycloparaffin series is another stable series, sometimes called the naphthalene series. These products have a lower heat energy content per unit weight than the paraffins, but they have a higher boiling point.

The aromatic series tend to dissolve or swell rubber fuel lines, rubber tank liners, and diaphragms. They have a high freezing point and a high density, and they produce a good deal of smoke when they burn, but the anti-detonation characteristics of this series is very good. Although some are blended into gasoline, they are found in larger amounts in diesel fuel.

The olefin series are the most unstable of the hydrocarbons used in gasoline. They combine with themselves through a process known as polymerization to produce gum-like residues. On their credit side, they are clean burning and have a high boiling point and low freezing point. Because of their unstable nature, they are not found in natural petroleum products, but are formed in the cracking process.

Blending

Because no one hydrocarbon series produces all of the desirable characteristics wanted in an aviation fuel, aviation gasoline is a blend of the various hydrocarbon series.

Also found in aviation gasoline are undesirable constituents such as sulfur compounds, which combine with other elements to form acids which promote corrosion and damage the fuel pumps, valves, metering systems, and even the engine itself.

Gums and varnishes form from the ageing of fuel and combustion of gasoline. They cause the piston rings and valves to stick. Some of these gums form during fuel storage, especially if the fuel is exposed to sunlight or to elevated temperatures. And it is the sulfur in this gum which gives old gasoline its characteristic sour "varnish" odor.

Specifications

Heat Energy Content

Aviation gasoline is required to have a minimum of 18,700 Btu per pound, but its nominal rating is 20,000 BTU's per pound.

Reid Vapor Pressure

Liquid gasoline does not readily combine with oxygen, so in order to burn, it must be vaporized, or evaporated. A liquid evaporates when the pressure of the escaping gasses is greater than the pressure of the air above the liquid, so a liquid may be made to evaporate by either lowering the pressure above it or by raising its temperature. The amount of pressure required to hold the vapors in a liquid is known as its vapor pressure, and it is expressed in pounds per square inch at a specific temperature.

Vapor pressure is measured in a Reid vapor pressure bomb. The fuel to be rated is enclosed in a container where its temperature can be accurately controlled, and the pressure of the vapors above the liquid is measured at the test temperature. The allowable range of Reid vapor pressure for aviation gasoline is from 5.5 to 7.0 psi at 100 °F.

If the vapor pressure of aviation gasoline is too low, the fuel will not vaporize properly and this will cause hard starting, especially in cold weather. But, if the vapor pressure is too high, the fuel will "boil" in the lines of the fuel system. Fuel vapors released in the lines by this boiling have a tendency to collect in high points and cause a vapor lock. A bubble of fuel vapor in the line, because of its compressibility, resists the flow of fuel from the tank to the carburetor or to the fuel pump. Most fuel pumps are not designed to pump vapor (a gas).

Critical Pressure And Temperature

When the fuel-air mixture in a cylinder reaches a certain pressure and temperature, it will explode rather than burn evenly, and this explosion is known as detonation.

In order to get the maximum amount of power and the lowest specific fuel consumption from an aircraft engine, the cylinder pressures must be raised as high as possible and this is usually done by increasing the compression ratio. But the maximum compression ratio is limited by the critical pressure of the fuel.

Octane Rating

In order to rate the fuel according to its critical pressure, a comparative rating system has been established. The detonation characteristics of two hydrocarbon fuels are used as references and a variable compression ratio test engine is used to establish the detonation characteristics. Iso-octane (C_8H_{18}), a member of the paraffin series of hydrocarbons, has a high critical pressure and desirable anti-detonation characteristics and is assigned a rating of 100. Normal heptane (C7H16), on the other hand, has a very low critical pressure and undesirable detonation characteristics, and so it is assigned a rating of zero.

For the rating test, the fuel is run in the test engine and the compression ratio is raised until a definite condition of detonation is produced. The test fuel is then switched out and a metering system is put into operation which feeds a mixture of iso-octane and normal heptane into the engine. The ratio of the octane and heptane is varied until the same detonation characteristics are obtained as were obtained with the fuel under test. If the blend of reference fuels is, for example, 80% octane and 20% heptane, the fuel is given an octane rating of 80.

The fuel-air mixture ratio determines the detonation characteristics of a fuel and, for a period of time, aviation gasoline was given a dual rating based on the mixture ratio. A test was run on the fuel using a rich mixture ratio, such as would be used for takeoff, and the octane rating was determined. Then the test was repeated, this time using a lean mixture ratio as would be used for cruise flight. The fuel was given a rating such as 80/87 which means that with a rich mixture, the fuel has a rating of 87 octane. With a lean mixture, however, its rating is 80 octane. This dual rating system has been superseded with a fuel grade rating, and the same fuel is now called grade-80 aviation gasoline.

When aircraft engines grew in size and power output, fuels were demanded that had anti-detonation characteristics that were better than those of iso-octane. In order to rate these fuels, various amounts of tetrethyl lead was added to the reference fuel to increase its critical pressure, and ratings greater than 100 were created. These ratings were not called octane numbers, but are performance numbers.

Fuel Additives

In order to get better anti-detonation characteristics from a particular aviation gasoline, tetraethyl lead, a heavy, oily, poisonous liquid is added. Grade 80 gasoline is allowed to have a maximum of 0.5 milliliter per U.S. gallon, and grade-100 gasoline is allowed to have as much as 4.6 milliliters per gallon. Tetraethyl lead allows engines to develop more power without detonation, but using a fuel with a lead content higher than the engine is designed to accommodate leads to problems of spark plug lead fouling and sticking valves.

In recent years, the economics of aviation fuel production have caused the petroleum industry to try to phase out the production of grade-80 gasoline, but the higher lead content of the grade-100 fuel makes it impractical for use in engines designed for the lower lead content grade 80 gasoline. To accommodate the lower lead engines and at the same time have a fuel with an octane rating high enough for the high compression engines, the petroleum industry has brought out a fuel called grade 100-LL. This low-lead 100-octane fuel has a maximum of two milliliters of lead per gallon, and it seems to be a workable compromise. The two milliliters of lead provide enough lubrication of parts requiring the lead, and at the same time its lead content is low enough that spark plug fouling and valve sticking is not a major problem.

Tetraethyl lead has a lower volatility than gasoline and under conditions of low power output or of uneven fuel-air distribution, some spark plugs may have their electrodes bridged over by a conductive lead oxide which completely shorts out the spark plug. A scavenging agent, ethylene dibromide, is added to the fuel to combine with the lead oxide and form lead bromide. This is more volatile than the oxide and it passes out the exhaust as a gas.

Fuel Grades

Aviation gasoline is manufactured in several grades, depending upon the octane or performance number and the amount of tetraethyl lead it contains. The various grades are dyed for identification.

Grade-80 aviation gasoline was formerly called 80/87 gasoline, and it is dyed red. It may contain up to 0.5 milliliter of tetraethyl lead per gallon.

There was at one time a 91/97 aviation gasoline that was dyed blue, but this grade of fuel has been phased out. In its place, grade 100-LL or 100-octane low-lead gasoline is available, and it is dyed blue. This grade of fuel is allowed to have up to two milliliters of lead per gallon.

The popular 100-octane fuel, which was formerly called 100/130, is dyed green and it may have up to 4.0 milliliters of lead per gallon. This was the main fuel for the military and the airlines before jet aircraft took over in both of these areas. But now, of course, this vast market for 100-octane aviation gasoline has disappeared and the petroleum industry is trying to fill the need for this fuel with 100-LL with its reduced lead content.

Large, high-powered engines such as the Pratt and Whitney R-4360 require a fuel with better anti-detonation characteristics than the 100/130 fuel had, and the 115/145 aviation gasoline was brought out. This fuel is dyed purple.

There is much the pilot can do to prevent contamination, many of which are discussed in the next section on remote area fuel handling.

Importance Of The Proper Grade Of Fuel

Aircraft engines are designed to operate with a specific grade of fuel and will not operate efficiently or safely if an improper grade of fuel is supplied to the engine.

The required grade of fuel must be placarded on the filler cap of the aircraft fuel tanks, and it is important to know that the required grade is being pumped into the tanks. The various grades of aviation gasoline are dyed for identification, and turbine fuel has a distinctive color and odor to distinguish it from gasoline.

If an improper grade of fuel has been inadvertently used, the following should be accomplished:

If The Engine Has Not Been Operated

1. Drain all of the improperly filled tanks.
2. Flush out all of the fuel lines.
3. Refill the tanks with the proper grade of fuel.

If The Engine Has Been Operated

1. Perform a compression check of all cylinders.
2. Inspect all of the cylinders with a borescope.
3. Drain the oil and inspect the oil screens.
4. Drain the entire fuel system, including all of the tanks and the carburetor.
5. Flush the entire system with the proper grade of fuel.
6. Fill the tanks with the proper grade of fuel.
7. Perform a complete engine run-up check.

Fuel Contamination

Fuel contaminants clog fuel strainers, decompose fuel lines, pump vanes and gaskets and displace fuel. This has come to be the cause of many aircraft accidents.

The two main hazards of the fueling process are (a) fuel contamination of the aircraft's fuel system and (b) static sparks. Both of these hazards can and have been lethal, so the processes involved should be understood by the pilot.

Contaminants can be classified into three categories:

Solids

Solid contaminants, including dust, dirt, leaves, bugs, lint off the sweater of the fueler, etc. In other words, anything solid that will not dissolve in gasoline but remains a solid in the tank. Over the years, these solids accumulate and build up in the tank. Depending on their density, they may remain forever on the bottom of the tank (until cleaned out) or, if lighter, these particles may be disturbed by turbulent sloshing and be taken up by the outflow of fuel into the fuel lines.

If the particle is large enough, it will be stopped by the screen in the gascolator (see figure 6-3). Also, some aircraft fuel tanks have a coarse finger screen in the tank outlet. It will prevent large objects from blocking fuel flow from the tank outlet. If the contamination is small enough to pass the screens, it may pass through the carburetor without notice. The fuel injection system, however, is less tolerant of small particle contamination. This, then, becomes one of the advantages of the carburetor for a bush plane.

If solid particle contamination in the fuel tank builds up over the years, it may become great enough to cause a power loss, especially if the flight is very turbulent, stirring up the sediment in the bottom of the tank so that enough of the solid contaminants pass through the fuel lines to the gascolator to plug up the screen to the extent that fuel flow becomes limited. Then the pilot will experience power loss due to excessive leaning of the mixture and finally, fuel starvation.

Have you taken a flashlight and looked down in the depths of your fuel tanks recently?

Surfactants

These partially soluble compounds are byproducts of the fuel processing or may be caused by fuel additives. They tend to adhere to other contaminants and cause them to drop out of the fuel and settle to the bottom of the tank as sludge.

Water

Though water has always been present in aviation fuel, it is now considered to be a major source of contamination, since modern airplanes fly at altitudes where the temperature is low enough to cause suspended and dissolved water to condense out of the fuel and form free water. This freed water can freeze and clog the fuel screens.

Water is slightly soluble in gasoline, and even more soluble in some other organic compounds that are completely soluble in gasoline (alcohol, for example). Therefore, since there is little control

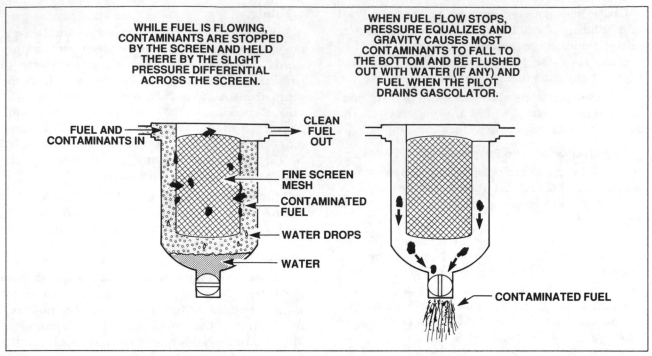

WHILE FUEL IS FLOWING, CONTAMINANTS ARE STOPPED BY THE SCREEN AND HELD THERE BY THE SLIGHT PRESSURE DIFFERENTIAL ACROSS THE SCREEN.

WHEN FUEL FLOW STOPS, PRESSURE EQUALIZES AND GRAVITY CAUSES MOST CONTAMINANTS TO FALL TO THE BOTTOM AND BE FLUSHED OUT WITH WATER (IF ANY) AND FUEL WHEN THE PILOT DRAINS GASCOLATOR.

FUEL AND CONTAMINANTS IN

CLEAN FUEL OUT

FINE SCREEN MESH

CONTAMINATED FUEL

WATER DROPS

WATER

CONTAMINATED FUEL

Figure 6-3. The gascolator in operation.

over what is in the fuel purchased, it is possible that there may be water in the fuel. This is especially true in warm weather because fuel will hold more water in solution if it is warm than it will if it is cold.

The above is a potentially dangerous situation because airplanes very often fly into a colder environment after departure. As the fuel in the tank cools off, water droplets form because the fuel can no longer hold that much water in solution. Fortunately, quantities of liquid water formed are not great. Usually the only way the pilot knows for sure that this is happening is because a few drops more of water can now be sumped from the tank sump and perhaps, once in awhile, an alert pilot who takes the trouble to catch the discharge from the gascolator will find a few drops of water in that sample, even though the aircraft has not been refueled.

More evidence of this readily shows up in a fleet of landplanes that are stored in a heated hangar in the winter in the north country. When the pilot sumps the tanks before the first flight of the day, no water is found. After a flight of an hour or so, the fuel cools and releases some water which sinks to the sump area, then cools more and freezes. The tank fuel sump valve is found to be frozen, when checked before the next flight!

Please keep this in perspective, though. In the last example, a temperature change from plus 50° F to minus 20° F is occurring so larger amounts of water are coming out of solution due to the large temperature change of the fuel, and even those amounts of water are very small. Small enough so the water remains in the sump of the fuel tank. Even if all of the water from this source found its way to the gascolator, it would be contained there and probably never reach the engine.

There are two other, more dangerous side effects of this phenomena of soluble water in fuel, which are **power loss due to ice crystals blocking the fuel strainer** and **fuel dye transference**.

Is It Snowing In Your Tanks?

If the aircraft flies into an air mass of a temperature below freezing, this phenomenon might be experienced. Here is what can happen: As the fuel cools off, it can hold less water in solution, so tiny water droplets form. Since they are small, the water droplets remain suspended in the fuel, then freeze as the temperature of the fuel descends below freezing. As it crystallizes, the droplet transforms into a beautiful snowflake!

These snowflakes are large enough to be stopped on the gascolator screen. When and if enough of them accumulate there, fuel flow is impeded and the engine fuel mixture is effectively leaned. Eventually, this leaning very likely will cause a power loss.

What can be done to prevent this? Prevention can be accomplished by adding the recommended amount of the recommended type of alcohol or "Prist®," which lowers the freezing point of the water that comes out of solution so that it is unlikely to freeze. The tiny water droplets simply pass through the engine with little effect.

If alcohol is not used, the pilot should be alert for symptoms of leaning that are unexplained. The "first alert" will come from the EGT gauge, with a rise in temperature. The pilot can richen the mixture, for awhile, by use of the mixture control. When that is no longer effective in keeping the EGT at the correct value, the carburetor heat will effectively richen the mixture because (a) less dense air decreases the amount of oxygen getting into the cylinder, thus increasing the fuel:air ratio, and (b) higher air temperature will do a better job of vaporizing the fuel, thus increasing the fuel:air ratio. When the application of carb heat no longer gets the job done, the pilot must decrease the throttle setting in order to keep the engine running. By good knowledge of what is happening, the pilot has kept full power (or nearly full power when carb heat is in use) much longer than the unaware pilot, and was forewarned sufficiently that, when the power loss comes, the aircraft has landed.

A more common problem is the one of dye transference. When the pilot acquires a sample of fuel in the little cup, a meniscus or line showing separation of fuel and water is looked for as indication of water in the fuel. The water is on the bottom because it is heavier. If there is no meniscus, there is no water, right? Wrong! What if it is ALL WATER? How does the pilot tell if the liquid in the cup is fuel or water?

Smell it? It will smell like fuel anyway. The cup smells like fuel. The back of my pickup smells like fuel because I have spilled fuel there in the past. Our nose is not a very useful test instrument in this case.

By the color of the fuel in the cup? No good. If you are using 100 or 100LL exclusively, the green or blue color may be present in the cup in either the fuel or the water because the dye may be soluble in water or in some of the compounds in the fuel that are soluble in water. If you are using autogas or Mexican 100 octane avgas, both are colorless or at least without dye. With any combination of the above, the dye will be diluted to the extent that identification of dye color may be iffy.

Water flyers have the perfect test kit very close at hand. They just put a drop or two of water off a finger dipped in the water under their feet in the cup. If the drops are visible at the bottom of the cup, the rest of the liquid in the cup is gasoline (or

kerosene or jet fuel, in which case the pilot's nose is the ideal test kit).

That brings us to the third classification of fuel contaminants: contaminants that are soluble in gasoline. These can be many different types of organic compounds that are liquids at normal ambient temperatures. The most common one seems to be jet fuel, as it is readily available at most airports. The best defense against this sort of contamination is a vigilant pilot or crewmember standing by or helping with the fueling process. As mentioned above, sumping tanks is a must, with a sniff test included, not to determine if the solution in the cup is gasoline or water, but to determine if the solution is contaminated with kerosene or jet fuel.

A rather clever device that helps me with the above mentioned problems is a fuel sampling cup that contains a small test ball restrained within the cup by a screen. The ball is constructed to have a very specific density so that it sinks in gasoline, floats in water and suspends in a mixture of 80% gasoline and 20% jet fuel!

Microorganisms

Airborne bacteria gather in the fuel, where they remain dormant until they come into contact with free water. The bacteria then grow at a prodigious rate as they live in the water and feed on the hydrocarbon fuel and on some of the surfacant contaminants. The scum which they form holds water against the walls of the fuel tanks and causes corrosion as well as clogging of fuel screens.

Draining a sample of fuel from the main strainers has long been considered an acceptable method of assuring that the fuel system is clean. But tests on several designs of aircraft have shown that this cursory sampling may not be adequate to assure that no contamination exists.

In one test reported to the FAA, three gallons of water were added to a half-full fuel tank, and after time was allowed for this water to settle, it was necessary to drain ten ounces of fuel before any water appeared at the strainer. In another airplane, one gallon of water was poured into a half-full fuel tank, and more than a quart of fuel had to be drained before the water appeared at the strainer. The tank sumps had to be drained before all of the water was eliminated from the system.

A commercial water test kit is available to test for water in aircraft fuel. This kit contains a small glass jar and a supply of capsules containing a grayish-white powder. A 100-cc sample of fuel is taken from the tank or from the fuel truck and put into the jar, and a capsule of powder is dumped into it. The lid is screwed on, and the contents are shaken for about ten seconds. If the powder changes color from gray-white to pink or purple, the fuel has a water content of more than 30 parts per million, and the fuel is not considered to be safe for use. This test is fail-safe, meaning that any error in performing the test will cause an unsafe indication to be given.

Protection Against Contamination

All fuel tanks are required to have their discharge protected by an eight- to 16-mesh finger screen. Downstream of this finger screen is the main fuel strainer, which usually is either a fine wire mesh or a paper-type element.

Fuel Handling

Remote Area Fueling

There are two main hazards for the operator/pilot when fueling. They are (a) static electricity generated spark ignition of the fuel while it is being transferred to the aircraft, and (b) contamination of the aircraft's fuel system. The following steps are appropriate when fueling at small airports and from fuel caches in the wilderness:

1. Investigate the vendor's filtration. If the filtration is aircraft quality, and the source is used by several aircraft each day, probably purchaser filtration is not necessary and will only slow the fueling process. If the vendor's filtration is an unknown or at all suspect, use of filtration provided by the purchaser is appropriate. Filtration systems available for use by the pilot are many and varied. The old standbys of the metal funnel and either a chamois or felt hat worked for years but have some disadvantages which include the fact that an already wet chamois will pass water and both the chamois and felt hat do dispense particles or fibers of chamois and felt into the tanks. Another objection to the chamois system is that the chamois will smell of gasoline in the aircraft cabin for hours after departing the fuel dock. I have, in recent years, developed confidence in the "Filter Funnel®." It stops free water, flows 8 GPM and is made of a conductive plastic. It dries quickly while I am paying the fuel bill and weighs only 9 oz. Stamped in the plastic of the funnel appear the words "not approved for aviation fuels," apparently the manufacturer's

attempted defense against this country's incredible liability history.

2. Before fueling the aircraft, my next step is to turn the aircraft's fuel selector valve from "both" to the "off" position. Two reasons for this:
   (a) fuel cannot transfer from the filled tank to the empty one while I am relocating to and filling the second tank. That way, I know the first tank filled is completely full.
   (b) if one or both of the tanks are contaminated with water, etc., the contaminant is confined to the tank(s). The rest of the ship's fuel system remains clean. This makes decontamination much easier and quicker. The selector valve must be left in the "off" position until the pilot has determined that no contamination has occurred (by sumping the tanks after fueling).

3. If the fueling is to be done with the dispensing hose reaching the aircraft, I prefer to do the job myself unless there is an experienced, well trained, caring individual who wishes to do the job. Care must be taken not to drag the fuel hose over the windshield as it will surely be scratched.

4. Touch the metal fuel hose nozzle against metal of the airplane before opening the fuel filler port to discharge any developed static electrical charge. A static charge can develop on a hose anytime the hose is moved, especially if it is dragged across or over something, or if something is moved over or dragged across the outside of the hose, or if there is flow inside the hose. This static electrical charge on the hose establishes a different electrical voltage on the hose than is on the airplane. When the hose and the airplane come in close enough proximity to each other, a spark will jump across the small space, moving electrons to equalize the voltage on both sides. Just like a spark from a spark plug, it is capable of igniting fuel-air vapors if they are present in the right mixture. The results are immediate and not pleasant.

5. Keep the nozzle in contact with the metal parts of the filler hole or the conductive plastic or metal of the filter funnel all during the fuel transfer. If there is not a low resistance (metal to metal or metal to conductive material to metal) electrical connection between the hose nozzle and the aircraft, an electrical charge may build up on the hose, ready to discharge (arc) just before the next metal to metal contact.

6. After fueling is complete and filler caps are firmly back in place, wait 15 minutes (time

enough to settle your bill with the supplier and let the filter funnel air out), sump each fuel tank to check for the three types of contamination listed above.

If water is found, continue sumping until no more water appears in the cup. Then rock the wings and lift the tail up and down several times and sump again. Repeat this procedure until no more water is found.

Now open the fuel selector valve. Next, sump the gascolator and inspect this sample for water. Continue sumping until the fuel is clear (cloudy fuel is an indication of microscopic water droplets dispersed in the fuel). To be very sure that all the water is out of the system, run the engine for 5-10 minutes, stop the engine and sample the gascolator again. Being of a suspicious nature, I would probably sample the gascolator and tank sumps again after the completion of the flight, to see if flight movements and turbulence caused any more water to settle out.

Under no circumstances should the gascolator be sumped until all fuel tanks have been sumped. To do so may pull water from the fuel tanks into fuel lines leading to the engine(s).

Static Electrical Discharge

There are several things the fueler can do to prevent the spark jump of an electrical discharge, including:

1. When possible, use of grounding cables from nozzle or refueler to aircraft and refueler to ground should be used. If not available, the following suggestions can help.

2. Don't let the static charge build up. Discharge it often, or better yet, continuously, by touching hose and or nozzle or jerry jug to the metal of the aircraft at a point far enough away from the filler hole, and upwind, so that no fuel vapors exist where the spark might jump. Allow the filler hose to lay up next to the metal skin of the aircraft somewhere along the length of the hose. Be sure to touch the nozzle of the hose to the metal of the aircraft before opening the filler port.

3. Keep the spark gap wide. When using a plastic funnel/filter, the hose nozzle is separated from the metal of the aircraft by a considerable distance. If the plastic funnel is capable of conducting an electrical or static charge, the funnel is constantly touching the metal filler cap so the charge can dissipate. If the plastic funnel does not conduct, then the spark gap

between nozzle and aircraft is kept too wide for a spark to jump. Therefore, the nozzle should be touching the plastic funnel during the fueling process, even if the funnel is not conductive or if you aren't sure about the funnel's conductivity. It is preferable not to use a non-conductive funnel as large voltage differentials can build up, making the non conductivity of the funnel questionable. It may be non-conductive at low voltages but at high voltages, unpredictability reigns.

The fact that there is rarely an incident of fuel ignition during fueling is a good indication that it is a pretty safe process. It is very safe if the pilot is aware of the two hazards of fueling. So, keep these hazards and recommended processes in mind each time you fuel up.

Pilots are sometimes required to fuel aircraft and to maintain the sophisticated fueling equipment found at larger airports. Each type of bulk fuel storage facility is protected from static electricity discharges and from contamination as much as is practical, and it is the responsibility of the operator of these facilities to assure that the proper grade of fuel is put into the fuel truck and that the truck is electrically grounded to the bulk fuel facility when it is being filled. All of the fuel filters should be cleaned before pumping, and all water traps must be carefully checked for any indication of water.

When the aircraft is fueled from a tank truck, it is the responsibility of the truck driver to position it well ahead of the aircraft and to be sure that the brakes are set, so there will be no possibility of the truck rolling into the aircraft. The sumps on the truck storage tanks should be checked and a record made of the purity of the fuel. A fully charged fire extinguisher should be mounted on the truck ready for instant use if the need should arise, and static bonding wires should be attached between the aircraft and the truck, with a ground connected between the truck and the earth. A ladder or stand should be used if needed, and a wing mat should be put in place to prevent damage to the aircraft.

The fuel nozzle must be free of any loose dirt which could fall into the fuel tank, and when inserting the nozzle into the tank, care must be taken not to damage the light metal of which the tank is made. Be sure that the end of the nozzle doesn't strike the bottom of the tank. When the fueling operation is completed, replace the nozzle cover and secure the tank cap. Remove the wing mat and return all of the equipment to the truck and roll the hose and grounding wire back onto their storage reels.

Fire Protection

All fueling operations should be done under conditions which allow a minimum possibility of fire. Never refuel an aircraft in a hangar, and defueling, as well, must be done in the open. Electrical equipment that is not absolutely necessary for the fueling operation should not be turned on, and fueling must not be done where radar is operating, as enough electrical energy can be absorbed by the aircraft to cause a spark to jump and ignite the fuel vapors.

If a fire should break out, it can be extinguished either with a dry powder or with a carbon dioxide fire extinguisher. Soda-acid or any water-type fire extinguishers should not be used, because fuel is lighter than water and it will float away, spreading the fire.

Fuel Metering Systems

Principles Of Fuel Metering

In order for an engine to develop its power most efficiently, the fuel must be mixed with exactly the correct weight of air. The volume of this mixture must be controllable by the pilot, and it must be uniformly distributed to all of the cylinders.

The mixture ratio between the fuel and the air must be variable in order to provide for either full power or for economy as the operating conditions require. Provision must be made to compensate the mixture ratio for variations in the air density caused by changing temperature and altitude.

Absolute dependability is essential for an aircraft fuel metering system, and the system must operate efficiently under conditions of moisture, dust, vibration, and engine heat.

Modern aircraft engines are being operated with cylinder pressures so high that any mismanagement of the fuel-air mixture ratio can cause detonation that can destroy an engine in a very few seconds.

The fuel metering systems used with aircraft engines have evolved from a very simple drip-type system in which liquid gasoline was dripped into a hot portion of the induction system that passed through or near the warm cooling water jacket where it was vaporized. Then the vapors were drawn into the cylinders.

The float carburetor that followed this primitive system has remained basically the same for the last sixty years or so. Float carburetors are simple and dependable, but they have some limitations that are primarily caused by non-uniform mixture distribution and susceptibility to carburetor icing.

These problems have been solved to a great extent by the pressure carburetor and fuel injection systems. The float carburetor, the pressure carburetor, and two different types of fuel injection systems will be discussed in some detail.

The Aircraft Float Carburetor

While studying how the carburetor and fuel injection systems operate, remember that the principal function of these devices is to sense the amount of air entering the engine at any moment and to meter into that air an amount of fuel that will provide the correct fuel:air ratio. These systems provide a uniform fuel:air ratio as the airflow varies. Most of these systems use airflow to produce a pressure differential that is proportional to the amount of airflow. The pressure difference then meters the correct amount of fuel into the airflow.

Airflow Sensing

Production Of Pressure Drop

All of the air burned in an engine must pass through the carburetor and specifically through the venturi. A venturi as shown in figure 6-4 is a specially shaped restriction placed in the main air passage. The principle of operation of the venturi is simply one of exchange of forms of energy.

Energy exists in two forms: potential, manifested as pressure; and kinetic, manifested as velocity. According to the law of conservation of energy, energy can neither be created nor destroyed, but its form can be altered. If energy is neither added nor taken away, any increase in kinetic energy will result in a decrease of potential energy. As the airflow for the engine passes through this venturi-shaped restrictor, its speed (kinetic energy) is increased, just as water increases its speed as it flows through shallow areas to form rapids. This increase in speed causes a corresponding decrease in potential energy (pressure).

Thus, the greater the airflow into the engine, the faster the air must flow through the venturi and the lower the pressure will be in the venturi as compared to ambient pressure.

Fuel Metering Force

Liquid fuel is delivered from the aircraft's tank to the carburetor through a fine mesh wire screen and into the float bowl or float chamber (see figure 6-4). A needle valve operated by the float keeps the fuel in the bowl at a constant level. As fuel is used from the bowl, the float tends to drop, opening the valve, allowing more fuel to enter the chamber.

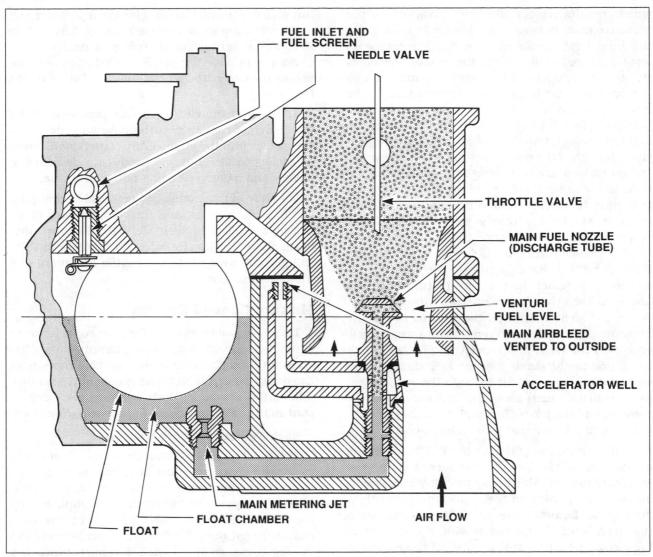

FUEL INLET AND FUEL SCREEN

NEEDLE VALVE

THROTTLE VALVE

MAIN FUEL NOZZLE (DISCHARGE TUBE)

VENTURI FUEL LEVEL

MAIN AIRBLEED VENTED TO OUTSIDE

ACCELERATOR WELL

MAIN METERING JET

FLOAT CHAMBER

FLOAT

AIR FLOW

Figure 6-4. Simplified aircraft updraft carburetor.

Located in the center of the venturi, with its end near the narrowest part of the venturi is the discharge tube, or main fuel nozzle. This nozzle is connected by the main fuel passage to the float bowl. The top of the discharge nozzle is slightly above the level of the fuel in the bowl. If the engine is not running, no air flows through the venturi so no pressure differential exists to raise the fuel level in the main discharge nozzle. When air is flowing in the venturi, a pressure differential between the venturi (low) and the float chamber (ambient — higher) exists. This pressure is known as the **fuel metering force** and will increase in strength as the airflow through the venturi increases.

The float bowl is vented to atmospheric pressure so the existing atmospheric pressure forces the fuel out through the discharge nozzle. Again: with no airflow, the pressure on the discharge nozzle is

exactly the same as that in the float bowl, and no fuel flows. As air flows through the venturi, the pressure at the discharge nozzle drops below that in the float bowl, causing fuel to flow out the nozzle.

The maximum amount of fuel which can flow out of the float bowl is limited by the main metering jet.

Air Bleed

The air bleed system of the float carburetor has three main functions, which are, (1) to provide a more even mixture over a wide range of airflows, (2) to improve fuel vaporization and (3) to decrease the size of the fuel metering force needed which decreases the amount of restriction necessary at the venturi.

One of the disadvantages of the venturi-float-bowl arrangement is the uneven fuel-air mixture which results as the air flow changes. As the pressure drop between the discharge nozzle and the float bowl increases due to an increase in airflow, more fuel flows and the mixture becomes richer. A wide open air bleed will produce exactly the opposite results. As the airflow through the venturi increases, airflow through the air bleed increases faster than the fuel flow through the main metering jet, and the mixture grows progressively leaner. If the conditions shown in these two configurations are combined into a single operation, the leaning tendency of one will cancel the enriching tendency of the other, and the fuel-air mixture will be essentially constant as airflow changes. This is done by restricting the main air bleed as shown in figure 6-4. The size of the air bleed is critical. Exactly enough air must be admitted to the fuel on the way to the discharge nozzle to keep the fuel:air mixture ratio constant. When the airflow through the venturi is low the pressure differential between the discharge nozzle and the float bowl is relatively small. There is a correspondingly small flow of fuel through the main metering jet and air through the airbleed jet. As the airflow increases, the pressure drop increases, and the flow of fuel and the flow of air both increase in an essentially constant ratio.

Another function of the air bleed is to aid in the atomization of the fuel. It introduces air into the stream of fuel, breaking it up into tiny bubbles, or an emulsion of air and fuel. This emulsion is less dense than liquid fuel and may be brought up to the lip of the discharge nozzle with less fuel metering force. The larger surface area of the emulsion also allows vaporization to begin before the fuel is sprayed into the venturi.

Again, the advantages of the main airbleed are:

1. Less pressure differential is required to move the fuel out of the discharge nozzle, so less restriction at the venturi is needed (less restriction of airflow to the engine).

2. The fuel:air ratio is more uniform over a wide range of airflows because the orifice size of the airbleed can be regulated.

3. Better fuel vaporization occurs which improves the uniformity of mixture flowing to all cylinders.

Airflow Limiter

In a reciprocating aircraft engine, all of the air used to combine with fuel must pass through the venturi of the carburetor. The throttle butterfly valve, figure 6-4, located downstream of the venturi, controls the amount of airflow into the engine,

but with the throttle in the wide open position, the venturi becomes the airflow limiting device. The size of the venturi is therefore critical, and is chosen to provide the proper air velocity, and the proper pressure drop for the volume of air required by the engine.

Any way of increasing the air pressure at the inlet, such as ram air or turbo-charging, increases the airflow into the engine. Any obstructions such as a clogged air filter will produce a decrease of airflow and a corresponding power decrease.

The RPM of the engine determines the pumping action of the pistons and thus the amount of air drawn through the venturi. With the same opening of the throttle butterfly valve, an increase in RPM will bring more air into the engine and more fuel from the float bowl.

Mixture Control System

There are two ways to meter the proper amount of fuel into the airflow: Varying the pressure drop across a fixed size orifice, and varying the size of the orifice while maintaining a constant pressure differential across it. Both methods are used to control the fuel-air ratio on float type carburetors.

Back Suction Mixture Control

The back suction mixture control, figure 6-5, varies the pressure in the float chamber between atmospheric and a pressure slightly below atmospheric. This pressure variation is accomplished by the use of a control valve located in the float chamber vent line. The float chamber is vented to the low pressure area near the venturi through a suction channel. This lowers the pressure in the float bowl. When the vent valve is opened, the pressure in the float bowl is raised to essentially atmospheric pressure, and a differential pressure exists across the main metering jet. This causes fuel to flow out the discharge nozzle. When the vent is closed, pressure in the float chamber decreases to a point essentially the same as the discharge nozzle. This lack of pressure differential stops the flow of fuel.

Variable Orifice Mixture Control

A more common way of varying the fuel-air ratio is to control the fuel flow by changing the size of the opening between the float bowl and the discharge nozzle. The float chamber has an unrestricted vent to maintain atmospheric pressure on the fuel in the float bowl. A needle valve is located in series with the main metering jet. When the valve shuts off the flow of fuel completely, the engine cannot run. This is the idle-cut-off position.

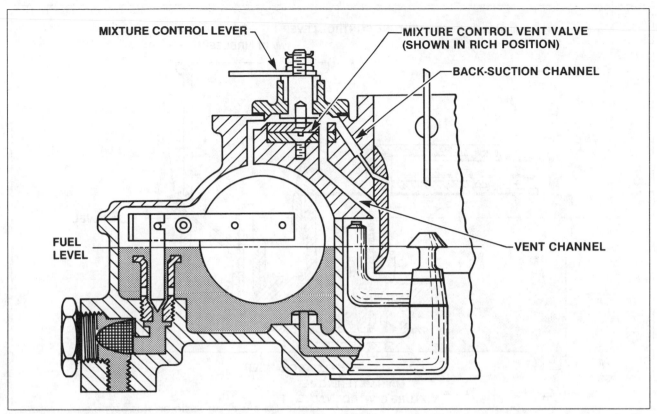

Figure 6-5. The back-suction mixture control leans the mixture by decreasing the pressure drop across the metering orifice by decreasing pressure in the float chamber.

When the valve is opened, fuel flows to the discharge nozzle and is metered by the valve as long as the area of the opening in the valve is smaller than the area of the main metering jet. When the mixture control valve is fully open, or in the full-rich position, the area of the opening of the mixture control is larger than the area of the main metering jet, and the jet limits the amount of fuel which can flow to the discharge nozzle. See figure 6-6.

Idle System

At engine speeds below about 1000 RPM, the airflow through the venturi is not sufficient to produce a pressure drop at the main discharge nozzle great enough to discharge the fuel, so an auxiliary system is provided. The throttle butterfly valve restricts the air which flows into the engine, and during idling is almost closed. The only air which flows into the engine must pass around the edge of the throttle valve disc. The velocity of the air at this point is naturally quite high, and the pressure at the edge of the valve, and above the valve, is low. In the wall of the throttle body, figure 6-7, where the butterfly valve almost touches, are two or three small holes or idle discharge ports. These ports are connected by an idle emulsion tube to a supply of fuel between the float bowl and

the discharge nozzle. Fuel rises from its level in the float bowl due to the low pressure above the throttle valve. The emulsion tube incorporates the idle metering jet and the idle air bleed. The upper idle discharge port is fitted with a tapered needle valve to control the amount of fuel-air emulsion allowed to flow from the discharge ports when the throttle valve is closed. The idle RPM of the engine is adjusted by varying the amount of throttle opening, by the throttle stop screw. The idle mixture is controlled by the needle valve. When the throttle is opened, the butterfly valve moves down, extending the area of low pressure over the secondary and tertiary idle discharge ports. This provides the fuel required for operating in conditions of off-idle, yet with not enough airflow to allow fuel to be drawn from the main discharge nozzle.

When the engine is operated at cruise or higher RPM, there is no low pressure above the throttle valve so air replaces fuel in the idle ports which then serve as an auxiliary air bleed, aiding the atomization of fuel at the higher flow rates.

Acceleration System

Between the time the idle system loses its effectiveness, and the time that there is sufficient airflow for the main metering system to operate,

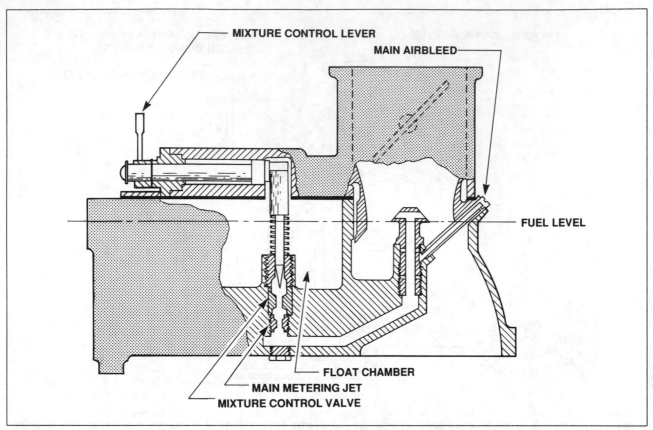

Figure 6-6. The needle-type mixture control controls the mixture by varying the size of the metering orifice.

there is a tendency for the engine to develop a "flat spot," or a point where there is insufficient fuel for continued acceleration. To overcome this condition, an **acceleration system** is installed.

The acceleration system may be as simple as the acceleration well of figure 6-4. In this simple system, an enlarged annular chamber around the main discharge nozzle at the main air bleed junction stores a supply of fuel during idling. When the throttle is opened suddenly, this fuel is readily available between the airbleed and the discharge nozzle to produce a rich mixture at the time the mixture would otherwise be too lean. Engines which require more fuel for this transition use a pump, such as the moveable piston type accelerator pump such as seen in figure 6-8 which shows a leather packing type piston, held against the walls of the pump bore by a coiled spring. The pump is actuated by a linkage from the throttle. When the throttle is closed, the piston moves upward, filling the cylinder with fuel from the float bowl through a ball type check valve. When the throttle is opened, the piston moves downward, closing the inlet check valve, and forcing the fuel out past the discharge check

valve into the airstream through the accelerator pump discharge nozzle. The piston is mounted on a spring loaded, telescoping shaft. When the throttle is opened, fuel is unable to immediately discharge because of the restriction of the nozzle, so the shaft telescopes, compressing the spring. The spring pressure sustains the discharge, providing a rich mixture during the transition period.

Power Enrichment System

Aircraft engines are designed to produce maximum power consistent with their weight. They are not, however, designed to dissipate all of the heat the fuel is capable of releasing, so some provision must be made to remove some of this heat. This is done by enriching the fuel-air mixture at full throttle. The additional fuel absorbs this heat as it changes into a vapor. Power enrichment systems are often called **economizer systems** because they allow the engine to operate with a relatively lean, economical mixture for all conditions other than full power. Pilots should be aware that many economizer systems only provide full power enrichment when the throttle is all the way open (last ¼" or so of throttle movement). Therefore, when takeoff power is required, throttle(s) should be opened fully.

100

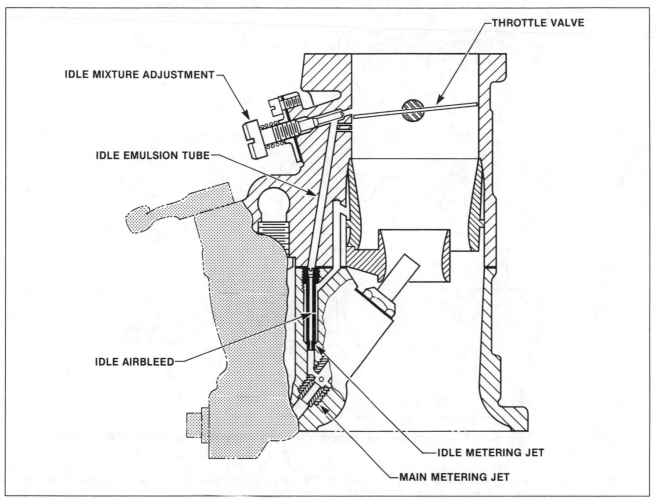

Figure 6-7. Fuel for idling is picked up after it passes through the main metering jet, brought up through the idle metering jet, mixed with air at the idle air bleed, and is discharged near the edge of the throttle valve.

Float Carburetor Preflight Inspection

At each preflight inspection, the pilot should determine that there is no fuel leaking from the carburetor. Leaking fuel is evidenced by fuel dye stains on the carburetor body or in the cowling below the carburetor.

Sumping of all drain points must be done after each refueling and after the aircraft has been tied down or stored overnight or after the aircraft has been flown in visible moisture (upper wing surfaces, operating in low pressure air are operating in air that is colder than ambient. This cools the upper wing surface, causing condensation which flows back along the upper wing surface to collect in fuel cap wells).

Carburetor Icing And Carburetor Heat Use

Carburetor ice means ice at any location in the induction systems of aircraft equipped with reciprocating engines. The term is traditional. It is used in aircraft accident records, even though many reciprocating engine installations have fuel injectors rather than carburetors per se. The term does not apply to icing associated with turbine engine installations.

Carburetor ice normally does not remain in evidence for very long after an accident occurs. Hence, there are probably many power loss incidents caused by this occurrence that can not be properly attributed to this phenomenon. Losses resulting from carburetor icing can be reduced by greater awareness and vigilance by pilots. The NTSB has long been concerned with carburetor icing as one of the unnecessary causal factors in general aviation accidents. Unlike mechanical failure over which the pilot has little in-flight control, carburetor icing can be avoided because the means to preclude it are usually readily available.

Since carburetor icing accidents are attributed to the pilot in virtually all cases, improved pilot

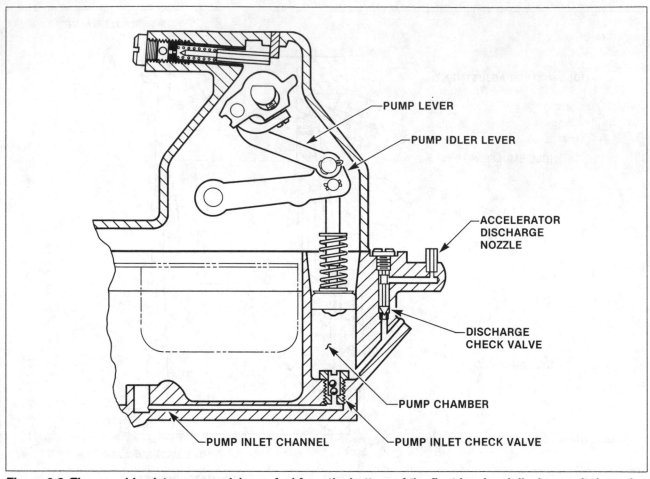

Figure 6-8. The movable piston pump picks up fuel from the bottom of the float bowl and discharges it through a pump discharge nozzle in the venturi.

awareness and attention becomes an even more important pilot-task in reducing incidents or accidents of this type.

Three categories of carburetor ice, as defined by the NTSB, are **impact ice, fuel ice and throttle ice**.

Impact ice is formed by the impingement of moisture-laden air at temperatures between +15° F and 32° F onto the elements of the induction system which are at temperatures below about 32° F. Under these conditions ice builds up on such components as the air scoop, heat valve, carburetor air screen, throttle valve and carburetor metering elements. Pilots should be particularly alert to such icing when they are operating in freezing rain, snow, sleet, rain or clouds. The ambient temperature at which impact ice can be expected to build up most rapidly is about +25° F when the super cooled moisture is still in a semi-liquid state, according to a study by the NTSB (1972).

Fuel ice forms at and downstream from the point at which fuel is introduced, when and if any air or fuel entrained moisture reaches a freezing temperature as a result of cooling of the mixture by fuel vaporization. This cooling process takes place in the aircraft induction system when the heat necessary for fuel vaporization is taken from the surrounding air, thus cooling the air. Then, since cooler air can hold less water vapor, the excess water is precipitated in the form of condensation. Further vaporization cooling freezes the condensate. When any structure, such as an adapter elbow or throttle plate, lies in the path of the water at time of freezing, ice accretion is initiated on that structure. If this condition continues and no anti-icing action is taken, the ice buildup will increase until the obstruction throttles the engine.

Visible moisture in the air is not necessary for fuel icing. Only air of high humidity is required. This fact, coupled with the fact that fuel icing can occur at ambient temperatures well above freezing, makes this type of icing very insidious. It can occur in no more than scattered clouds, or even in bright sunshine with no sign of rain.

The usual range of ambient temperatures at which fuel icing may be expected to occur is 40 °F to 80 °F although the upper limit may extend to as high as 100 °F. A temperature of around 60 °F should be regarded as the most suspect. The minimum relative humidity generally necessary for fuel icing is 50 percent, with the icing hazard increasing as the humidity level increases (NTSB, 1972).

Throttle ice is formed at or near a partly closed throttle when water vapor in the induction air condenses and freezes due to the expansion cooling and lower pressure as the air passes the restriction imposed by the throttle. This temperature drop normally does not exceed 5 °F. When ambient temperature is above 37 °F the pilot need not be concerned with throttle icing as long as only air passes the throttle (systems such as pressure carburetors and fuel injection systems usually inject fuel into the air at a point downstream of the throttle).

When there is fuel:air mixture passing the throttle, any ice formation would be attributable to water vapor freezing from the cumulative effects of the fuel ice and throttle ice phenomena. Icing at the throttle then can occur at ambient temperatures much higher than 37 °F.

Carburetor Ice Formation And Prevention

Any one or combination of these ice-forming situations may cause loss of power by restriction of induction flow and interference with an appropriate fuel:air ratio. One reason it can be important to use carburetor heat as an anti-icer rather than a deicer lies in the fact that carburetor heat available declines as carburetor ice forms in the induction system, decreasing power and exhaust gas temperature. It is best to take action early to guard against a buildup of carburetor ice before deicing capability is lost.

Carburetor air heaters in small aircraft are usually of the exhaust pipe cuff type. The exhaust-heated air is directed into the carburetor air duct as desired by the heat valve so that, with full carburetor heat, the normal cold air duct is essentially closed off and only the intake air passing through the exhaust pipe heat cuff enters the carburetor.

It should be realized that partial carburetor heat can be worse than none at all under certain conditions. For example, the fuel:air mixture temperature might be 20 °F with no heat applied, which normally would not be so conducive to ice formation as if the temperature were brought up to 33 °F by partial application of carburetor heat. In the cold air case, water in the form of ice crystals would pass through the carburetor to be vaporized by engine heat with little consequence. Partial carburetor heat would melt the ice crystals which would re-freeze upon contact with cold carburetor parts near regions of low pressure and temperature. So, the general rule is, for engines with no carburetor air temperature (CAT) instrumentation, use full heat whenever any heat is applied. CAT, if installed, allows the pilot to properly modulate application of heat so that the negative effects of full carburetor heat may be avoided.

Negative Effects Of Carburetor Heat

Notwithstanding the importance of using carburetor heat when necessary, the importance of guarding against undue overuse should be recognized. Carburetor heat increases induction air temperature, causing it to be less dense. This decreases the number of oxygen molecules that can be packed into each cylinder, thus decreasing power output, increasing the fuel:air ratio and increasing cylinder temperatures needed to get the same power output, possibly increasing the detonation hazard at high power outputs.

There are exceptions to the rule that carburetor heat application results in lower power. In extremely cold ambient air temperatures (below +15 °F), the use of carburetor heat may increase power because of increased fuel vaporization. At very cold temperatures, fuel may pass through the engine because it wasn't vaporized in time for the power stroke. Symptoms of poor fuel vaporization in cruise flight are a rough running engine that will afterfire (fuel burning in the cylinder or exhaust stack during the exhaust stroke) if the mixture is leaned. Application of carburetor heat will smooth out the engine. If temperatures are very cold and humidity is low, cruise operation with partial carburetor heat may be necessary in order to lean normally. If CAT is installed, I have used heat to bring the CAT to –5 °C, then leaned, enjoying many hours of arctic flight behind a happy engine. CAT temperatures just above 0 °C are not advisable as ice crystals in the air may melt and refreeze in the carburetor.

Induction air temperature instrumentation (CAT) is a very useful instrument, promoting safe engine operation under conditions of extremes of high humidity and very high or very low temperatures.

Engine starting in very cold air temperatures suggests the use of carburetor heat to properly vaporize the fuel. Heat from the engine compartment (assuming the engine is warm or has been preheated) is drawn into the carburetor instead of the very cold outside air.

During operation in cold air, carburetor heat provides even another benefit. Warmed air decreases the chance of thermal shock during closed throttle operations. Arctic pilots select carburetor heat on before shutting off the engine with the mixture control to prevent cold air from passing through the engine after combustion stops but the engine has not yet stopped turning, thus decreasing thermal shock.

Pressure Carburetors

Recognizing the limitations of the simpler float carburetor, steps have been taken to overcome them with the pressure carburetor. These limitations are essentially:

Susceptibility to icing.

Uneven fuel-air mixture distribution.

Critical at high density altitudes due to vapor lock.

Affected by gravity (attitude critical).

Metering a function of air volume, not mass.

The pressure carburetor utilizes a closed fuel system, one in which the fuel is not open to the atmosphere at any point from the tank to the discharge nozzle, as it is in the bowl of the float carburetor. Fuel leaves the tank under pressure from the boost or auxiliary pump, goes through the filter and the engine pump to the carburetor. Here fuel is metered and directed to the discharge nozzle.

Pressure of the fuel delivered to the metering jet is controlled by the volume and density of air flowing into the engine. In this way, the fuel flow becomes a function of the mass air flow.

The Bendix PS7BD, figure 6-9, is typical of the modern pressure carburetors used on light reciprocating engine aircraft, and is the unit discussed here.

Air Metering Force

Air flows into the engine, passing first through the inlet air filter and into the carburetor throttle body then through the venturi, past the throttle valve and discharge nozzle into the intake manifold. As the air enters the carburetor body, some of it flows into the channel around the venturi where its pressure (impact pressure) increases due to its decrease in velocity. This impact pressure is directed into chamber A of the computer or regulator unit. Any change in the velocity or density of air entering the carburetor changes the pressure in chamber A.

Air flowing through the venturi produces a low pressure proportional to the velocity of air entering the induction system. This low pressure is directed into chamber B of the regulator where it operates on the opposite side of the diaphragm from the impact air pressure. These two air forces work together to move the air diaphragm proportional to the volume of air entering the engine.

The mass or weight of air is a function of the air density and to modify the effect of air volume to reflect its density, an automatic mixture control is placed in the vent line between the two air chambers. When air density decreases (high altitude or high temperature), the automatic mixture control opens the vent, decreases the pressure drop across the diaphragm, and lowers the metering force.

Fuel Metering Force

Fuel enters the carburetor from the engine pump under a pressure of approximately 9 to 14 pounds per square inch, and passes through a fine mesh wire screen in chamber E on its way to the poppet valve.

The amount the poppet valve opens is determined by a balance between the air metering force and the regulated fuel pressure. The air metering force moves the diaphragm to the right, opening the poppet valve.

When the poppet valve opens, the fuel flows from chamber E into D and exerts a force on the fuel diaphragm moving it back enough to allow the spring to close the poppet valve. The fuel pressure in chamber D is, in this way, regulated to be proportional to the mass of air flowing into the engine.

When the engine is idling with not enough airflow to produce a steady air metering force, the large coil spring in chamber A forces the diaphragm over and opens the poppet valve to provide the fuel pressure required for idling.

Fuel from chamber D, regulated but unmetered, flows through the main metering jet and through the idle needle valve. For all conditions other than idle, this valve is off its seat enough that its

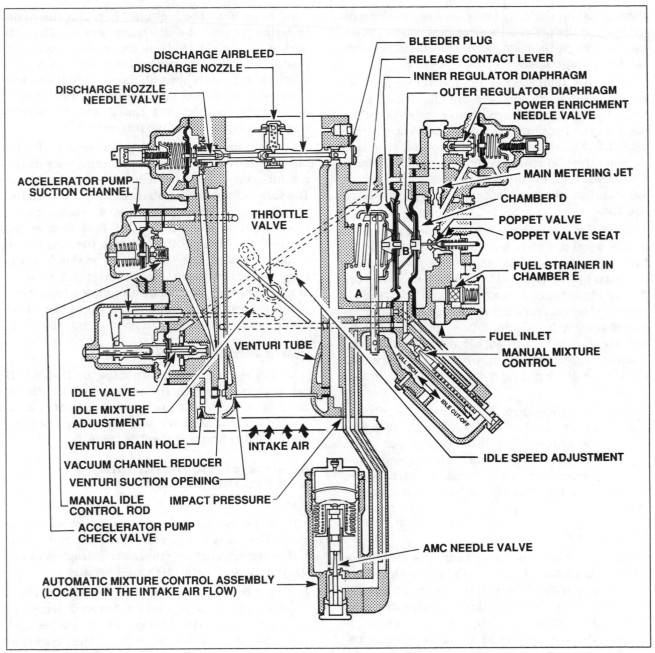

Figure 6-9. A Bendix PS7BD pressure carburetor with an airflow-type enrichment valve and an automatic mixture control.

opening is larger than the main metering jet, so no metering is done by the idle valve. Fuel flows to the discharge valve, and the discharge nozzle. Air from the impact annulus is mixed with the fuel in the nozzle to produce a spray for better vaporization.

The spring loaded diaphragm type discharge valve provides a fast and efficient cut-off when the mixture control is placed in the Idle Cut-off position. When the fuel pressure drops low enough, a spring forces the needle valve onto its seat, stopping all flow of fuel from the nozzle. This valve also

provides a constant pressure downstream of the metering jet; so the variable pressure from the regulator will force a flow through the jet proportional to the mass air flow.

Mixture Control System

As altitude increases, the air density becomes less, and unless a correction is made for this, the mixture will enrich and the engine will lose power. To maintain an essentially constant fuel-air mixture, the pilot must decrease the weight of the fuel flowing to the discharge nozzle. This is done by decreasing the pressure differential across the air metering diaphragm by opening the bleed between the two air chambers.

When the pilot wishes to stop the engine, he pulls the mixture control to the Idle Cut-off position. The mixture control needle valve is pulled back so the pressures in chambers A and B are essentially equalized. When the control is in this position the idle spring is depressed and its force is removed from the diaphragm, closing the poppet valve and shutting off all of the fuel to the metering sections.

An automatic mixture control relieves the pilot of the necessity of regulating the mixture as altitude changes. A brass bellows filled with helium and attached to a reverse tapered needle varies the flow of air between chambers A and B as the air density changes. An inert oil in the bellows damps out vibrations. This automatic mixture control, by varying the amount of air bleed between the two chambers, maintains a pressure differential across the air diaphragm appropriate for any air density.

Since mixture control is automatic, the mixture control in the cockpit will have only two (idle cut-off and auto-run) or three (idle cut-off, auto rich and auto lean) positions. This provides the pilot a strong clue that pressure carburetors equipped with automatic mixture control are installed.

Idle System

The idle system of the Bendix PS carburetor controls both the idle air and the idle fuel. All of the air which flows into the engine during idle must flow around the almost closed throttle valve. The amount this valve is held away from closing is controlled by the idle speed adjustment, an adjustable stop on the throttle shaft extension. The amount of fuel allowed to flow during idling is regulated by the amount the idle fuel valve is held off its seat by the control rod, as it contacts a yoke on the throttle shaft extension.

Acceleration System

A single diaphragm pump is used on most carburetors of this type to provide a momentarily rich mixture at the main discharge nozzle when the throttle is suddenly opened.

The accelerating pump is located between the idle valve and the discharge nozzle, with one side of the diaphragm vented to the manifold pressure, downstream of the throttle. The other side of the

diaphragm is in the fuel line, between the main metering jet and the discharge nozzle. The coil spring in the air side compresses when the manifold pressure is low and fuel fills the pump. When the throttle is opened, the manifold pressure increases; the spring pushes the diaphragm over and forces the fuel out the discharge nozzle, momentarily enriching the mixture.

Some pumps used with this series of carburetors have a divider in the fuel chamber with a combination check and relief valve and a bleed. The valve allows a rapid discharge of fuel when the throttle is first opened, but soon seats, and a lesser, but sustained flow of fuel discharges through the pump bleed. When the throttle is suddenly closed, the decrease in manifold pressure causes a rapid movement of the pump diaphragm, and the check valve closes to prevent the pump starving the discharge nozzle. This would cause the mixture to go momentarily lean.

Power Enrichment System

Two types of systems are used in Bendix PS carburetors to enrich the mixture under conditions of full throttle; one system operates as a function of the airflow, and the other operates mechanically from the throttle valve.

Advantages and Disadvantages
Of Pressure Carburetors

Icing

Pressure carburetors are less susceptible to icing than float carburetors because, with float carburetors, a greater temperature drop occurs in the vicinity of the throttle valve, because:

1. The temperature drop due to venturi effect (lowered pressure causes lowered temperature) is greater in the float carburetor because its venturi is more severe than the venturi in the pressure carburetor or fuel injection system. In the two latter systems the venturi-caused pressure drop does not need to be large because the fuel metering force can be amplified by the size of the air diaphragm and fuel metering force does not directly supply fuel pressure as it does in the float carburetor.

2. The temperature drop near the throttle valve caused by fuel vaporization occurs only in the float carburetor because it is the only one of the three systems that vaporizes fuel upstream of the throttle valve. The fuel injection system vaporizes fuel in the warm area of the induction system very near the cylinder intake valve just outside of the cylinder.

The pressure carburetor is less susceptible to icing than the float carburetor but somewhat more susceptible than fuel injection systems because fuel vaporization still occurs in the carburetor body. From personal experience I can tell you that ice will form in pressure carburetors (almost instantaneously under the right ambient conditions), so carburetor heat is still needed during the severe conditions conducive to carburetor ice formation.

Pressure carburetors can suffer ice formation but, since the temperature drop near the throttle is much less than that of the float carburetor, icing will take place over a much narrower range of ambient conditions. Pilots should remain vigilant for signs of carburetor ice formation with all three systems.

Fuel Injection Systems

Uneven fuel:air mixture distribution, which is a major problem with conventional carburetors, has become of increasing importance as the horsepower requirements of modern aircraft engines have continued to go up. The modern aircraft engine operates with such high cylinder pressures that an inadvertently lean mixture on one cylinder can cause detonation that can destroy the engine. Because of uneven fuel distribution, carbureted engines cannot enjoy maximum leaning and do not have all cylinders working evenly. Therefore, fuel injected engines produce slightly more power and use less fuel than carbureted engines of equal displacement and compression ratio. Continuous flow fuel injection systems deposit a continuous flow of fuel into the induction system near the intake valve just outside of the cylinder. The fuel vaporizes in this warm environment then is sucked into the cylinder during the intake stroke. Continuous flow systems are much simpler than direct injection systems yet are very effective.

Two such continuous flow systems are discussed here. The first, the Bendix RSA, uses a venturi and air diaphragm to develop a fuel metering force, very much like the pressure carburetor does. The second, the Teledyne-Continental system, determines the amount of fuel to deliver to the engine without the use of a venturi.

Bendix Fuel Injection System

The Bendix RSA fuel injection system is a good example of a system that utilizes a venturi-developed fuel metering force. The line diagram of figure 6-10 will help follow the flow in this system.

Air Metering Force

The impact tubes P, in the inlet of the throttle body, sense the total (dynamic plus static) pressure of the air entering the engine and the venturi O senses its velocity. These two forces combine to provide a force that moves the air diaphragm (between chamber I and J) proportionally to the amount of air ingested into the engine.

Fuel Metering Force

Fuel from the engine driven fuel pump C enters the fuel control through the filter E and the mixture control valve F. The pressure from this fuel acts on the diaphragm to cause it to close the ball valve T. Fuel for the engine operation flows through the main metering jet G and the throttle fuel valve H into chamber L of the regulator M. This metered fuel opens the ball valve.

Metered Fuel Flow

The actual metering is done by the pressure drop across the orifices in the fuel injector nozzles. In this system, fuel metering force is done by the position of the ball valve in its seat. See figures 6-11 and 6-12. The inlet pressure is held relatively constant by the engine driven fuel pump and the outlet pressure is controlled by the balance between the fuel and air metering forces. When the throttle is opened, the air metering force increases; this opens the ball valve and lowers the pressure in chamber L. The pressure in chamber K is greater than that in L, by the drop across the fuel control, and tends to close the valve. The balance between the air and fuel forces therefore holds the valve off its seat a stabilized amount for any given airflow.

Flow Divider

After the fuel leaves the regulator, it flows through a flexible hose to the flow divider, figure 6-13, located on top of the engine in a central location. The injector nozzles are connected to the flow divider by 1/8" stainless steel tubing. A pressure gage in the cockpit reads the pressure at the outlet of the flow divider. This is actually the pressure drop across the injector nozzles and is directly proportional to the fuel flow through the nozzles. For all flow conditions other than idle, the restriction of the nozzles causes a pressure to build up in the metered fuel lines which influences the fuel metering force. Under idle flow conditions, the opposition caused by the nozzles is so small that metering would be erratic. To prevent this, a spring holds the flow divider valve closed to oppose

107

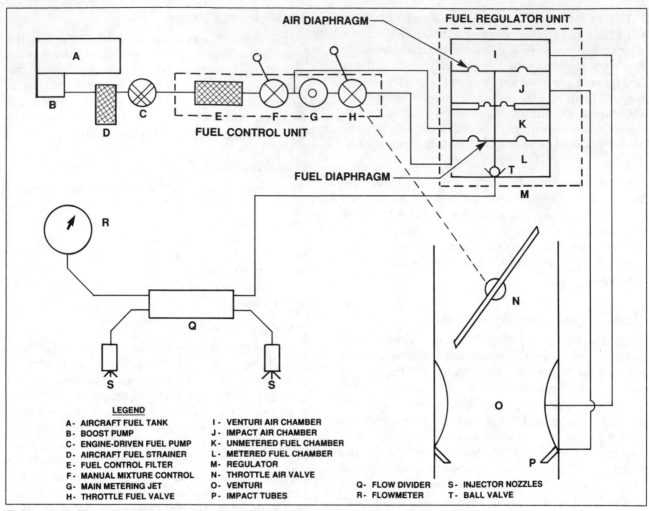

AIR DIAPHRAGM
FUEL REGULATOR UNIT
FUEL CONTROL UNIT
FUEL DIAPHRAGM

LEGEND

A- AIRCRAFT FUEL TANK
B- BOOST PUMP
C- ENGINE-DRIVEN FUEL PUMP
D- AIRCRAFT FUEL STRAINER
E- FUEL CONTROL FILTER
F- MANUAL MIXTURE CONTROL
G- MAIN METERING JET
H- THROTTLE FUEL VALVE

I - VENTURI AIR CHAMBER
J - IMPACT AIR CHAMBER
K- UNMETERED FUEL CHAMBER
L- METERED FUEL CHAMBER
M- REGULATOR
N- THROTTLE AIR VALVE
O- VENTURI
P- IMPACT TUBES

Q- FLOW DIVIDER   S- INJECTOR NOZZLES
R- FLOWMETER   T- BALL VALVE

Figure 6-10. Flow diagram for the Bendix RSA fuel injection system of figure 6-11.

the fuel flow until metered fuel pressure becomes sufficient to off-seat it.

For idle fuel flow, the flow divider opens only partially and thus serves the double function of creating the downstream pressure for the fuel control and dividing the fuel to the cylinders for these extremely low flow conditions. When the mixture control is placed in the Idle Cut-Off position, the flow divider provides cut-off for the fuel.

Injection Nozzles

This system also uses an air bleed type nozzle. These nozzles screw into the cylinder head near the intake port. See figure 6-14. Each nozzle consists of a brass body which incorporates a metering orifice, an air bleed hole, and an emulsion chamber. Around this body is a fine mesh metal screen and a pressed steel shroud. These nozzles are calibrated to flow within plus or minus 2% of each other and are interchangeable between engines and cylinders. An identification mark is

stamped on one of the hex flats of the nozzle opposite the air bleed hole. When installing a nozzle in a horizontal plane, the air bleed hole should be positioned as near the top as practical to minimize fuel bleeding from the opening immediately after engine shut down.

Idle System

When there is a low airflow through the engine such as is encountered during idling, the air metering forces are not sufficient to open the ball valve enough for idle fuel to flow. The air diaphragm is between two springs which hold the ball valve off its seat until the airflow becomes sufficient. The constant head spring, figure 6-11, pushes against the air diaphragm and forces the ball valve off its seat. This maintains a constant head of pressure across the fuel control. As the airflow increases, the air diaphragm moves over, compressing the constant head spring until the diaphragm bushing makes solid contact with the

108

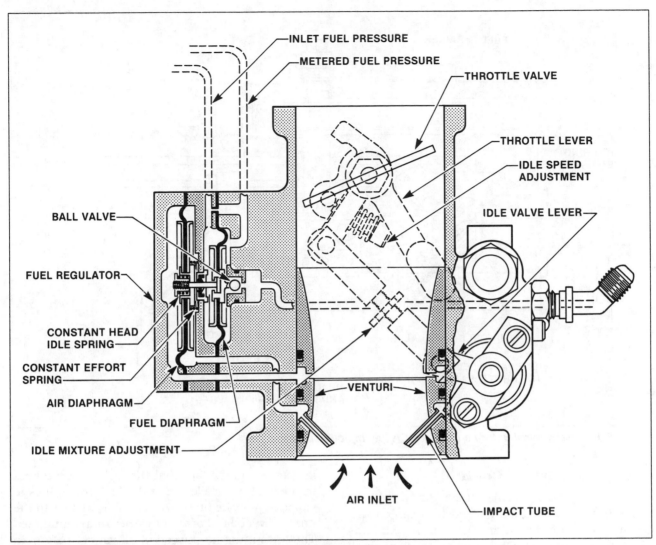

Figure 6-11. Regulator unit of a Precision Airmotive RSA fuel system.

ball valve shaft. Beyond this point the ball valve acts as though it is directly connected to the diaphragm. A smooth transition between idle and cruise RPM is provided by the use of the constant effort spring working between the air diaphragm and the housing. This spring, in effect, preloads the air diaphragm, giving it an initial loaded position from which to work.

As with any fuel metering system, this one controls the idle RPM by limiting the amount of air allowed to flow past the throttle valve, and the idle mixture by the amount of fuel allowed to flow to the discharge nozzles.

A spring loaded screw, figure 6-11, contacts a stop on the throttle body to limit the amount the throttle air valve can close. An adjustable-length rod connects the throttle air valve to the throttle fuel valve and controls the amount the throttle fuel valve remains open. Adjustment of this length determines the idle mixture ratio, and ultimately the idle manifold pressure.

Manual Mixture Control

A spring loaded, flat plate type valve in the fuel control, figure 6-12, is moved by a linkage from the cockpit to regulate the amount of fuel that can flow to the main metering jet. When the mixture control is placed in the idle cut-off position, the passage to the main metering jet is completely closed and no fuel can flow to the jet. In the full rich position, the opening afforded by the mixture control is larger than the metering jet, and the jet limits the flow. In any intermediate position, the opening is smaller than the main jet and the mixture control becomes the flow limiting device.

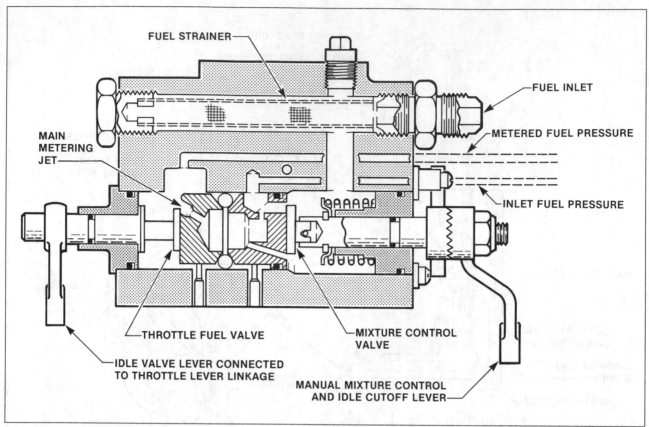

FUEL STRAINER

FUEL INLET

METERED FUEL PRESSURE

MAIN METERING JET

INLET FUEL PRESSURE

THROTTLE FUEL VALVE

MIXTURE CONTROL VALVE

IDLE VALVE LEVER CONNECTED TO THROTTLE LEVER LINKAGE

MANUAL MIXTURE CONTROL AND IDLE CUTOFF LEVER

Figure 6-12. Fuel control unit of a Bendix RSA fuel injection system.

Automatic Mixture Control

A reverse tapered needle attached to a bellows, figure 6-15, varies the air bleed between the air chambers of the regulator. The bellows contains helium to sense density changes and a small amount of inert oil to dampen vibrations. As the air density decreases, either from an increase in either altitude or temperature, the pressure inside the bellows causes it to expand. This increases the bleed across the air diaphragm which decreases the air metering force and leans the mixture.

Starting Procedure

Engines equipped with this injection system are started by first placing the mixture control in the idle cut-off position and opening the throttle about ⅛ of the way. Turn on the master switch and the boost pump. Move the mixture control to the Full Rich position until there is an indication of flow on the flow meter and return the mixture control to the idle cut-off. Turn the ignition on and engage the starter. As soon as the engine starts, move the mixture control to the Full Rich position.

Fuel injected engines have the reputation of being difficult to start when they are hot. This is

largely due to the fact that the high temperatures in the engine nacelle cause the fuel in the lines to vaporize and the lines from the flow divider to the nozzles will be full of vapor rather than liquid fuel. The lines must be purged of all vapors before an effective start can be made.

The Teledyne-Continenta Fuel Injection System

The Continental fuel injection system meters its fuel as a function of the engine RPM and throttle position. It does not use air flow as a metering force. A special engine driven pump, an integral part of the system, produces the fuel metering pressure. A line drawing flow diagram, figure 6-16, should help with a mental arrangement of the components of this system.

Components

Injection Pump

Fuel enters the system into a swirl chamber where vapor and air bubbles are ejected back to the fuel tank for venting. See figure 6-17. From the swirl chamber, fuel is drawn into the engine-driven pump which is the heart of this fuel injection

110

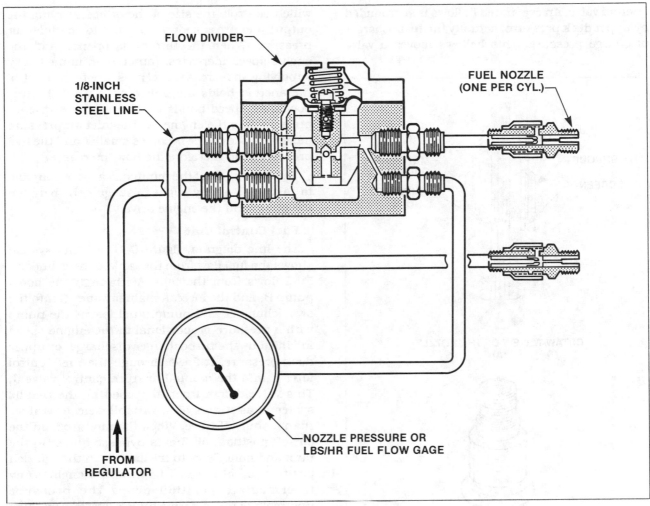

Figure 6-13. Flow divider for a Bendix RSA fuel injection system.

system. It is basically a vane type, constant displacement pump with special features that allow it to produce an output pressure which varies with the engine speed. If a passage containing an orifice bypasses the pump mechanism, figure 6-17, the output pressure will vary according to the speed of the pump. The size of the orifice will determine the pressure for any given speed. If its size is increased, the output pressure will decrease. This system works well for flows in the cruise or high power range, but when the flow is low, as in idling, there is not enough restriction to maintain a constant output pressure. An adjustable pressure relief valve is therefore installed in this line. During idle the output pressure is determined by the setting of the relief valve and the orifice has no effect, while at the high power end of operation, the relief valve is off its seat and the pressure is determined by the orifice.

Any fuel injection system must have vapor free fuel in its metering section, and provision is made

in the pump to remove all vapor from the fuel and return it to the tank.

Fuel enters the pump through a chamber where the vapor is swirled out of the liquid and collects in the top. Some fuel from the pump outlet returns to the tank through a venturi arrangement on the top of this chamber and produces a low pressure which attracts the vapors and returns them to the aircraft fuel tank. A final feature in this pump is a by-pass check valve around the pump so fuel from the boost pump may flow to the fuel control for starting. As soon as the engine pump pressure becomes higher than that of the boost pump, the valve closes and the engine pump takes over.

Turbocharged engines have a unique problem during acceleration. If the fuel flow increases before the turbocharger has time to build up to speed and increase the airflow proportionately, the engine may falter from an overly rich mixture. Pumps for these engines have the simple orifice replaced with a variable restrictor controlled by an aneroid valve. An evacuated bellows is surrounded by upper deck pressure, actually the turbocharger discharge pressure. This bellows moves a valve which controls the size of the orifice, varying the output fuel pressure proportional to the inlet air pressure. When the throttle is opened and the engine speed increases, rather than immediately supplying an increased fuel pressure to the control the aneroid holds the orifice open until the turbocharger speed builds up and increases the air pressure into the engine. As the inlet air pressure increases, the orifice becomes smaller and the fuel pressure, and therefore the flow, increases.

The drive shaft of the pump has a loose coupling to take care of any slight misalignment between the pump and the engine drive.

Fuel Control Unit

The line diagram, figure 6-16, of this system shows the fuel flow from the tank to the cylinders. Fuel flows from the tank A, through the boost pump B, and the aircraft main strainer C into the swirl chamber and pump. Fuel leaves the pump with a pressure proportional to the engine speed modified by the turbocharger discharge, or upper deck pressure. It flows through the fuel control filter I, into the manual mixture control valve J. This valve differs from that used by the Bendix systems, as it acts as a variable selector rather than a shut-off valve. When the mixture is in the cut-off position, all fuel is bypassed back to the tank and none flows to the engine. In the full rich position, all of the fuel flows to the engine. Any intermediate position drops the pressure upstream of the metering orifices by routing some of the fuel back to the tank and some to the engine. A metering plug with a precision orifice limits the maximum amount of fuel that may flow

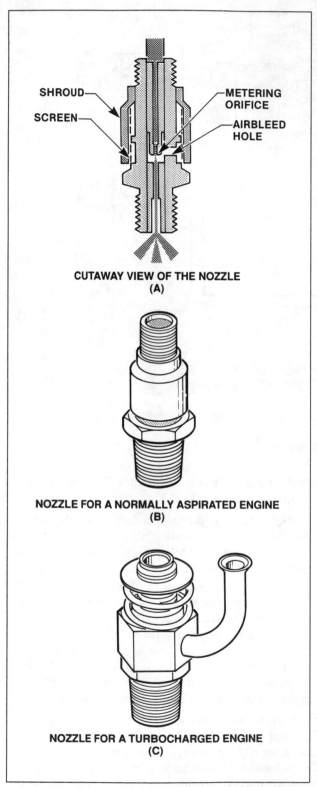

Figure 6-14. Fuel nozzles for a Bendix RSA fuel injection system.

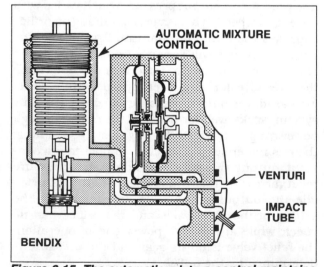

Figure 6-15. The automatic mixture control maintains the mixture ratio as the pilot sets it by controlling the pressure differential across the air diaphragm.

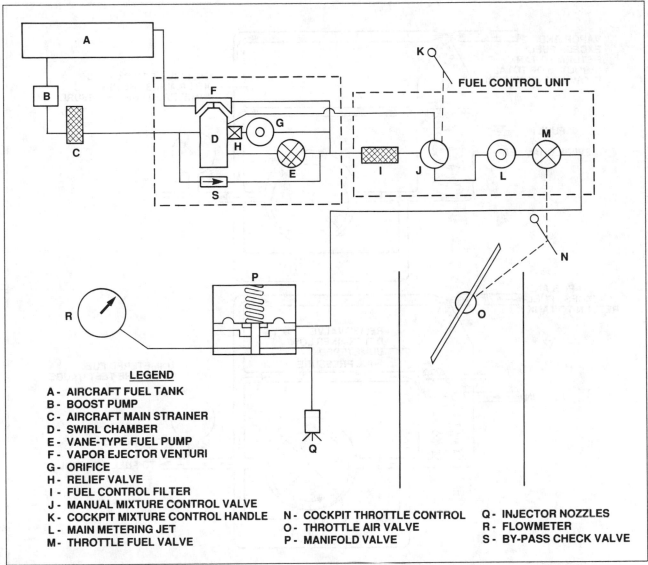

LEGEND

A - AIRCRAFT FUEL TANK
B - BOOST PUMP
C - AIRCRAFT MAIN STRAINER
D - SWIRL CHAMBER
E - VANE-TYPE FUEL PUMP
F - VAPOR EJECTOR VENTURI
G - ORIFICE
H - RELIEF VALVE
I - FUEL CONTROL FILTER
J - MANUAL MIXTURE CONTROL VALVE
K - COCKPIT MIXTURE CONTROL HANDLE
L - MAIN METERING JET
M - THROTTLE FUEL VALVE

N - COCKPIT THROTTLE CONTROL
O - THROTTLE AIR VALVE
P - MANIFOLD VALVE

Q - INJECTOR NOZZLES
R - FLOWMETER
S - BY-PASS CHECK VALVE

Figure 6-16. The Teledyne-Continental fuel injection system.

into the engine under full throttle, full rich conditions. The throttle control N in the cockpit controls both the air valve 0, similar to that used in any carburetor, and the fuel valve M, essentially a variable orifice which determines the amount of fuel allowed to flow to the engine for any given pressure.

Fuel Manifold Valve

After the fuel leaves the throttle fuel valve, it flows through a flexible fuel line to the manifold valve which is usually located on the top of the engine (figure 6-18). This device, sometimes referred to as the "spider", is very similar to the flow divider of the Bendix system. This valve serves two basic functions: it distributes the fuel evenly to all of the cylinders, and it provides a positive

shut-off when the mixture control is placed in the idle cut-off position.

When fuel pressure rises, the diaphragm lifts the valve off its seat but a spring loaded poppet inside the valve stays on its seat until the fuel pressure has opened the valve completely so it will do no metering. Then the poppet opens and allows fuel to flow to the nozzles. The opposition caused by this valve provides a constant pressure downstream of the jets for metering at idle. Above idle, the valve is fully open, and offers no opposition. When the mixture control is placed in the Idle Cut-off position, the fuel pressure drops and the spring closes the manifold valve to provide a positive shut off of the fuel to the nozzles. If the poppet should become plugged or sticky, erratic or rough idling will result.

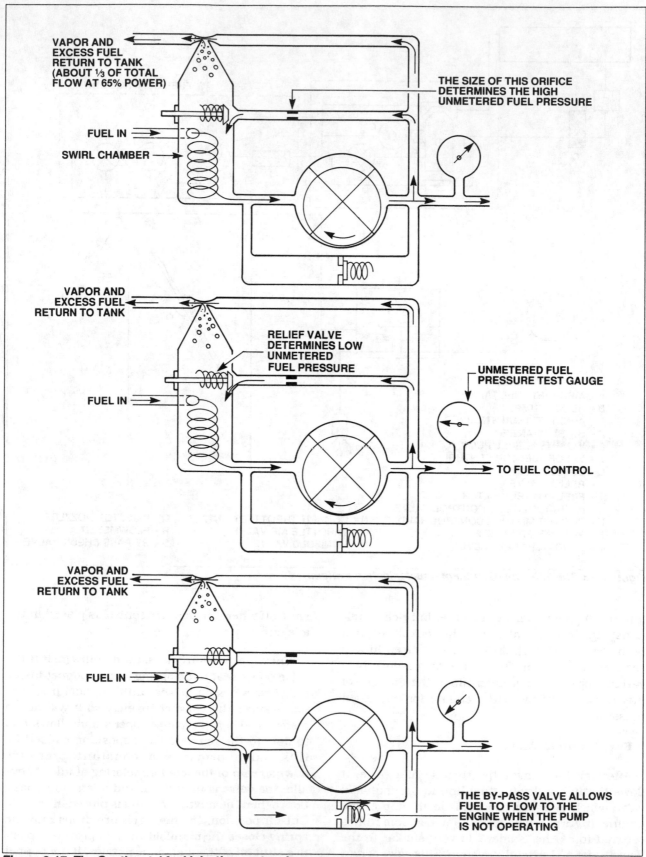

VAPOR AND EXCESS FUEL RETURN TO TANK (ABOUT ⅓ OF TOTAL FLOW AT 65% POWER)

THE SIZE OF THIS ORIFICE DETERMINES THE HIGH UNMETERED FUEL PRESSURE

FUEL IN

SWIRL CHAMBER

VAPOR AND EXCESS FUEL RETURN TO TANK

RELIEF VALVE DETERMINES LOW UNMETERED FUEL PRESSURE

UNMETERED FUEL PRESSURE TEST GAUGE

FUEL IN

TO FUEL CONTROL

VAPOR AND EXCESS FUEL RETURN TO TANK

FUEL IN

THE BY-PASS VALVE ALLOWS FUEL TO FLOW TO THE ENGINE WHEN THE PUMP IS NOT OPERATING

Figure 6-17. The Continental fuel injection system incorporates a constant displacement pump with controls in its relief system which makes its output pressure proportional to the engine RPM.

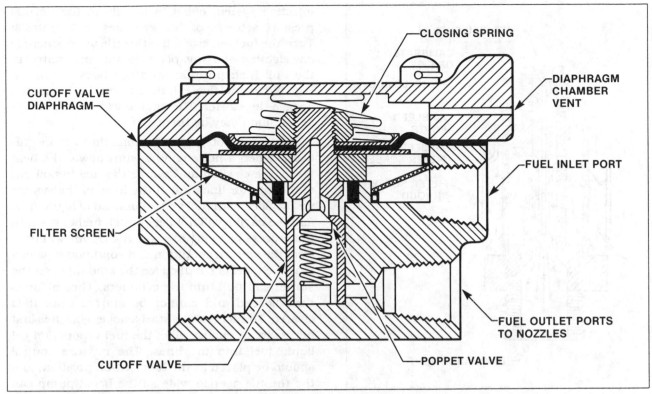

Figure 6-18. Fuel manifold valve for a Teledyne-Continental fuel injection system.

Injector Lines

The nozzles are connected to the manifold valve by six stainless steel lines, one eighth inch in diameter and all the same length in order to insure that fuel pressures at the injector nozzles will all be equal.

Injector Nozzles

Possibly the simplest, yet one of the most important components of a fuel injection system, is the injector nozzle, figure 6-19. One nozzle in each cylinder head provides the point of discharge for the fuel into the induction system. Fuel flows from the manifold valve into the nozzle, through a calibrated orifice into the intake valve chamber of the cylinder head. Air is drawn through a screen into the injector nozzle where it mixes with the fuel and forms an emulsion to improve fuel vaporization. The sheet metal shroud, pressed onto the nozzle to protect the screen, should never be removed. Plugged or partially plugged nozzles can be a source of rough operation and should be one of the first components checked when troubleshooting the system. Nozzles must be cleaned by soaking them in lacquer thinner or acetone and blown out with clean compressed air in the direction opposite regular flow. Wires, drills or other cleaning devices should never be used to

remove obstructions from the orifice. If soaking and blowing out does not clear the obstruction, the nozzle should be replaced.

There are three sizes of nozzle orifices in use with this system, each identified by a letter stamped on the nozzle. The A size nozzle will flow a given amount of fuel for a given pressure. A set of B size nozzles flows one half gallon more fuel per hour with the same pressure and the C size flows yet another half gallon per hour. When the engine was calibrated in its factory run-in, the proper size nozzles were installed and this size should be kept in the engine from then on.

Different models of engines require different styles of nozzles. Some engines require long nozzles to inject the fuel farther into the intake chamber; others function best with short nozzles. The nozzles differ in appearance, but their function is the same. Nozzles used with turbocharged engines have their shrouds vented to the turbocharger side of the throttle valve. If this were not done, increasing the manifold pressure above atmospheric would blow fuel out the bleed holes.

Starting Procedure

Fuel injected engines have the reputation of being difficult to start especially if the engine is hot. Starting an engine equipped with a Continental

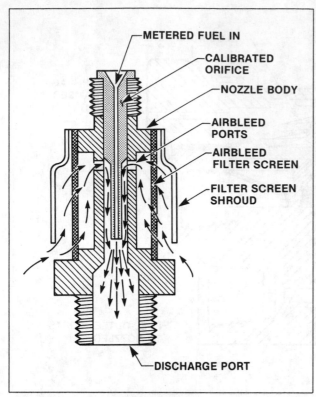

- METERED FUEL IN
- CALIBRATED ORIFICE
- NOZZLE BODY
- AIRBLEED PORTS
- AIRBLEED FILTER SCREEN
- FILTER SCREEN SHROUD
- DISCHARGE PORT

Figure 6-19. Fuel injector nozzle for a Teledyne-Continental fuel injection system.

injection system consists mainly of the normal prestart settings of the switches and controls: Turn the fuel on, crack the throttle approximately one eighth of the way, place the mixture control in the Full Rich position, and the boost pump on high. When fuel flow is indicated on the flow meter, engage the starter. The engine will start, and the boost pump may be turned off.

Starting a hot engine is sometimes more difficult. When a hot engine is shut down, the heat inside the cowling may cause the fuel to boil out of the injector lines. All of the lines in the system will be filled with fuel vapors instead of liquid fuel, and the engine will not get enough fuel to start. In the process of trying to start, the engine will pass from the starved condition to a condition in which too much fuel is supplied for the amount of air the starter can pull into the cylinders. The engine is then flooded and cannot be started until it is cleared. To successfully start a hot engine, it is first necessary to remove all of the fuel vapors and get liquid fuel into the lines. The mixture control should be placed in the Idle Cut-off position, and the throttle opened wide so the boost pump can operate at full pressure. Turn the boost pump on high and allow fuel to circulate through the pump for fifteen or twenty seconds. Turn the boost pump off and place the mixture control in Full Rich. Close the throttle to the correct position for starting, and engage starter. The lines up to the fuel pump and the pump itself are now full of liquid fuel, and a normal start can be made.

Additional Reading

1. IAP, Inc. (1985); Aircraft Fuel Metering Systems; ISBN 0-89100-057-7.
2. IAP, Inc. (1983); Powerplant Section Textbook, Chapter 6, Aircraft Fuel Metering Systems; ISBN 0-089100-251-0.
3. Bent & McKinley (1978); Aircraft Powerplants; Fourth Ed.; Chapters 4-6; McGraw Hill; ISBN 0-07-004792-8.
4. NTSB, 1972; Carburetor Ice in General Aviation; a letter of findings.

Study Questions And Problems

1. What provides the low pressure at the discharge nozzle of a float carburetor?

2. What limits the maximum amount of fuel which can flow from the float bowl for any given pressure differential?

3. Name two functions of an air bleed on a float carburetor.

4. Why are float carburetors more susceptible to icing (at least two reasons)?

5. By NTSB classification, what are the three categories of carburetor ice? Name and describe each.

6. Under what conditions should partial carburetor heat not be used?

7. Under what conditions would the use of carburetor heat be detrimental?

8. When should carburetor heat be used on the ground (several answers)?

9. What is the purpose of the acceleration system on an aircraft carburetor?

10. Why is a power enrichment system also called an economizer system?

11. Is the fuel-air mixture enriched or leaned when the main air bleed is restricted?

12. Does application of carburetor heat richen or lean the mixture? How and why? (more than one reason).

13. What are the advantages of the pressure carburetor over the float carburetor?

14. What are the disadvantages of the pressure carburetor?

15. What does the fuel flowmeter actually measure on the fuel injection systems discussed in this chapter?

16. What is the purpose of the air bleed in an injector nozzle?

17. What are the functions of the spider?

18. What are the advantages of a fuel injection system, as a fuel metering system?

19. What are the disadvantages of a fuel injection system, as a fuel metering system?

20. What types (NTSB categories) of icing do fuel injection systems avoid?

21. What type of icing is avoided by the Teledyne-Continental fuel injection system that is not avoided by the Bendix system?

Chapter VII
Power Management

RPM And MAP

The pilot's selection of power from the engine is done mostly by changing MAP or RPM, or both. **Manifold absolute pressure** (MAP) and RPM are the two main variables in the production of power. Figure 7-1 presents the relationship of MAP, RPM and power. It shows that, if RPM remains constant and MAP is increased, power will be increased. Also, if MAP is held constant and RPM is increased, power will be increased.

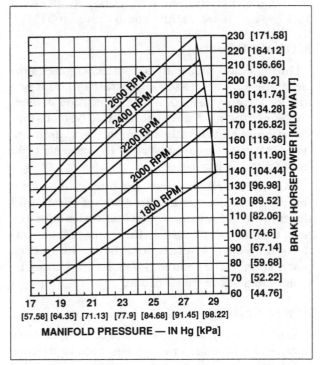

Figure 7-1. Bhp vs. intake manifold pressure vs. revolutions per minute.

Other Factors Affecting Power

Humidity

Water vapor in the air will cause a significant reduction of power output from reciprocating engines. If the humidity of the air is high, more water molecules displace oxygen molecules in a cylinder full of air, producing two power-reducing effects: (1) less O_2 is available to burn fuel to produce power, and (2) the fuel:air ratio, already richened for full power production on takeoff, becomes even richer and causes a further decrease in power production (see figure 6-1).

Power losses of 5% are not unusual in warm air having 100% relative humidity. It should be pointed out that a 5% decrease in power output may cause much larger decreases in performance, depending on the amount of excess thrust horsepower (ETHP) being produced at the time. A 5% decrease in power may produce ETHP decreases of 100% if operating in a flight condition where ETHP was only equal to that 5% decrease!

Since warm air holds more water as vapor, the effect of humidity on power decreases as temperature decreases. Also for this reason, humidity effects become inconsequential at high altitudes.

Temperature

Temperature affects the density of the air. More O_2 molecules will be present in a cylinder full of cold air so more engine power can be expected from cold air. As a general rule, for each 6° C below standard temperature, one percent more horsepower is produced than at standard temperature and one percent should be subtracted for each 6° C the temperature is above standard.

Very cold temperatures introduce another effect on engine power — poor fuel vaporization in the carburetor which creates poor fuel distribution to the cylinders. The reader is referred to the discussion of carb air temp in ch.6 and total air temp in ch.17.

Mixture

The fuel:air mixture has a very significant effect on power production. This effect is well illustrated in figure 6-1.

Ambient Pressure

If the throttle is fixed in the open position, MAP will vary with amount of (1) **ambient pressure**, and (2) **friction loss** in the intake manifold.

If the engine is not running, pressure in the intake manifold (MAP) will be equal to ambient pressure. So, before engine start, a glance at the MAP gauge will confirm that it is reading correctly if the pilot knows the ambient pressure.

The altimeter setting from a nearby station produces the ambient pressure, corrected to sea level, at that station. To obtain the ambient pressure at the field elevation where the MAP gauge is being checked, subtract one inch of Mercury for each 1000 feet of field elevation from the altimeter setting.

Example:

Given: altimeter setting: 30.14″ Hg
        field elevation: 800 feet.

        pressure altitude: 580 feet: PA =
        800 feet – ([30.14 – 29.92] × 1000)

Ambient Pressure =
Altimeter Setting – (1″ Hg × Field Elevation/1000 ft)
        = 30.14″ – (1″ × 800/1000)

        = 30.14″ – .80″

        = 29.34″ Hg

The manifold pressure gauge should read approximately 29.3″ Hg before engine start under the conditions of the above example.

As can be seen from this example, although pressure altitude is the standard of altimetry with which nearly all altimetry problems are computed, in the case of converting altimeter settings to ambient pressure, field elevation is used instead of pressure altitude because the altimeter setting itself was derived by converting barometric pressure at field elevation to barometric pressure at sea level by adding about 1.06 inches per 1000 feet of field elevation to the embient pressure.

After engine start, with engine idling steadily, MAP will be low because the pilot has closed the throttle which greatly restricts airflow into the cylinders. This, in turn, restricts RPM. During idle, MAP will probably be about 8-12″ Hg and a large pressure drop could be measured across the throttle valve, if pressure measuring instruments were placed on both sides of the throttle valve.

When the throttle valve is opened fully for takeoff, the large pressure drop across the throttle is removed so the MAP will rise to nearly equal the ambient pressure.

Friction Loss

During the takeoff run from a field elevation of 800 ft MSL, with throttle completely open, MAP will probably read about 27″ Hg. Why not 29.3″ Hg as it was before engine start? Because now the air in the intake system is moving at high velocity, creating friction and drag as it moves through the air filter and encounters the bends and skin fric-

tion caused by the geometry of the intake system. As with any fluid dynamics, the faster the air flows and the more twisted is the path the air must travel, the more will be the **friction loss** which appears as a pressure loss.

A typical intake system will cause 2″ to 3″ Hg pressure loss at takeoff RPM and open throttle. So takeoff manifold pressure at the example airfield will be:

MAP = ambient pressure - intake system friction loss

MAP = 29.3″ Hg – 2″ to 3″ Hg
    = 26.3″ to 27.3″ Hg

To see if you understand the above discussion and to fix it better in your mind, do study question #2. The MAP gauge, read before engine start should indicate the ambient pressure of about 23.1″ Hg and takeoff MAP should be about 21″ Hg.

Altitude Effect

During climb (normally aspirated engine), ambient pressure decreases at the rate of about one inch of Mercury per 1000 feet of altitude gained. Therefore, the pilot will observe, if RPM (constant speed prop) and throttle setting are not changed, that MAP will decrease about 1″ Hg for each 1000 feet of altitude gained.

To maintain climb power, the pilot must adjust the throttle every few hundred feet of climb until the throttle is fully open. **Critical altitude** is the term used to describe the altitude where the throttle is fully open in order to achieve the desired power setting. The desired power setting cannot be maintained above this altitude.

Power Settings

Most normally aspirated reciprocating engines in general aviation aircraft are designed to produce continuous power with economical operation and long life if operated at or below 75% of their sea level rated power. Full power (takeoff) should be used when needed but not for continuous operation. 100% power is only available at sea level and with full throttle. With full throttle, power decreases with altitude until an altitude is reached where 75% power is the maximum power available. This usually occurs at altitudes of 6000 - 10,000 feet, depending on the amount of friction loss in the intake system and the RPM selected by the pilot (75% power critical altitude).

The airframe manufacturer selects an engine which will give adequate cruise performance in that particular airframe using 55%-75% power.

So, today's aircraft are usually designed to be operated at the following power settings:

Takeoff Power = 100% or as limited by altitude (full throttle)

Climb Power = 75%

Fast cruise = 65-75%

Normal cruise = 65%

Economy cruise = 55-65%

Maximum range power = 40-50% (best miles/gallon)

Maximum endurance power = 35-45% (longest time aloft/gallon)

Figure 7-2 shows some of the relationships of altitude, power setting and range.

Figure 7-3 is the classic airspeed vs. altitude chart showing the effect of different power settings. For the aircraft equipped with a normally aspirated engine, the following observations can be made from this graph:

1. Maximum true airspeed is achieved at sea level.

2. Maximum speed decreases as altitude increases.

3. Critical altitude can be read directly from this chart. For example, critical altitude for 75% power at 2400 RPM is 6500 feet and for 2700

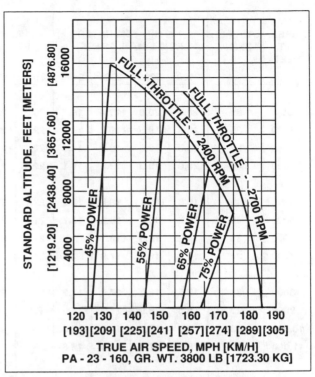

Figure 7-3. Airspeed vs. altitude and power.

RPM, it is about 7800 feet (will vary between airplane models).

4. At a constant power setting, the airplane will fly faster at higher altitudes (the air is thinner — less parasite drag). However, this is only true at power settings where parasite drag is the dominant drag form. At 40% power or less, induced drag is dominant. Since induced drag increases with altitude, the slope of the % power lines becomes negative indicating that, at constant (low) power settings the aircraft will fly slower as altitude increases.

Once percent power is selected for the flight by the pilot, MAP and RPM can be chosen from the charts provided by the manufacturer. If, for example, 65% power is selected for a flight at 6000 feet, the aircraft's handbook will provide several choices of MAP and RPM that will produce 65% power. Power setting charts take many different formats. Two are presented in figures 7-4 and 7-5.

The Lycoming table of figure 7-4 gives four choices of RPM, with corresponding values of MAP that will produce 65% power. Which one should the pilot use? It seems reasonable that the lowest approved RPM will result in the least wear (friction) because the piston doesn't have to travel as far in its cylinder in one hour at 2100 RPM as it does at 2400 RPM. So the lowest approved RPM might be

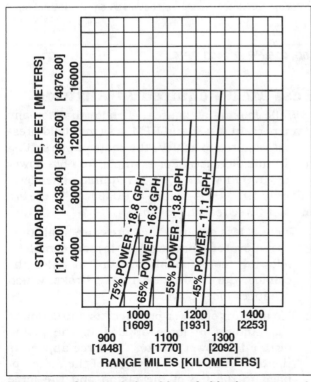

Figure 7-2. Some relationships of altitude, power setting and range.

PRESS. ALT. FEET	STD. ALT. TEMP. °F	129 HP - 55% RATED RPM & MAN. PRESS.				153 HP - 65% RATED RPM & MAN. PRESS.				175 HP - 75% RATED RPM & MAN. PRESS			200 HP - 85% RATED RPM & MAN. PRESS		
		2100	2200	2300	2400	2100	2200	2300	2400	2200	2300	2400	2200	2300	2400
SL	59	20.8	20.0	19.4	18.7	23.2	22.4	21.7	21.0	24.6	23.9	23.1	27.2	26.4	25.5
1000	55	20.5	19.8	19.2	18.5	22.9	22.2	21.5	20.8	24.3	23.6	22.9	26.9	26.1	25.3
2000	52	20.3	19.5	19.0	18.3	22.7	21.9	21.2	20.6	24.1	23.4	22.6	F.T.	25.8	25.0
3000	48	20.0	19.3	18.8	18.1	22.4	21.7	21.0	20.4	23.8	23.1	22.4		F.T.	24.7
4000	45	19.8	19.1	18.5	17.9	22.1	21.4	20.8	20.2	23.5	22.8	22.1			F.T.
5000	41	19.5	18.9	18.3	17.7	21.9	21.2	20.5	20.0	23.2	22.6	21.9			
6000	38	19.3	18.6	18.1	17.85	21.6	21.0	20.3	19.7	F.T.	22.3	21.7			
7000	34	19.1	18.4	17.9	17.36	21.3	20.7	20.1	19.5	—	F.T.	21.5			
8000	31	18.8	18.2	17.7	17.2	21.1	20.5	19.9	19.3	—	—	F.T.			
9000	27	18.6	18.0	17.5	17.0	F.T.	20.2	19.7	19.1						
10,000	23	18.3	17.7	17.2	16.8	—	F.T.	19.4	18.9						
11,000	20	18.1	17.5	17.0	16.6	—		F.T.	F.T.						
12,000	16	17.8	17.3	16.8	16.4										
13,000	13	F.T.	17.0	16.6	16.2										
14,000	9	—	F.T.	16.4	16.2										
15,000	6	—	—	F.T.	15.8										
16,000	1				F.T.										

NOTE: To maintain constant power, correct manifold pressure approximately 0.17" Hg for each 10 °F variation in carburetor air temperature from standard altitude temperature. Add manifold pressure for air temperatures above standard; subtract for temperatures below standard.

Figure 7-4. Power setting table — Avco Lycoming 0-540-J3A6D, 235 HP @ 2400 RPM.

a good choice because of less wear. On the other hand, pressures in the cylinder will be lower at the higher RPM settings resulting in less stress on internal parts.

Actually, all settings between 2100 and 2400 RPM are approved for this engine. Probably little, if any, difference in wear and stress will be apparent at overhaul time if the engine is always operated within the approved range. Usually, the pilot chooses a value of RPM which produces minimum vibration and cabin noise level.

To produce 65% power at 6,000 feet in the engine represented by the chart in figure 7-4, 21.6" Hg is used at 2100 RPM and 19.7" Hg will work at 2400 RPM. If the pilot desires to use 2250 RPM, interpolation of the table suggests using 20.65" Hg. The MAP gauge can't be read that accurately, so the pilot sets the throttle while reading the MAP gauge as closely as possible.

Less MAP Required At Altitude

While looking at figure 7-4, note that for any given percent power and RPM, less manifold pressure is required to achieve the same power setting as altitude increases. For example, for 65% power using 2300 RPM, 21.7" Hg is required at sea level but only 19.7" Hg is needed at 9000 feet. There are at least three reasons for this:

1. There is less exhaust back pressure at altitude so removal of exhaust gasses is more complete, allowing more fuel:air mixture to enter the cylinder during the next intake stroke, which produces more power.

2. The amount of power produced is a function of the amount of force applied to the piston during the power stroke. The force applied is the difference in the strength of the force applied to the top of the piston and the force applied to the bottom of the piston. This latter

force is the force of ambient pressure (the crankcase is vented) which decreases as altitude increases. Therefore, to produce the same amount of power at high altitude, less cylinder pressure is required so less MAP is needed.

3. Colder air at high altitude is denser, so less pressure (MAP) is needed to move the same number of oxygen molecules into the cylinder. To understand this, it might help to consider two similar aircraft, one at high altitude and one at low altitude, each at 21″ MAP. The pressure in each manifold is the same but the air in the high aircraft's manifold is colder and thus denser.

Engine Operation

A number of precautions must be observed by the pilot in the operation of any reciprocating engine. The larger and more powerful the engine is, the more important the precautions become and the more expensive are the repairs if the engine is abused.

The most important rule for the pilot to follow is: **keep all operations within the limitations established by the manufacturer.** Monitoring of oil pressure and temperature, cylinder head and exhaust gas temperature, RPM, MAP and fuel flow must be frequent in order to always insure operation within the limitations.

The following rules generally apply when operating most general aviation reciprocating engines:

1. Make throttle, propeller RPM and mixture control changes slowly and smoothly. Quick movements of the engine controls impose large stresses on all moving parts in an engine and are usually indications of an inexperienced pilot who has not yet learned to anticipate the need for power setting changes in time to make those changes slowly and smoothly. Try to make the changes so that no one else in the airplane knows a change was made!

2. In order to avoid momentary high cylinder pressures and internal stresses, keep RPM high during power changes. In other words, when reducing power, reduce MAP first. When increasing power, increase RPM first. The power management quadrant in most airplanes is arranged so that the throttle is on the left, RPM control is in the middle and mixture control is on the right. A good habit to get into is to work from right to left across the quadrant when

CRUISE PERFORMANCE
PRESSURE ALTITUDE 6000 FEET

CONDITIONS:
2650 Pounds
Recommended Lean Mixture
Cowl Flaps Closed

NOTE:
For best fuel economy, operate at the leanest mixture that results in smooth engine operation or at peak EGT if an EGT indicator is installed.

RPM	MP	20 °C BELOW STANDARD TEMP −17 °C			STANDARD TEMPERATURE 3 °C			20 °C ABOVE STANDARD TEMP 23 °C		
		% BHP	KTAS	GPH	% BHP	KTAS	GPH	% BHP	KTAS	GPH
2500	23	- - -	- - -	- - -	75	136	10.0	72	136	9.6
	22	73	132	9.7	70	132	9.4	68	132	9.1
	21	68	128	9.1	66	128	8.8	63	128	8.6
	20	63	123	8.6	61	123	8.3	59	123	8.1
2400	24	- - -	- - -	- - -	77	137	10.2	74	138	9.9
	23	75	133	10.0	72	134	9.6	70	134	9.3
	22	70	130	9.4	68	130	9.1	66	130	8.8
	21	66	126	8.8	63	126	8.6	61	125	8.3
2300	24	77	134	10.2	74	135	9.8	71	136	9.5
	23	72	131	9.6	70	132	9.3	67	132	9.0
	22	68	127	9.1	65	128	8.8	63	127	8.5
	21	63	123	8.5	61	123	8.3	59	123	8.0
2200	24	74	132	9.9	71	133	9.5	69	133	9.2
	23	70	129	9.3	67	129	9.0	65	129	8.7
	22	65	15	8.8	63	125	8.5	61	125	8.2
	21	61	121	8.3	59	120	8.0	57	120	7.8
2100	23	67	126	8.9	64	126	8.7	62	126	8.4
	22	62	122	8.5	60	122	8.2	58	122	7.9
	21	58	118	8.0	56	117	7.7	54	117	7.5
	20	54	113	7.5	52	112	7.3	50	110	7.0
	19	50	108	7.0	48	106	6.8	46	103	6.6

PRESSURE ALTITUDE 10,000 FEET

CONDITIONS:
2650 Pounds
Recommended Lean Mixture
Cowl Flaps Closed

NOTE:
For best fuel economy, operate at the leanest mixture that results in smooth engine operation or at peak EGT if an EGT indicator is installed.

RPM	MP	20 °C BELOW STANDARD TEMP −25 °C			STANDARD TEMPERATURE −5 °C			20 °C ABOVE STANDARD TEMP 15 °C		
		% BHP	KTAS	GPH	% BHP	KTAS	GPH	% BHP	KTAS	GPH
2700	20	72	136	9.7	70	136	9.3	67	136	9.0
	19	67	131	9.0	65	131	8.7	62	130	8.4
2600	20	70	134	9.4	68	134	9.0	65	133	8.8
	19	65	129	8.8	63	128	8.5	61	128	8.2
	18	60	123	8.2	58	123	7.9	56	121	7.7
2500	20	68	132	9.1	66	132	8.8	63	131	8.5
	19	63	127	8.5	61	126	8.3	59	125	8.0
	18	58	121	8.0	56	120	7.7	54	119	7.5
	17	54	115	7.4	52	113	7.2	50	110	7.0
2400	20	66	130	8.9	63	129	8.6	61	129	8.3
	19	61	124	8.3	59	124	8.0	57	123	7.8
	18	56	119	7.7	54	118	7.5	52	115	
2300	20	64	127	8.6	61	127	8.3	59	126	8.0
	19	59	122	8.0	57	121	7.8	55	119	7.5
	18	54	116	7.5	52	114	7.3	51	112	7.1
	17	50	109	7.0	48	106	6.8	46	103	6.6
2200	20	61	125	8.3	59	124	8.0	57	123	7.8
	19	57	119	7.8	55	118	7.5	53	116	7.3
	18	52	113	7.3	50	111	7.0	49	108	6.9
2100	20	59	122	8.0	57	121	7.8	55	119	7.5
	19	55	116	7.5	52	115	7.3	51	112	7.1
	18	50	110	7.0	48	107	6.8	47	104	6.6

Figure 7-5. Cruise performance chart for 6000 and 10,000 ft for a high-performance general aviation airplane.

Figure 7-6. A Cessna 180 is wearing its "winter kit" which includes airflow limiting devices for engine air intake, engine cooling and oil cooler.

Figure 7-7. Engine blankets conserve engine warmth during cold weather operations in the north country.

increasing power and left to right when decreasing power.

3. Always be mindful of thermal shock. Reducing power suddenly from high power settings and cylinder head temperatures rapidly cools the exhaust ports, setting up stresses in the differentially cooling metal of the cylinder heads, resulting in stress cracking of those parts, often in the exhaust port area. On let down (descent), reduce power by increments to allow the engine to cool slowly. Proper cowl flap management is important to prevent rapid engine cooling. When operating in very cold temperatures, use of "winter kits" to decrease airflow into and over the engine is important to engine operation (see figure 7-6) as is proper use of carburetor heat to reduce cold air flow through the engine.

4. During a prolonged power-off glide, such as is done when practicing forced landings from altitude, "clear" the engine periodically by advancing the throttle to a medium power position for a few seconds until the engine runs smoothly, then reduce power. This helps keep the bottom spark plugs from fouling, warms the engine and exhaust so that carburetor heat is available, thus preventing carburetor ice.

5. Use full rich mixture setting when operating near or at full power. Also, it is a good idea to move

the throttle fully forward as some automatic enrichment systems only work when the throttle is fully open. At high power settings, the engine needs the added fuel for internal cooling.

6. Be mindful of the possibility of carburetor or induction system icing. For a full discussion of this topic, see chapter 6.

7. Before starting a cold engine, it should be preheated and the propeller moved through several "blades" to distribute oil to the wear-surfaces. See figure 7-7. Some pilots like to prime the engine before or while pulling the propeller through. Properly done, the engine will start "on the first blade" which reduces starter and battery wear. The above procedure requires that some precautions be taken. There should be a competent pilot at the controls during this procedure, or at least the aircraft tail should be secured, mags checked "off" and mixture checked in idle-cutoff position. The propeller should be swung as if the engine will start because if the P-leads to the magnetos are discontinuous, the mags will be "hot", even with the mag switches off, so the engine could start. If it does start, it will run only for a moment, until the primed fuel is used up if the mixture control is in the idle-cutoff position.

Figure 7-8. When hand-starting by "propping," proper techniques must be used.

"Propping" To Start An Engine

There are two schools-of-thought on this subject. The more modern one seems to be: don't do it — its too dangerous. The "old school" thinking is that every pilot should learn and know how to prop-start an engine properly. Properly is the key word here. There is a proper procedure which, if not followed, does indeed make the procedure dangerous. Have a qualified flight instructor teach you the proper procedure.

Most fuel injected engines can't be prop-started unless there is enough battery power available to prime the engine and provide spark-boost. Some fuel injected engines used in remote areas are fitted with hand primers and impulse coupling magnetos so they can be hand started. Three bladed propeller equipped engines aren't propped because the blades are closer together so there is less time between blades to get hands out of the blade path.

Stopping Procedure

The engine should be allowed to cool from normal operating temperatures before stopping. Usually, taxi time takes care of this requirement (except for helicopters and some turbine equipped aircraft). A **P-lead continuity check** should be done before shutdown while the engine is idling by moving both mag switches to the off position then quickly on. The cylinders should stop firing during the brief moment the switches are off. If the engine continues to run, at least one P-lead is discontinuous. The aircraft should not be left unattended on the ramp until the prop is labelled "DANGER-MAG SWITCHES ON".

Following a successful P-lead continuity check, the engine should be stopped by moving the mixture control to the idle cutoff position. If temperatures are very cold (below +15 °F) it is a good idea to pull carburetor heat on before stopping the engine to avoid some of the tremendous temperature change that takes place in the cylinder and exhaust system when "the fire goes out" and the engine is still turning over.

In certain situations, seaplane pilots stop the engine with the magneto switches instead of the mixture control. This procedure provides "instant-off" — remember, they have no brakes so the engine must be stopped at precisely the right instant when executing a difficult approach to a dock. Also, stopping the engine with the mag switches doesn't interrupt the flow of fuel to the cylinders, so the engine is primed and ready to start "on the first blade" (even hot fuel-injected engines) in case the approach to the dock is missed and there is an airplane at a dock a short distance downwind!

Why isn't this procedure used by landplane pilots? Because it is considered safer if the engine is stopped without fuel in the next cylinder to fire, in case the propeller is moved. Also, taxiing with rich mixture may cause carbon particles to adhere to the walls of the cylinder head. These "glowers" can ignite the fuel:air mixture after the mags are shut off, causing **dieseling** (engine running without ignition, either forward or backward). Note the discussion about "parking the prop" near the end of chapter 5.

Study Questions And Problems

1. A pilot uses an altimeter setting from a nearby weather station that is located within a mile of the airplane but is situated nearly 600 feet higher than the field elevation. Explain why no error is introduced to the calculations by the elevation difference when computing ambient pressure in order to check the MAP gauge before engine start.

2. Altimeter setting at Flagstaff, Arizona, field elevation 7000 feet, is 30.12″ Hg. What should MAP be, before engine start and during takeoff with fully opened throttle?

3. At Death Valley airport, 211 feet below sea level, what would the MAP gauge read with engine off with an altimeter setting of 30.42″ Hg?

4. What is the difference between detonation and preignition?

5. What is critical altitude?

6. Why does MAP increase when RPM is decreased, when there is no change of throttle position?

7. Let's imagine for a moment that the aircraft is flying in clouds and its static system is plugged with ice, so the altimeter is not working properly. The pilot increases prop RPM to takeoff RPM and opens the throttle fully. The MAP gauge reads 18″ Hg. What is the aircraft's approximate altitude?

8. What MAP is required to produce 55% power at 2300 RPM at 7500 feet in the engine of figure 7-4? In the engine of figure 7-5? (standard temperature).

9. What power setting (combination of RPM and manifold pressure) will allow the highest possible 75% power critical altitude for the engine of figure 7-1?

10. What is your best estimate of what the critical altitude of Question #9 will be?

Chapter VIII
Supercharging And Turbocharging

Supercharging and turbocharging are two methods used to provide the engine with higher manifold pressures than can be achieved with normal aspiration. Superchargers and turbochargers compress the fuel:air mixture after it leaves the carburetor or the induction air before it is mixed with the fuel. **Superchargers** are driven directly from the engine, using some of the power produced by the engine. **Turbochargers** are driven by the hot, high velocity exhaust gasses that are being expelled from the engine. The term **turbo-supercharger** is most correctly applied to an engine system that has both a turbocharger that boosts air pressure before the carburetor and a supercharger that further boosts pressure of the fuel:air mixture. These latter systems are quite rare within the general aviation fleet.

A **normally or naturally aspirated engine's** cylinders are filled with air due to the force of atmospheric pressure, which becomes less as altitude increases. Volumetric efficiency is not very good for these engines. The "boosted" engine receives air that is pushed into the cylinder under pressure, resulting in higher volumetric efficiencies and the ability to produce power at higher altitudes.

Flying High

Flying at high altitudes has advantages, including:

1. Fly above icing conditions, turbulence and high terrain.

2. Utilize beneficial winds aloft.

3. Enjoy a lower traffic density at higher altitudes.

4. Better speed. The aircraft flies faster at altitude given the same power setting.

Can you think of other reasons to fly high?

As seen in figure 7-3, engine power in a normally aspirated engine decreases as atmospheric pressure decreases. These engines produce maximum power at sea level and power decreases with increased altitude. Maximum airspeed is highest at sea level and decreases with increased altitude due to similar changes in available power.

The 65% power curve of figure 7-3, however, tells a different story. As the normally aspirated engine climbs to altitude with power maintained at 65%, true airspeed *increases* but only until the altitude

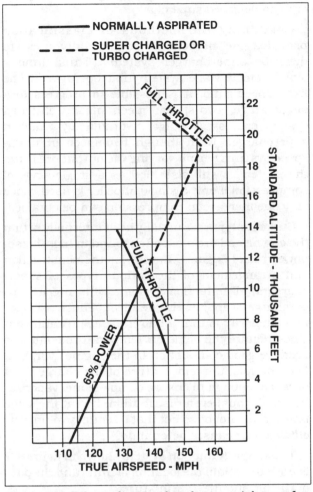

Figure 8-1. Effect of supercharging on cruise performance at high altitude.

is reached where 65% power can no longer be maintained.

If some additional air could be "pumped" into this engine's manifold, allowing 65% power to be maintained to higher altitudes, the airplane would fly faster as it climbed higher, according to the extension of the 65% power line seen in figure 8-1, because drag is reduced in the thinner air of high altitude.

Superchargers

Superchargers usually compress the fuel/air mixture after it leaves the carburetor, while turbochargers usually compress the air before it is mixed with the metered fuel from the carburetor or fuel injection system. Each increase in the pressure of the air or fuel/air mixture in an induction system is called a **stage**. Superchargers can be classified as single-stage, two-stage, or multi-stage, depending on the number of times compression occurs. Superchargers may also operate at different speeds. Thus, they can be referred to as single-speed, two-speed, or variable-speed superchargers.

Combining the methods of classification provides the nomenclature normally used to describe supercharger systems. Thus, from a simple single-stage system that operates at one fixed speed ratio, it is possible to progress to a single-stage, two-speed, mechanically clutched system or a single-stage, hydraulically clutched supercharger. Even though two-speed or multi-speed systems permit varying the output pressure, the system is still classified as a single-stage of compression if only a single impeller is used, since only one increase in compression can be obtained.

Superchargers are usually built internally within the engine and are most often found in high-horse-power radial engines. Except for the construction and arrangement of the various types of superchargers, all induction systems with internal superchargers are nearly identical. The reason for this similarity is that all aircraft reciprocating engines require the same air temperature control to produce good combustion in the engine cylinders. For example, the temperature of the charge must be warm enough to ensure complete fuel vaporization and, thus, even distribution; but at the same time it must not be so hot that it reduces volumetric efficiency or causes detonation.

The simple induction system shown in figure 8-2 is used to explain the location of units and the path of the air and fuel/air mixture.

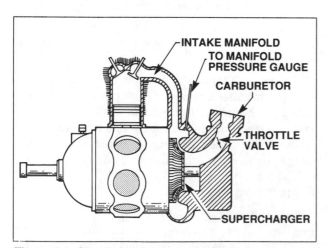

Figure 8-2. Simple induction system.

Air enters the system through the ram air intake. The intake opening is located so that the air is forced into the induction system, giving a slight ram effect. The air passes through ducts to the carburetor. The carburetor meters the fuel in proportion to the air and mixes the air with the correct amount of fuel. MAP is controlled by the throttle to regulate the flow of air. In this way, the power output of the engine is controlled.

Impeller Speeds

The gear ratio of the impeller gear train varies from approximately 6:1 to 12:1. Impeller speed on an engine equipped with a 10:1 impeller gear ratio operating at 2,600 RPM would be 26,000 RPM. This requires that the impeller unit be a high-grade forging, usually of aluminum alloy, carefully designed and constructed. Because of the high ratio of all supercharger gear trains, considerable acceleration and deceleration forces are created when the engine speed is increased or decreased rapidly. This necessitates that the impeller be splined on the shaft.

Single-Stage, Two-Speed Supercharger Systems

Some aircraft engines are equipped with internally driven superchargers which are single-stage, two-speed systems. The impeller in such systems can be driven at two different speeds by means of clutches within the engine.

This type of unit is equipped with a means of driving the impeller directly from the crankshaft at a ratio of 10:1, which is accomplished by moving the control in the cockpit, thereby applying oil pressure through the high-speed clutch and thus locking the entire intermediate gear assembly. This is called "high blower" and is used above a specified altitude ranging from 7,000 to 12,000 ft. Below these levels, the control is positioned to release the pressure on the high-speed clutch and apply it to the low-speed clutch.

In effect, this gives two engines in one. It improves the power output characteristics over a range of operating conditions varying from sea level to approximately 20,000 ft. Naturally, a device of this kind complicates and increases considerably the initial and maintenance costs of the engine. A higher grade fuel is also required to withstand the additional pressures and, in some cases, higher temperatures created within the combustion chamber due to more complete fuel charging of the cylinder. The addition of this unit also complicates the operation of the powerplant because it requires more attention and adds to the variables which must be controlled.

Superchargers On Horizontally Opposed Engines

Superchargers have been used on high horsepower horizontally opposed engines such as the Lycoming GSO-480 series. Figure 8-3 shows an exploded view of the parts of such an installation.

Since the power required to drive the supercharger is taken from the crankshaft, the net gain in horsepower obtained by supercharging is reduced. The amount of supercharging that can be done is restricted by the temperatures produced (compressing a gas increases its temperature) to avoid the problems of preignition and detonation. These problems are at least partially solved by turbocharging.

Turbocharging

Turbochargers are designed to deliver compressed air to the inlet of the carburetor or fuel/air control unit of an engine. Turbochargers derive their power from the energy of engine exhaust gases directed against some form of turbine.

The typical turbocharger is composed of three main parts:

(1) The compressor assembly.

(2) The exhaust gas turbine assembly.

(3) The pump and bearing casing.

These major sections are shown in figure 8-4. In addition to the major assemblies, there is a baffle between the compressor casing and the exhaust-gas turbine that directs cooling air to the pump and bearing casing, and also shields the compressor from the heat radiated by the turbine. In installations where cooling air is limited, the baffle is replaced by a regular cooling shroud that receives its air directly from the induction system.

The compressor assembly (A of figure 8-4) is made up of an impeller, a diffuser, and a casing. The air for the induction system enters through a circular opening in the center of the compressor casing, where it is picked up by the blades of the impeller, which gives it high velocity as it travels outward toward the diffuser. The diffuser vanes direct the airflow as it leaves the impeller and also converts the high velocity of the air to high pressure.

Motive power for the impeller is furnished through the impeller's attachment to the turbine wheel shaft of the exhaust-gas turbine. This complete assembly is referred to as the **rotor**. The rotor revolves on the ball bearings at the rear end of the pump and bearing casing and the roller bearing at the turbine end. The roller bearing carries the radial (centrifugal) load of the rotor, and the ball bearing supports the rotor at the impeller end and bears the entire thrust (axial) load and part of the radial load.

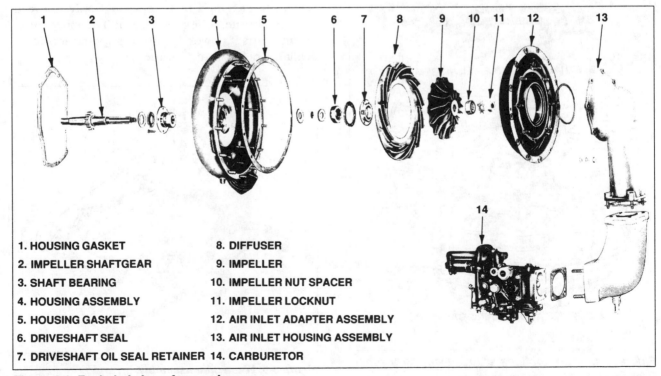

1. HOUSING GASKET	8. DIFFUSER
2. IMPELLER SHAFTGEAR	9. IMPELLER
3. SHAFT BEARING	10. IMPELLER NUT SPACER
4. HOUSING ASSEMBLY	11. IMPELLER LOCKNUT
5. HOUSING GASKET	12. AIR INLET ADAPTER ASSEMBLY
6. DRIVESHAFT SEAL	13. AIR INLET HOUSING ASSEMBLY
7. DRIVESHAFT OIL SEAL RETAINER	14. CARBURETOR

Figure 8-3. Exploded view of supercharger.

The exhaust gas turbine assembly (B of figure 8-4) consists of the turbine wheel (bucket wheel), nozzle box, butterfly valve (wastegate), and cooling cap. The turbine wheel, driven by exhaust gases, drives the compressor impeller. The **nozzle box** collects and directs the exhaust gases onto the turbine wheel, and the **wastegate** regulates the amount of exhaust gases directed to the turbine by the nozzle box. The **cooling cap** controls a flow of air for turbine cooling.

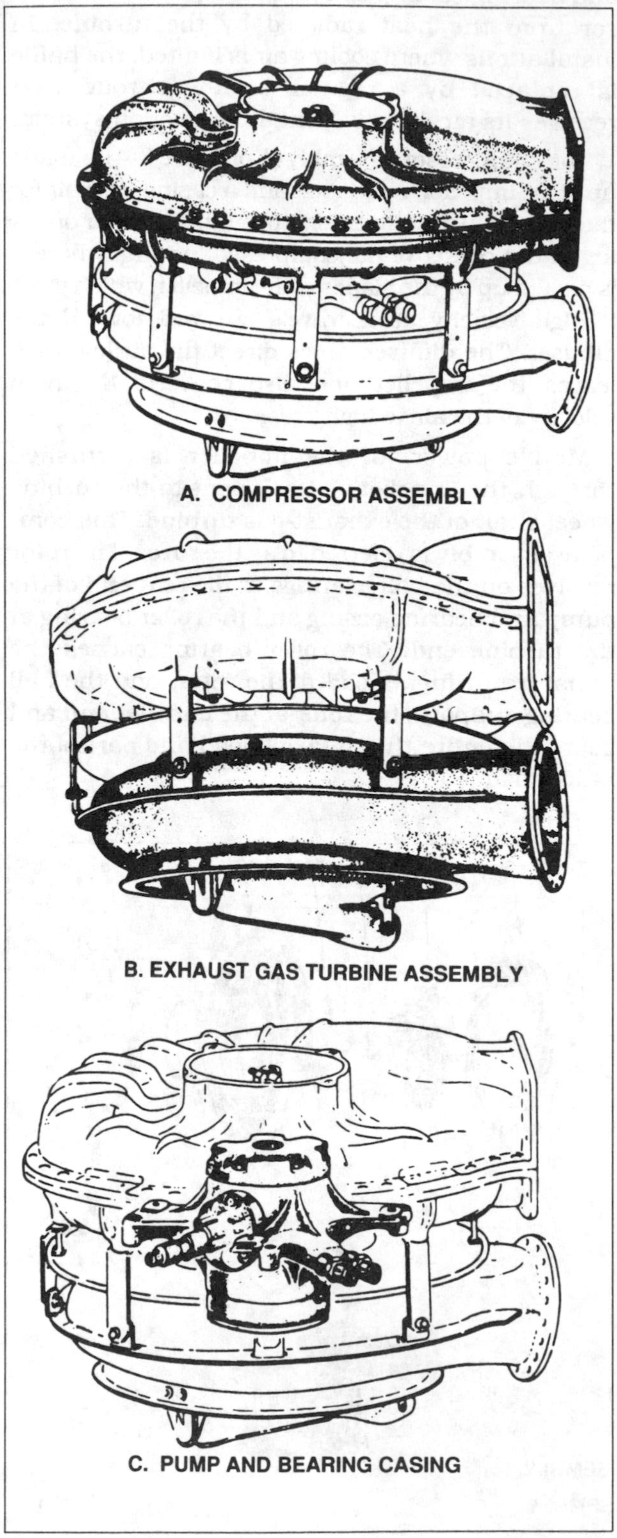

The **wastegate** (figure 8-5) controls the volume of the exhaust gas that is directed onto the turbine and thereby regulates the speed of the rotor (turbine and impeller) and the output of the compressor impeller.

If the wastegate is completely closed, all the exhaust gases are "backed up" and forced through the nozzle box and turbine wheel. If the wastegate is partially closed, only part of the exhaust gas is directed to the turbine. The nozzles of the nozzle box allow the gases to expand and reach high velocity before they contact the turbine wheel. The exhaust gases, thus directed, strike the cup-like buckets, arranged radially around the outer edge of the turbine, and cause the rotor (turbine and impeller) to rotate. The gases are then exhausted overboard through the spaces between the buckets. When the wastegate is fully open, nearly all of the exhaust gases pass overboard through the tailpipe and do not drive the turbine wheel.

An increasing number of engines used in light aircraft are equipped with turbocharger systems. On some small aircraft engines, the turbocharger system is designed to be operated only above a certain altitude; for example, 5,000 ft., since maximum cruise power without supercharging is available below that altitude.

A. COMPRESSOR ASSEMBLY

B. EXHAUST GAS TURBINE ASSEMBLY

C. PUMP AND BEARING CASING

Figure 8-4. Main sections of a typical turbosupercharger.

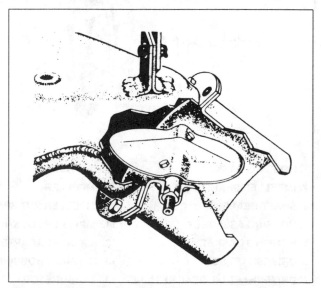

Figure 8-5. Wase gate assembly.

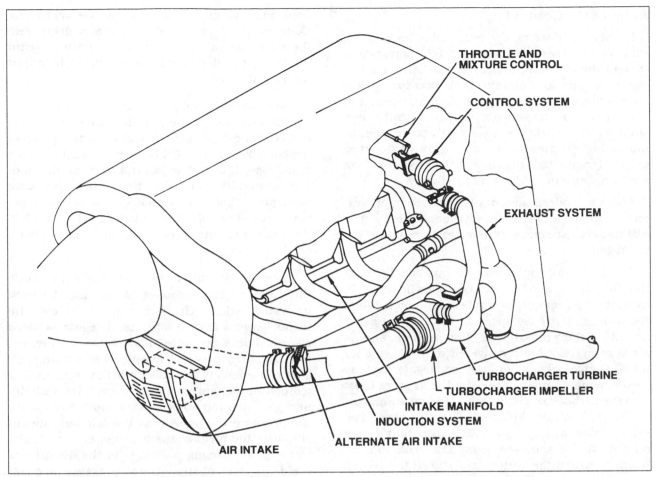

Figure 8-6. Turbocharger induction and exhaust systems.

Induction Air System

The location of the air induction and exhaust systems of a typical turbocharger system for a small aircraft is shown in figure 8-6.

The filtered ram-air intake admits filtered air which permits compressor suction to automatically admit alternate air (heated engine compartment air) if the induction filter becomes clogged. On some aircraft, the alternate air door can be operated manually in the event of filter clogging.

Wastegate control is either manual (pilot controlled) or automatic.

Manual Wastegate Control

Manual control is accomplished usually with a vernier control in the cockpit connected by cable directly to the wastegate. Manual control systems are often designed for retrofit installations. Since these engines are not designed for MAP's above 30″ Hg (one atmosphere), takeoff and climb to about 5000 feet is accomplished with the wastegate control set in the open position so that no boost of MAP takes place. Arriving at 5000 feet, for example, with throttle open and climb RPM set, the pilot will observe the MAP dropping about one inch for each 1000 feet of altitude gain. With throttle open, the pilot begins to close the wastegate in order to maintain climb MAP setting. As altitude increases, the wastegate must be closed more to achieve more boosting to maintain climb MAP. This process continues until the wastegate is completely closed. When the wastegate is closed, **critical altitude** for this MAP setting has been reached. Further climb will result in a reduction of MAP as the turbocharger isn't able to produce enough boost differential pressure from the thin air above critical altitude.

For descent, the process is reversed. The pilot must gradually open the wastegate as the aircraft descends, in order to keep MAP from increasing above the desired setting during descent. When the wastegate is fully open, throttle is used for further MAP control.

It is apparent that this type of system requires a high pilot workload and nearly constant monitoring during climbs and descents.

Automatic Control

Numerous control systems have been used, from a relatively simple but not totally satisfactory system that uses a pressure controller to control the wastegate to maintain **deck pressure** (pressure between the discharge of the compressor and the engine throttle valve) at some constant value, such as 28-30" Hg for a **normalized engine**, or much higher pressures for a **ground boosted engine** (engine that requires MAP in excess of sea level pressure to produce its rated power).

Today's modern automatic turbocharger systems usually utilize a **density controller** and a **differential pressure controller** to regulate the wastegate.

By regulating the wastegate position between the "fully open" and "closed" positions (figure 8-7), a constant power output can be maintained. When the wastegate is fully open, all the exhaust gases are directed overboard to the atmosphere, and no air is compressed and delivered to the engine air inlet. Conversely, when the wastegate is fully closed, a maximum volume of exhaust gases flows into the turbocharger turbine, and maximum turbocharging is accomplished. Between these two extremes of wastegate position, constant power output can be achieved below the maximum altitude at which the system is designed to operate.

A **critical altitude** exists for every possible power setting, and if an attempt is made to fly the aircraft at or above this altitude without a corresponding decrease in the power setting, the wastegate will be automatically driven to the fully closed position in an effort to maintain a constant power output. Thus, the wastegate will be almost fully open at sea level and will continue to move toward the closed position as the aircraft climbs, in order to maintain the preselected manifold pressure setting.

When the wastegate is fully closed (leaving only a small clearance to prevent sticking) the manifold pressure will begin to drop if the aircraft continues to climb above the critical altitude. If a higher power setting cannot be selected, the turbocharger's critical altitude has been reached. Beyond this altitude, the power output will continue to decrease.

The position of the wastegate valve is controlled by oil pressure. Engine oil pressure acts on a piston in the wastegate controller (figure 8-7) which is connected by linkage to the wastegate valve. When oil pressure is increased on the piston, the wastegate valve moves toward the "closed"

position, and engine **deck pressure** increases. Conversely, when the oil pressure is decreased, the wastegate valve is moved by the spring toward the "open" position, and deck pressure and output power is decreased.

The position of the piston attached to the wastegate valve is dependent on bleed oil which controls the engine oil pressure applied to the top of the controller piston. Oil is returned to the engine crankcase through two control devices, the **density controller** and the **differential pressure controller**. These two controllers, acting independently, determine how much oil is bled back to the crankcase, and thus establish the oil pressure on the piston.

The **density controller** *limits* maximum MAP, thus preventing **overboost** by limiting the deck pressure while the aircraft is below the turbocharger's critical altitude. It regulates bleed oil only if deck pressure (density) becomes excessive. The pressure- and temperature-sensing bellows of the density controller react to pressure and temperature changes. The bellows, filled with dry nitrogen, maintains a constant density by allowing the pressure to increase as the temperature increases. Movement of the bellows repositions the bleed valve, causing a change in the quantity of bleed oil, which changes the oil pressure on top of the wastegate piston. **Overboost** (excessive MAP) is prevented by the density controller, but can also be prevented by a simpler device called the **overboost control valve** which is a spring activated pressure relief valve. The overboost control valve senses and reacts (opens) only to pressure, so does not directly protect the engine from excessively high power outputs due to maximum manifold pressure and very cold intake air temperatures. (See figure 8-7.)

The **differential pressure controller** *regulates* the position of the wastegate valve to maintain a pre-set pressure differential across the throttle (usually 2-3" Hg) so that if the pilot opens the throttle, there will be a response. One side of the diaphragm in the differential pressure controller senses air pressure upstream from the throttle **(deck pressure)**; the other side samples pressure on the cylinder side of the throttle valve **(MAP)** (see figure 8-7). At the "wide open" throttle position, when the density controller controls the wastegate, the pressure across the differential pressure controller diaphragm is at a minimum and the controller spring holds the bleed valve closed. At "part throttle" position, the air differential is

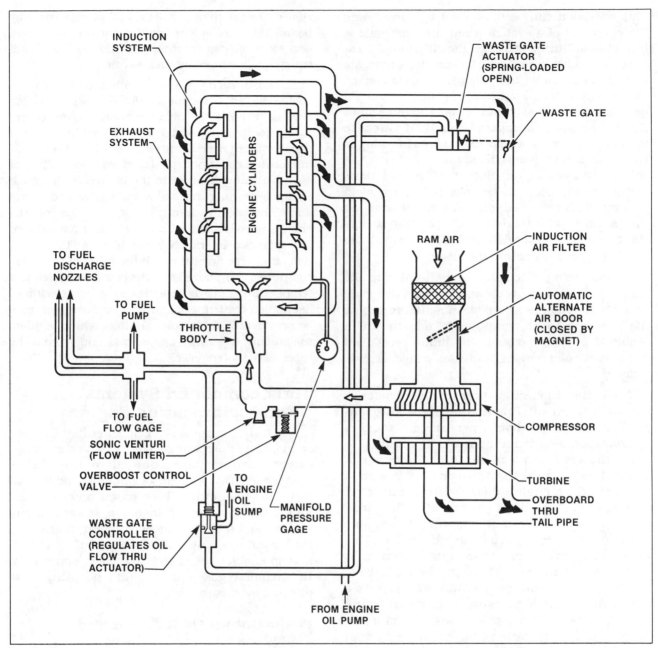

Figure 8-7. Typical turbocharger system for a general aviation aircraft. The turbocharger not only supplies the compressed air for the engine induction system, but it also provides air for cabin pressurization.

increased, opening the bleed valve to bleed oil to the engine crankcase and repositioning the wastegate piston.

Thus, the two controllers operate independently to control turbocharger operation at all positions of the throttle. Without the overriding function of the differential pressure controller during part-throttle operation, the density controller would position the wastegate valve for maximum power. The differential pressure controller, by maintaining a small pressure difference across the throttle, allows pilot control of MAP that is no different than the throttle

response of a normally aspirated engine, by continually repositioning the wastegate over the whole operating range of the engine. At all times, the density controller guards against overboosting the engine. Should overboost occur, the density controller can override the differential pressure controller and partially open the wastegate.

The differential pressure controller reduces the unstable condition known as "bootstrapping" during part-throttle operation. **Bootstrapping** is an indication of unregulated power change that results in the continual drift of manifold pressure.

This condition can be illustrated by considering the operation of a system when the wastegate is fully closed. During this time, the differential pressure controller is not modulating the wastegate valve position. Any slight change in power caused by a change in temperature or RPM fluctuation will be magnified and will result in MAP change since the slight change in temperature or RPM will cause a change in the amount of exhaust gas flowing to the turbine. Any change in exhaust gas flow to the turbine will cause a change in MAP, and thus, power output and will be reflected in manifold pressure indications. **Bootstrapping**, then, is an undesirable cycle of turbocharging events causing the manifold pressure to drift in an attempt to reach a state of equilibrium.

Bootstrapping is sometimes confused with the condition known as **overboost**, but bootstrapping is not a condition which is detrimental to engine life. An **overboost** condition is one in which manifold pressure exceeds the limits prescribed for a particular engine and can cause serious damage.

Thus, the differential pressure controller is essential to smooth functioning of the automatically controlled turbocharger, since it reduces bootstrapping by reducing the time required to bring the system into equilibrium. There is still a great deal more throttle sensitivity with a turbocharged engine than with a naturally aspirated engine. Rapid movement of the throttle can cause a certain amount of manifold pressure drift in a turbocharged engine. This condition, less severe than bootstrapping, is called **overshoot**. Overshoot can be a source of concern to the pilot or operator who selects a particular manifold pressure setting only to find it has changed in a few seconds and must be reset. Since the automatic controls cannot respond rapidly enough to abrupt changes in throttle settings to eliminate the inertia of turbocharger speed changes, overshoot must be controlled by the operator. This can best be accomplished by slowly making changes in throttle setting, accompanied by a few seconds wait for the system to reach a new equilibrium. Such a procedure is effective with all turbocharged engines, regardless of the degree of throttle sensitivity.

The Intercooler

As air moves through the turbocharger compressor, its temperature is raised because of compression. If the hot air charge is not properly cooled, the temperature of the fuel:air mixture may be too high, causing *preignition* and/or *detonation* and exhaust gas temperature may be too high, causing valve burning and warping.

The air in turbo-equipped induction systems is cooled by an **intercooler** (figure 8-8), so called because it cools the air after compression (or between compression stages), thus making it more dense without decreasing its pressure and eliminating most of the problems described above. The hot air flows through tubes in the intercooler in much the same manner that water flows in the radiator of an automobile. Fresh outside air, separate from the charge, is collected and piped to the intercooler so that it flows over and cools the tubes. As the induction air charge flows through the tubes of the intercooler, heat is removed and the charge is cooled to a temperature that the engine can tolerate without detonation occurring. Control for the cooling air may be provided by intercooler shutters which regulate the amount of air that passes over and around the tubes of the intercooler.

Turbocompound Systems For Reciprocating Engines

The turbocompound engine consists of a conventional, reciprocating engine in which exhaust-driven turbines are coupled to the engine crankshaft. This system of obtaining additional power is sometimes called a **power recovery turbine** (PRT) system. It is not a supercharging system, and it is not connected in any manner to the air induction system of the aircraft. The PRT system enables the engine to recover power from the exhaust gases that would be otherwise directed overboard.

Power Output Of Turbocharged Engines

Figure 8-9 compares cruise power settings of two engines of the same displacement and rated power. It is apparent that the turbocharged engine requires much higher MAP to produce similar power output. The turbocharged engine has a lower compression ratio because the compressor adds a first stage of compression. This is a main reason for the difference in MAP. This turbocharged engine is a ground boosted engine, permitting 38 to 43″ Hg MAP for takeoff, so it also has a larger, stronger crankshaft, special exhaust valves and guides to withstand the higher exhaust gas temperature (EGT) and oil jets in the crankcase that direct oil at the bottom side of the pistons for additional cooling.

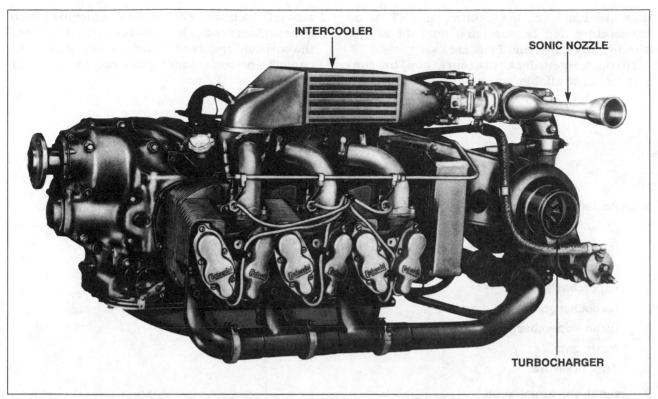

Figure 8-8. *Note that the high pressure air from the turbocharger compressor flows through an intercooler on its way to the cylinders. The decreased temperature of the induction air improves power output and decreases engine wear. Note the sonic venturi seen at the upper rear (right) of the engine. This device limits the amount of air that can be taken from the manifold to pressurize the cabin, thus insuring adequate induction air for the engine.*

Pilots should keep in mind that the turbocharger functions as an *amplifier*, increasing outside pressure by several times (typically 3:1). The power stroke in the cylinder is also an amplifier, increasing pressures many times. A small change of throttle or RPM setting produces large changes in the powerplant. Therefore, the pilot must make power changes slowly and smoothly to allow the modulating (controlling) systems to function without having to deal with "surprises." Also, turbocharged engines operate under maximum stress. Be alert to guard against mixture settings

POWER SETTING TABLE (CRUISE) - LYCOMING MODEL IO-540-C4B5, 250 HP ENGINE							
NORMAL CRUISE APPROX 210 HP		**INTERMEDIATE CRUISE APPROX 190 HP**		**ECONOMY CRUISE APPROX 175 HP**		**LONG RANGE CRUISE APPROX 140 HP**	
RPM	**MAP**	**RPM**	**MAP**	**RPM**	**MAP**	**RPM**	**MAP**
2400	26.0	2200	26.0	2200	24.0	2100	21.0
		2300	25.0	2300	23.2	2200	20.0
		2400	24.0	2400	22.4	2300	19.3

POWER SETTING TABLE (CRUISE) - LYCOMING MODEL TIO-540-C1A, 250 HP ENGINE							
TURBO CRUISE APPROX 232 HP		**INTERMEDIATE CRUISE APPROX 200 HP**		**ECONOMY CRUISE APPROX 173 HP**		**LONG RANGE CRUISE APROX 140 HP**	
2400	34.0	2300	31.0	2200	28.0	2100	25.0
		2400	30.0	2300	27.0	2200	24.0
		2500	29.0	2400	26.0	2300	23.0

Figure 8-9. Power setting table (cruise).

135

that are too lean, overheating of CHT or oil temperature and be sure that only the recommended lubricants and fuels are used.

Turbocharger deck pressure can be considered an available source of pressurized air, some of which can be drawn off for other uses such as deice boots and cabin pressurization providing the turbocharger is of sufficient size to produce enough pressurized air for the engine and other systems.

Study Questions And Problems

1. Define and describe the function of the following:

> ground boosted engine
>
> normally aspirated engine
>
> normalized engine
>
> supercharger
>
> turbocharger
>
> turbo-supercharger
>
> waste gate
>
> deck pressure

2. describe the function of:

> density controller
>
> differential pressure controller

3. In an automatic turbocharger, what provides the force to open the wastegate? To close it?

4. What is the purpose of an alternate air valve and how does it work?

5. What is critical altitude?

6. What is the purpose of an intercooler and how does it work?

7. Describe the operation of a manually controlled turbocharger.

8. What engine pressure is sensed in an automatic turbocharger control system to prevent overboost?

9. Describe how the differential pressure controller maintains the correct boost at partial throttle settings?

Chapter IX
Pressurization And High Altitude Operations

Chapter 8 listed some good reasons for flying above 10-12,000 feet. But when operations are carried out above those altitudes, some additional concerns must be dealt with. **Dysbarism** is the term which indicates *any physiological disorder brought about by changes in pressure.* Because pilots and passengers alike are affected, those in command of the aircraft must be knowledgeable about the many aspects of flight at high altitudes, including an understanding of how the atmosphere changes as altitude increases, how the human respiratory and circulatory systems function and are affected by altitude, how oxygen systems function, how aircraft pressurization systems work and what laws affect operations at higher altitudes.

The following pages contain much of what the pilot needs to know, but there is a lot to learn. As a study aid, I suggest the reader make notes in figure 9-2 so that, as a summary, it can be retained in the pilot's mind.

Altitude Physiology

Humans have a remarkable ability to adapt to their surroundings. The human body makes adjustments for changes in external temperature, acclimates to barometric pressure variations from one habitat to another, compensates for motion in space and postural changes in relation to gravity, resists toxic agents and diseases, and performs all these adjustments while meeting changing energy requirements for varying amounts of physical and mental activity. The human body does adjust to acute and chronic reductions in its oxygen supply by increasing respiration, chemical changes in the blood, and by increasing production of red blood cells; however, a complete absence of oxygen will cause death in approximately five to eight minutes.

In aviation, the demands upon the compensatory mechanisms of the body are numerous and of considerable magnitude. The environmental changes of greatest physiological significance involved in flight are: marked changes in barometric pressure, considerable variation in temperature, and movement at high speed in three dimensions.

The advances in aeronautical and mechanical engineering in the past decade have resulted in the development of highly versatile aircraft. Since we are essentially ground creatures, we must learn how to adjust to or protect ourselves from the low pressures and temperatures of flight, and the effects of acceleration on the body. Low visibility with its concomitant problems of disorientation, and problems related to the general physical and mental stress associated with flight, should also be considered. Humans cannot operate these machines at full capacity without physical aids, such as a supplemental supply of oxygen and pressurized cabins for use at altitudes starting as low as 5,000-10,000 feet.

We must overcome the handicaps imposed by nature on an organism designed for terrestrial life. In particular, the limiting factors in adjustment of the human body to flight must be appreciated. The extent to which these limiting factors are alleviated by available equipment must be understood clearly.

An effort is made in the following pages to outline some of the important factors regarding physiological effects of flight, and to describe the devices and procedures that will contribute to the safety and efficiency of those who fly.

The Atmosphere

Perhaps the primary problem of flight related to physiology has to do with the fact that the pressure of the gases in the atmosphere change during ascent and descent. Thus, it is essential that the pilot have an understanding of the gases found in the atmosphere and their effects upon the body. There are other factors such as temperature change which also need to be understood in order to protect ourselves from potential hazard.

Composition Of The Atmosphere

The atmosphere is a mixture of gases. It is composed primarily of nitrogen (N_2) and oxygen (O_2). The gases other than nitrogen and oxygen are so low in percentage that they are considered to be negligible from the standpoint of pressure changes and oxygen requirements. Therefore, pilots

TABLE OF U.S. STANDARD ATMOSPHERE					
FEET	IN. OF HG	MM OF HG	PSI	C°	F°
0	29.92	760.0	14.69	15.0	59.0
2,000	27.82	706.7	13.66	11.0	51.9
4,000	25.84	656.3	12.69	7.1	44.7
6,000	23.98	609.1	11.77	3.1	37.6
8,000	22.23	564.6	10.91	−0.8	30.5
10,000	20.58	522.7	10.10	−4.8	23.4
12,000	19.03	483.4	9.34	−8.8	16.2
14,000	17.58	446.5	8.63	−12.7	9.1
16,000	16.22	412.0	7.96	−16.7	1.9
18,000	14.95	379.7	7.34	−20.7	−5.1
20,000	13.76	349.5	6.76	−24.6	−12.3
22,000	12.65	321.3	6.21	−28.6	−19.4
24,000	11.61	294.9	5.70	−32.5	−26.5
26,000	10.64	270.3	5.22	−36.5	−33.6
28,000	9.74	237.4	4.78	−40.5	−40.7
30,000	8.90	226.1	4.37	−44.4	−47.8
32,000	8.12	206.3	3.99	−48.4	−54.9
34,000	7.40	188.0	3.63	−52.4	−62.0
36,000	6.73	171.0	3.30	−55.0	−69.7
38,000	6.12	155.5	3.00	−55.0	−69.7
40,000	5.56	141.2	2.73	−55.0	−69.7
42,000	5.05	128.3	2.48	−55.0	−69.7
44,000	4.59	116.6	2.25	−55.0	−69.7
46,000	4.17	105.9	2.05	−55.0	−69.7
48,000	3.79	96.3	1.86	−55.0	−69.7
50,000	3.44	87.4	1.70	−55.0	−69.7
55,000	2.71	68.8	1.33	TEMPERATURE REMAINS CONSTANT	
60,000	2.14	54.4	1.05		
63,000	1.95	46.9	.907		
64,000	1.76	44.7	.86		
70,000	1.32	33.5	PSF 113.2		
74,000	1.09	27.7	77.3		
80,000	.82	20.9	58.1		
84,000	.68	17.3	47.9		
90,000	.51	13.0	35.9	See Study Question #14	−56.6
94,000	.43	10.9	29.7		−50.0
100,000	.33	8.0	22.3		−40.1

Figure 9-1. The standard atmosphere.

usually refer to the percentages of nitrogen and oxygen as approximately 80% and 20% respectively, rather than their true percentages of 78.08% and 20.95%.

Nitrogen — 78.08%

This gas is responsible for the major portion of the total atmospheric pressure or weight. The gas itself is inert as far as the human body is concerned. Once the body becomes saturated, the same amount of nitrogen is exhaled as was inhaled. In other words, the body utilizes none of the nitrogen, although it is saturated with it.

Oxygen — 20.95%

This gas is essential for life. When the body is deprived of oxygen, death follows a short time later. Each time we breathe, about 21% of that breath is oxygen. In the lungs this gas is absorbed into the blood stream and is carried by the blood to all parts of the body. It is used to burn or oxidize food material for the production of heat, kinetic, electrical (nerve) energy, as well as the production of special chemical compounds that are concentrated ready-to-use energy forms.

A human body rapidly exposed to an altitude of 45,000 feet clothed in everyday street apparel could become unconscious in 9-12 seconds, may be reduced to a vegetable in a few minutes, and perhaps be dead a few minutes later. The dangerous element here is the reduced oxygen pressure found at this level. Since air is a mixture of gases, it behaves as such and is subject to those laws which govern gases.

Respiration And Circulation

When the human organism is exposed by aerial flight to various stresses, both physiological and psychological, all body functions are affected. However, the areas of the body which are affected most directly are the respiratory and circulatory systems. Therefore, it is important for the individual to be familiar with the actions and limitations of human respiratory and circulatory systems.

Respiration

The Concept Of Respiration

Respiration is defined as the *exchange of gases between the organism and its' environment*. The more obvious features of this process are the absorption of oxygen from the atmosphere and the elimination of some carbon dioxide from the body.

Cells in the body require oxygen for the burning of food material and the production of heat and energy. Carbon dioxide (CO_2) is produced from these reactions and excesses must be removed from the body. The lungs receive oxygen from the atmosphere which diffuses through the lungs into the blood. The blood at the same time releases carbon dioxide into the lungs and it is then expelled to the outside air. The oxygen absorbed by the blood is transported to nearly every cell in the body.

The respiratory system is made up of the lungs, a series of conducting tubes called bronchi, the windpipe or trachea, and the mouth and nose. Air first enters the nasal passages where it is warmed, moisturized, and filtered. It then passes down the throat through the windpipe into the bronchial tubes and into the lungs. Once inside the lungs, the large tubes branch into many thousands of smaller tubes. Located at the very end of each individual branch is an air sac. These air sacs (alveoli) are very small, the total number in the lungs is estimated to be three hundred million. Tiny blood vessels (capillaries) surround the thin moist walls of each air sac. Because of these thin walls (1/50,000th of an inch) gases can diffuse back and forth into and out of the blood which flows constantly through the capillaries.

The chest cavity is surrounded by the ribs on the sides and separated from the abdominal cavity by a large flat sheet of muscle (diaphragm). Since the chest is a closed cavity with only one opening to the outside, any change in total volume ventilates the airspaces in the lungs. The chest size is altered by muscular action which raises and lowers the diaphragm and by contraction and relaxation of muscles between the ribs. **Inspiration** is the active phase of lung ventilation. **Expiration** is a passive phase resulting from the relaxation of chest muscles and diaphragm. These two phases are reversed when crew are breathing from a pressure-demand O_2 system at altitudes above 35,000 feet.

Movement Of Gases In The Respiratory System

External respiration is the *exchange of gases between the lungs and the surrounding atmosphere*.

Dalton's Law states that *the total pressure of a mixture of gases is equal to the sum of the pressures of each gas in that mixture*. Each gas exerts its own pressure, depending on the percentage of that gas in the mixture. The pressure of each gas in the mixture is expressed as the **partial pressure** (p) of

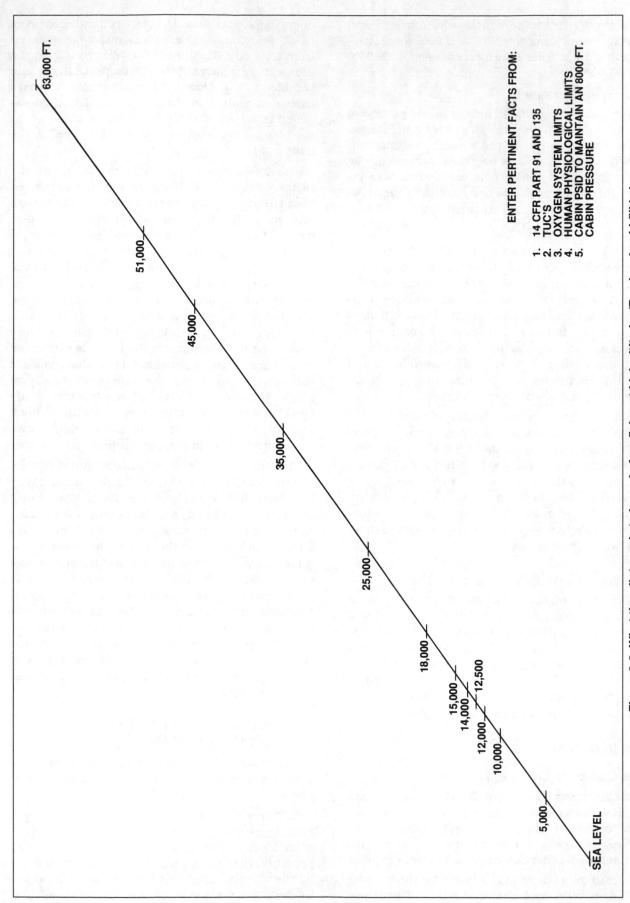

Figure 9-2. What the pilot needs to know before flying at high altitude. (Reader should fill in.)

SEA LEVEL
5,000
10,000
12,000
12,500
14,000
15,000
18,000
25,000
35,000
45,000
51,000
63,000 FT.

ENTER PERTINENT FACTS FROM:

1. 14 CFR PART 91 AND 135
2. TUC'S
3. OXYGEN SYSTEM LIMITS
4. HUMAN PHYSIOLOGICAL LIMITS
5. CABIN PSID TO MAINTAIN AN 8000 FT.
    CABIN PRESSURE

that gas, i.e., pO_2 (for oxygen). The application of this law to aviation is that even though the percentage of oxygen in the atmosphere doesn't change with altitude, its pO_2 will decrease proportionately as atmospheric pressure decreases. This reduction of pO_2 by ascent to altitude explains "altitude (or mountain) sickness," a form of hypoxic hypoxia.

The pO_2 forces oxygen through the air sacs into the blood stream. The partial pressure of oxygen at ground level is about 21% of the total atmospheric pressure. When a breath of air is taken into the lungs, it would be expected that the partial pressure of oxygen in the lungs to be about 159 mm Hg. However, the lungs also contain other gases which exert a fairly constant pressure (water vapor 47 mm Hg, and carbon dioxide 40 mm Hg). Therefore, these gases reduce the partial pressure of the oxygen entering the air sacs to about 102 mm Hg at ground level. See figure 9-5.

Graham's Law states that *a gas of high pressure exerts a force toward a region of lower pressure, and that if an existing membrane separating these regions of unequal pressure, is permeable or semipermeable, the gas of higher pressure will pass (or diffuse) through the membrane into the region of low pressure.* This process, which occurs in milliseconds, continues until the unequal regions are nearly equal in pressure. This law explains the transfer of oxygen, carbon dioxide and other gases from one part of the body to another.

Due to the function of Graham's Law, the high partial pressure of oxygen (102 mm Hg) now diffuses through the air sac wall and enters the blood stream. This will raise the partial pressure of oxygen in the blood from approximately 40 mm Hg to approximately 100 mm Hg. At the same time this is happening, the high pressure of carbon dioxide (47 mm Hg CO_2) in the blood will cause some of the CO_2 to diffuse into the air sac where the CO_2 pressure is about 40 mm Hg.

Internal respiration is the exchange of gases between the blood and the body cells.

The same principle that applies to external respiration is found in this phase of respiration except that the partial pressures of the gases involved are now reversed. High partial pressure oxygen diffuses from the circulatory system into the body tissues and excess partial pressure of carbon dioxide diffuses out of the tissue into the blood stream for return to the heart and lungs, some of which is to be exhaled from the body.

Circulation

The **circulatory system** is concerned with the transportation of blood throughout the body. Blood carries food and oxygen to the tissues and waste material, including carbon dioxide, from the tissues to the organs of excretion which include the lungs, kidneys, liver, sweat glands, and skin. Blood has additional functions of maintaining body heat and fluid balance.

Structure

The segments of the body which comprise the circulatory system are the heart, arteries, veins, and capillaries.

The **heart** is a pumping organ capable of forcing the blood through the vessels as tissue requirements dictate. The interior of the heart is divided into right and left halves and each half has two cavities.

Arteries are the vessels which normally carry oxygenated blood away from the heart. The elastic walls are muscular and strong, permitting the arteries to vary their carrying capacity. Small arteries connect large arteries to capillaries. The capillaries convey blood from the arteries to the veins. They are very small, thin walled, and usually form a network in the tissues in which the exchange of all gases between tissues and blood takes place.

Veins are the vessels which carry deoxygenated blood back to the heart. They have thinner walls and are less elastic than the corresponding sized arteries. When blood enters the veins from the capillaries it is under low pressure. Therefore, some method is necessary to get the blood back to the heart, especially from the lower regions of the body. The muscles around the veins produce a milking action of the veins forcing the blood toward the heart. Backflow of blood is prevented primarily by one-way valves located in the veins.

Control Of The Heart

Carbon dioxide has an important effect on the heart. An increased concentration of CO_2 results in acceleration of the heart. In exercise there is not only a strong heart beat but also a faster rate because of the increased amount of carbon dioxide and need for oxygen in the blood. A decrease in carbon dioxide tends to slow the heart rate until the body again reaches its normal CO_2 quantity. A deficiency of oxygen causes an increase in the heart rate and has a definite effect upon the heart's contraction. This rate increase will transport a greater volume of blood to the lungs to obtain oxygen thus alleviating tissue deficiency. The heart has influence on blood pressure in that the faster the rate and the greater the force of the heart beat, the higher will be the arterial pressure. Because the heart is a pump, any change in its action will affect the pressure of the fluid pumped.

Composition Of The Blood

Blood is made up of two parts, plasma and solids. Approximately 90% of the plasma is water, in which many substances are dissolved or suspended. The solid part of the blood is composed mainly of white and red blood cells. White blood cells are composed primarily of substances which act as antibodies to assist in fighting disease and infections in the body. The red blood cells are formed in the bone marrow and contain a colored substance called **hemoglobin**. Hemoglobin is an iron-containing compound. The iron in the hemoglobin molecule is responsible for the chemical affinity of hemoglobin for oxygen. About 95% of the oxygen is transported by hemoglobin and the remainder is in simple solution. It can readily be seen that a person who is anemic, or for any other reason does not have enough functioning red blood cells, will begin to suffer from lack of oxygen at relatively low altitudes. The blood of the average person contains about 15 grams of hemoglobin per 100 ml (milliliters). Each gram of hemoglobin is capable of combining with 1.34 ml of oxygen so the blood could contain 20 ml of oxygen per 100 ml of blood or 20 volumes percent if it were completely saturated. Normal arterial saturation is about 95-97 percent and the oxygen content is 19 volumes percent. The ability of hemoglobin to take up or release oxygen is not a linear function of the partial pressure of oxygen. However, the relationship is well defined and is shown in figure 9-3. Venous or return blood has a normal oxygen tension of 40 mm Hg, contains 14 volumes percent of oxygen, and is 65-75% saturated.

Hypoxia

Hypoxia is one of the most important physiological problems for those who fly at high altitudes. One factor which tends to make it dangerous is its insidious onset. Any aviator who flies above 12,000 feet in an unpressurized aircraft without supplemental oxygen is a potential hypoxia case. Successful handling of this matter calls for complete knowledge of the causes, effects, prevention, treatment, and personal symptoms of hypoxia. Only then will the pilot be able to fly safely at high altitude.

Hypoxia can be defined as *a lack of sufficient oxygen in the body cells or tissues.* It is caused by an inadequate supply of oxygen, inadequate transportation of oxygen, or inability of the body tissues to use oxygen. The forms of hypoxia are divided into four major groups.

Hypoxic (altitude) hypoxia is caused by *an insufficient partial pressure of oxygen in the inspired air.* This reduction of oxygen pressure becomes apparent in the physiological deficient zone of the atmosphere as discussed previously. There are also several pathological conditions which interfere with normal ventilation of the lungs and these conditions may cause hypoxic hypoxia, or predispose a person to an extreme case of hypoxia (i.e., strangulation, drowning, pneumonia, etc.).

Hypemic (anemic) hypoxia may be defined as *hypoxia caused by a reduction in the oxygen carrying capacity of the blood.* It is caused by a reduction in total circulating hemoglobin or red blood cells as in anemia or of dysfunctional oxyhemoglobin as found in carbon monoxide poisoning or excessive smoking.

Stagnant hypoxia is *an oxygen deficiency in the body due to poor circulation of the blood.* It is caused by a failure of the circulatory system to pump blood and adequate oxygen to the tissues. Failure of the circulatory system may be seen in cases of shock when the total circulating blood volume is

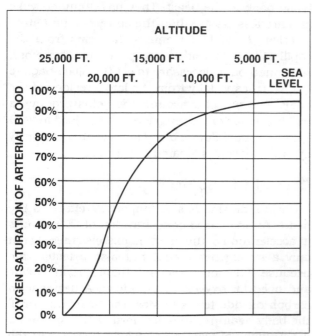

Figure 9-3. Oxygen saturation of arterial blood vs. altitude.

reduced. In flying, this type of hypoxia may be caused by positive pressure breathing for long periods of time or excessive head to foot type "G" forces.

Histotoxic hypoxia is *hypoxia caused by the inability of the body cells to utilize oxygen.* It is caused by a failure of the tissues to use oxygen efficiently because of impairment of cellular respiration. Alcohol, the type we buy at the bar, and certain other drugs are directly responsible for the disruption of cellular respiration.

Symptoms Of Hypoxia

All of the above types of hypoxia could occur while flying. The most important single hazardous characteristic of hypoxia is that if the (untrained) crewmember is hypoxic and engrossed in flight duties, the flyer may not notice the effects of hypoxia due to inadequate knowledge or training. Therefore, in order to detect hypoxia, each pilot must know their own symptoms. An altitude chamber flight gives an individual the opportunity to experience symptoms of hypoxia under controlled conditions. Since individuals differ so much in their reactions to hypoxia, it is impossible to group the symptoms in a specific order. Therefore, the order of appearance of symptoms on the list below is a random one.

1. An increased breathing rate, headache, fatigue.
2. Light-headed or dizzy sensations, listlessness.
3. Tingling or warm sensations, sweating.
4. Poor coordination, impairment of judgment.
5. Loss of vision or reduced vision, sleepiness.
6. **Cyanosis.** (Discoloration at the fingernail beds, lips and ear lobes).
7. Behavioral changes, feeling of well being, (euphoria).

Effective Performance Time (EPT) Or Time Of Useful Consciousness (TUC)

This is *the amount of time in which a person is able to effectively or adequately perform flight duties with an insufficient supply of oxygen.* At altitudes below 30,000 feet the TUCs may differ considerably from the time of total consciousness (the time it takes to pass out). Above 35,000 feet the times become shorter and eventually coincide for all practical purposes, with the time it takes for blood to circulate from the lungs to the head.

Average TUC for flying personnel without supplemental oxygen is shown in figure 9-4.

20,000 FEET	30 MINUTES OR MORE
22,000 FEET	5 TO 10 MINUTES
25,000 FEET	3 TO 5 MINUTES
28,000 FEET	2.5 TO 3 MINUTES
30,000 FEET	1 TO 2 MINUTES
35,000 FEET	30 TO 60 SECONDS
*40,000 FEET	15 TO 20 SECONDS
*45,000 FEET	9 TO 15 SECONDS

NOTE: BASED ON INTERRUPTION OF O_2 SUPPLEMENT AFTER BEING AT ALTITUDE RATHER THAN ASCENDING.

*TIMES WOULD BE 50% LESS THAN SHOWN TO BE REPRESENTATIVE OF A SUDDEN DECOMPRESSION TO THESE INDICATED ALTITUDES (REVERSE DIFFUSION).

Figure 9-4. Average TUC for flying personnel without supplemental oxygen.

There are a variety of factors which will determine TUC, some of which are:

1. Altitude. TUC decreases at higher altitudes.
2. Rate of Ascent. In general, the faster the rate, the shorter the TUC.
3. Physical Activity. Exercise decreases TUC considerably.
4. Day-to-Day Factors. Physical fitness, diet, rest, drugs, smoking, illness, and other factors may change one's ability to tolerate hypoxia from day to day, therefore, changing one's TUC.

Methods To Combat Hypoxia

Hypoxic hypoxia is avoided at altitude by pressurization of the cabin to 10,000 feet or below, or proper use of an oxygen system. If either fails, a realization or recognition of the pilot's hypoxia symptoms is required and should be followed by immediate use of supplemental oxygen to combat hypoxic hypoxia successfully. Crewmembers must check their oxygen system periodically to assure an adequate supply and proper function. Frequency of checks will depend upon the cabin altitude of the aircraft. In most cases, recovery is rapid and the individual usually regains full faculties within fifteen seconds after oxygen is administered. In aircraft where supplemental oxygen is not available, an emergency descent to altitudes of 12,000 feet or below generally will have the same effect as supplemental oxygen. Because of the need to recognize one's own symptoms of hypoxia as the only and final defense against the insidious onset of hypoxia due to an undetected failure of pressurization or oxygen system, it is strongly recommended that any pilot planning to fly above 10,000 feet obtain high altitude certification in the altitude chamber. In 1991, 14CFR (FAR) Part 61 was amended to require this training for pilots in command of aircraft capable of flight above 25,000 feet. For information on high altitude physiology training, contact:

FAA Airman Education Section (AAC-142)
Civil Aeromedical Institute
P.O. Box 25082
Oklahoma City, OK 73125
(405) 686-4837

Flight Physiology Group
Center for Aerospace Sciences
University of North Dakota
P.O. Box 8216
Grand Forks, ND 58202
(701) 777-2791

Oxygen Equipment

The development of oxygen equipment has necessarily paralleled the progress in aircraft performance. Without protection from the physiological problems at altitude, the human element becomes the limitation on how fast, how high, and how well an aircraft can perform. Oxygen equipment is just one area of development which has enabled flight in the hostile environment above 12,000 feet. The proper and effective use of oxygen equipment includes the methods of checking the oxygen equipment prior to and during flight.

Oxygen Storage

Aircraft operators who fly routinely either pressurized or unpressurized at altitudes in excess of 10,000 feet commonly employ portable or fixed oxygen installations. Fixed installations consist of containers affixed within the aircraft and serviced through an exterior fuselage valve.

Light aircraft operators who normally fly below 10,000 feet usually use portable O_2 equipment consisting of a container, regulator, mask outlet, pressure gauge, etc., as an integral unit which may be taken aboard the aircraft each time a flight is contemplated at high altitude. Portable equipment, in order to avoid weight and bulk problems, is limited in oxygen supply duration. Typical breathing time for four people at 18,000 feet is in the range of 1½ hours using a 22 cubic foot container. Fixed O_2 installations usually offer much longer duration times or service for more people or both. Actual times will depend upon size of oxygen containers in the system, and the number of people using the system. See figure 9-6.

Oxygen Storage Methods

Gaseous oxygen is stored in the containers at a pressure of 1800-2200 PSI. This is termed a high pressure system. The high pressure system is used very extensively in general aviation and commercial aviation.

Latest development in oxygen systems for aircraft make use of chemical action. **Solid state oxygen** systems employ chemical action to form O_2 and have come into use in new jumbo jet transports. They have weight, duration, and storage advantages not found in systems currently in use.

Regulators And Masks

Continuous Flow

The continuous flow O_2 regulator provides a flow of 100% oxygen. The rate of flow is usually measured in liters per minute and must increase as altitude increases. Flow rate may be automatic or manually controlled by turning a valve to alter the flow rate. Several regulators are offered which employ an altitude sensing aneroid to change the flow rate automatically.

Continuous flow masks utilize an oronasal face piece to receive the oxygen flow. The face piece does not usually have an air tight or oxygen tight face seal. This permits the user to exhale around the face piece or through small face piece ports or openings designed to dilute the oxygen with ambient air.

Continuous flow masks in use today make use of a *rebreather* bag. This bag is attached to the mask and enables the wearer to reuse a part of the exhaled oxygen. Usually, there is a device in the oxygen hose which enables the wearer to see that oxygen is flowing through the system.

The design of continuous flow systems limits the altitude at which the system can be used. These systems are normally limited to use up to 25,000 feet if the mask described above is used, or 18,000 feet if a nasal cannula is used. If a nasal cannula is used, there must be a standby face mask available for each cannula used in case the user experiences nasal congestion. FAR 23.1447 discusses restrictions to cannula use.

In order to accommodate personnel flying in the higher altitude ranges, several continuous flow masks are now being marketed that provide an air tight seal with exhalation valves which convert the rebreather bag into a reservoir bag. Very careful

144

attention to system capabilities is required in use of this type equipment above 25,000 feet even though it has been certificated to 41,000 feet.

Demand And Pressure Demand Systems

Figure 9-5 shows lung (alveolar) O_2 pressure and blood-hemoglobin percent O_2 saturation at altitude with and without each oxygen system. 90% saturation is considered minimum for adequate crew function. The demand regulator, as the name implies, operates to furnish oxygen only when the user inhales or *demands* it. A lever may also be employed to enable the regulator to automatically give either a mixture of cabin air and oxygen or 100% oxygen. This is referred to as the automix lever. The regulator is set up to give varying amounts of oxygen to the user depending upon the altitude attained.

The demand mask is designed to accommodate an air tight and oxygen tight seal to the face. This mask is expected to retain all of the oxygen thus inhaled into the mask by the user and not be diluted by entry of outside air. The demand regulator and mask provides a higher altitude capability than most continuous flow systems. It may be safely used to altitudes of 35,000 feet.

Pressure demand regulators are designed to furnish oxygen on inhalation either as a mixture of air and oxygen or 100% O_2. This regulator also provides a positive pressure application of oxygen to the mask face piece enabling the users lungs to be pressurized with oxygen. This is of great benefit at extreme altitudes (35,000-45,000 feet). The oxygen pressure flow may be either manually controlled or function automatically on some regulators at a certain altitude through aneroid action.

The pressure demand masks are designed to create an air tight and oxygen tight seal. The inhalation and exhalation valves are specially designed to permit oxygen pressure build up within the mask face piece and thus supply oxygen under pressure to the lungs.

It is essential that demand and pressure demand masks be properly suspended by an adequate head harness and that the masks be afforded tension adjustments in order for the user to obtain a leak proof seal to the face. The higher you fly the more critical this adjustment becomes.

Preflight Oxygen Equipment Check

Prior to flight, a person should locate the oxygen mask, practice donning it, and adjust head harness to fit; locate and check function of oxygen pressure gauges, flow indicators and connections; and check quantity of oxygen in the system. The mask should be donned and the O_2 system should be checked for function.

A physical check of the mask and tubing to spot any cracks, tears, or deterioration would also be indicated. If a person is using a mask connection to an individual regulator, check for regulator condition and lever or valve positions as required by that particular system.

General Rules For Oxygen Safety

Do not inspect oxygen equipment with greasy hands. Do not permit accumulation of oily waste or residue in the vicinity of the oxygen system. Avoid the use of chapstick, lipstick, vaseline and other combustibles while using O_2.

Do not use surplus oxygen equipment unless it is inspected by a certified FAA inspection station and approved for use.

Some military components use oxygen containers stressed for a pressure of 450 PSI (low pressure). Needless to say, a hazard exists if a person attempts to put 1800-2200 PSI O_2 pressure in this type container. High pressure O_2 containers should be marked to indicate *1800 PSI* before attempting to fill the container to this pressure. Most individuals do not possess the equipment necessary to fill an aircraft oxygen container from another source of high pressure oxygen. It is recommended that oxygen system servicing be done at FAA certified stations such as are located at some fixed base operations, terminal complexes, etc.

After any use of oxygen, careful attention should be given to ascertain that all flow is shut off before lighting cigarettes, etc.

Oxygen systems must be engineered to protect the individual to the maximum anticipated flight altitude of the aircraft. Before purchasing any oxygen equipment, it is recommended that the distributor be briefed on such factors as peak altitude to be flown, number of persons who will use the oxygen system, expected oxygen breathing duration, range of the aircraft, etc., so a proper oxygen system can be designed. Do not make any modification to the system without consulting the supplier or distributor. Oxygen system duration is shown in figure 9-6.

Do not place portable oxygen containers in the aircraft unless they are fastened securely to ensure against displacement in the event of turbulence, unusual aircraft attitudes, etc.

Figure 9-5. Oxygen requirements.

AIRCRAFT ALTITUDE	BAROM. PRESS. mm Hg	H2O PRESS. (BODY)	TRACH. PRESS. mm Hg	%O2 INSP. AIR	TRACH. PRESS. pO2	AVEOLAR pCO2 mmHg	AVEOLAR pO2 mmHg	%O2 SAT. Hb.	%SUPPLEM. O2 REQUIRE. INSP. AIR	TRACH. PRESS. pO2	%O2 SAT. Hb.	WITH CONSTANT FLOW O2 SYSTEM
Sea Level	760	47	713	.21	149	40	103	96%	(.21) 21%	149	96%	NTPD*
5,000 ft.	632	47	585	.21	122	38	78	94%	(.250) 25%	149	96%	.5 LPM
10,000 ft.	523	47	476	.21	100	36	61	90%	(.31) 31%	149	96%	1.0 LPM
15,000 ft.	429	47	382	.21	80	33	46	70%	(.40) 40%	149	96%	1.5 LPM
20,000 ft.	349	47	302	.21	63	30	33	62%	(.49) 49%	149	96%	2.0 LPM
25,000 ft.	282	47	235	.21	49				(.63) 63%	149	96%	2.5 LPM

Sat. Hb. = Saturation of hemoglobin

CONTINUOUS FLOW OXYGEN EQUIPMENT — TOTALLY INADEQUATE ABOVE THIS ALTITUDE

Rebreather Type Equip. Not Recomm. Above 25,000

WITH DILUTER DEMAND OXYGEN EQUIPMENT

AIRCRAFT ALTITUDE	BAROM. PRESS. mm Hg	H2O PRESS. (BODY)	TRACH. PRESS. mm Hg	%O2 INSP. AIR	TRACH. PRESS. pO2	AVEOLAR pCO2 mmHg	AVEOLAR pO2 mmHg	%O2 SAT. Hb.	%SUPPLEM. O2 REQUIRE. INSP. AIR	TRACH. PRESS. pO2	%O2 SAT. Hb.	WITH CONSTANT FLOW O2 SYSTEM
30,000 ft.	225	47	178	.21	37	40	103	96%	(.84) 84%	149	96%	3.0 LPM
35,000 ft.	179	47	132	.21	28	39	93	95%	(1.00) 100%	132	95%	3.5 LPM
40,000 ft.	141	47	94	.21	20	35	59	87%	(1.00) 100%	94	87%	4.0 LPM

4.1 LPM IS MAXIMUM FLOW MAXIMUM CERT. ALTITUDE OF 41,000 ft.

WITH PRESSURE DEMAND OXYGEN EQUIPMENT

AIRCRAFT ALTITUDE	BAROM. PRESS. mm Hg	H2O PRESS. (BODY)	TRACH. PRESS. mm Hg	%O2 INSP. AIR	TRACH. PRESS. pO2	AVEOLAR pCO2 mmHg	AVEOLAR pO2 mmHg	%O2 SAT. Hb.	%SUPPLEM. O2 REQUIRE. INSP. AIR	TRACH. PRESS. pO2	%O2 SAT. Hb.	WITH CONSTANT FLOW O2 SYSTEM
40,000 ft.	141	47	94	.21	20	Inadequate						
				(× 1.00)	94	35	59	87%	100%	94	87%	Must be a Phase Sequential Type Mask.
				(+ 8mm Hg)	102	36	66	92%	100% + Pres	102	92%	
42,000 ft.	128	47	81	.21	17							
				(× 1.00)	81	33	48	71%	100%	81	71%	
				(+ 16mm Hg)	97	36	61	90%	+ Pos. Pres	97	90%	
45,000 ft.	111	47	64	.21	13							
				(× 1.00)	64	30	34	62%	100%	64	62%	
				(+ 33mm Hg)	97	36	61	90%	+ Pos. Pres	97	90%	

(*NORMAL TEMPERATURE, PRESSURE, DRY 70 °F, PRESSURE = 760 mm Hg, P_{H_2O} = O)

CYL VOL CU FT	*NUMBER OF PEOPLE USING											
	1	2	3	4	5	6	7	8	9	10	11	12
22	151	75	50	37	30	25	21	18	16	15	13	12
49	336	167	111	83	66	55	47	41	37	33	30	27
66	448	222	148	111	89	74	63	55	49	44	40	37

*FOR DURATION TIME WITH CREW USING DILUTER DEMAND QUICK-DONNING OXYGEN MASK WITH SELECTOR ON 100% OR NORMAL MODE, INCREASE COMPUTATION OF "NUMBER OF PEOPLE USING" BY TWO PERSONS (E.G., WITH FOUR PASSENGERS AND A CREW OF TWO, ENTER THE TABLE AT "8")

Figure 9-6. Oxygen system duration ~ minutes.

Oxygen cylinders must be hydrostatically tested every five years or they cannot legally be refilled. The date of the last test is stamped in the metal of the bottle.

Hyperventilation

Respiratory controls of the body react to the amount of carbon dioxide (CO_2) found in the blood stream. In a physically relaxed state, the amount of carbon dioxide in the blood stimulates the respiratory control centers and breathing rate is stabilized at about 12 to 16 breaths per minute. When physical activity occurs, the body cells use more oxygen and more carbon dioxide is produced. Excessive carbon dioxide enters the blood and subsequently the respiratory center responds to this, and breathing increases in depth and rate to remove the oversupply of carbon dioxide. Once the excess CO_2 is removed, the respiratory center causes the breathing rate to change back to normal.

The same process is involved when a maximum effort is made to hold the breath. While the breath is being held, the body cells continue to manufacture carbon dioxide which enters the blood. The amount in the blood finally becomes so great that, in spite of conscious efforts, the respiratory center overrides it and breathing is resumed.

Hyperventilation Symptoms And Treatment

Hyperventilation, or over breathing, is a disturbance of respiration that may occur in individuals as a result of emotional tension or anxiety. Under conditions of emotional stress, fright, or pain, lung ventilation may increase, although the carbon dioxide output of the body cells remains at a resting level. As a result, carbon dioxide is washed out of the blood. This results in an excessive loss of carbon dioxide from the lungs, lowering the partial pressure of carbon dioxide below the normal 40 mm Hg. The most common symptoms are dizziness, hot and cold sensations, tingling of the hands, legs, and feet, tetany,

nausea, sleepiness, and finally unconsciousness. Unconsciousness is due to the respiratory centers overriding mechanism to regain control of breathing by reducing blood flow to the brain (Cerebral Stagnant Hypoxia).

After becoming unconscious, the breathing rate will be exceedingly low until enough carbon dioxide is produced to stimulate the respiratory center. Hyperventilation also occurs as a result of the body's normal compensatory response to hypoxia. However, excessive breathing does little good in overcoming hypoxia.

Should symptoms occur which cannot definitely be identified as either hypoxia or hyperventilation, the following steps should be taken:

1. Check oxygen equipment immediately and put the regulator auto-mix lever on 100% oxygen (demand or pressure demand system). Continuous flow system — check oxygen supply and flow mechanism.

2. After three or four breaths of oxygen, the symptoms should improve markedly, if the condition experienced was hypoxia. (Recovery from hypoxia is extremely rapid.)

3. If the symptoms persist, consciously slow the breathing rate until symptoms clear and then resume breathing at a normal rate. Breathing can be slowed by breathing into a bag, or talking aloud.

Several aircraft accidents have been traced to probable hyperventilation. It is recommended that you induce hyperventilation by voluntarily breathing several deep breaths at an accelerated rate. You will begin to get some of the symptoms mentioned previously. Once you experience several of these symptoms, return to your normal rate of breathing. After you become familiar with the early warnings your body gives you, the likelihood of an accident caused by hyperventilation will be reduced. ***Caution: Do not hyperventilate while alone or in a standing position. You may fall and injure yourself.***

147

Dysbarism

Dysbarism is a term that describes any physiological disorder caused by changes in pressure. In addition to disorders already discussed, increased or decreased pressure can cause undesirable effects on the body caused by gas expanding or evolving within the body which may be divided into two groups:

Trapped Gas

During ascent and descent, free gas expands or contracts in body cavities. A body cavity's inability to equalize with pressure changes may cause abdominal pain, toothache, or pain in ears and sinuses.

Evolved Gas

This condition is produced by the low atmospheric pressure of high altitude, primarily above 30,000 feet. However, such problems have occurred as low as 18,000 feet. Gases coming out of solution in the blood and other body tissues may be responsible for such conditions as bends, chokes, paresthesia, and central nervous system problems. These gases or bubbles consist mainly of nitrogen, with some oxygen, carbon dioxide, and water vapor.

The formation of these bubbles is explained by **Henry's Law**, which states that *the amount of gas in solution varies directly with the pressure of the gas over the solution; or, when the pressure of a gas over a certain liquid decreases, the amount of gas dissolved in the liquid will also decrease (or vise versa).* This gas law affects the body in that gases, primarily nitrogen, will come out of solution when the body is exposed to reduced atmospheric pressure. This occurs because the pressure of nitrogen is reduced proportionately as the total atmospheric pressure is reduced. This "evolved" gas phenomenon may lead to disorders similar to the "bends" which a deep sea diver experiences as a result of rapid ascent to the surface (going from an area of high external pressure, therefore high N_2 saturation, to an area of lower external pressure and lower nitrogen saturation, possibly resulting in nitrogen bubble formation throughout the body).

Trapped Gases — Cause, Effects, Prevention, And Treatment

Ear Block — Barotitis Media

The ear is composed of three sections: the outer ear, the middle ear, and the inner ear. The **outer ear** includes the auditory canal, which ends at the **eardrum**. The eardrum is a thin membrane about 0.004 inches thick. The **middle ear** is located within the temporal bone of the skull and is separated from the outer ear by the eardrum. A short slit-like tube that connects the middle ear cavity and the back wall of the throat is called the **eustachian tube**. The inner ear is used for both hearing and certain equilibrium senses.

During ascent or descent, air must escape or be replenished through the eustachian tube to equalize the pressure in the middle ear cavity with that of the atmosphere. If one is unable to equalize this pressure because of a head cold or an infection in the tubes or other causes of eustachian tube closure, pain and discomfort will result.

Normally, there is little difficulty equalizing pressure during descent because this can be accomplished by swallowing, yawning, or tensing the muscles of the throat at intervals. During sleep, the rate of swallowing slows down. For this reason it is advisable to awaken sleeping passengers prior to descent for the purpose of permitting them to ventilate their ears. Infants should be given a bottle or pacifier. Small children should avoid difficulty by chewing gum. If the preceding actions fail to equalize the pressure, the person experiencing ear block should **valsalva**. The valsalva procedure is done by closing the mouth, holding the nose and blowing. This will force air up the eustachian tube and into the middle ear. This is not a dangerous procedure and should not be delayed until the pressure in the ears becomes painful, otherwise it may be extremely difficult to open the eustachian tube. Painful ear block generally occurs when descent is made too rapidly. If the valsalva maneuver won't relieve this pain, ascent to higher altitude is recommended to equalize pressures, making it easier to "clear" the ears. This should be followed by a slower descent. During the second descent, close attention should be given to the prompt use of clearing techniques. Prudent use of nasal inhalants may also prove to be very helpful but should be used sparingly due to their compounding effect with hypoxia.

After a flight in which 100 percent oxygen is used, the valsalva procedure should be accomplished several times to ventilate the middle ear. This is recommended because the middle ear will be filled with pure oxygen, which is then gradually absorbed by the tissue of the middle ear. This causes a reduction in pressure which may become painful later in the day or night.

Sinus Block — Barosinusitis

The sinuses present a condition in flight similar to that of the middle ear. The sinuses are air filled, rigid, bony cavities lined with mucous membrane. They connect with the nasal cavity by means of one or more small openings. When these openings into the sinuses are normal, air passes out of and into these cavities without difficulty at any moderate rate of ascent or descent. If the openings of the sinuses are obstructed by the swelling of the mucous membrane lining, ready equalization of pressure becomes difficult. When the maxillary sinuses are affected, the pain will probably be felt on either side of the nose, in the cheek bones. Maxillary sinusitis may produce pain referred to the teeth of the upper jaw and may be mistaken for a toothache. When the frontal sinuses are affected, the pain will be located above the eyes and usually is quite severe. This type of sinus problem is the most common.

Equalization of pressure to relieve pain in the sinuses is best accomplished by use of the valsalva procedure, and/or inhalants previously mentioned in conjunction with ear block. Reversing the direction of pressure change by climbing may be necessary to clear severe sinus blocks.

Toothache — Barodontalgia

A toothache may occur at altitude. The pain may or may not become more severe as altitude is increased, but descent almost invariably brings relief. The toothache often disappears at the same altitude at which it was first observed on ascent.

Common sources of this difficulty are abscesses, mechanically imperfect fillings, inadequately filled root canals, and pulpitis.

Anyone who experiences a toothache at altitude should see a dentist without delay for examination and treatment. Maxillary sinus discomfort may be misinterpreted as a toothache.

Gastrointestinal Pain

Gastrointestinal pain is the discomfort caused by the expansion of gas within the digestive tract during ascent into the reduced pressure found at altitude. Fortunately, the symptoms are not serious in most individuals. In flights above 25,000 feet, enough distention may occur to produce severe pain.

The stomach, small and large intestines, normally contain variable amounts of gas with pressure approximately equivalent to that of the ambient atmosphere. The chief sources of this gas are swallowed atmospheric air and, to a lesser extent, gas formed as a result of digestive processes. As gases in the stomach and intestines expand during ascent, relief is ordinarily obtained by belching or by passing flatus.

Gas pains of even moderate severity may result in lowered blood pressure. Shock will be the eventual result if relief from distention is not obtained. Immediate descent from altitude should be made in order to obtain relief.

Evolved Gases — Cause, Effects, Prevention, And Treatment

Evolved gases cause **decompression sickness** and are due to the same things that cause bends in caisson workers or deep sea divers. The formation of gas bubbles within the body resembles the release of bubbles in a carbonated beverage, such as soda pop or beer, when the cap is removed.

Nitrogen, always present in body fluids, comes out of solution and forms bubbles if the pressure on the body drops sufficiently. Fatty tissue contains many times more nitrogen than other tissue, making an overweight person more susceptible to evolved gas decompression sicknesses. The action of these bubbles on various tissues of the body are thought to cause various types of evolved gas sicknesses.

The bends is characterized by pain in and about the joints, and may be mild at the onset. Failure to descend from altitude following the onset of bends may result in deep, gnawing, and penetrating pain which becomes intolerable in severity. Ordinarily, the pain is progressive and becomes worse if ascent is made to a higher altitude. Severe pain can cause loss of muscular power of the extremity involved, and if the pain is allowed to continue, collapse may result. The pain may diffuse from the joint over the arm or leg as a whole, or over the entire area of a long bone. Joints, such as those of the knee and shoulder, are most frequently affected.

Chokes, the common term for symptoms referable to the chest, probably are caused in part by blocking of the smaller pulmonary blood vessels by innumerable small bubbles. This, at first, may cause a deep burning sensation underneath the sternum. As the condition progresses, the pain may become more severe, may be stabbing in character, and may be markedly accentuated upon deep inhalation. It is necessary to take short breaths to avoid distress. There is an uncontrollable desire to cough, but the cough is ineffective and nonproductive. Finally, there is a sensation of suffocation, breathing becomes progressively more shallow, and there may be cyanosis. Immediate descent is imperative when these symptoms occur. This condition, if allowed to progress, frequently results in collapse and unconsciousness. Fatigue and weakness, as well as soreness in the chest, may persist for several hours after descent to ground level.

Another sign of decompression sickness is called **paresthesia**. Symptoms of paresthesia are tingling, itching, and cold and warm sensations. These symptoms are thought to be caused by the occurrence of bubbles locally or in the central nervous system where they may involve nerve tracts leading to the affected areas in the skin. A mottled red rash may appear on the skin and, more rarely, a welt accompanied by a burning sensation. Skin manifestations of paresthesia are not in themselves incapacitating or critical, but rather, symptoms of a dangerous condition which must not be ignored.

Central nervous system disturbances (CNS) include visual disturbances which are similar to the symptoms of hypoxia or the pulling of positive "G's." Lines or spots before the eyes may be seen. Some parts of the field of vision may disappear or blur. Dull and persistent headaches are commonly associated with visual symptoms. Other comparatively rare effects are partial paralysis, sensory disturbances, and aphasia, all of which are usually transient. These nerve manifestations occur primarily during or after descent and are generally relieved upon reaching ground level, although in rare cases they may persist for a long period of time.

Complications

One of the outcomes of decompression sickness may be **shock**. It consists of faintness, dizziness, nausea, or even loss of consciousness, accompanied by pallor and sweating. This reaction is the body's protest against a disturbance in the circulation of blood. This form of shock may be experienced during the time spent at high altitude. However, symptoms occasionally persist even after return to the ground or may reappear several hours after landing. This is why decompression sickness should be treated with respect. The shock reaction makes it dangerous. Prompt recognition of symptoms and treatment, usually by recompression (descent), results in usual cure.

Persons exhibiting symptoms of decompression sickness should remain quiet, keep the affected area immobile, and descend. The condition could prove to be crippling and the resultant lack of crew

efficiency caused by a severe attack can be extremely dangerous. Pain is usually relieved after descent. Pilots should contact the closest aviation medical examiner (AME) in the event of an occurrence of this type. He will know how to treat the problem, and he will know where the closest recompression chamber is located. No one should fly again or venture to higher altitudes without a thorough checkup by a physician experienced in decompression sickness physiology.

Scuba Diving And Flying

Many flying personnel and passengers are also diving enthusiasts. SCUBA (Self Contained Underwater Breathing Apparatus) is an exhilarating hobby; however, recent research has shown that diving and flying shortly thereafter can have marked effects on an individual. The SCUBA diver uses compressed air in the breathing tanks. When dives are made to a depth of approximately 30 feet, (a pressure of two atmospheres) the body will absorb about twice as much nitrogen as it had at the surface. Generally the diver has no problems when returning to the surface, provided the recommended steps as prescribed in the diving manual have been followed. However, the problem is compounded if that individual decides to fly in an aircraft shortly after diving. A person flying an aircraft to altitude in excess of 8,000 feet following SCUBA diving is in the same predicament the non-diver is when flying at 40,000 feet unpressurized. In other words, the body is subject to altitudes where evolved gases can and do occur.

The recommended waiting time before going to flight altitudes of 8,000 feet is at least 12 hours after diving which has not required controlled ascent (non-decompression stop diving), and at least 24 hours after diving which has required controlled ascent (decompression stop diving). The waiting time before going to flight altitudes above 8,000 feet should be at least 24 hours after any SCUBA diving. The altitudes given here are intended to be maximum possible exposure pressure altitudes and not pressurized cabin altitude because of decompression possibilities (failure of the cabin pressurization system).

Pilots should query passengers picked up at locations where diving is done to ascertain time-since-last-dive and dive types. Passengers going into these areas should be made aware of flight limitations after diving.

If you or a member of your flight experience evolved gas type decompression sickness such as bends, CNS, or chokes, you should contact the National Diving Accident Network, Duke University, North Carolina, (919) 684-8111 or the School of Aero-Space Medicine, Brooks AFB, Texas, at (512) 536-3278. Delays may cause the situation to worsen and collapse may occur. There are numerous treatment facilities over the world where these manifestations can be treated by skilled personnel and return to normal will be expedited if prompt action is taken. The local flight surgeon or aviation medical examiner or the above listed facilities will have current information about these treatment facilities.

Cabin Pressurization And Decompression

Cabin pressurization is the maintenance of a cabin altitude lower than the actual flight altitude. This is accomplished by compressing air in the aircraft cabin. Pressurized aircraft reduce the physiological problems at altitude and increase the effectiveness and comfort of the aircrew member and passengers. However, when pressurized to an altitude lower than actual flight, the possibility of sudden loss of this pressurization exists. If all aboard are prepared for this eventuality, the effects should be minor, unless the **rapid decompression** happens at extreme altitudes.

Advantages Of Cabin Pressurization Systems

Flights to altitude may be made without the use of supplemental oxygen. However, the oxygen ceiling, in relation to cabin altitude, must still be observed. Additional precautions should be taken to see that oxygen equipment is readily available in the event of a sudden loss of pressure.

Decompression sicknesses are prevented or made less serious because the body is not exposed to extremely low barometric pressure.

Heating, cooling and ventilation can be controlled more closely.

Cabin pressurization is especially advantageous in transport aircraft on long flights at medium altitudes. Eliminating the need for an oxygen mask contributes to the comfort of all crew members and increases their efficiency by enabling them to move more freely about the aircraft.

Physiological Effects Of Decompression

Decompression is defined as the inability of the aircraft's pressurization system to maintain its designed pressure schedule. This can be caused by a malfunction in the pressurization system or structural damage to the aircraft. Physiologically, decompressions fall into two categories:

Explosive decompression is defined as a change in cabin pressure faster than the lungs can decompress. Therefore, it is possible that lung damage may occur. Normally, the time required to release air from the lungs where no restrictions exist, such as masks, etc., is 0.2 seconds. Most authorities consider any decompression which occurs in less than 0.5 seconds as explosive and potentially dangerous.

Rapid decompression is defined as a change in cabin pressure where the lungs can decompress faster than the cabin. Therefore, there is no likelihood of lung damage, provided the airway is open. Persons startled during the initial stage of rapid decompression close their glottis as a normal startle reaction. Doing so is very dangerous as the gases in the lungs are expanding rapidly. **Air embolism** may result due to rupture of the tiny air sac membrane, allowing air bubbles to enter the blood. It is often fatal or causes permanent damage. This is why the author believes that some physiological briefing/training should be required for anyone planning flight in a small cabin aircraft at very high altitudes. They must be trained to "say ahh" when they detect symptoms of an RD, just as a diver is trained to "blow bubbles" constantly during a free ascent.

Experiments have been conducted in altitude chambers where normal, healthy persons were decompressed from 5,000 feet to 35,000 feet in 0.7 seconds. There were no ill effects from these decompressions. Almost all large pressurized aircraft decompress at a relatively slow rate (10 seconds or more). Thus, anyone exposed will, in all probability, experience a rapid decompression rather than an explosive decompression. This is not necessarily true for small volume pressurized aircraft. The use of safety belts becomes much more critical when seated near exits or windows.

During a decompression there may be noise, and for a split second one may feel dazed. The cabin air will fill with fog, dust, and flying debris. Fog occurs due to the rapid drop in temperature and the change of relative humidity. Normally, the ears clear automatically. Belching or passage of intestinal gas may occur. Air will rush from the mouth and nose due to the escape of air from the lungs, and may be noticed by some individuals.

The primary danger of decompression, besides air embolism, is hypoxia. Unless proper utilization of oxygen equipment is made when the aircraft is flying above 30,000 feet, unconsciousness will occur in a very short time. The period of useful consciousness is considerably shortened when a person is subjected to a rapid decompression. This is due to the rapid reduction of pressure on the body. Thus, oxygen in the lungs is exhaled rapidly. This in effect reduces the partial pressure of oxygen in the blood and may reduce the effective performance time by 1/3 to 1/2 its normal time. It is for this reason the oxygen mask should be worn on the face when flying at very high altitudes, regardless of the cabin pressure altitude.

Another potential hazard of high altitude decompressions is the possibility of evolved gas decompression sickness. Exposure to windblast and extremely cold temperatures are other hazards.

Rapid descent from altitude is indicated if these problems are to be minimized. Automatic visual and aural warning systems should be included in the equipment of all pressurized aircraft so that slow decompressions will not occur and overwhelm the occupants before being detected.

Cabin Pressurization Systems

The basic requirements to have cabin pressurization include:

1. A cabin structure, strong enough to withstand the normal twisting and flexing forces of flight while serving as a pressure chamber, with seals for doors, windows and transitory controls (control cables, wires, etc. that must pass in and out of the pressure chamber).

2. A source of compressed air for pressurization and ventilation capable of large quantities of air flow at maximum differential (10-15 pounds of air per minute for a 6-8 place aircraft).

3. A means of regulating the temperature of the air flowing into the cabin.

4. A means of regulating the pressure and rate of pressure change in the cabin.

5. A means of controlling the rate of outflow of air from the cabin.

Cabin Pressure Sources

Today, most general aviation reciprocating engine aircraft utilize some air from their turbochargers. The compressed air destined for the cabin passes through a venturi, called a **flow control unit** to limit flow from the compressor so as not to starve the engine and to increase the volume of air flowing to the cabin (ambient air is mixed with the compressed air). If air flow velocities in the venturi are designed to reach

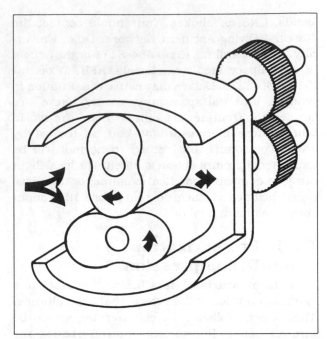

Figure 9-7. Schematic drawing of a Roots-type cabin compressor.

trans-sonic speeds, a sonic shock wave forms, restricting airflow. This device is referred to as a **sonic nozzle**.

Larger pressurized aircraft may utilize bleed air from the engine supercharger if the compressed air does not yet have fuel mixed with it, or a separate supercharger or Roots-type compressor driven by the engine (see figure 9-7). Turbine engine equipped aircraft utilize bleed air from the compressor section of the engine.

Cabin Temperature Control

When the air exits the flow control unit it is at a pressure above but near the cabin differential pressure and is still hot from pressurization. This air flows to a bypass valve which modulates flow, sending part of the warm air through the **air-to-air heat exchanger**, or **intercooler** (not the same one found in the engine intake system) where ambient air flowing through the heat exchanger cools the pressurized air. Depending on the position of the bypass valve, which is thermostatically controlled, different proportions of cool and warm air flow on into the **muffler** which insures quiet operation of the system and of the air entering the cabin through the **mixing plenum** which combines new air from the muffler with recirculated cabin air from the vent blower.

On the ground and under extreme conditions of ambient air temperature, additional heating of the

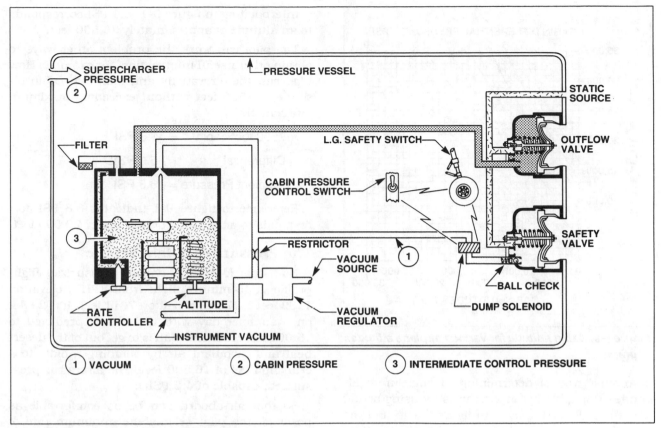

Figure 9-8. Pressurization control system.

air is done by electric heating elements and additional cooling is provided by a referigeration system.

The mixing plenum is also often a distribution point, sending air forward and aft, up and down to floor ducts, overhead eye vents, cockpit vents and defroster vents.

Cabin Pressure Control

Since it is impractical to build a pressure vessel that is absolutely air tight and because the occupants need good ventilation (air flow), pressurization is accomplished by pumping in more air than is needed and letting it leak out. Some **uncontrolled leakage** always occurs but **controlled leakage** through the outflow valve determines the cabin pressure altitude and rate of change of cabin altitude. See figure 9-8.

The **cabin altitude (pressure) controller** functions in one of two modes: **isobaric control** and **differential control**. During **isobaric control**, the regulator controls cabin pressure in order to climb or descend the cabin and to maintain the cabin altitude selected by the pilot. If the aircraft climbs so high that the selected altitude cannot be maintained without exceeding the maximum design pressure differential, the controller becomes a **dif-** **ferential controller** limiting cabin pressure so the maximum differential is not exceeded. While the controller is in isobaric control mode, the cabin rate of climb or descent to the selected altitude is also controlled.

Pressure Differential

Each aircraft has a design limited maximum allowable pressure differential between inside (the cabin) and outside (ambient) pressure.

For example, the King Air C90A differential is 5.0 PSID ± 0.1 PSID (pounds per square inch differential). As can be seen from figure 9-9, any aircraft with a maximum pressure differential of 5.0 PSID can fly at altitudes up to 23,000 feet with a cabin altitude of 8,000 feet, and with a cabin altitude of 10,000 feet, about 26,000 feet can be reached.

The greater the PSID allowable, the higher the aircraft can fly while maintaining an acceptable cabin altitude.

In many cases, maximum pressure differential is the limiting factor in determining the aircraft's **service ceiling**.

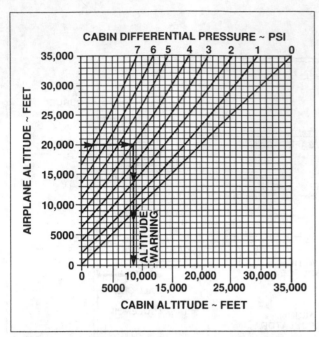

Figure 9-9. Cabin altitude for various airplane altitudes graph.

Another way of determining the maximum altitude attainiable by an aircraft is by using actual pressures. Referring back to figure 9-1 to find the ambient pressure in PSI for any altitude, the King Air's maximum operating altitude can be found:

Cabin pressure @ 10,000 feet   = 10.1 PSI

Maximum differential   = − 5.0 PSID

Minimum ambient (outside) pressure = 5.1 PSI

Interpolating in figure 9-1, 5.1 PSI corresponds to an altitude of approximately 26,500 feet.

The pilot can work the problem other ways to obtain other useful information, for example: How high can the aircraft fly and maintain cabin altitude at 7500 feet without exceeding maximum differential?

7500 foot cabin   = 11.8 PSI

Differential   = −5.0 PSID

Ambient Pressure =   6.8 PSI

Reference to figure 9-1 indicates 6.8 PSI corresponds to an altitude of just under 20,000 feet.

The Pressurized Flight Profile

Now, lets look at a typical pressurized flight profile: a 45 minute flight from KPHX, elevation 1132 feet to KFLG, elevation 7011 feet. It is 2 P.M. on a summer day with turbulence predicted to 15,000 feet MSL. The plan is to get out of the desert heat and turbulent air by climbing rapidly to a cruising level of 16,500 feet. The aircraft is pressurized, capable of 4.2 PSID.

So that all aboard can be as comfortable as possible, the pilot selects the minimum rate of pressure change (climb) consistent with all parameters of the flight which, with respect to **cabin altitude**, starts at 1132 feet (unless pressure variation from standard is extreme, variations in pressure altitude are not significant in these calculations). The cabin climbs to 7500 feet

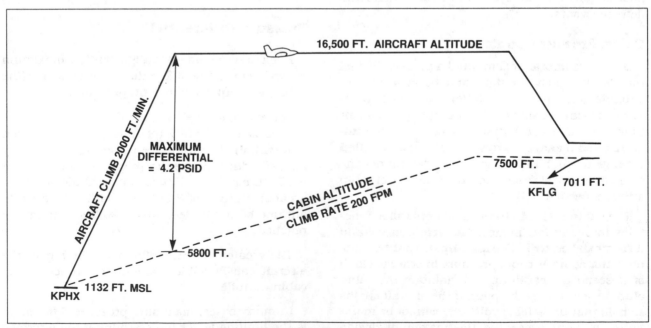

Figure 9-10. Flight profile showing cabin and pressurized aircraft altitudes.

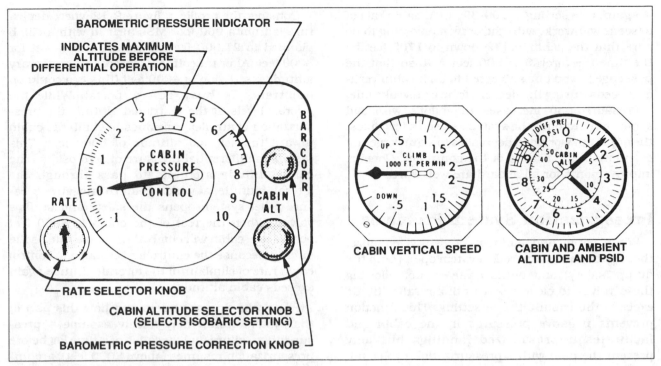

Figure 9-11. Pressurization panel.

(about 500 feet above destination field elevation so that PSID = 0 while the aircraft still has a few hundred feet to descend, assuring that the cabin will be depressurized at touchdown). The cabin altitude profile should start to climb after takeoff (see figure 9-10) and reach 7500 feet *before* arrival at KFLG, so the rate of climb needs to be: (done in the pilot's head or with a flight computer)

$$ROC = \frac{7500 - 1100 \text{ feet}}{40 \text{ min}} = \frac{6400 \text{ feet}}{40 \text{ min}} = 160 \text{ FPM}$$

Generally, rates of climb not exceeding 500 FPM climbing and 300 FPM descending are rarely uncomfortable for passengers. For this flight, the pilot sets the **rate control** (figure 9-11) for about 200 FPM, as this setting will get the cabin to 7500 feet well before arrival and should be very comfortable for all aboard unless someone is suffering from sinus and nasal congestion (this condition will be more of a problem on the return to KPHX). The **isobaric (altitude) control** should be set for 7500 feet.

With a near maximum effort climb (to get up to cool, smooth air) of 2000 FPM, the cruise altitude of 16,500 feet should be reached in 7.5 - 8 minutes at which time the cabin altitude should be 2700-2800 feet, so the maximum differential pressure would be:

2,800 feet = 13.3 PSI

16,500 feet = <u>7.8 PSI</u>

Differential = 5.5 PSID

This exceeds the maximum PSID of this aircraft's pressurization system by a significant amount. If this profile was actually flown, the system would go to differential control at 4.2 PSID, causing the cabin to climb for awhile at a faster rate to maintain 4.2 PSID. Knowing this, the pilot must climb the cabin a little faster or the aircraft slower so that when the aircraft reaches 16,500 feet, which is, as can be seen at a glance looking at the flight profile in figure 9-11, the moment of maximum differential, the cabin altitude must be up to:

16,500 feet = 7.8 PSI

Differential = <u>4.2 PSID</u>

12.0 PSI = 5800 feet

The return trip to KPHX will require the isobaric setting of the pressurization controller to be about 1600-1700 feet (500 feet above the airport). Descent rate can be computed as:

$$\frac{7000 - 1700 \text{ feet}}{40 \text{ min}} = \frac{5300 \text{ feet}}{40 \text{ min}} = 133 \text{ FPM}$$

Again, rate setting of 200-300 FPM cabin rate of descent will work, with the crew monitoring to be sure that the cabin will be down to 1700 feet by the time the aircraft is 500 feet AGL so that the passengers won't be subjected to high cabin vertical speeds during the descent from cruise altitude.

Pressurization increases the pilot's workload somewhat but once the system is used a few times, the apparent workload decreases to a quick mental calculation and setting of the controller. Comfort improvements for all aboard are well worth it!

Pressurization System Problems

When cabin and ambient pressures equalize, the negative-pressure relief system opens both the dump (safety) and outflow valves, not allowing these valves to close again until aircraft altitude exceeds the cabin altitude setting. This function prevents negative pressures in the cabin and facilitates unpressurized landings but may present the pilot with a pressurization problem if ATC places a low-altitude hold on the aircraft during climbout.

Consider the profile of figure 9-12 where departure is from a 900 foot MSL airfield with a climb planned to 21,000 feet, cabin altitude is set for 8000 feet. After takeoff, ATC imposes a temporary altitude restriction of 4000 feet (this is a common occurrence at busy TCA airports). While the aircraft levels at the restricted altitude the pressurization controller continues to climb the cabin toward the selected altitude of 8000 feet. This results in differential pressure of zero psid as the cabin altitude attempts to pass through the aircraft altitude, at which point the negative pressure relief system opens the safety and outflow valves. When the rest of the climb to 8000 feet occurs, the cabin will climb at the same rate as the aircraft because the controller is unable to control cabin rate of climb until the aircraft altitude again exceeds cabin altitude.

Pilots who do not understand how this part of the system works may falsely assume a pressurization failure and, because of a time lag before pressurization resumes (above 8000 feet), are unable to associate the cause of the problem with their failure to readjust the cabin altitude to a

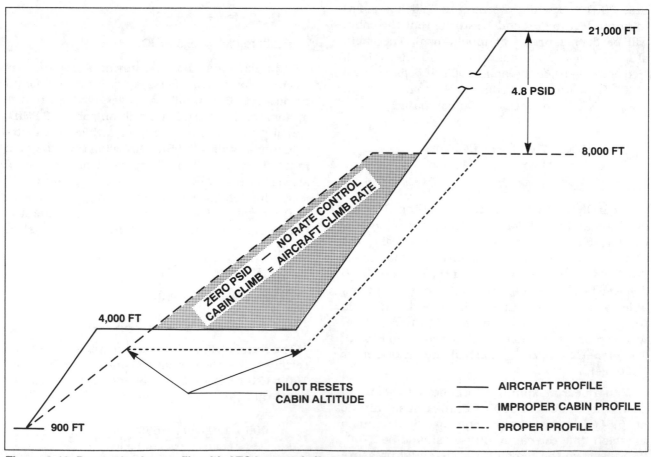

Figure 9-12. Pressurization profile with ATC imposed climb restriction.

value below the assigned aircraft altitude when the altitude restriction is received, thus failing to maintain a positive differential pressure, as shown by the "proper profile" in figure 9-12.

The **dump valve** or **safety valve** (figure 9-8) can be opened by the pilot by causing the cabin and ambient pressure to equalize as on final approach to landing or as in the scenario above, or with a switch in the cabin. Opening the dump valve is indicated if there is a pressurization failure causing the pressure to increase uncontrolled to maximum differential, or if there is windshield damage or smoke to be cleared from the cabin. In the event that cabin pressure needs to be reduced to zero, the pilot has the choice of opening the dump valve or shutting off the inflow of air to obtain a **leakdown** of pressure which is usually a less severe decrease in cabin pressure.

Most real (not pilot induced) pressurization problems will appear soon after liftoff. The three most common symptoms are:

1. Rapid pressurization towards maximum psid,
2. Lack of pressurization (cabin climb = aircraft climb rate) and
3. Leakdown (cabin altitude decreases at notable rate).

The first two symptoms are generally caused by controller, control system or outflow valve malfunctions. The third one is often caused by air inflow problems.

Both (or all) outflow valves should open and close freely as directed by the controller. If one or both are stuck closed, cabin altitude will descend rapidly after takeoff. Possible sources of this problem include: a stuck preset solenoid, a cracked pressurization controller, a diaphragm failure in the controller, disconnected or leaking plumbing, a cracked or dirty outflow valve or a failed diaphragm in the outflow valve.

Unusual or excessive pressure changes (**pressure bumps**) are usually caused by dirty, sticking outflow valves. The filters of this system can get dirty causing large differences in cabin climb vs. descent with no change in the rate knob setting. Heavy smoking in the cabin, dusty conditions or failure to keep the cabin very clean all contribute to these symptoms, requiring more frequent maintenance of the filters and valves.

If any of the above symptoms occur, the crew should consult the pressurization system malfunction procedures for the aircraft in use, starting with a check of all control settings.

Additional Reading

1. Airframe and Powerplant Mechanics Airframe Handbook (EA-AC 65-15A), Chapter 14; IAP, Inc., Publ.
2. A&P Airframe Technician Textbook (EA-ITP-A2), 1991; Chapter 16; IAP, Inc., Publ.

Study Questions And Problems

1. What is the highest altitude an aircraft with a 9.0 PSID pressurization system can fly while maintaining an 8,000 foot cabin altitude? Prove your answer using the alternate method. Show the graph you expanded from figure 9-9 to accommodate this aircraft. (Hint: For the alternate method, construct a very accurate graph of altitude vs. pressure from table 9-1 to help you interpolate for altitudes not shown in the table.)

2. For the flight from KPHX to KFLG described on pages 154-156, what rate of ascent must the pilot set in the controller if the isobaric setting is 8000 feet and the aircraft is climbing at 2200 FPM to 28,000 feet? (Hint: Do Question #3 simultaneously.)

3. What PSID must the pressurization system of question #2 be capable of?

4. What is the cabin rate of climb of the aircraft, in the text of the chapter, that is enroute from Phoenix to Flagstaff as it passes through 15,500 feet (remember, the controller is in differential mode because the pilot set the controller for a cabin climb rate of 200 FPM).

5. On the return flight from Flagstaff described in the text of the chapter, what would be the maximum PSID acheived if the pilot waited to the last minute to descend at 2000 FPM, assuming the pressurization system could handle any differential pressure?

6. Describe the emergency treatment, performed at altitude, for someone suffering from hyperventilation and for someone suffering from hypoxic hypoxia.

7. Define the terms expiration, respiration, circulation, hypemic hypoxia.

8. Why is the partial pressure of oxygen less at the alveolar level in the lung than in ambient air? (at least three reasons)

9. Define dysbarism, trapped gas, evolved gas, valsalva.

10. What is the function of the eustachian tube? What disorder occurs while flying if it malfunctions?

11. How long should one wait before flying after SCUBA diving?

12. List the three basic types of oxygen equipment and the altitude limitations of each.

13. Why should pilots avail themselves of a "chamber ride" or FAA high altitude physiological training?

14. What is the temperature of the standard atmosphere at 90,000, 94,000 and 100,000 feet in degrees Celsius?

Chapter X
Electrical Principles

Electricity, hydraulics and pneumatics all have a major commonality: they each use power or force which is produced in one location, transported, then used at another location. That ability to do work is transported by moving electrons, in the case of electricity, hydraulic fluid (a fluid liquid) in the case of hydraulics, and air (a fluid gas) in the case of pneumatics. Examples: electrical power, generated by the alternator in the engine compartment, lowers the flaps. Hydraulic pressure, generated by a hydraulic pump driven by the engine, retracts the landing gear at the pilot's command. Ice accumulated on the leading edge of the wing is broken and dislodged by inflation of the leading edge boots with air pressure generated with engine power. Each of these systems are discussed in the following chapters.

The simplest powered airplane would not fly without **electricity** which this book defines as *the flow of electrons* or *the presence of flowing electrons* or *the ability (potential) of electrons to flow*. This movement of electrons is necessary for electrical heaters and motors to work, radios to transmit and receive communication and navigations signals, electrically operated flaps and landing gear to operate, CRT's (cathode-ray tubes), computers, and many instruments, autopilots, etc. to work. The more complex an aircraft is, the larger is the role of electricity in operating and controlling the aircraft.

Fly-by-wire flight controls work because electrons carry messages of pilot control input to the device that moves the flight control and its systems. The pilot must gain a working knowledge of this system power source in order to use it well and in order to be able to compete well with other pilots when attending familiarization classes on larger, more complex aircraft. For example, the electrical systems diagrams book for the DC-10 is a book about the same size as this one.

Any pilot who understands this and the next two chapters well and is willing to continue to learn should have no trouble understanding advanced aircraft electrical systems. Several sources are listed at the end of this chapter which will provide further detail about all of the topics of these chapters on electricity.

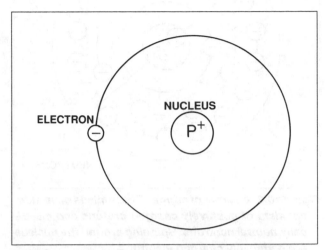

Figure 10-1. Diagram of a hydrogen atom.

Electron Flow

The **electron** is *a very small (its mass is ¹⁄₁₈₃₇th that of a proton), electrically charged particle that exists as a part of every atom*. An **atom** is *the smallest particle of any matter that is able to exhibit the properties and characteristics of that matter*. Atoms are made up of numerous building blocks called sub-atomic sized particles. The three largest that make up almost all of the mass of an atom are the proton, the neutron and the electron. The **proton** is 1837 times greater in mass than the electron and has an equal, but opposite, electrical charge. The **neutron** has about the same mass as the proton and is without electrical charge.

The simplest atom, hydrogen, has one proton in the nucleus and one electron orbiting it (just as a satellite orbits the earth). See figure 10-1. All atoms of hydrogen have this makeup. Different elements (hydrogen, helium, copper, aluminum, gold, etc.) differ mainly in the number of electrons, protons, and neutrons in their respective atoms.

For example, copper has a nucleus made up of 29 protons and 36 neutrons (figure 10-2). Surrounding the nucleus and spinning in four rings, or "shells," are 29 electrons. This combination gives what is called a balanced atom, as there are exactly the same number of positive charges (protons) as there are negative charges (electrons). The neutrons have no electrical charge, and so

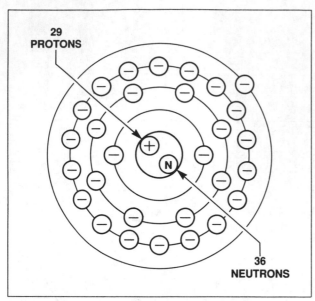

29
PROTONS

36
NEUTRONS

Figure 10-2. An atom of copper. The nucleus of an atom consists of positively charged protons and electrically neutral neutrons. Spinning around the nucleus are negatively charged electrons.

they do not enter into the development of electrical potential.

All matter contains energy, and energy in the atom causes the electrons to spin around the nucleus. As they spin, centrifugal force tends to pull them away from the nucleus. But there is an electrostatic field within the atom that produces a force which exactly balances this centrifugal force and holds the electrons a specific distance away from the nucleus, much like gravity holds satellites in orbit about the earth against their centrifugal force.

The electrons spin around the nucleus in rings or shells, and when energy is added to an atom, such as is done when the material is heated, the orbiting radius, or the distance between the electrons and the proton, is increased, and the bond, or force of attraction, between the proton and the electron is decreased, making such electrons more loosely held.

Ions

An electrical force outside the atom can attract electrons away from the outer ring and leave the atom in an unbalanced condition (less electrons than protons). These unbalanced atoms are called ions. For example, copper has one electron in its outer ring, and if a positive force is applied to the atom, this outer ring electron, which is negatively charged and loosely held, will be drawn from the atom, leaving it with more protons than electrons.

It then becomes a positive ion and will exhibit an attractive force to electrons from nearby balanced atoms. Electrons constantly move about within conductive materials from one atom to another, in a continuous but random fashion.

Conductors And Insulators

Materials which have atoms with loosely held electrons are said to be conductive—that is, electrons can be made to move through a conductive material in response to a force. **Insulators** are made up of materials which are composed of atoms which hold their outer electrons more tightly so that electrons can't flow in response to an electromotive force. Since the atoms of each element (copper, iron, etc.) are different, how tightly the outer electrons are held varies, so the resistance to electron flow exhibited by each element varies. Metals are all considered to be quite good conductors.

Electromotive force is the force that causes electrons to flow. Imagine a region where there are many more positive ions than electrons and another region where there are more electrons than protons. Each of these regions are electrically charged—one positively, the other negatively, thus producing an electrical force field.

The concept that opposites attract applies. Negatively charged particles (electrons) are attracted to the positively charged region. Arriving there, they cancel, or neutralize, some of the positive charges and weaken the force.

It may assist the reader to think only about electrons (negatively charged particles). A region of excess electrons (negative pole of a battery) creates a strong negative charge while an area of strong positive charges can be thought of as having a deficiency of electrons. If these opposite charges are brought into proximity with each other, an **electrostatic field** develops as shown in figure 10-3.

Electron flow occurs when a conductor such as copper is connected across a source of electrons (figure 10-4). An electron is attracted from an atom by the electromotive force (EMF) of the source. The atom which lost the electron has now become a positive ion and pulls an electron away from the next atom. This exchange continues until the electron that left the conductor is replaced by one from the negative terminal of the source, then is repeated as long as EMF *(voltage or electrical pressure)* exists.

Electron movement takes place within the conductor at about the speed of light; that is, about

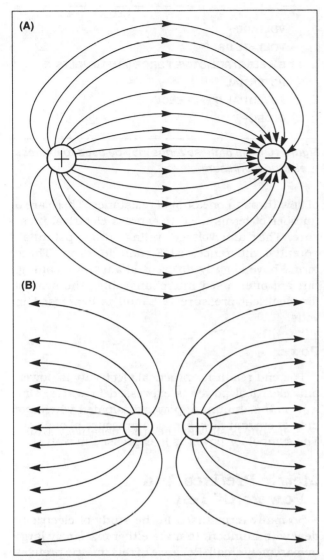

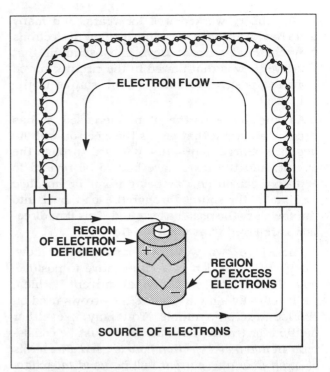

Figure 10-4. When an electron is attracted from the conductor by the positive charge of the source, it leaves a positive ion. This ion attracts an electron from an adjoining atom. This exchange continues through the conductor until an electron is furnished by the negative terminal of the source to replace the one that was taken by the source.

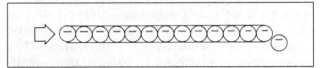

Figure 10-5. When one electron is forced into the conductor, it immediately forces an electron out of the opposite end of the conductor.

Figure 10-3(A)—Lines of force leave the charged body at right angles to its surface, and since they are polarized alike, the lines spread apart. They enter the oppositely charged body at a right angle to its surface. (B)—Charged bodies reject lines of electrostatic force from other bodies having the same charge.

186,000 miles per second. This is not saying that a single electron moves through the conductor from one end to the other at this speed, but because of the domino effect, an electron entering one end of the conductor will almost immediately force an electron out of the other end (figure 10-5).

Effects Of Electron Flow

As electrons flow, useful effects are produced. The flow causes magnetic and electrical fields to surround the conductor. The greater the amount of flow, the stronger will be the fields. Also, as

electrons are forced to flow, the opposition to their flow (called **resistance**) produces heat within the conductor.

Direction Of Flow

The effects of electricity were observed long before there was any knowledge of electrons, and in explaining what was seen, a wrong assumption was made: electricity appeared to follow the rules of hydraulics, in that there was pressure, flow, and opposition, and there was a definite relationship between the three. Since the flow of electricity could not actually be observed, it was only natural to assume that it flowed from a high level of energy to a lower level or, in electrical terms, from positive to negative.

This theory worked well for years, and many texts have been written calling the flow of electrons **conventional current flow**, and assuming that whatever it was that flowed in the circuits moved from the positive terminal of the source to the negative.

As knowledge of the atom increased, it has become apparent that is was the electron with its *negative* charge that actually moved through the circuit, and the texts have had to be revised to explain electron flow as being from the *negative* terminal of the source through the load, back into the source of the *positive* terminal. This flow direction is referred to as **electron flow**.

There are two ways we can consider flow: electron flow, which is from negative to positive, and the flow of "conventional current," which, while actually a myth, follows the arrows used on semiconductor symbols. You may use either method for tracing flow, but you must be consistent. *In this book, electron flow is used. The terms electron flow and current will be used interchangeably.*

Units Of Electrical Measurement

Quantity

The electron is such an extremely small particle of electricity that an enormous number of them are required to have a measurable unit. The **coulomb** is the basic unit of electrical quantity and is equal to 6.28 billion, billion electrons. This is most generally written as 6.28×10^{18}, which means that the number of electrons is 628 followed by 16 zeros. The symbol for quantity is Q.

Flow

When one coulomb flows past a point in one second, there is a flow of one **ampere**, or one amp. **Rate** of electron flow is called **current**, and its symbol is I, expressed in amps.

Resistance

The **ohm** is the standard unit of resistance, or opposition to current flow, and is the resistance at a specific temperature of a column of mercury having a specified length and weight. More practically, it is the resistance through which a pressure of one volt can force a flow of one ampere. The symbol for resistance is R.

Pressure

The **volt** is the unit of electrical pressure and is the amount of pressure required to force one amp

```
VOLTAGE
VOLTAGE DROP
E = ELECTROMOTIVE FORCE (EMF) = VOLTS
POTENTIAL
POTENTIAL DIFFERENCE
IR DROP
```

Figure 10-6. All of these terms may be used to express electrical pressure.

of flow through one ohm of resistance. There are a number of terms used to express electrical pressure. They are: voltage, voltage drop, potential, potential difference, EMF, and IR drop. These terms have slightly different shades of meaning, but are often used interchangeably. The symbol for electrical pressure is E and is expressed in volts.

Power

The end result for practical electricity is power, and electrical power is expressed in **watts**. One watt is the amount of power dissipated when one amp of current flows under a pressure of one volt. The symbol for power is P, expressed in watts.

Metric Prefixes And Powers Of Ten

So many terms used in the study of electricity deal with numbers that are either extremely large or extremely small. Because of this, metric prefixes have been adapted to them.

For example, the emergency frequency used for aircraft communications is 121,500,000 **hertz**, or cycles per second. This number is large and unwieldy, so we can divide it by one million and use the metric prefix mega-. The number becomes 121.5 megahertz.

These prefixes also help us in dealing with very small numbers. For example, the basic unit of capacitance, the farad, is much too large for practical use in aircraft electronics, and one of the commonly used capacitors has a capacity of 0.000,000,000,002 farad. A number such as this is awkward to work with, and its use encourages errors. A much more convenient way to express this same unit is to use the term two picofarads, or 2pf. This term, by the way, has previously been called a micromicro farad, and you may still see it referred to in this way. It may be written 2 mmf or 2μμf, with the Greek letter μ (mu) used.

Scientific Notation

Multiplying and dividing very large and very small numbers is made easier by the use of powers of ten. In this method of handling numbers, convert every number into a number between one and ten by moving the decimal the proper number of places in the correct direction. For example, 0.000,000,002 can be converted into 2.0 by moving the decimal to the right nine places. Since the number is smaller than one, the number two will have to be multiplied by a *negative* power of ten. When 0.000,000,002 is converted into a power of ten number, it becomes 2×10^{-9}.

Numbers larger than one are converted in exactly the same way, except they are multiplied by a *positive* power of ten. One coulomb contains 6,280,000,000,000,000,000 electrons. This number is easier to work with when it is converted to 6.28×10^{18} by moving the decimal to the left 18 places and multiplying 6.28 by ten, 18 times.

Numbers that have been converted into powers of ten may be multiplied or divided by performing the required work on the numbers, and then adding the powers of ten (the exponents), to multiply, or subtracting the powers, to divide.

$$0.0025 \times 5,000 = 2.5 \times 10^{-3} \times 5 \times 10^{3} = 12.5$$

$$0.125 \times 0.5 = (1.25 \times 10^{-1}) \times (5 \times 10^{-1}) = 6.25 \times 10^{-2} = 0.0625$$

$$5,000,000 \div 250,000 = 5 \times 10^{6} \div 2.5 \times 10^{5} = 2 \times 10^{1} = 20$$

$$0.125 \div 0.5 = 1.25 \times 10^{-1} \div 5 \times 10^{-1} = 0.25 \times 10^{0} = 0.25$$

The above discussions are summarized for you in figure 10-7.

Static Electricity

Electricity may be classified in two types, **current** and **static**. In current electricity, the electrons move through a circuit and perform work, either by the magnetic field created by their movement, or by the heat generated when they are forced through a resistance. Static electricity, on the other hand, rarely serves a useful purpose, and is more often a nuisance rather than a useful form of electrical energy.

We cannot see electricity, but the effects of both types are easy to observe. If we have a couple of balls of pith wood, we can suspend them by

MULTIPLIER	PREFIX	SYMBOL	NOTATION
1,000,000,000,000	tera	t	1×10^{12}
1,000,000,000	giga	g	1×10^{9}
1,000,000	mega	M	1×10^{6}
1,000	kilo	k	1×10^{3}
100	hecto	h	1×10^{2}
10	deka	fk	1×10^{1}
0.1	deci	f	1×10^{-1}
0.01	centi	v	1×10^{-2}
0.001	milli	m	1×10^{-3}
0.000,001	micro	μ (mu)	1×10^{-6}
0.000,000,001	nano	n	1×10^{-9}
0.000,000,000,001	pico	p	1×10^{-12}

Figure 10-7. Management of very large or small numbers.

threads and observe the effects of static electricity as we charge them.

Rub a glass rod with a piece of wool or fur and the rod will pick up extra electrons, and will

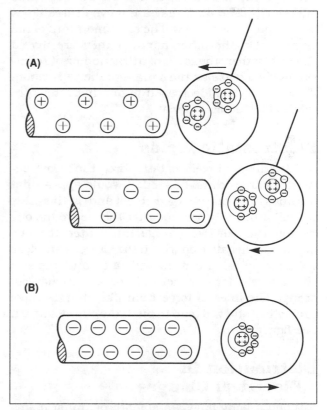

Figure 10-8(A)—An uncharged pith ball will be attracted to a rod that has either a positive or a negative charge. (B)—Once the charged rod has contacted the ball, the ball will assume the same charge as the rod and it will be repelled by the rod.

163

become negatively charged. Hold the rod close to a ball which has a neutral charge—it will be attracted to the rod. But as soon as the ball touches the rod, electrons will flow to the ball and give it a negative charge. Now, it will be repelled by the rod. See figure 10-8.

If the glass rod is rubbed with a piece of silk, the glass will give up electrons to the silk, and the rod will have a deficiency of electrons, or will become positively charged. When it is held near one of the balls, it, too, will attract the ball, and once it has touched it, the ball will be repelled. This is the same thing that happens with the first ball, but there is a difference. One of the balls is charged *negatively*, and the other is charged *positively*. When they are brought close together, they will be attracted to each other.

This demonstrates that there are two types of charges and that like charges repel each other, while unlike charges attract. The strength of the repelling and attracting force varies as the square of the distance between the two charges. For example, if the distance between the two balls is doubled, the force of attraction will be reduced to one-fourth ($2^2 = 4$). If they are moved three times as far apart, the force will be only one-ninth of the original. On the other hand, if they are moved closer together, the force of attraction or of repulsion will *increase* as the square of the separation. If the distance is decreased to one-half, the force will be four times as great.

Electrostatic Fields

If the lines of electrostatic force that exist between charges could be seen, it would appear that the lines leave one charged object and, in this case, we will arbitrarily assume that they leave the one having the positive charge and enter the one having the negative charge. If the charges are close together, all of the lines will link, and the two charges will form a neutral, or an uncharged, group. The lines of force from like charges repel each other and will tend to push the charges apart. See figure 10-3.

Distribution Of Electrical Charges

When a body having a smooth or uniform surface is electrically charged, the charge will distribute evenly over the entire surface but if the surface is irregular in shape, the charge will concentrate at the points or areas having the sharpest curvature.

Figure 10-9. Static wicks (dischargers) provide sharp points from which static charges are dissipated into the air before they can build up to a high potential on the control surface.

This explains the action of static dischargers used on many aircraft control surfaces. As the airplane flies through the air, friction causes a large static charge to build up on the aircraft's surfaces. The control surfaces are connected to the airframe structure by hinges which do not provide a particularly good conductive path, so the differential charge builds up in strength. **Static wicks** are attached to the control surface to dissipate this charge. They have sharp points on which the static charges concentrate and will discharge into the air before they can build up on the smooth surface sufficiently high to jump across the hinges causing hinge bearing damage and radio interference. See figure 10-9.

Figure 10-10. Bonding straps provide a low-resistance path between the control surface and the aircraft structure to prevent the buildup of a static charge on the control surface.

Figure 10-11. Airplanes and fuel trucks should be grounded (connected electrically) together to neutralize the charge of static electricity before the fueling nozzle is put into the tank.

As a further aid in preventing radio interference, the control surface is bonded to the structure. This means that a flexible metal braid **bonding strap** is attached to both the control surface and the structure to act as a good conductor, so the charge can be neutralized as it forms. See figure 10-10.

Static electricity is of real concern during the fueling operation of an aircraft. As the aircraft flies, friction between the air and the surface builds up a large static charge which cannot readily bleed off upon landing, because the rubber tires insulate the aircraft from the ground. Fuel trucks and fueling pits are grounded to the earth, so if the first contact with the aircraft is with the fuel nozzle in the open filler neck, a spark can jump in the explosive fumes and cause a serious fire. To prevent this, the aircraft is connected to the fuel truck and to the ground by the grounding cables which are provided on all fuel trucks for this purpose (figure 10-11).

Magnetism

Magnetic Characteristics

One of the most useful devices in both the production and use of electricity is the magnet. First discovered in the province of Magnesia in Asia Minor, a form of iron oxide demonstrated a strange property: when it was suspended in the air or floated on a chip, it would always turn in a northerly direction. This strange stone was used by early seafarers as a leading stone, or "lodestone" to aid in navigation.

A **magnet** is a body that has the property of attracting iron and producing magnetic fields ex-

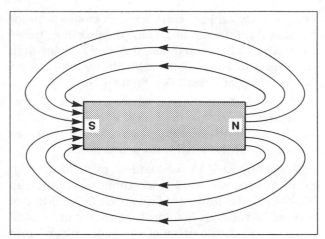

Figure 10-12. Lines of magnetic force form complete loops, leaving the magnet at its north pole and returning at its south pole.

ternal to itself. It is these magnetic fields that are of interest and use.

Lines of magnetic force, or **flux**, are always complete loops that leave the magnet at right angles to its surface at the north pole, and since they are all polarized in the same direction, they repel each other and spread out. They draw closer together as they re-enter the magnet at the south pole, at right angles to its surface, and travel through the magnet to complete the loop. See figure 10-12.

There is no insulation against lines of flux, as they will pass through any material; but if it is important that a device be protected from magnetic fields, it can be entirely surrounded by a soft iron shield. Iron, as we will see, has a very high

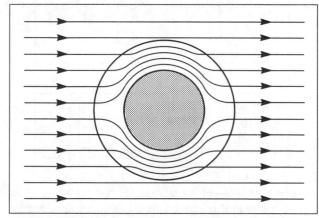

Figure 10-13. The only way to shield an object from lines of magnetic force is to enclose it in a shield made of a highly permeable material. The lines of force will flow through the shield and bypass its center.

165

permeability and provides a much easier path for the flux than the air, and since the flux lines travel the path of least resistance, they will flow through the iron and leave an area inside the shield that has no magnetic field. See figure 10-13.

The ends of the magnet where the lines of force leave and return are the poles, and are called the north- and south-seeking poles or, more commonly today, just the north and south poles.

If we were able to see inside a piece of unmagnetized iron, we would see that it contains an almost infinite number of magnetic fields oriented in such a random fashion that they cancel each other. Now, if this piece of iron were placed in a strong magnetic field, all of these little fields, or domains, as they are called, would align themselves with the strong field, and the iron would become a magnet, having a north and south pole, and lines of magnetic flux would encircle it. See figure 10-14.

The domain theory of magnetism is supported by the fact that each magnet has both a north and south pole, regardless of the size of the magnet. If we break a bar magnet in two, each half will demonstrate the characteristics of the original magnet, and if we break each of these halves in two, all of the small pieces will still be magnets having north and south poles, with lines of magnetic flux surrounding them (figure 10-15).

Soft iron has a very low **retentivity**, meaning that as soon as the magnetizing force is removed, the domains will lose their alignment, the fields

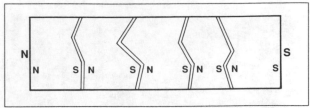

Figure 10-15. *Regardless of how small the pieces of a magnet may be broken, each piece will have a north and a south pole.*

will cancel each other, and the iron will no longer act as a magnet. Hard steel and some of the alloys of iron using aluminum, nickel, cobalt, and molybdenum have very high retentivities, and will retain the alignment of their domains long after the magnetizing force has been removed. It is materials of this type that are used for permanent magnets in aircraft magnetos, instruments, and radio speakers.

The number of lines of flux that loop through the magnet give an indication of its strength. One line of flux is called one **maxwell**. The **flux density** is the number of lines of flux for a unit area, and it is measured in gausses, with one **gauss** representing a density of one maxwell per square centimeter.

Lines of flux always follow the path of least resistance as they travel from the north pole to the south. They will even travel a longer distance if the traveling is easier. The measure of the ease with which the lines of flux can travel through a material or medium is measured in terms of **permeability**. Air is used as a reference and is given the permeability of one. Flux can travel through iron much more easily than through air, since it has a permeability of 7,000, and some of the extremely efficient permanent magnet alloys have permeability values as high as 1,000,000.

Lines of flux are invisible, but if we place a magnet under a piece of paper and sprinkle iron filings over it, they will form a definite pattern showing the lines they follow. They pass directly between the poles of a horseshoe magnet, but, if we place a piece of soft iron above the poles, the lines will enter the iron and flow through it to the south pole. This is because iron has so much greater permeability than air (figure 10-16).

It is the characteristic of the lines of flux to pass through a permeable material that explains the attraction of a piece of iron to a magnet. Remember that the lines of flux always seek the path of least resistance between the poles, and since air has a very low permeability compared to iron, if a piece

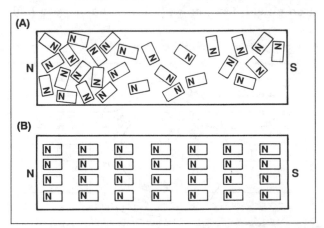

Figure 10-14(A)—*In an unmagnetized material, all of the individual magnetic fields, or domains, are arranged in a random fashion and cancel each other. (B)—When the material is magnetized, all of the domains are aligned, and the material has a north and south pole, just like that of the individual domains.*

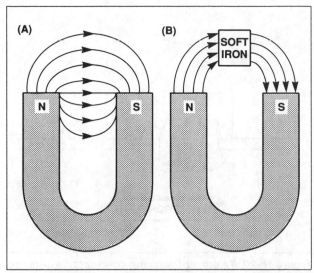

Figure 10-16(A)—Lines of magnetic flux are assumed to leave the north pole of a magnet at right angles to its surface and travel to the south pole, where they enter at right angles to its surface. (B)—The flux lines always seek the path of least resistance, even traveling longer distances if they can travel through a material with a high permeability.

of iron gets within the field of a magnet, the lines of flux will travel through it rather than through the air around it. The lines of flux want to link the poles with the shortest possible loops, so they will

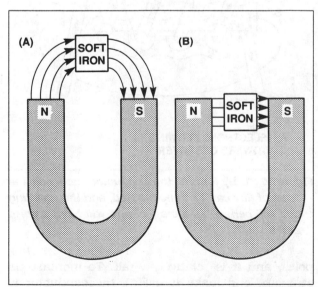

Figure 10-17(A)—The lines of flux pass through a material having a high permeability. In their effort to keep all of the force loops as short as possible, a force is exerted on the soft iron to pull it into the center of the magnetic field. (B)—When the soft iron is centered between the poles, it will resist any attempt to lengthen the lines of force.

exert a strong pull to get the piece of iron centered between the poles as the iron is pulled in closer, more lines of flux can pass through it and the pull will be stronger. When it is centered, it will resist any force that tries to lengthen the lines of flux by pulling the iron out of the field.

Almost any magnetic material regardless of its retentivity, will lose some of its magnetic strength if its lines of flux must pass through the air. Because of this, magnets whose strength is critical are stored with keepers of soft iron linking the poles to provide a highly permeable path for the flux.

Magnetism and lines of magnetic flux follow the same rules as charges of static electricity. Like poles repel each other, and the force of repulsion follows the inverse square law. This means that if the distance between the poles is doubled, the force of repulsion will be reduced to one-fourth. Unlike poles attract each other, and the force of attraction is squared as the distance is decreased. Halving the separation increases the force of attraction four times.

Electromagnetics

Though the effects of magnetism had been observed for centuries, it was not until 1819 that the relationship between electricity and magnetism was discovered. The Danish physicist Hans Christian Oersted discovered that the needle of a small compass would be deflected if it was held near a wire carrying electric current. This deflection was caused by an invisible magnetic field surrounding the wire (figure 10-18).

The effect of this field can be seen if iron fillings are sprinkled on a plate that surrounds a current-carrying conductor. The filings will arrange themselves in a series of concentric circles around the conductor (figure 10-19).

By observing the effect of the field and remembering what has been learned about electron flow, one can understand a great deal about electromagnets, and about the way they serve us in many useful ways.

Electrons are negative charges of electricity that can be forced to flow through a conductor. As they travel, they produce the forces we have just observed. The greater the amount of flow, the stronger will be the magnetic field.

The **left-hand rule** gives the direction of these lines of flux, and by knowing their direction, one can determine the direction of rotation of motors and the polarity of electromagnets. If one grasps

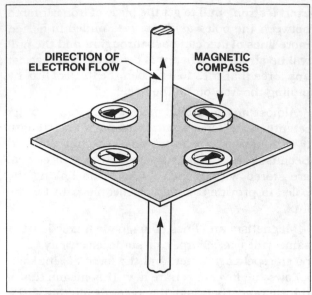

Figure 10-18. *The relationship between magnetism and electricity was discovered when it was found that the needle of a magnetic compass was deflected when it was placed near a current-carrying conductor.*

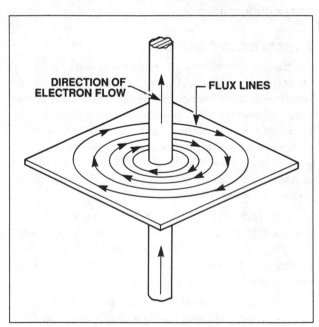

Figure 10-19. *Lines of magnetic force encircle a current-carrying conductor.*

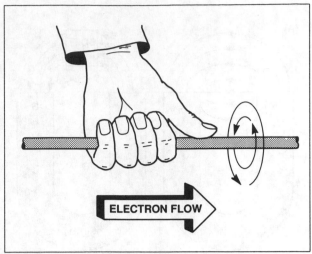

Figure 10-20. *If a current carrying conductor is grasped with the left hand, with the thumb pointing in the direction of electron flow (from negative to positive), the lines of force will encircle the conductor in the same direction as the fingers are pointing.*

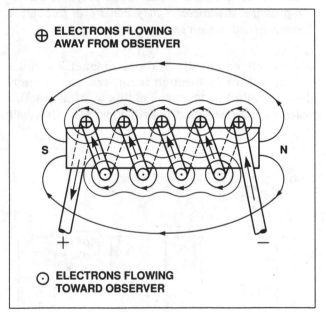

Figure 10-21. *By forming the conductor into a coil, the lines of flux can be concentrated, and the resulting coil will behave as a magnet and will have a north and a south pole.*

the conductor in the left hand with the thumb pointing in the direction of electron flow from negative to positive, the fingers will encircle the conductor in the direction of the lines of flux (figure 10-20).

Now, this magnetic field around the conductor does not serve much of a practical purpose because, while it has direction, it does not have any

poles, and it is relatively weak. To increase its strength, and make it useful, the conductor is wound into the form of a coil. When this is done, the lines of flux are concentrated and the coil attains the characteristics of a magnet.

In figure 10-21, the electrons flow into the coil from the right, and as the conductor passes over the top of the coil, the electrons are flowing away from us, as seen by the cross representing the tail of the arrow. Below the coil, the electrons flow toward us, as is seen from the dot representing the head of the arrow. When electron flow is away from us, the lines encircle the conductor in a counterclockwise direction, and when they come toward us, the field circles the conductor clockwise.

The lines of flux around each turn aid or reinforce the flux around every other turn, and there is a resultant field that enters the coil from the left and leaves it from the right. When the same terminology is used for permanent magnets, one finds that in this electromagnet, the pole on the right is the north pole, since the lines of flux are leaving it, and the pole of the left where the lines enter, is the south pole.

From what has just been seen, one can apply another left-hand rule—this one, the left-hand rule for coils. If one grasps a coil with the left hand in such a way that the fingers wrap around the coil in the direction of electron flow, the thumb will point to the north pole formed by the coil (figure 10-22).

The strength of an electromagnet is determined by the **number of turns** in the coil and by the **amount of current** flowing through it. This strength, or the magnetomotive force produced by the coil, is measured in **gilberts**. One gilbert is the amount of magnetomotive force produced by one **maxwell** (one line of flux) flowing through a mag-

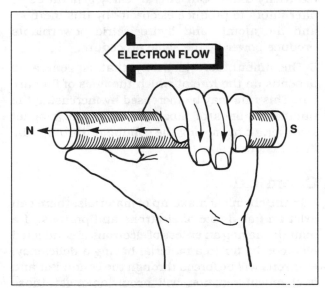

Figure 10-22. *If a coil is grasped with the left hand in such a way that the fingers encircle it in the same direction as the electron flow (from negative to positive), the thumb will point to the north pole of the electromagnet formed by the coil.*

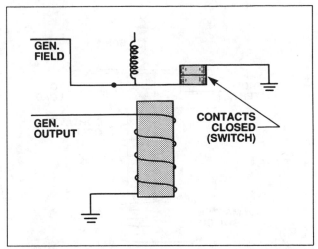

Figure 10-23. *An electromagnetically operated switch with a fixed core is called a relay.*

netic circuit having one unit of **"reluctance,"** the opposition of the circuit to the flow of magnetic flux. One gilbert is produced by a coil having $10/4\pi$ ampere-turns. More simply stated, an electromagnet having one ampere-turn will produce a magnetomotive force of 1.256 gilberts.

We can further increase the strength of an electromagnet by concentrating the lines of flux. This is accomplished by using some highly permeable material, such as soft iron, for the core.

There are two types of electromagnets used in practical applications, those having **fixed cores**, and those with **movable cores**. Fixed-core electromagnets are used in such devices as voltage regulators, where current proportional to the generator output voltage flows through the coil.

When enough current flows, the magnet will become strong enough to pull the contacts open and lower the generator output (figure 10-23).

Movable-core electromagnets are called **solenoids**. A soft iron core is held out of the center of the coil by a spring, but as soon as current flows in the coil, the magnetic field will pass through the core and, in its efforts to keep the loops of force as small as possible, will overcome the spring force to pull the core into the center of the coil, thus closing the switch (figure 10-24).

Terms Relating To Magnetism And Electromagnetism:

ampere-turn The amount of magnetomotive force produced by an electromagnet of one turn when a current of one ampere flows through it. One ampere turn produces a magnetomotive force of 1.256 gilberts.

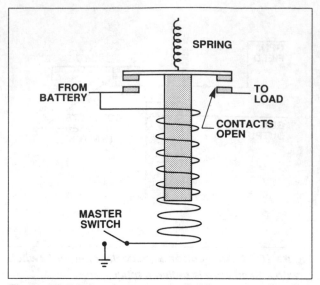

Figure 10-24. An electromagnetically operated switch with a movable core is called a solenoid. The application illustrated is that of the aircraft's master solenoid.

flux Lines of magnetic force

gauss A measurement of flux density equal to one line of flux per square centimeter

gilbert The amount of magnetomotive force required to maintain one maxwell in a magnetic circuit having one unit of reluctance. One gilbert = $10/4\pi$ ampere turns.

maxwell One line of magnetic flux.

permeability A measure of the ease with which lines of magnetic flux can pass through an object or body.

reluctance The opposition within a magnetic circuit to the flow of lines of magnetic flux.

residual magnetism The magnetism that remains in a material after the magnetizing force has been removed.

retentivity The property of a magnetic material to retain its magnetism after the magnetizing force has been removed.

Sources Of Electrical Energy

Energy cannot be created nor destroyed, but within rather wide guidelines, conversion of energy from one form into another is possible. This ability to exchange forms makes electrical energy all the more valuable to us. As we will see shortly, we daily convert mechanical, thermal, chemical, and light energy into electricity, and fortunately, electricity can be converted back into all of these forms of energy.

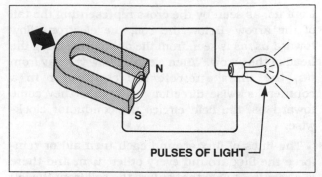

Figure 10-25. The amount of electricity generated by electromagnetic induction is determined by the rate at which the conductor cuts through lines of magnetic flux.

In the single-engine trainer aircraft, there are numerous sources of electrical energy. As you read this section, see how complete a list you can make of electrical sources that exist in the aircraft you fly. Remember, these are not sources of electrons (the electrons are always there in any conductor), but sources of EMF or voltage (the pressure or force needed to *move* those electrons).

Electromagnetic Induction

Lines of magnetic flux pass between the poles of a magnet, and if a conductor is moved through these lines of flux, they will transfer to the conductor and force electrons to flow through it. This is the principle used to generate most of our electricity today. Our aircraft carry generators or alternators to produce electricity by this method, and the atomic and hydroelectric powerplants produce power by the same procedure.

The amount of electricity that is generated depends on the *rate* at which the lines of flux are cut. This rate may be increased by increasing the number of lines of flux by making the magnet stronger, or by moving the conductor through the lines faster.

Chemical

In the chemical make-up of materials, there can exist an imbalance of electrons and protons. If a material having an excess of electrons is connected by a conductor to a material having a deficiency, electrons will be forced through the conductor and a chemical reaction will occur amongst those materials.

For example, if a piece of aluminum and a piece of copper are immersed in a solution of hydrochloric acid and water, and the two pieces of metal are connected by a piece of wire, electrons

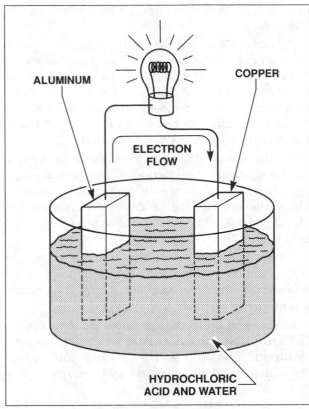

ELECTRON FLOW

ALUMINUM

COPPER

HYDROCHLORIC ACID AND WATER

Figure 10-26. Electrons will flow between two dissimilar metals when they are connected by a conductor and are both immersed in an electrolyte.

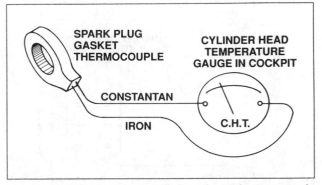

Figure 10-27. Electrons will flow in a thermocouple made of certain dissimilar metals when there is a temperature difference between the two junctions.

will leave the aluminum and flow to the copper. The electrons which leave the aluminum are replaced by the negative chlorine ions from the acid. When the chlorine combines with the aluminum, it eats away part of the metal and forms a gray powdery material on the surface. Positive hydrogen ions will be attracted to the copper where they are neutralized by the electrons that came from the aluminum, and bubble to the surface as free hydrogen gas. More on this in the discussion of batteries.

Thermal (Heat)

When certain combinations of wire, such as iron and constantan or chromel and alumel, are joined into a loop with two junctions, a **thermocouple** is formed. An electrical current will flow through the wires when there is a *difference* in the temperature of the two junctions. A **cylinder head temperature** measuring system has one junction held tight against the engine cylinder head by a spark plug, while the other junction is in the relatively constant temperature of the instrument panel.

Pressure

Crystalline material such as quartz has the characteristic that when it is bent or deformed by a mechanical force, an excess of electrons will accumulate on one surface, leaving the opposite surface with a deficiency. Such material is called **piezoelectric**. This feature is made use of in crystal microphones and phonograph pickups. The fact that this interchange between mechanical and electrical energy is reversible makes crystals useful for producing alternating current for radio transmitters. A piece of crystal has only one natural frequency at which it will vibrate, and if it is excited by pulses of electrical energy, it will vibrate at this frequency. As it vibrates, it produces between its faces an alternating voltage having an accurate frequency (figure 10-28).

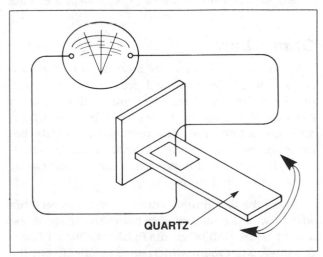

QUARTZ

Figure 10-28. A small electrical potential difference will be built up across the faces of certain crystalline materials when they are bent or otherwise subjected to mechanical pressure. Such material is termed piezoelectric.

171

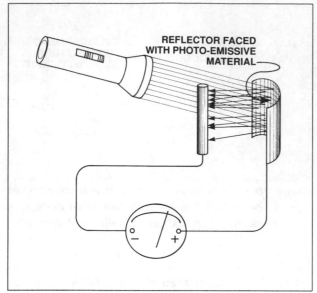

Figure 10-29. A photoemissive material will emit electrons when it is struck by light.

Light

Light is a form of energy, and when it strikes certain materials that are photoemissive, electrons will be knocked off of it and forced to flow in a circuit and do work (figure 10-29). Switches may be controlled by light-sensitive devices* to turn airport lights on at dark and off at dawn, and sensitive light measuring meters are used in photography to determine the amount of light available, so that the proper exposure can be made.

*Do you have one of these in your airplane? I do!

Ohm's Law

In our study this far, we have seen that a concentration of electrons will produce an electrical pressure that will force electrons to flow through a circuit. By assigning values to the pressure, flow, and opposition, the relationship that exists between them can be understood, and can accurately predict what will happen in a circuit under any given set of conditions.

It was the German scientist George Simon Ohm who proved the relationship between these values, and in 1826, published his findings. **Ohm's law** is the basic statement which says in effect that *the current that flows in a circuit is directly proportional to the voltage (pressure) that causes it, and inversely proportional to the resistance (opposition) in the circuit.* The units used make this relationship easy to see: one volt of pressure will cause one ampere

of current to flow in a circuit whose resistance is one ohm.

By convention, voltage is represented by the letter E, current by the letter I, and resistance by the letter R. A statement of Ohm's law in the form of a formula is, therefore, E = I × R. This simple formula can be algebraically rearranged to find the current, the formula becomes I = E/R, and resistance may be found by the formula R = E/I.

For those who are not comfortable with doing a simple algebraic transformation, the formula may be obtained by placing your thumb over the symbol of the quantity to be determined on figures 10-30 or 10-31. The formula becomes those symbols still visible. Try it!

Power in an electrical circuit is measured in watts, and one **watt** is the amount of power used in a circuit when one amp of current flows under a pressure of one volt.

Power is equal to voltage times current, P = I × E. Figure 10-31 shows an easy way to find current by dividing power by voltage, I = P/E, and to find the voltage by dividing power by the current, E = P/I.

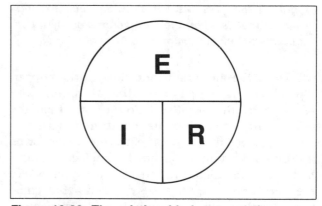

Figure 10-30. The relationship between voltage, current, and resistance.

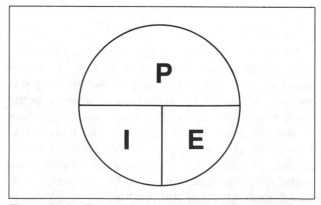

Figure 10-31. The relationship between power, current, and voltage.

Mechanical Power In Electrical Circuits

Power, as is remembered from basic physics, is the time-rate of doing work, and the practical unit of measurement in the English system is the **horsepower**, which is the amount of power required to do 33,000 foot-pounds of work in one minute, or its equivalent of 550 foot-pounds of work in one second. Seven hundred forty-six watts is the electrical equivalent of one horsepower.

If using a 24-volt electric hoist, one can find the amount of current needed to raise a 1,000-pound load six feet in 30 seconds. Neglecting efficiency of the hoist motor:

$$1,000 \times 6 = 6,000 \text{ foot–pounds}$$

$$\frac{6,000}{30} = 200 \text{ foot–pounds per second}$$

$$\frac{200}{550} = 0.364 \text{ horsepower}$$

$$746 \times 0.364 = 271.5 \text{ watts}$$

$$I = \frac{P}{E} = \frac{271.5}{24} = 11.31 \text{ amps}$$

A 250 watt landing light requires 20.8 amps at 12 volts:

$$I = P/E = \frac{250}{12} = 20.8 \text{ amps}$$

Advantage Of Alternating Current Over Direct Current

Alternating Current

Alternating current which continually changes its values of voltage and current and periodically reverses its direction of flow has some advantages over direct current, in which the electrons flow in one direction only. AC is much easier to generate in the large quantities needed for our homes and industries and for large transport aircraft.

More important, though, is the ease with which the values of current and voltage can be changed to get the most effective use of electrical energy. For example, in our homes and shops, most AC electricity has a pressure of about 115 volts, and if a kilowatt of power is needed, almost nine amps of current must flow. The current flowing in a conductor determines the amount of heat generated, and therefore the size of conductor needed. So, to get the same amount of power with less current, smaller conductors can be used with higher voltage which will save both money and weight.

In order to use the smallest conductor possible for cross-country transportation of electrical power, the voltage of the electricity carried in our transmission lines is boosted up to several thousand volts. For example, at 15,000 volts, we will need to transmit only 0.067 amp for one kilowatt of electrical power. Before the electricity is brought into our homes or shops, it is transformed down to a usable value of around 115 volts, so it will be safer and more convenient to handle. Between the generation of alternating current and its final use, the voltage and current will be changed many times. The transformers that do this changing are quite efficient and very little energy is lost.

Generating Alternating Current

We have seen in our study of direct current electricity that there is a close relationship between magnetism and electricity. Any time electrons flow in a conductor, a magnetic field surrounds the conductor, and the strength of this field is determined by the amount of electron flow. We have also found that when a magnetic field is moved across a conductor, electrons are forced to flow in it, and the amount of this flow is determined by the *rate* at which the lines of magnetic flux are cut by the conductor. Increasing the number of lines of flux by making the magnet stronger or increasing the speed of movement between the conductor and the magnet will increase the amount of electron flow.

If a conductor, wound in the form of a coil, is attached to an electrical measuring instrument, and a permanent magnet is moved back and forth through the coil, the meter will deflect from side to side (figure 10-32). This indicates that the electrons flow in one direction when the magnet is moved into the coil and reverse and flow in the opposite direction when the magnet is withdrawn. This is alternating current, AC.

The AC electricity with which we are most familiar has been generated by a rotary generator in which the conductor (in the form of a coil) is rotated inside a magnetic field. If we were to watch on an oscilloscope the changing values of the voltage as the coils are rotated, we would see that the voltage starts at zero, rises to a peak, and then

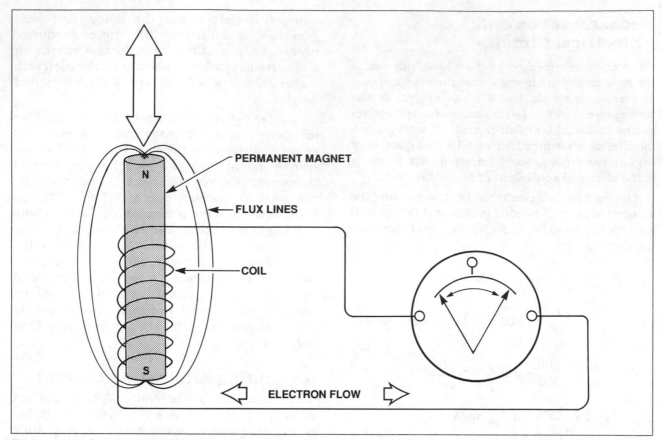

Figure 10-32. Current will flow in a conductor when lines of magnetic flux from a permanent magnet cut across it.

drops back off to zero. As the coils continue to rotate, the voltage builds up in the opposite direction to a peak and then back to zero. There is one complete cycle of voltage changes for each complete revolution of the coil (figure 10-33).

The wave form of alternating current or voltage produced by a rotary generator is called a sine wave, because the voltage or current changes according to the sine of the angle through which the generator has rotated.

Alternating Current Term And Values

A **cycle** is one complete sequence of voltage or current changes from zero through a positive peak to zero, then through a negative peak, back to zero where it can start over and repeat the sequence.

The time required for one cycle of events to occur is known as the **period** of the alternating current or voltage.

The number of complete cycles per second is the **frequency** of the AC, and it is expressed in **hertz**. One hertz is one cycle per second. The frequency of the alternating current produced by a generator is determined by the number of *pairs* of magnetic poles in the generator and the speed in revolutions per minute of the rotating coils. Frequency may be found by the formula:

$$\text{Frequency (Hz)} = \frac{\text{Poles}}{2} \times \frac{\text{rpm}}{60}$$

The frequency of commercial alternating current in the USA is 60 hertz, while in some foreign countries it is 50 hertz. The AC power used in most aircraft is 400 hertz, because the inductive reactance at this frequency is high enough to allow smaller transformers and motors to be used efficiently.

For a complete discussion of AC power, see 'Additional Reading,' item 1, section 3.

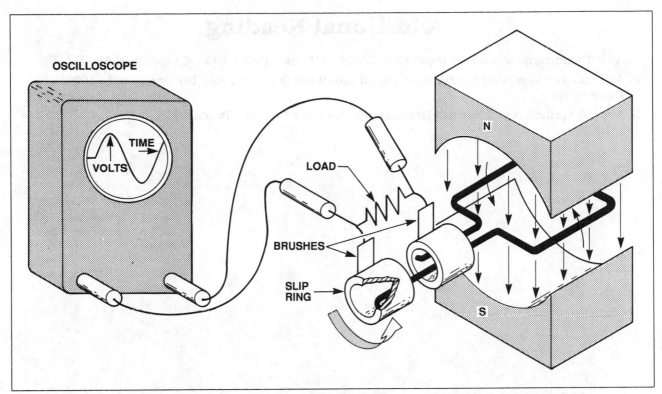

Figure 10-33. Alternating current is produced in a conductor when it is rotated within a magnetic field.

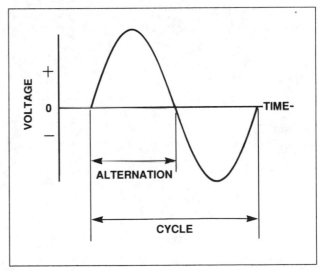

Figure 10-34. Sine wave values.

Additional Reading

1. A&P Technician General Textbook (EA-ITP-G2); IAP, Inc., Publ.; 1991; Chapters 3, 5.

2. Airframe & Powerplant Mechanics General Handbook (EA-AC65-9A); IAP, Inc., Publ.; 1976; Chapter 8.

3. Aircraft Ignition And Electrical Components (EA-IGS); IAP, Inc., Publ.; 1985.

Study Questions And Problems

1. Why use electricity in a simple airplane?

2. What is electricity? Check some other sources to see if they agree with this chapter. Do they, in your opinion?

3. Define:
   atom
   electron
   ion
   current
   voltage
   resistance
   Ohm's Law
   conductor
   insulator

4. Express the following numbers in scientific notation:
   a) 2,400,000
   b) 126
   c) .0034
   d) 2,360

5. Express the following in megaHertz:
   a) 122800000 cycles per second
   b) 5680000 cycles per second
   c) 1040000 cycles per second
   d) 1220 cycles per second (express megaHertz in scientific notation).

6. Define:
   ampere
   ohm
   volt
   watt
   static discharge wick

7. What is the purpose of a bonding strap?

8. What does the term "permeability" mean, with respect to magnetism?

9. What is the difference between a relay and a solenoid?

10. Name at least seven sources of electrical energy in a typical single-engine aircraft.

11. A 2500-pound airplane is requiring 22.5 amps for its avionics and 18 amps for its landing and other lights (12-volt system). If all of this electrical demand was shut off, how much would the rate of climb increase? Assume 80% propeller efficiency, 92% alternator efficiency.

Chapter XI
Electrical Components

A discussion of the components that are typically found in aircraft electrical systems is appropriate before learning about the system in the airplanes.

The aircraft electrical system can be classified into the following parts:

1. *The electrical energy source*—Typically the battery is the source until the alternator or generator is operating, then the latter becomes the source because it produces a greater voltage.

2. *The circuit*—This is the transportation and distribution system for the electrical energy (moving electrons). A pathway from the source through the active component and back to the source must exist in order for electrons to be able to flow.

3. *The active component*—The site where the work is done is evidenced by (a) work being done, (b) heat being generated, and (c) a voltage drop occurs. It is the location of significant electrical resistance.

The components are presented below according to this classification.

Batteries

A **battery** is a device composed of one or more cells in which chemical energy is converted into electrical energy. The chemical nature of the battery components provides an excess of electrons at one terminal and a deficiency at the other. When the two terminals are joined by a conductor, electrons flow, and as they flow, the chemical composition of the active material changes. In a primary cell, the active elements become exhausted and cannot be restored, while in a secondary cell, electricity from an external source will restore the active material to its original, or charged, condition.

Primary Cells

The most commonly used primary cell, or dry cell, as it is called, is of the **carbon-zinc type**. In this cell, a carbon rod is supported in a zinc container by a moist paste containing ammonium chloride, manganese dioxide, and granulated carbon. When a conductor joins the zinc case and the carbon rod, electrons leave the zinc and flow to the carbon. The zinc is left with positive ions which attract negative chlorine ions from the ammonium chloride electrolyte and form zinc chloride on the inside of the can (figure 11-1). Zinc chloride is actually corrosion, and it eats away the zinc, eventually causing the battery to fail.

The positive ammonium ions that are left in the electrolyte are attracted to the carbon rod, where they accept the electrons that have just arrived from the zinc. As they are neutralized, they break down into ammonia and hydrogen gases which are absorbed into the moist manganese dioxide.

Carbon-zinc cells produce one and a half volts regardless of their size, but the size of the cell determines the amount of current it can supply.

Leakage, which at one time was a problem with carbon-zinc batteries, is minimized by the use of effective seals and by enclosing the zinc can inside of a steel jacket.

When longer life is needed, **alkaline cells** are used, rather than the less expensive carbon-zinc variety. In the modern alkaline cell, the center electrode is a zinc rod supported in a carbon container by a moist electrolyte of potassium hydroxide. All of this is housed inside a steel case, with an insulating disc isolating the center electrode from the case (figure 11-2).

Potassium hydroxide has a lower resistance than ammonium chloride, and these cells will produce more current than carbon-zinc cells and so are excellent for tape recorders and other devices which contain motors. In order for alkaline cells to be interchangeable with carbon-zinc cells, their polarity has to be reversed. This is done by mounting the cell in an insulated steel outer case. The negative center terminal of the cell bears against the outer case by a spring, and the case of the cell contacts only the center conductor of the outer shell. By doing this, both types of cells have negative outer cases and positive center terminals.

Another type of alkaline cell is the **mercury cell** which is used in some hearing aids, cameras, and for other applications where a high capacity for small size overcomes their higher cost.

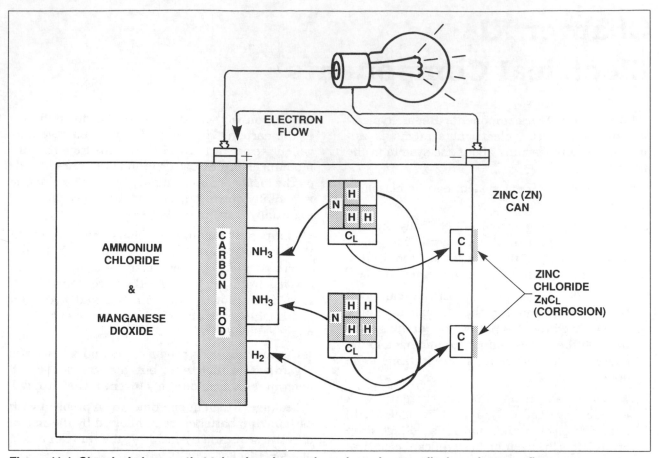

Figure 11-1. Chemical changes that take place in a carbon-zinc primary cell when electrons flow.

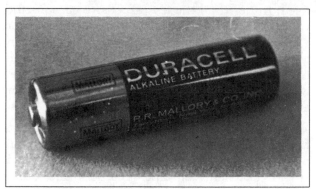

Figure 11-2. An alkaline cell uses a carbon container and a zinc center electrode with potassium hydroxide as an electrolyte.

Figure 11-3. A mercury cell is used when a high capacity is needed in a very small cell.

A pellet of mercuric oxide is placed inside a steel container, and a porous insulator or separator is placed over the pellet. A roll of extremely thin corrugated zinc is placed over the paper which is saturated by a solution of potassium hydroxide which serves as the electrolyte. The cell is then closed with a steel cap that contacts the zinc, and is insulated from the container. The container is the positive terminal, and the cap is the negative. In applications where it is necessary to have a negative case, the same procedure is used to reverse the polarity as was used by the other alkaline cell (figure 11-3).

Secondary Cells

It is possible to construct a cell which, instead of consuming or destroying one of its elements as it is used, will convert the element into a chemical that can be changed back into its original material by flowing electricity into it. Cells of this type are called secondary cells, and do not produce electrical energy, but merely store it. It is for this reason that batteries made up of these cells are called storage batteries.

Lead-acid Batteries

The most commonly used storage battery for aircraft is the lead-acid battery.

Construction

The positive plate is made up of a grid of lead and antimony filled with lead peroxide. The negative plate uses a similar grid, but its open spaces are filled with spongy lead in which an expander material is used to prevent its compacting back into a dense form of lead, losing much of its surface area.

The cells are made up of a group of positive plates, joined together and interlaced between a stack of negative plates. Porous separators keep the plates apart and hold a supply of electrolyte in contact with the active material.

The battery container was formerly made of hard rubber, but is now made of a high-impact plastic which has individual compartments for the cells (figure 11-4).

Figure 11-4. A typical lead-acid aircraft battery.

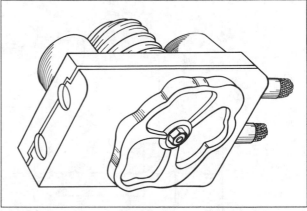

Figure 11-5. A lead weight actuated valve allows the filler plug to vent when upright, but seals when tipped.

Connector straps join the cells and provide the external terminals of the battery. A cover seals the cells in the case, and holes in the cover provide access to the cells for servicing them with water and for checking the condition of the electrolyte.

In aircraft batteries, these cell openings are closed with vented screw-in-type caps which have lead weights inside them to close the vent when the battery is tipped. This prevents the electrolyte from spilling in unusual flight attitudes (figure 11-5).

An electrolyte consisting of a mixture of sulfuric acid and water covers the plates and takes an active part in the charging and discharging of the cell.

The reaction at the plate covered with spongy lead, when the cell is being discharged, is as follows:

$$Pb + SO_4^{--} \rightarrow PbSO_4 + 2\ e^-$$

The insoluble lead sulfate is deposited in the spongy lead, and the two electrons are given up to the plate, which is connected to the negative electrode. These electrons flow through the electric circuit to the positive terminal which is connected to the lead dioxide plate, where the reaction is:

$$PbO_2 + SO_4^{--} + 4\ H^+ + 2\ e^- \rightarrow PbSO_4 + 2\ H_2O$$

See figures 11-6 and 11-7.

The complete chemical reaction is:

$$Pb + PbO_2 + 2\ H_2SO_4 \underset{charge}{\overset{discharge}{\rightleftarrows}} 2PbSO_4 + 2\ H_2O$$

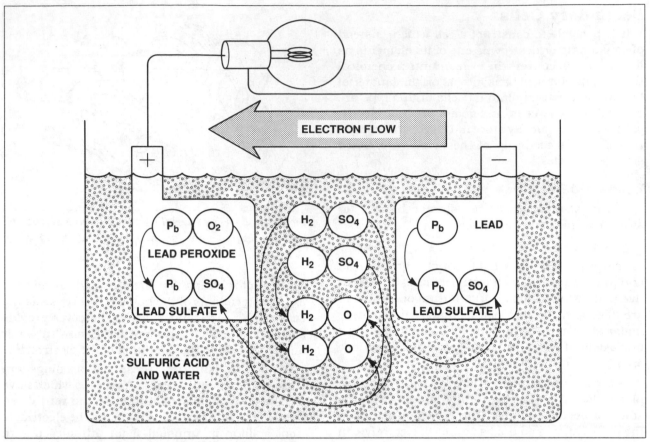

Figure 11-6. Chemical changes that take place during discharge.

As the cell delivers current, the two electrodes approach an identical condition by the deposit of lead sulfate on both. At the same time sulfuric acid is withdrawn from the electrolyte, which in consequence diminishes in density. The latter fact permits the condition of the cell to be judged by measuring the density of the sulfuric acid with a simple hydrometer. See figure 11-7.

Capacity

The capacity of a battery is *a measure of* its ability to produce a given amount of current for a specified time, and is expressed in **ampere-hours**. One ampere-hour of capacity is the amount of electricity that is put into or taken from a battery when a current of one ampere flows for one hour. Any combination of flow and time that moves the same amount of electricity is also one ampere-hour. For example, a flow of one-half amp for two hours or two amps for one-half hour is each one ampere-hour.

The amount of active material, the area of the plates, and the amount of electrolyte determine the capacity of a battery. But also of importance in rating a battery is the rate at which the current flows.

Standard Rate—Five-hour Discharge

A standard rating used to specify the capacity of a battery is the five-hour discharge rating. This is the number of ampere-hours of capacity of the battery when there is sufficient current flow to drop the voltage of a fully charged battery to 1.75 volts per cell at the end of five hours.

High Rate—Twenty-minute Discharge And Five-minute Discharge

Two ratings are used to specify the capacity of a battery when the current is taken out at a high rate. Both ratings will produce lower current rates than the five-hour rating. Figure 11-8 shows the relationship between the capacity at the five-hour rating and that for the twenty-minute and the five-minute rating for typical 12- and 24-volt aircraft batteries.

Battery capacity is important to the pilot when it comes time to start the engine in very cold weather (temperature decreases battery rated output) and when a generator or alternator fails in flight.

Rated capacity indicates the battery's ability to perform when it is new. Actual capacity may be quite different and depends on:

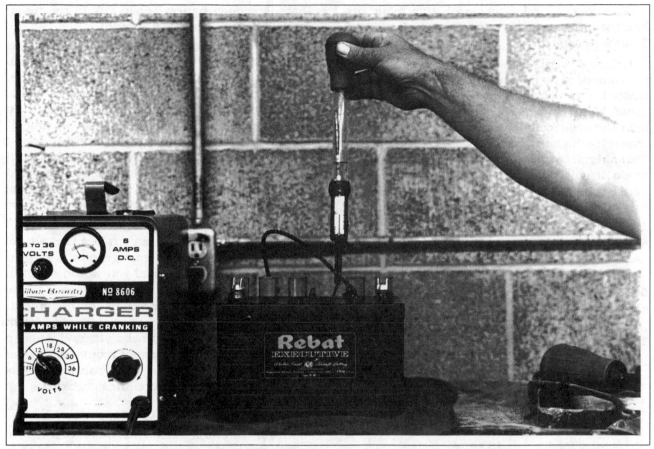

Figure 11-7. Checking electrolyte density with a hydrometer

1 *Battery temperature*—cold temperatures decrease battery performance.

2. *Battery state of charge*—A partially discharged battery will exhibit degraded performance.

3. *Battery condition*—Affected by age, service history, g forces applied (hard landings, turbulence).

Battery condition is degraded by:

Age—Each charge-discharge cycle decreases the porosity (surface area) of the plates and results in some lead reactants moving by gravity to the bottom of the case. Addition of water that is not distilled introduces contaminants that react to form inert compounds, taking reactants out of action or contaminating them. If the plates are exposed to the air, oxides form which permanently contaminate the exposed portion of the plates. Full discharge and recharging adversely alters the physical condition of the plates. Finally, sufficient sludge accumulates in the bottom of the case to short circuit the bottom of the plates causing failure. Thus, the battery condition deteriorates with time and service history.

G forces stress the plate attach points. This can be caused by poor landing or even rough roads during transport on a new battery, hard landings and even strong turbulence. Plates may break loose and end up in the bottom of the case, thus decreasing capacity or causing an internal short-circuit.

BATTERY VOLTAGE	PLATES PER CELL	DISCHARGE RATE *					
		5-HOUR		20-MINUTE		5-MINUTE	
		A.H.	AMPS	A.H.	AMPS	A.H.	AMPS
12	9	25	5	16	48	12	140
24	9	17	3.4	10.3	31	6.7	80
* Battery is considered discharged when closed-circuit voltage drops to 1.2 volts per cell.							

Figure 11-8. Relationship between ampere-hour capacity and discharge rate.

How does the pilot cope with this many unknowns with respect to battery performance? Some suggestions:

1. Install a good quality battery of a capacity recommended by the maintenance manual. Ask a trusted, experienced mechanic which battery is giving his customers the best service. If you are flying into cold country or into remote areas, purchase the highest capacity battery permitted. If your aircraft can be started by propping, learn how to do it safely from an experienced instructor before you go.

2. In the case of an alternator or generator failure in flight, reduce electrical load to the minimum needed, then the remaining time before electrical system failure can be estimated by multiplying the amp-hour capacity of the battery by 50% (condition factor) and dividing the amps of current in use into it.

Example: Alternator fails at 15:00Z on a single-engine aircraft with a 35 amp-hour rated battery. Load is reduced to the following:

Strobe	3 amps
1 Navcom	5 amps (average)
Transponder/encoder	5 amps
Total	13 amps

$$35 \text{ amp-hours} \times 50\% = \frac{17.5 \text{ A–H}}{13 \text{ amps}} = 1 + \text{hours}$$

Momentarily used items such as fuel pump, landing lights, etc. can be subtracted from the battery capacity before the continuous use items are, for example:

Landing light: 16 amps used for 1 minute

$$16 \text{ amps} \times \frac{1}{60} \text{ hr} = .27 \text{ A–H}$$

$$(17.5 \text{ A–H}) - (.27 \text{ A–H}) = 17.23 \text{ A–H}$$

Then: $\dfrac{17.23 \text{ A–H}}{13 \text{ amps}} = 1.3$ hrs

There are numerous modifying factors to the above. The IFR aircraft should get out of IFR conditions as soon as possible. Many modern aircraft have all electrically driven gyros whose operational times are thirty minutes or less. Know your system!

For example, I often train pilots in an amphibian that uses electrical power to extend the landing gear. The emergency extension (without power) requires 380 strokes of a hand pump! Upon alternator failure most pilots find it prudent to extend the gear right after the power failure and hope

there is enough battery left to talk to the tower, rather than the other way around!

The above example assumes that the pilot has asked a mechanic to list the current flow for each electrical item in the aircraft. If this hasn't been done (and in most cases it should have been done but wasn't—there is no regulation requiring it), an approximation can be made on the basis that (1) the pilot knows what circuit breaker operates what equipment, and (2) the total load on a circuit breaker or fuse doesn't exceed 80% of its rated capacity (which is marked in view of the pilot). So, a circuit breaker marked 5 amps should not have more than 4 amps flowing through it.

I hope the reader/pilot can now see the value of pre-planning for minor emergencies. Pre-plan what reduced electrical load you would select in the event of alternator or generator failure in the aircraft you fly for conditions of day/night, IFR/VFR, congested/non-congested area (eight conditions, right? Do you need others?). Make a list of "ON" items, their current requirements and estimated battery life. You may wish to add this list to your emergency procedures checklist.

Nickel-cadmium Batteries

Nickel-cadmium batteries have come into popularity in the past few years because of their ability to accept high charge rates, and to discharge at equally high rates without the voltage drop associated with lead-acid batteries (figure 11-9). These batteries can operate in temperatures ranging from –65 degrees to 165 degrees F, but, naturally, the extremes of this range limit the output of these batteries.

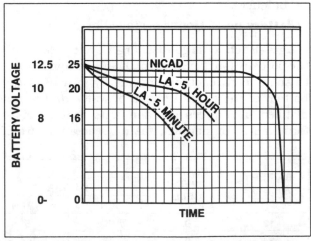

Figure 11-9. Discharge voltage related to percentage of charge for typical nickel-cadmium and lead-acid batteries.

184

The excellent voltage producing capability of the nickel-cadmium battery during discharge, as compared to the lead-acid battery (see figure 11-9) and somewhat lower susceptibility to very low temperatures are its main advantages over the lead-acid battery. However, it is more expensive, requires expensive maintenance because it develops a "memory" which must be periodically erased and is subject to thermal runaway, requiring sophisticated monitoring systems (more on these last two subject in a moment). Therefore, nickel-cadmium aircraft battery use is largely limited to turbine engine aircraft.

Construction

The base of the nickel-cadmium battery is its plate material or **plaque**. Powdered nickel is fused or sintered about a fine mesh nickel screen to form an extremely porous material. This plaque is formed into plates.

Plaque is impregnated with nickel hydroxide, electrochemically deposited within the pores to form the **positive plates**.

Cadmium hydroxide is electrochemically deposited in the pores of plaque to form the **negative plates**.

Nickel tabs are welded onto one corner of each plate, and these plates are assembled into a core assembly. Terminals are welded to the tabs of both the positive and negative plates to form the **core assembly**.

The positive plates are meshed with the negative plates to form a stackup. Between these plates is interlaced a multi-layer **separator** consisting of continuous strips wider than the plates are tall in order to completely insulate them. The entire stackup is bound with a plastic binder to form a compact assembly.

A **cover and vent assembly** is attached to the plate stackup to provide a foundation for the cell. The filler cap can be removed for electrolyte servicing, but when in place it serves as a vent for the gases liberated on charging.

A polystyrene or nylon (polyamid) **case** houses the cell. The separator extends all the way to the bottom of the case for rigidity.

Unlike the lead-acid battery, the cells of a nickel-cadmium battery are individual units, and may be serviced individually. Individual cells in their plastic cases are assembled into 12 or 24 volt batteries. These batteries are housed in epoxy coated steel cases. Twelve-volt batteries use nine or ten cells, while twenty-four volt batteries have either nineteen or twenty cells linked in series.

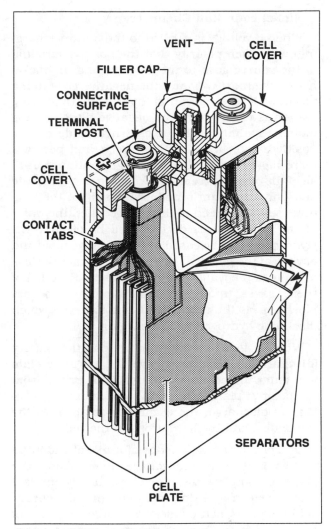

Figure 11-10. Typical nickel-cadmium cell.

Electrolyte

A 30% by weight solution of potassium hydroxide (KOH) and pure water serves as the electrolyte for nickel-cadmium batteries. Since this electrolyte serves only as a conductor and does not enter into the chemical changes, as it does with a lead-acid battery, its specific gravity does not change with the condition of charge of the cell. The specific gravity is about 1.24 to 1.30 at 80 degrees F.

Potassium hydroxide is a strong alkali, or base, and can do as much or more damage to your skin, eyes, or clothing as acid. If any of this electrolyte is spilled, it may be neutralized with a solution of boric acid and water, or with vinegar. Nickel-cadmium batteries must never be stored or transported in proximity to lead-acid batteries because the chemical reaction of their respective electrolytes, should they get together, would be severe.

Nickel-cadmium Chemistry

When a voltage is applied to the battery (charging) with such polarity that the positive terminal of the source goes to the plate having the nickel hydroxide (positive), and the negative goes to the plate with the cadmium hydroxide (negative), oxygen is driven from the negative plate, leaving metallic cadmium. The nickel hydroxide on the positive plate accepts this oxygen and becomes more highly oxidized. The charge continues until all of the oxygen is removed from the negative plate and only cadmium remains. Gassing of the cells occurs near the end of the charge when the water in the electrolyte is decomposed by electrolysis. Hydrogen gas is released at the negative plate and oxygen at the positive. This gassing causes the loss of some of the water in the electrolyte. Reaction of these gases upon re-combination may be explosive, as is the case of lead-acid batteries which discharge hydrogen while charging.

When a load is connected across the battery (discharging), electrons leave the negative plate and enter the positive, and oxygen is driven from the positive plate to be recovered by the negative. During discharge, the electrolyte is absorbed by the plates, and the level drops in the cell.

One of the characteristics of a nickel-cadmium battery is its constant voltage up to almost the point of complete discharge; another is the fact that the electrolyte does not enter into the chemical changes which cause the charge. Because of these two facts, there is no simple way to determine the condition of charge of the battery. The only way to know exactly how much charge it has, is to know how many ampere-hours of charge have been put in.

Thermal Problems

The desirable characteristics of being able to discharge at a high rate, and to accept a charge at an equally high rate gives the nickel-cadmium battery a characteristic which is equally undesirable. The nickel-cadmium battery has a very low internal resistance, and its voltage and this resistance vary as an *inverse function* of temperature. This means an increase in cell temperature will cause a decrease in voltage and internal resistance.

The temperature rise that triggers thermal problems can come from heat generated by a fast discharge (engine start), from high ambient temperatures, a poorly vented installation, or from a breakdown of the separator material. If the cellophane separator material becomes perforated,

oxygen can migrate from the positive plate to the negative. It will there unite with the cadmium to generate enough heat to decrease the internal resistance to a dangerous point.

Batteries installed in most airplanes are subjected to a constant voltage charging source, the generator. These generators have a high current producing capability.

Turbine starters require a lot of electrical energy. The battery provides this, and it usually does it without much strain. If the start has been difficult, a high current has been taken from the battery for a prolonged period of time. The battery, especially the center cells, gets warm. These cells, being warmer than the other cells, have a lower voltage and a lower internal resistance. When the engine starts and the generator gets on line, a high current is put back into the battery. Normally as this current is put back into the battery it will drop off rapidly as the battery regains its charge, but if some of the cells are unbalanced because of temperature, the current will continue to rise, causing more heat. This rise in temperature causes a further rise in current, and a condition develops in which the battery accepts all of the charging current the source is capable of producing. This condition is known as **thermal runaway**, and can cause so much heat that the battery may explode or melt its way out through the bottom of the aircraft. This problem and high cost causes operators to look on nickel-cadmium batteries with mixed emotions.

Battery Servicing

Inspection In The Airplane

Unlike the lead-acid battery, which is usually forgotten except for an occasional check for corrosion and electrolyte level, a nickel-cadmium battery should be carefully monitored for its condition. The Federal Aviation Administration has issued a requirement for inflight monitoring capability for nickel-cadmium batteries. Any unusual temperatures encountered, or any high or low current or voltage indications on the cockpit instruments serve as a warning flag to carefully check the battery.

The battery container should be free of any contamination. The ventilation system should be clear and operating. The pad in the sump jar should be saturated with a boric acid solution. At least every fifty hours, if everything is going right, and more often if trouble has been reported, the battery should be inspected for indication of spewed electrolyte, as indicated by a white powder on the top of the battery. This powder is potassium carbonate which is formed where the electrolyte has combined with carbon dioxide.

Typically every 100 hours, nickel-cadmium batteries must be **deep cycled** because they develop a "memory." If the battery is discharged numerous times to 65% charge, eventually the battery will not discharge much below that value and will fail during a difficult start. Its capacity is also decreased if one or more of the cells is out of balance with the others. To remedy this situation the battery must undergo **deep cycling** which is a shop process that starts with complete discharge. Then the battery is allowed to remain in this condition for a given period of time, and then charged back to 140% of its ampere hour capacity with a constant current charge. This is called equalization of deep cycling. Then the battery is discharged at a lower rate than before, until the individual cell voltage is down somewhere around 0.2 volts. When this voltage is reached, the cell is considered to be completely discharged. Each is then individually short circuited with a shorting strap. The battery is allowed to sit for three to eight hours in this shorted condition to allow all of the cells to reach a completely discharged condition.

After the rest period, the battery is placed on a constant current charge at its five-hour rate, for seven hours. This will put 140% of its rated ampere-hour capacity into the battery. In the last five minutes of charge, the voltage of each individual cell is measured. Any cell not having a voltage somewhere between 1.55 and 1.80 volts at a temperature between 70 and 80 degrees F is considered faulty and must be replaced.

Before the battery is returned to service, a **final check** is made of the torque of all the intercell connection hardware, the level of the electrolyte, and leakage current between the cells and the case. Any time a battery is serviced, a complete record should be made of the battery condition. This record will enable spotting troubles before they become serious. This entire process, done every 100 hours, may cost in excess of $1,000.

In-flight Nickel-cadmium Monitoring Systems

Nickel-cadmium batteries with their possibility of thermal problems require careful monitoring in flight. The FAA has issued a requirement for equipment to be installed in aircraft using this type of battery or auxiliary

Figure 11-11. A warning light indicates a cell temperature of 150 degrees F.

power unit (APU) starting to warn the flight crew of impending problems. These warning devices may be of either the temperature, current, or voltage sensing type.

Temperature sensing system—A typical temperature monitoring device uses a temperature sensor to measure the temperature of a central intercell link. When the temperature of this link reaches 140 degrees F, a warning light on the panel illuminates to tell the pilot a problem exists, and that the battery should be taken off of the generator line. If the temperature rises to 160 degrees F, a second light comes on telling the pilot that there is a problem and a landing should be done as soon as possible. A more complex system gives the pilot a direct readout of the battery temperature and activates a warning light when the intercell temperature reaches 140 degrees F (figure 11-11).

Current sensing system—Another warning device works on the principle that a thermal problem cannot exist if the charging current is monitored closely. The pilot is unable to detect a dangerous condition by monitoring the normally installed load meter, so an automatic warning device senses the current through a shunt in the charging circuit and flashes a warning to the pilot when the charge exceeds a predetermined amount.

Voltage sensing system—A third type of warning device operates on the principle that a dangerous thermal problem cannot exist if all of the cells are balanced voltage-wise. A continual sampling of the voltage of all of the cells is taken and a warning flashed to the pilot if any cell falls outside of a minimum and a maximum voltage range. This sampling is done at approximately four-second intervals. If more than one cell is out of tolerance, the flashing becomes more rapid. To back up this system, a temperature sensor is placed between the central cells of the battery to warn the pilot of an over-temperature condition.

A few words about jump-starts and APU use for the pilot. If the battery won't start the engine because it has been discharged (if your partner left the master switch on, he or she should owe you a steak dinner for your inconvenience and for the fact that the battery's life span has been decreased), an APU or set of battery jumper cables are needed.

Before attaching any of the above to the airplane, *be certain* that all avionics are shut off as well as any other electrical circuits that are not needed. This is to protect from the possibility of **reversed polarity**—where the negative terminal of the auxiliary battery is connected to the positive terminal of the aircraft system which will destroy many delicate semiconductor devices and may cause SFA (smoke, fire and anxiety).

Of course, avionics must be off (isolated from the aircraft's electrical system) during any engine start to protect voltage sensitive elements from momentary high voltage peaks that occur when very high current flows are stopped suddenly, as is the case when the starter motor is disengaged.

In addition, the danger of improper voltage from an APU (which may be capable of 12, 24 and/or 32 volts) exists. Be certain that the person operating the APU is qualified and understands what the voltage requirements are for your aircraft.

Generators

As we reviewed in Chapter 10, any time a conductor is moved in a magnetic field, it cuts across the lines of flux and a voltage is generated in the conductor; this voltage causes a current to flow. In figure 11-12(A), when the conductor is moved out from the magnet, the lines of flux will encircle the conductor as shown.

Simple AC Generator

A single conductor in a magnetic field is not an efficient producer of electricity, so to improve the output, the conductor is formed into a loop with ends attached to slip rings (figure 11-12) with

brushes connecting the rings to the load. Now, if this loop is rotated in a clockwise direction, the side B of the loop in figure 11-12(A) will move downward as it passes in front of the north pole of the magnet. And, according to the left-hand rule, the lines of flux will encircle the conductor in such a direction as to cause the electrons to flow as shown.

Side A of the coil is, in the meantime, moving upward, across the south pole of the magnet, and the direction of the lines of flux causes the electrons to flow away from the slip rings. This flow gradually decreases until, in the position shown in figure 11-12(B), both conductors are traveling parallel to the lines of flux and no lines are being cut; thus there will be no voltage generated, and no current will flow.

Now, in figure 11-12(C), the conductors are again moving across the face of the magnet, and the maximum number of lines is being cut. A voltage is again generated in the wire, but this time the polarity is opposite to that in figure 11-12(A), and the electrons flow in the opposite direction. The output of this generator is a **sine wave**; that is, one whose amplitude varies as the sine of the angle through which the coil has rotated.

The amount of current produced in generators such as this is dependent on three things: the number of lines of magnetic flux, the number of conductors cutting the flux, and the speed at which the conductors move through the flux. In essence, the amount of current depends on the rate at which the lines of flux are cut.

Simple DC Generator

Rather than attaching the ends of the moving coil to slip rings, we can produce direct current by attaching the ends to a split ring. This must be done in such a way that the side of the coil passing in front of the north pole, figure 11-13, will always be connected to the negative brush, and the side moving across the south pole will feed the positive brush. AC is generated in the armature coil, but the split-ring commutator causes electron flow through the load to be in the same direction at all times.

Figure 11-14 shows the output of a single-coil DC generator. In position A, the conductors are moving parallel to the magnetic field, and no lines are being cut, so no voltage is being generated. In position B, the black side of the coil is moving downward across one of the poles, and the white side moves upward across the opposite pole. Maximum voltage is generated and maximum current

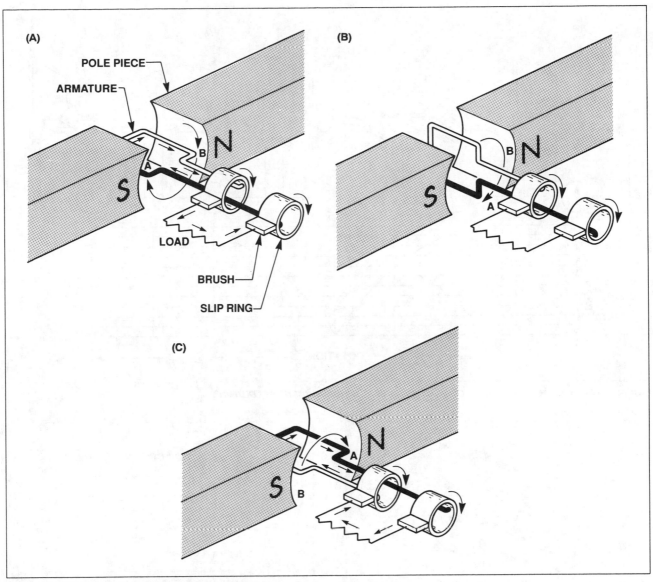

Figure 11-12. Principles of an AC generator.

flows through the load. In position C, the conductors are again moving parallel to the field and no current is generated. In D, the white side of the coil is moving downward and the black side upward. The flow in the conductor is opposite to that in position B, but since the white side is not feeding the black brush, the flow in the load circuit is the same as it was in B.

A generator such as this one has alternating current in the conductor, but because of the action of the split ring, called a **commutator**, the output is pulsating direct current. A single coil produces the pulsations shown in figure 11-14, and in order to get an output voltage with less pulsation, more coils are added to the armature. The output of an armature with three coils would be similar to that shown in figure 11-15.

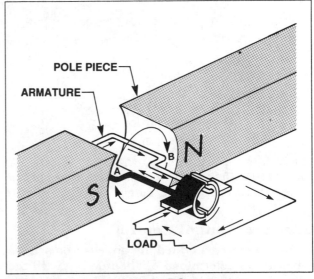

Figure 11-13. Principles of a DC generator.

189

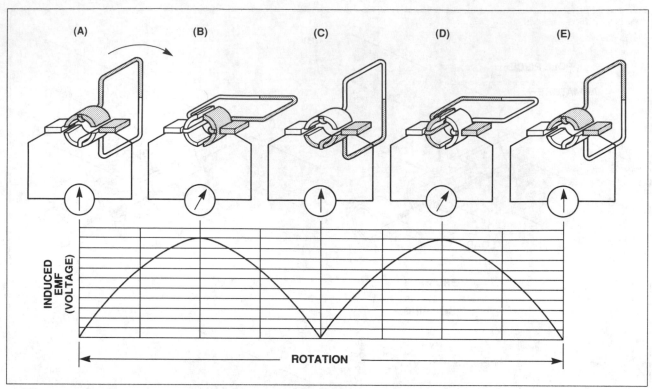

Figure 11-14. The output voltage of a DC generator is pulsating direct current.

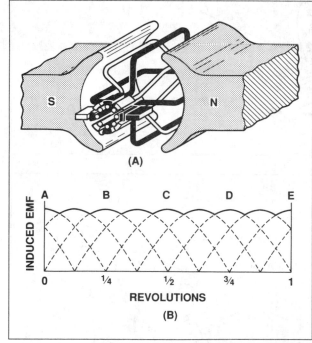

Figure 11-15. The use of more than one armature coil produces a smoother DC output.

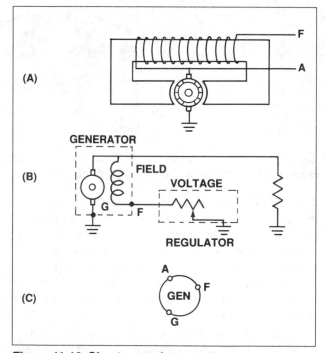

Figure 11-16. Shunt-wound generator.

Generator Output Control

The voltage generated in an electromagnetic generator is determined by the *rate* at which the lines of flux are cut. This rate is a function of the number of conductors, which is fixed by the construction of the generator, by the speed or rotation (the RPM of the generator), or by the strength of the magnetic field. *Practical voltage control is accomplished by varying the strength of the magnetic field.* Figure 11-16(A) is a schematic of a simple shunt-wound generator, in which the field flux is provided by electromagnets whose coils are in parallel with the armature output. If a variable resistor is installed in such a way that the current through the field coils can be varied, the output of the generator can be controlled (figure 11-16(B)).

Figure 11-16(C) is a typical symbol used to indicate a generator and its three connections to the aircraft electrical systems.

Control Of Aircraft DC Generators

An aircraft DC generator, as we have just seen, is self-excited; that means that the current used to produce electromagnetism in the field comes from the armature (figure 11-16(A)). If there were no control in the field circuit, the generator would run away with itself, producing more and more voltage and current and burn out its windings. As the generator starts to turn, a voltage begins to build up because of **residual magnetism** in the field frame. This voltage causes current to flow in the field coils and increases the voltage. As voltage rises, field current increases and, by a bootstrapping action, the voltage will continue to rise as indicated by the dashed curve in figure 11-17.

When the voltage required by the system is reached, the voltage regulator, which acts as a variable resistor in the field circuit, will automatically decrease the field current to prevent the voltage from rising above the regulated level. The voltage is held relatively constant by controlling the amount of field current allowed to flow.

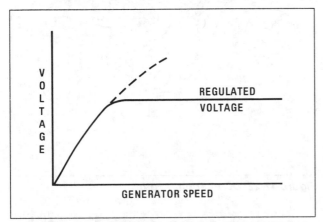

Figure 11-17. The voltage regulator allows voltage to rise to its regulated value, then holds the output voltage constant.

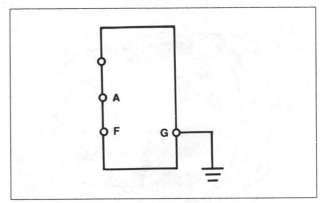

Figure 11-18. Symbol for voltage regulator for a generator.

Voltage regulators for generators are generally of the vibrator-type or carbon-pile type. For a complete explanation of how these voltage regulators work, see any of the listed texts in the 'Additional Reading' section.

Figure 11-18 shows the symbology for a voltage regulator. F is the field connection to the generator. Voltage and current supplied by the regulator to the generator through this terminal controls generator output. The regulator senses generator output by sensing system voltage at terminals A and G.

DC Alternators And Their Control

For many years direct current for use in aircraft was produced in a generator. The armature produced alternating current which was rectified by the brushes and the commutator. The main reason for using this inefficient way of producing electrical energy was the lack of non-mechanical rectifiers of sufficiently small size, which could stand the high temperature and vibration existing in a generator. When semiconductor diodes with high current rating and small size became available, there was a switch to the more efficient alternator.

Now we often see these devices referred to as *AC generators* or *AC alternators*, and they actually do generate AC in their windings. Before the AC leaves the housing, however, it is converted into direct current.

The alternator, such as the one shown in figure 11-19, has a number of advantages over the generator. In the first place, perhaps one of the most important advantages of the alternator is the fact that load current is generated in the stator or stationary winding and does not have to flow through brushes to the load. Figure 11-20 illustrates the three-phase stator commonly used

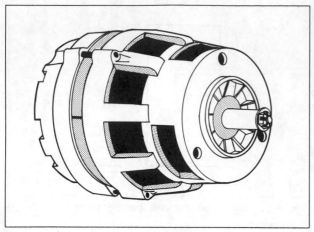

Figure 11-19. Typical alternator used in general aviation.

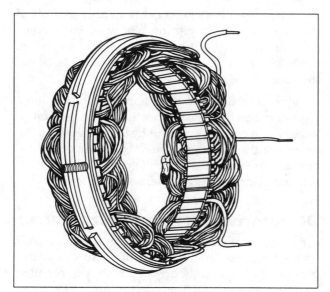

Figure 11-20. DC alternator stator.

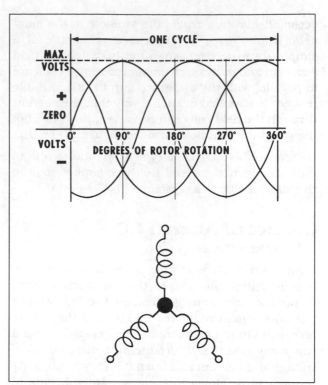

Figure 11-21. The three-phase AC output of the alternator stator is rectified by a six-diode, full-wave rectifier of figure 11-23.

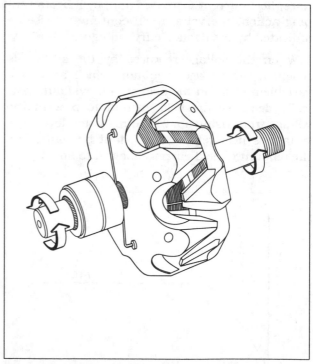

Figure 11-22. DC alternator rotor.

in aircraft alternators. The coils in this stator are connected as three windings, joined together to form a Y. There are seven pairs of poles in the rotor and seven coils in each leg of the Y, so, as the rotor turns within the stator, three phases of alternating current are generated (figure 11-21).

The rotor of an aircraft alternator is made up of two soft iron end pieces with intermeshing poles pressed onto the shaft, on either side of a drum-type exciter coil. The ends of the coil are attached to two slip rings, and the entire rotor is supported in the housing with either ball or needle bearings. The rotor shown in figure 11-22 has fourteen poles, or seven pairs, and because of their intermeshing configuration, the poles alternate, north and south. DC from the voltage regulator energizes the drum-type coil in the core of this rotor. The magnetic field produces alternating north and south poles in the soft iron, interlacing fingers formed by the two end-pieces.

192

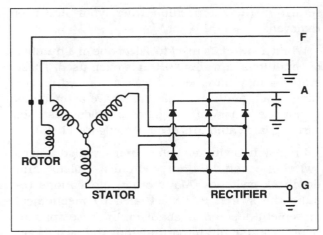

Figure 11-23. DC alternator circuit.

The rectifier of an alternator is made up of six silicon diodes, three mounted in the end frame, and the other three pressed into an insulated heat sink. Figure 11-23 is a schematic of an aircraft alternator showing the three-phase Y-connected stator, the single-phase rotor, and the six rectifier diodes. A capacitor is often placed at the output of the rectifier to protect the diodes from voltage surges.

It is possible to build a completely "brushless" alternator which does away with the stator slip rings. Current needed to produce the magnetic field in the rotor is transferred to the rotor inductively. For additional readings about alternators, see the publications listed as additional readings.

Alternators, like generators, must be controlled (regulated). If allowed to bootstrap, the typical small aircraft alternator, driven by a powerful engine, is capable of producing several hundred volts. **Note:** Circuit breakers are current sensitive devices, not voltage sensitive, so they provide no protection against high voltages.

Alternator regulators are either vibrator-type or transistorized (solid-state). See figure 11-24 for

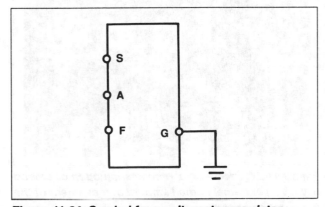

Figure 11-24. Symbol for an alternator regulator.

symbology. The S terminal is connected to +12 volts through the alternator switch which provides current to operate the solid-state circuitry of the regulator.

Circuit Control Devices

These devices control the flow of electrons through a circuit. **Switches, relays** and **solenoids** stop or permit the flow of electrons. **Fuses** and **circuit breakers** will open the circuit (stopping flow) if too much current is passing through them. **Resistors** increase the resistance of a circuit, decreasing current flow. **Capacitors** are utilized for many purposes, such as to stop DC flow yet permit AC current to flow, therefore filtering or removing an AC "ripple" from DC. **Inductors** (coils) are used to pass DC and stop AC flow and to produce magnetic fields such as in a solenoid. Diodes have many purposes including serving as a one-way "valve" which allows electrons to move in only one direction through the conductor. A **transformer** permits voltage of AC current (not DC) to be increased or decreased. A **rectifier** consists of a circuit and components that convert AC into DC. An **inverter** converts DC to AC.

A limited discussion of each of the above components in circuit control devices follows. Readers that would like to explore further into the myriad of ways these devices can be utilized to accomplish work for man's purposes are encouraged to explore the subjects further in the 'Additional Reading' sources which are a good place to start delving deeper into the subject.

Control Devices

Switches

Switches used to control (allow or disallow) the flow of electrons in most aircraft circuits are of the enclosed **toggle** or **rocker** type. These switches are actuated by either moving the bat-shaped toggle or by pressing on one side of the rocker (figure 11-25).

If the switch controls only one circuit and has only two connections and two positions, open and closed, it is called a single-pole, single-throw, or SPST switch. These may be used to control lights and other circuits that require either turning them on or off.

If the switch has three connections and either two or three positions, it is a single-pole, double-throw, SPDT switch. These switches are used to either raise or lower the landing gear, and may or

Figure 11-25. Most of the switches found in modern aircraft are either of the toggle or the rocker type.

may not have a neutral position in the middle. They may have physical limiting which requires that the handles be depressed or pulled out in order to change the switch position.

Some switches are used to control more than one circuit and may be either the single- or double-throw type. The double-pole, single-throw, DPST switch could be used to control both the battery and the generator circuit so they would both be turned on and off at the same time. A double-pole, double-throw, DPDT switch controls two circuits and has either two or three positions. The two circuits controlled are kept electrically isolated from each other in this switch.

Both toggle and rocker switches may have one or both of their positions spring-loaded so they will

return to the off position when your finger is removed.

When a switch is used to select one of a number of conditions, a **wafer switch** is often used. These switches may have several wafers stacked on the same shaft, and each wafer may have as many as twenty positions. Wafer switches are seldom used for carrying large amounts of current (figure 11-26).

Precision switches, sometimes called micro switches, include limit switches, uplock and downlock switches. Many other applications require the switch to be actuated by mechanical movement of some mechanism. In these applications it is usually important that the switch actuate when the mechanism reaches a *very definite and specific* location. These switches require an extremely small movement to actuate. Snap-acting switches of the type pioneered by Micro Switch division of Honeywell find a wide use in these applications. An extremely small movement of the plunger trips the spring so it will drive the contacts together. When the plunger is released, the spring snaps the contacts back open (figure 11-27).

Relays and **solenoids** are used when it is necessary to open or close a circuit carrying a large amount of current from a remote location and with a small switch. An example is the starter circuit for an aircraft engine. The starter motor requires a great deal of current and the cable between the battery and the starter must be as short as possible. But the starter switch must be located inside the cockpit and is often incorporated in the ignition switch. A solenoid, or contactor, is used for this application. A small amount of current is used

Figure 11-26. Wafer switches are used when it is necessary to select any of a large number of circuit conditions.

Figure 11-27. Precision switches snap open or closed with an extremely small amount of movement of the operating control.

Figure 11-28. Master and starter solenoid switches control large amounts of current, but they are operated by a very small current that flows through the coil (small wire in foreground).

to energize an electromagnet which closes the contacts in the circuit carrying the large amount of current.

If the electromagnet has a fixed core that attract a movable armature, the switch is called a **relay**, but if the core is movable, and is pulled into the hollow coil, it is called a **solenoid**.

Protective Devices

A **fuse** is a fusible link made of a low-melting-point alloy enclosed in a glass tube and is used as a simple and reliable circuit protection device. The load current flows through the fuse, and if it becomes excessive, will melt the link and open the circuit. Some fuses are designed to withstand a momentary surge of current, but will melt if the current is sustained. These slow-blow fuses have a small spring attached to the link so when the

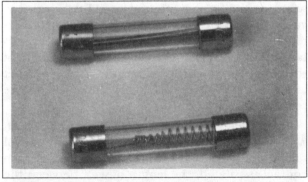

Figure 11-29. The heat caused by excess current melts the fusible link and opens the circuit. Slow-blow fuses will stand a momentary excess of current, but sustained overload will soften the link and the spring will pull it apart.

Figure 11-30. Most circuit protection for modern aircraft is provided by circuit breakers that may be reset in flight. At the top of the picture is a "pullable" circuit breaker that may be used as a switch as well as a circuit breaker.

sustained current softens the link, the spring will pull it in two and open the circuit (figure 11-29).

It is often inconvenient to replace a fuse in flight, so most aircraft circuits are protected by **circuit breakers** that will automatically open the circuit if the current becomes excessive, but may be reset by moving the operating control, which may be a toggle, a push-button, or a rocker. If the *excess current* was caused by a surge of voltage, or by some isolated and non-recurring problem, the circuit breaker will remain in and the circuit will operate normally. But if an actual fault such as a short circuit does exist, the breaker will trip again, and it should be left open.

Aircraft circuit breakers are of the trip-free type which means that they will open the circuit irrespective of the position of the operating control With this type of breaker, it is impossible to hold the circuit closed, if an actual fault exists.

There are two operating principles of circuit breakers. **Thermal breakers** open the circuit when the excess current heats an element in the switch and it snaps the contacts open. The other type is a **magnetic breaker** which uses the strength of the magnetic field caused by the current to open the contacts.

Automatic reset circuit breakers that open a circuit when excess current flows, but automatically close it again after a cooling-off period, are usually not used in aircraft circuits.

Resistors

It is often necessary in electrical circuits to control current flow by varying voltage. We will discuss this in more detail in circuit arrangement, but here we will discuss the physical characteristics of the devices used to do this.

Resistors may be inserted into circuits to drop voltage by converting some of the electrical energy into heat. And these resistors may be classified as **fixed** or **variable**.

The great majority of fixed resistors used to control small amounts of current are made of a mixture of carbon (a conductor) and an insulating material. The relative percentage of the two materials in the mix determines the amount of resistance a given amount of the material will have. Small amounts of material are used to dissipate small amounts of power and, for more power, more material is used. Composition resistors are normally available in sizes from ⅛ watt

Figure 11-31. Wire-wound resistors are used when there is a great deal of power that must be dissipated as heat.

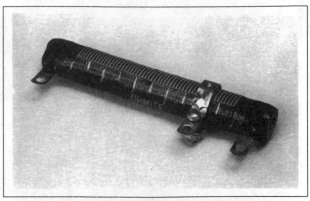

Figure 11-32. Wire-wound resistors may have a portion of the wire exposed and incorporate a movable tap.

up to two watts. The larger the physical size of the resistor, the more power it will dissipate.

Most modern resistors are of the axial-lead type; that is, the leads come directly out of the ends of the resistor. The ohmic value of this type of resistor is indicated by three or four bands of color around one end.

When more power needs to be dissipated than can be handled by a composition resistor, special resistors made of resistance wire wound over hollow ceramic tubes are used. Some of these resistors are tapped along the length of the wire to provide different values of resistance, and others have a portion of the wire left bare, so a metal band can be slid over the resistor, allowing it to be set to any desired resistance. When the screw is tightened, the band will not move from the selected resistance (figures 11-31 and 11-32).

When it is necessary to change the amount of resistance in a circuit, **variable resistors** may be used. These may be of either the composition or the wire-wound type. In the composition resistor, the mix is bonded to an insulating disk, and a wiper, or sliding contact, is rotated by the shaft to vary the amount of material between the two terminals. We saw earlier in the discussion of resistance that the resistance varies with the length of the conductor, and the farther the sliding contact is from the fixed contact, the greater will be the resistance.

Resistance wire may be wound around a form which is shaped so that the sliding contact will touch the wire at the edge of the form. As the contact is rotated, the length of wire between the terminals varies and the resistance changes (figure 11-33).

Variable resistors having only two terminals, one at the end of the resistance material, and the other, the sliding contact, are called **rheostats** and

Figure 11-33. Variable resistors allow the amount of resistance in a circuit to be changed by rotating the shaft.

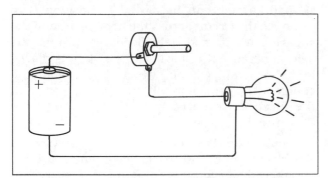

Figure 11-34. Rheostats are used to vary the amount of resistance in a circuit.

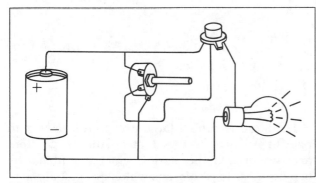

Figure 11-35. Potentiometers are used as voltage dividers in a circuit.

are used to vary the amount of resistance in a circuit (figure 11-34).

If the resistor has three terminals, one for either end of the resistance material, and one for the slider, it is called a **potentiometer**, and may be used as a voltage divider, of which we will say more in the section on circuits (figure 11-35).

Inductors

As the amount of current flow changes (as with AC), the magnetic field expands or contracts, and as it does, the flux cuts across the conductor and induces a voltage into it. According to Lenz's law, the voltage that is induced into the conductor is of such a polarity that it opposes the change that caused it. For example, as the voltage begins to rise and the current increases, the expanding lines of flux cut across the conductor and induce a voltage into it that opposes, or slows down, the rise. When the current flow in the conductor is steady, lines of flux surround it, and since there is no change in the amount of current, these lines do not cut across the conductor, and so there is no voltage induced into it. When the current decreases, the lines of flux cut across the conductor as they collapse, and they induce into it a voltage that opposes the decrease.

When a conductor carries alternating current, both the amount and the direction of the current continually change, and so an opposing voltage is constantly induced into the conductor. This induced voltage acts as an opposition to the flow of current.

Factors Affecting Inductance

Inductance opposes a change in current by the generation of a back voltage. All conductors have the characteristic of inductance, since they all generate back voltage any time the current flowing in them changes. The amount of inductance is increased by anything that concentrates the lines of flux, or causes more of the flux to cut across the conductor. If the conductor is formed into a coil, the lines of flux surrounding any one of the turns cut not only across the conductor itself, but also across each of its turns, and so it generates a much greater induced current to oppose the source current. Thus, a coil is often called an **inductor** or **choke**.

There is much more to the workings of inductors that the reader can explore (again, start with the additional readings). The pilot should at least know that inductance and inductors exist, create problems (induced current back flow as the current induced magnetic field collapses when DC current is shut off, causing switch contact arcing, etc.) and are very useful components in electrical circuits (create magnetic fields for relays, passes DC but restricts AC and many more useful tasks).

Energy Stored In Electrostatic Fields

Electrical energy, as we have seen, may be stored in the magnetic field which surrounds a conductor through which electrons are moving. It may also be stored in *electrostatic* fields caused by an accumulation of electrical charges that are not moving, but are static. The electromagnetic field strength is determined by the amount of *current* flowing in the conductor, but the strength of the electrostatic field is determined by the amount of pressure, or voltage, of the static charges.

A **capacitor**, sometimes called a **condenser**, is a device that stores electrical energy in the electrostatic fields that exist between two conductors that are separated by an insulator, or **dielectric**. Let's consider the circuit in figure 11-36 where two flat metal plates are arranged so they face each other, but are separated by an insulator. One of the plates is attached to the positive terminal of the power source and the other to the negative terminal. When the switch is closed, because of the battery EMF, electrons will be drawn from the plate attached to the positive battery terminal and will flow to the plate attached to the negative terminal. There can be no flow across the insulator, but the plates will become charged. If the voltmeter reading were taken across points C and D, it would be found to be exactly the same as that taken across points A and B. Current flow would be indicated by the ammeter during the time the plates are being charged, but when they become fully charged, no more current will flow.

In the circuit of figure 11-37, a power source is connected to a capacitor through a resistor, which limits the amount of current that initially flows

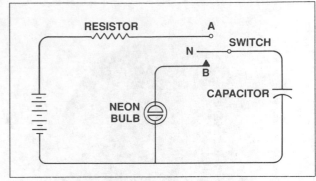

Figure 11-37. The resistor limits the rate at which the capacitor will charge when the switch is in position A. When it is in position B, the capacitor will discharge through the neon bulb, causing it to flash.

into the capacitor, but does not prevent the voltage across the capacitor rising to that of the source.

When the voltage across the capacitor rises to the voltage of the source, no more current will flow. Now the switch can be moved to position B to complete the circuit across the neon light, and the capacitor will immediately discharge through the light and cause it to flash. If the switch is placed in its neutral position, when the capacitor is charged, the capacitor will remain charged until the electrons eventually leak off through the dielectric.

A **capacitor** is a device which will store an electrical charge, and its capacity is measured in farads, with one farad the amount of capacity that will hold one coulomb of electricity (6.28×10^{18} electrons) under a pressure of one volt.

$$C = \frac{Q}{E}$$

C = capacity in farads

Q = charge in coulombs

E = voltage in volts

The farad is such a large unit that it is seldom used in practical circuits. Instead, most capacitors are measured in microfarads which are millionths of a farad, or in picofarads which are millionths of millionths of a farad. Picofarads have formerly been called micro-micro farads and may still be referred to in this way in some texts. The Greek letter mu (μ) is used to represent the prefix micro.

1 microfarad (μf) = 1×10^{-6} farad

1 picofarad (Pf or $\mu\mu$f) = 1×10^{-12} farad

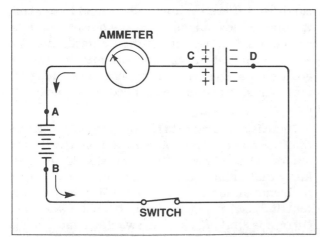

Figure 11-36. Current flows only when the capacitor is charging or discharging.

198

Figure 11-38. A magneto capacitor is a paper capacitor sealed in a metal container.

The capacity is affected by three variables: the area of the plates, the separation between the plates, and the dielectric constant of the material between the plates (how good an insulator it is).

It only stands to reason that the larger the plates, the more electrons can be stored. One very common type of capacitor has plates made of two long strips of metal foil separated by waxed paper and rolled into a tight cylinder. This construction provides the maximum plate area for its small physical size (figure 11-38).

Capacitors do many wonderful things for us in airplanes and in most electronic circuits. There are hundreds of them in your radios, TV sets, glass cockpits, etc. There is much more to know about them than is presented here but do try to remember that capacitors can store electrons for quick use (strobe flash) and for when there are too many electrons flowing in the circuit (an opening switch that is arcing). Effectively, capacitors stop DC flow and allow AC current to pass. They are indeed a useful device—you will see a few in diagrams of most aircraft electrical systems. See figure 11-39 for another interesting example of capacitance.

Transformers

When alternating current flows in a conductor, the changing lines of flux radiate out and cut across any other conductor that is nearby, and anytime they cut across a conductor, they generate a voltage in it even though there is no electrical connection between the two. See figure 11-40. This voltage is said to be generated by **mutual inductance**, and is the basis for **transformer** action that is so important to us in our use of alternating current, as it allows us to change the values of AC voltage and current in circuits.

Consider in figure 11-41 two coils of wire, a primary and a secondary, wound around a common core, but not connected electrically.

When an alternating current flows in the primary, a voltage will be induced into the secondary, and current will flow in it, lighting the bulb.

Since we can consider these windings to be purely inductive, the current in the primary winding lags the source voltage by 90°. The voltage induced into the secondary winding will be greatest when the change in current is the greatest, and it is therefore 90° out of phase with the current in the primary. So when the two phase shifts are added, the voltage in the secondary winding will be 180° out of phase with the voltage in the primary.

The amount of voltage generated in the secondary winding of a transformer is equal to the

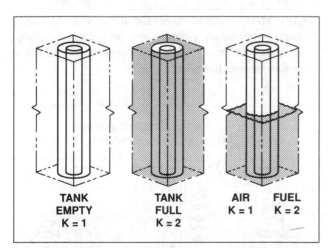

Figure 11-39. The capacitance-type fuel gauging system uses capacitors as probes in the fuel tank. When the tank is empty, the dielectric is air. When the tank is full, fuel is the dielectric. As the tank is filled, the dielectric changes proportionately, changing capacitance which is measured and indicated on the fuel gauge. See Chapter 16 for a full explanation of this system.

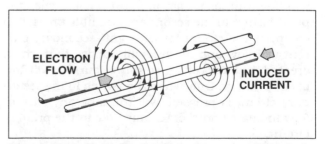

Figure 11-40. Mutual induction causes a voltage to be induced into a conductor not electrically connected to the conductor through which the source current flows.

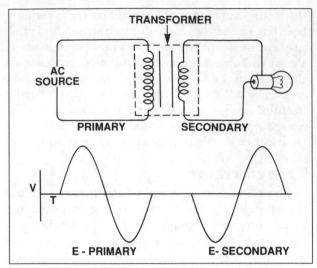

Figure 11-41. *The voltage induced into the secondary of a transformer by mutual induction is 180 degrees out of phase with the source voltage.*

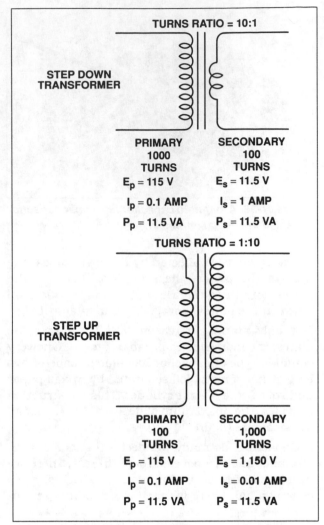

Figure 11-42. *Voltage increases directly as the turns ratio between the primary and secondary windings of a transformer, while the current decreases in proportion to the turns ratio.*

voltage in the primary, *times the turns ratio between the primary and the secondary windings.* For example, if there are 100 turns in the primary winding and 1,000 turns in the secondary, the turns ratio is 1:10, and if there are 115 volts across the primary, there will be 1,150 volts across the secondary. A transformer does not generate any power, so the product of the voltage and the current in the secondary must be the same as that in the primary. Because of this, there must be a flow of one ampere in the primary winding to produce a flow of 100 milliamps in the secondary. Actually, transformers aren't 100% efficient so some power is converted to heat and not all of the current is apparent in the secondary.

A transformer may have its primary winding connected directly across the AC power line, and as long as there is an open circuit in the secondary winding, the back voltage produced in the coil will block the source voltage enough that there will be almost no current flowing through the primary winding. But, as we see in figure 11-43, when the push button in the secondary circuit is pressed to complete the circuit for the light, secondary current will flow and its flux will oppose that which created the back voltage so source current will flow in the primary. Only when the switch in the secondary circuit is pressed, will there be any current flow indicated on the AC ammeter in the primary circuit.

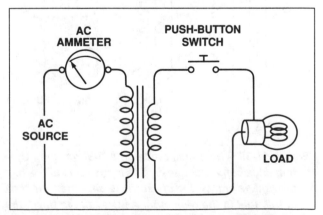

Figure 11-43. *Primary current flows only when the secondary circuit is completed.*

Types Of Practical Inductors

Chokes

It is often necessary to install an inductor, or a choke, in a circuit to impede the flow of alternating current of a particular frequency while not affecting the flow of AC below that frequency.

When alternating current in the power frequencies of 50 or 60 hertz is rectified, or changed into direct current, the output is in the form of pulsating direct current, and if an inductor is placed in series with the load, the changing current will induce a back voltage that will tend to smooth out the pulsations, or ripples. Chokes of this type have laminated iron cores and often have an inductance of more than one henry.

Figure 11-44. Forms of commonly used inductors.

Transformers

We can get almost any voltage of alternating current by using a transformer. The primary winding is designed to accept the voltage and frequency of the power source, and there may be one or more secondary windings needed for the particular application.

Step-up Or Step-down Transformers

When there are more turns in the secondary than in the primary, the transformer is called a step-up transformer, but if the secondary has fewer turns, it is called a step-down transformer. Step-down transformers are often used to get the high current necessary for operating some motors.

In order to increase voltage of DC current, it must first be converted to AC, then stepped up, then converted back to DC.

Inverters

Inverters convert DC to AC. One way to make an inverter is to drive an AC generator with a DC motor, thus producing a sine wave. Solid-state, low-power inverters are available for converting 12 VDC to 117 VAC, so the car or airplane 12-volt system can power an electric shaver or small electric drill.

An inverter is used in some aircraft systems to convert a portion of the aircraft's DC power to AC. This AC is used mainly for instruments, radio, radar, lighting, and other accessories. These inverters are usually built to supply current at a frequency of 400 hertz, but some are designed to provide more than one voltage; for example, 26-volt AC in one winding and 115 volts in another.

There are two basic types of inverters: the rotary and the static.

There are many sizes, types, and configurations of **rotary inverters**. Such inverters are essentially AC generators and DC motors in one housing. The generator field, or armature, and the motor field, or armature, are mounted on a common shaft which will rotate within the housing. One common type of rotary inverter is the permanent magnet inverter (figure 11-45(A)).

Permanent Magnet Rotary Inverter

A permanent magnet inverter is composed of a DC motor and a permanent magnet AC generator assembly. Each has a separate stator mounted within a common housing. The motor armature is mounted on a rotor and connected to the DC supply through a commutator and brush assembly. The motor field windings are mounted on the housing and connected directly to the DC supply. A permanent magnet rotor is mounted at the opposite end of the same shaft as the motor armature, and the stator windings are mounted on the housing, allowing AC to be taken from the inverter without the use of brushes.

Static Inverters

In many applications where continuous DC voltage must be converted to alternating voltage, static inverters are used in place of rotary inverters or motor generator sets. The rapid progress being made by the semiconductor industry is extending the range of applications of such equipment into voltage and power ranges which would have been impractical a few years ago. Some such applications are power supplies for frequency-sensitive military and commercial AC equipment, aircraft emergency AC systems, and conversion of wide frequency range power to precise frequency power.

The use of static inverters in small aircraft also has increased rapidly in the last few years, and the technology has advanced to the point that static inverters are available for any requirement filled by

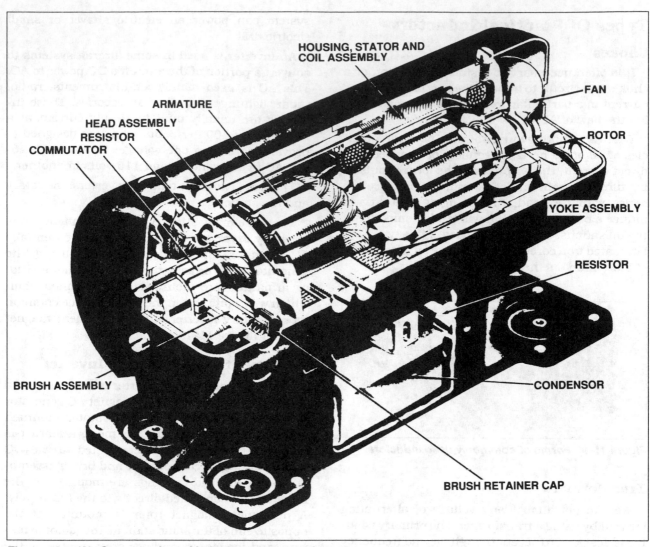

Figure 11-45(A). Cutaway view of inductor-type rotary inverter.

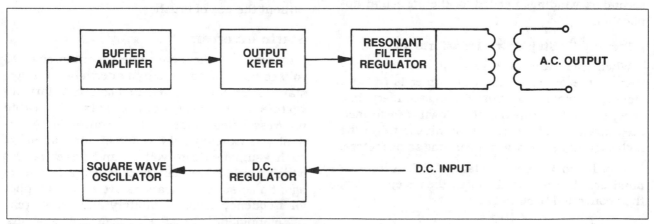

Figure 11-45(B). Regulated sine wave static inverter.

rotary inverters. For example, 250 VA emergency AC supplies operated from aircraft batteries are in production, as are 2,500 VA main AC supplies operated from a varying frequency generator supp-ly. This type of equipment has certain advantages for aircraft applications, particularly the absence of moving parts and the adaptability to conduction cooling.

Static inverters, referred to as solid-state inverters, are manufactured in a wide range of types and models, which can be classified by the shape of the AC output waveform and the power output capabilities. One of the most commonly used static inverters produces a regulated sine wave output. A block diagram of a typical regulated sine wave static inverter is shown in figure 11-45(B). This inverter converts a low DC voltage into higher AC voltage. The AC output voltage is held to a very small voltage tolerance, a typical variation of less than 1 percent with a full input load change. Output taps are normally provided to permit selection of various voltages; for example, taps may be provided for a 105-, 115-, and 125-volt AC outputs. Frequency regulation is typically within a range of one cycle for a 0-100 percent load change.

Variations of this type of static inverter are available, many of which provide a square wave output.

Since static inverters use solid-state components, they are considerably smaller, more compact, and much lighter in weight than rotary inverters. Depending on the output power rating required, static inverters that are no larger than a typical airspeed indicator can be used in aircraft systems. Some of the features of static inverters are:

1. High efficiency.
2. Low maintenance, long life.
3. No warmup period required.
4. Capable of starting under load.
5. Extremely quiet operation.
6. Fast response to load changes.

Static inverters are commonly used to provide power for such frequency-sensitive instruments as the attitude gyro and directional gyro. They also provide power for autosyn and magnesyn indicators and transmitters, rate gyros, radar, and other airborne applications.

Diodes

The Solid-state Diode

Solid-state electronics ushered in a revolution in communications, control, and in actually almost all aspects of electricity. One small piece of an insulating material such as germanium or silicon is "doped" with an element having an excess of electrons and is called an N-material. Another piece of the same material is doped with an element having a deficiency of electrons, and this is

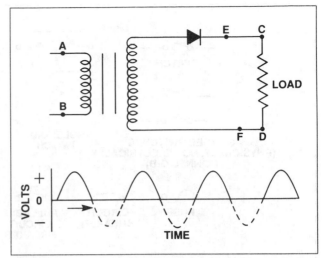

Figure 11-46. A semiconductor diode half-wave rectifier with its output waveform. Points A-B voltage is the full AC sine wave form. The diode blocks flow in one direction (dashed line), so the voltage measured at points C-D show the load getting a pulsating DC (all electrons flowing in the same direction). Adding a capacitor from E to F would smooth the pulsations or ripple in the DC.

called the P-material. These two pieces are fused together and they act as a check valve. Electrons can easily pass from the N-material to the P-, but they cannot flow from the P- to the N-.

When a solid-state diode is placed in the circuit such as that in figure 11-46, the alternating current is changed into pulsating direct current, producing a circuit called a **half-wave rectifier**, which is discussed below.

Today, diodes come in a vast array of types and sizes to do many tasks. Specialized diodes amaze us with what they can do. For example, the **zener diode** is a special silicon junction diode with an unusual characteristic wherein it stops current flow in the reverse direction, but only up to a specific voltage, called the *threshold voltage*, above which current can flow with little resistance from the zener diode. Therefore, this special diode can be used to limit, or regulate, voltage in a circuit. See figure 11-47 for the zener diode symbol and the other symbols you will use to understand the aircraft's electrical system in the next chapter.

Half-wave Rectifier

A half-wave rectifier circuit uses a single diode in series with the voltage source, which is the secondary of the transformer in figure 11-41, and the load. Electrons can flow only during the half-cycle when the cathode, represented by the bar

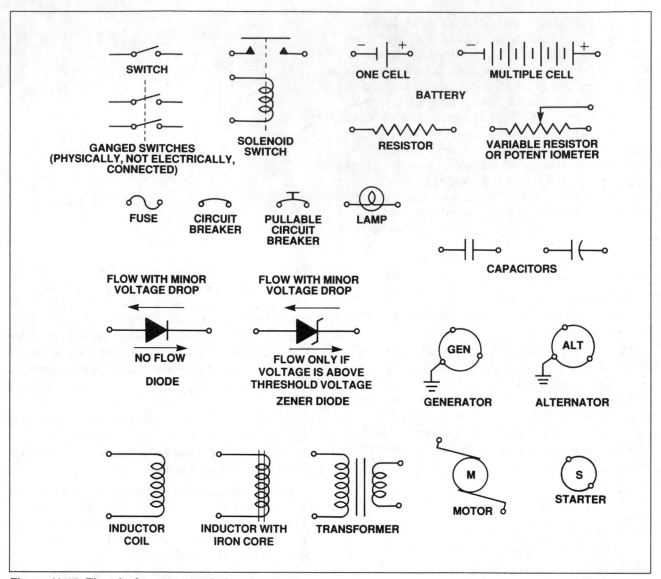

Figure 11-47. Electrical system symbols.

across the arrowhead, is negative. The output waveform of this type of rectifier is one-half of the alternating-current wave, and because of this, it is a very inefficient type of circuit.

Full-wave Rectifier

In order to change both halves of the AC cycle into DC, a transformer with a center-tapped secondary winding and two diodes can be used.

To give you a chance to think with conventional flow theory, let's follow the flow of *conventional* current (figure 11-48), which is from the positive side of the source, through the load to the negative side. Conventional current, with direction opposite that of electron flow, follows the direction of

the arrowheads in the diode symbols. During the half-cycle when the top of the secondary winding is positive, current flows through diode D_1, and passes through the load from the top to the bottom. This causes the top of the load to be positive. After leaving the load, the current flows into the secondary winding at the center tap which is negative during this half-cycle. During the next half-cycle, the bottom of the secondary winding will be positive and the center tap will be negative with respect to the bottom. Current will flow through diode D_2 and through the load resistor in the same direction it passed during the first half-cycle. The output waveform is pulsating direct current, whose frequency is twice that of the pulsating DC produced by a half-wave rectifier.

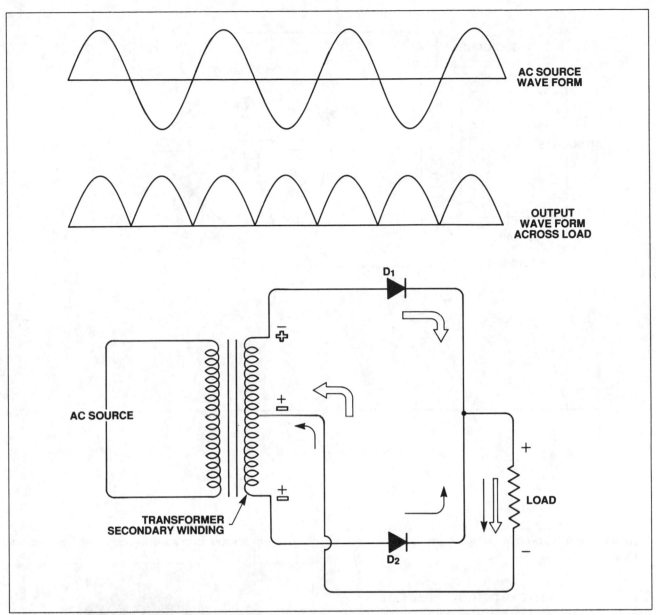

Figure 11-48. A two-diode, full-wave rectifier with its output waveform. All of the AC source is converted to DC with a less severe, more efficient pulse than that of the half-wave rectifier. Most battery chargers use a circuit like this.

Bridge-type Full-wave Rectifier

The two-diode full-wave rectifier requires a transformer that will give the desired output voltage across just one-half of the secondary winding. This is inefficient, and this inefficiency may be overcome by the bridge-type full-wave rectifier, which uses the entire secondary winding. To do this, four diodes instead of two are used. This type of rectifier circuit is commonly used.

Conventional current flow can be traced through the load in figure 11-49. During the half-cycle when the top of the transformer secondary is positive, the current flows through diode D_1 and through the load resistor from the right to the left, and then down through diode D_2 and back to the bottom of the secondary winding, which is negative. During the next half-cycle, the polarity of the secondary has reversed, and current flows through diode D_3, through the load in the same direction as before, and up through diode D_4 to the top side of the transformer, which at this time is negative. The output waveform is similar to that produced by the two-diode full-wave rectifier, but the voltage is quite a bit higher because with this circuit the entire secondary winding is used.

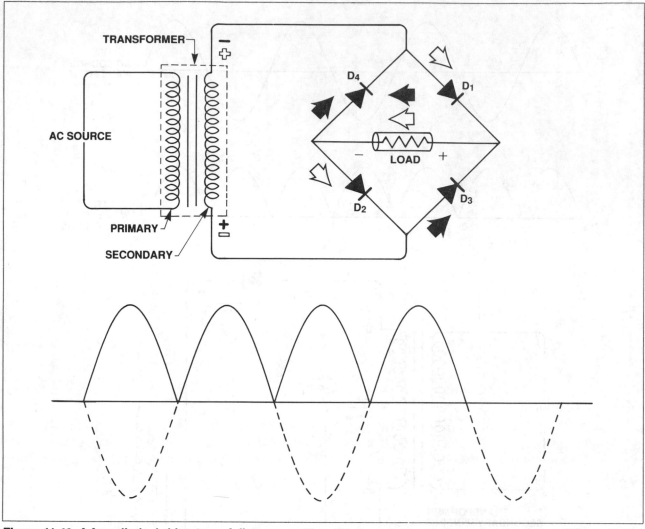

Figure 11-49. A four-diode, bridge-type, full-wave rectifier with its output waveform showing conventional current flow.

Full-wave Three-phase Rectifier

The most familiar six-diode three-phase full-wave rectifier is the one found in the DC alternator, which is rapidly replacing the generator as a device for producing direct-current electricity for modern aircraft. This alternator uses a three-phase stator and six silicon diodes arranged as those in figure 11-51.

Let's examine the current flow through the load resistor for one complete cycle of all three phases. Remember, we are tracing conventional current which is opposite of electron flow, and follows the direction indicated by the arrowheads in the diode symbols. In that portion of the cycle when the output end of phase A is positive, current leaves it and flows through diode D_1 and down through the load, making its top end positive. After leaving the load, the current flows through diode D_2 and coil

Figure 11-50. Most aircraft direct current is produced by a three-phase alternator.

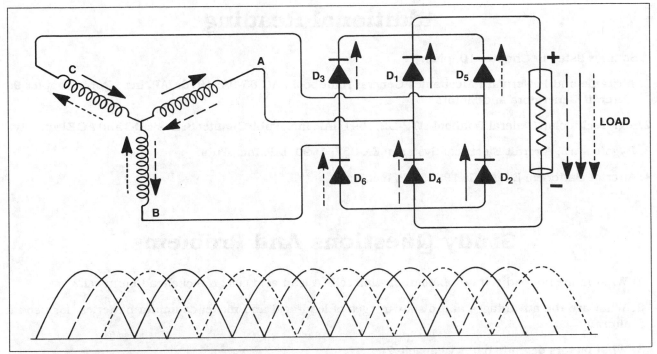

Figure 11-51. Three-phase, six-diode, full-wave rectifier and its output waveform.

C, whose output lead is negative. Now, as the alternator field rotates, it causes the output end of the coil B to become positive and the output of coil A to be negative. Current flows out of B, through diode D_3, the load, diode D_4, and back through coil A. Continued rotation causes the output of coil C to be positive and B to be negative, so the current leaves C, passes through diode D_5, the load and diode D_6 and back into coil B.

The output waveform of this rectifier gives a very steady direct current as the current from the three phases overlaps and there is never a time when the current drops below system voltage.

Additional Reading

Sources listed in Chapter 10, plus:

1. Airframe and Powerplant Mechanics General Handbook; AC 65-9A; 1976; IAP, Inc., Publ; Chapter 9: Aircraft Generators and Motors

2. A&P Technician General Textbook; ITP-G2; 1991; IAP, Inc., Publ; Chapter 3: Basic DC and AC Electricity.

3. Bygate, J.E.; Aircraft Electrical Systems; EA-357; 1990; IAP, Inc., Publ.

4. Aircraft Batteries; EA-AB-1; 1985; IAP, Inc., Publ.

Study Questions And Problems

1. What is the no-load voltage of a carbon-zinc cell? A lead-acid cell? A nickel-cadmium cell?

2. What are the advantages and disadvantages of lead-acid and nickel-cadmium batteries, for use in aircraft?

3. What factors degrade battery capacity?

4. What is battery capacity?

5. An aircraft is equipped with a 25 amp-hour battery. The alternator fails at 00:00Z on a night VFR flight. The pilot determines the minimum electrical load will be navigation lights (4 amps), instrument lights (1.5 amps), turn coordinator (1 amp) and one nav-com (5 amps in receive mode). The pilot can probably expect the electrical equipment to keep working until at least what time? (Show work.)

6. Make a list of electrical devices you would leave on after an alternator failure in the aircraft you regularly fly, the load of each, total load and expected duration before battery failure. Assume VFR day and night conditions. State type of aircraft and sources for each device's load.

7. What precautions must the pilot take before an APU or jump-start?

8. What electrical principal allows a generator to work? Explain it.

9. How is the output of a generator and alternator controlled?

10. Describe the function of each of the following and give the symbol for each, except the last two.
    Switch
    Relay
    Solenoid
    Fuse
    Circuit breaker
    Pullable circuit breaker
    Resistor
    Capacitor
    Diode
    Inductor
    Transformer
    Rectifier
    Inverter

Chapter XII
Aircraft Electrical Systems

Before looking into the aircraft electrical system, let's get familiar with some very simple, basic electrical circuits. The first page of Chapter 11 indicated that a circuit must have at least three parts, called (1) the source, (2) the electron transportation and distribution system, and (3) the **load**—the location(s) where the work is done.

Figure 12-1 shows a very simple circuit—that of a one cell flashlight. The source is the 1.5 volt dry cell battery. The distribution system consists of a wire or conductor connecting the positive terminal of the battery to the switch, a conductor connecting the other pole of the switch to the light bulb filament which has a high resistance to electron flow, creating heat sufficient to make the filament glow, and a conductor from the filament to the battery's negative terminal.

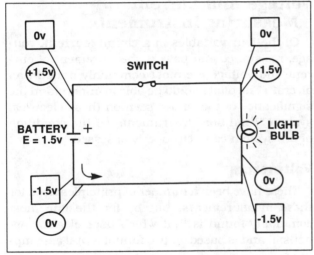

Figure 12-1. The circuit of a one cell flashlight.

The above describes the pathway of electrons but the description started at the positive terminal and described the pathway in the direction opposite that of electron flow, again illustrating the idea of "conventional current theory." The arrow in figure 12-1 indicates the direction of electron flow.

Ohm's law ($E = IR$) and the power relationship ($P = IE$) can be used to determine the current flow and the power dissipated as heat and light by the filament (the **load**). Dividing both sides of the Ohm's law equation by R:

$$\frac{E}{R} = \frac{IR}{R}, \quad \text{we get} \quad I = \frac{E}{R} = \frac{1.5\,V}{100\,watts} = .015\,amps$$

Then,

$$P = IE = .015\ amps \times 1.5\,volts = .0225\ watts$$

$$= 2.25 \times 10^{-2}\ watts$$

The circled values of voltage indicate the voltage at all parts of the circuit if the zero-voltage reference point is taken as the negative terminal of the battery. Values in the square boxes indicate voltages if the zero-voltage reference point is assumed to be the positive terminal of the battery. In either case, it can be seen that the voltage drops across the filament, where the work is being done. *There is always a voltage drop across a resistance when work is being done.*

Figure 12-2 presents two ideas; the first is the use of the ground symbol. Consider all ground symbols to be electrically connected together by a very low resistance conductor. So, all grounded

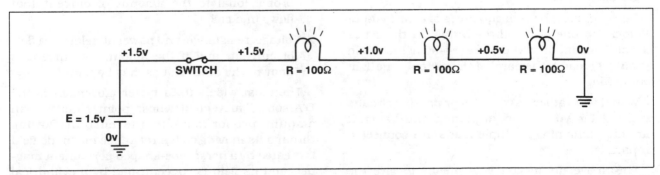

Figure 12-2. A series circuit.

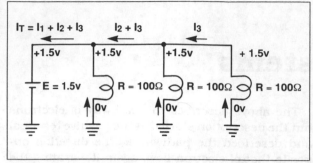

Figure 12-3. A parallel circuit (electron flow).

points are electrically attached together (usually by attaching to the metal frame of the airplane or other vehicle). The second idea presented is that of a series circuit.

Series And Parallel Circuits

In a **series circuit** all electrons must flow through all active components. In figure 12-2, all current must flow through all three light bulbs. In the **parallel circuit** of figure 12-3, a portion of the current flows through each filament. In this case, since the three filaments are equal in resistance, one-third of the electrons flow through each filament. The pathway of electron flow is from the source (always) out the negative terminal into the ground, then a portion of the electrons flow through each filament, then back to the source at the positive terminal.

The amount of voltage dropped across the filament in a series circuit is always less than the total. The value of the voltage drop depends on the amount of resistance of the individual load and the amount of resistance of other loads in the series circuit. In this example, since all resistances are equal, the voltage drops across each of the filaments are equal. Voltage drops across the loads in a parallel circuit are equal to the total voltage produced by the source. Figure 12-4 analyzes the current and power of these sample series and parallel circuits.

Consider also the consequences of a burned-out filament in one of the three bulbs. In the series circuit, current flow would cease—all lights would go out whereas in the parallel circuit, only one light would fail.

Would you rather have series or parallel circuits working for you in an airplane? Most aircraft circuits containing multiple loads are connected in parallel.

The above discussion was meant to give the reader an intuitive feeling for what occurs in rela-

---

SERIES (FIGURE 12-2)

$$I = \frac{E}{R} = \frac{0.5\,V}{100\,\Omega} = .005\,A$$

$$P = .005\,A \times .5\,V = .0025\,W = 2.5\,\text{MILLIWATTS}$$

PARALLEL (FIGURE 12-3)

$$I = \frac{E}{R} = \frac{1.5\,V}{100\,\Omega} = .015\,A$$

$$P = .015\,A \times 1.5\,V = .0225\,W = 22.5\,\text{MILLIWATTS}$$

Figure 12-4. Current and power for each individual resistor in a series and a parallel circuit.

tively simple DC circuits such as are found in light aircraft. For more details regarding circuit analysis, see the 'Additional Reading' section at the end of this chapter.

Voltage And Current Measuring Instruments

Of the four variables in a circuit (current, voltage, resistance and power), measurement of current and voltage are most commonly done in an aircraft. The pilot should be able to understand the significance of the values seen on these electrical system "condition" instruments so the condition of the system can be properly assessed.

Voltmeters

There have been a number of principles used for these measurements, but by far the one most commonly found is that which uses electromagnetism, and is based on two fundamental assumptions:

1. The strength of an electromagnetic field is proportional to the amount of current that flows in a coil.

2. Voltage, resistance, and power all relate to a flow of current, and if the amount of current is known, the other values may be found.

The most widely used meter movement is the D'Arsonval movement whose pointer deflects an amount proportional to the current flowing through its moving coil. A reference magnetic field is created by a horseshoe-shaped permanent magnet, and its field is concentrated by a cylindrical keeper in the center of the open end.

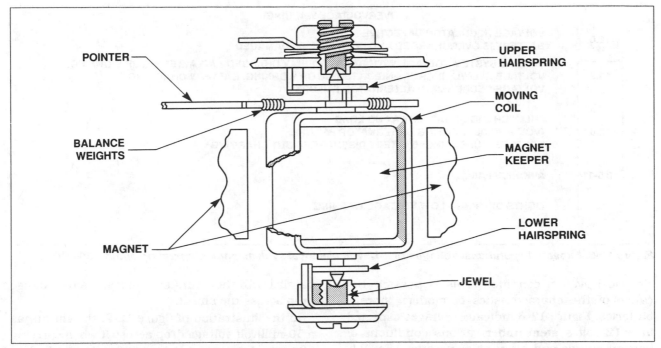

Figure 12-5. The moving coil of a D'Arsonval-type meter.

Surrounding the keeper and supported by hardened steel pivots riding in smooth glass jewels, is a coil through which the current to be measured flows. The current enters and leaves the coil through calibrated hairsprings, one surrounding each of the pivots. Current flowing through this coil creates a magnetic field whose polarity is the same as that of the permanent magnet, and thus the two fields oppose each other. The opposing force rotates the coil on its pivots until the force of the hairspring exactly balances the force caused by the magnetic fields.

Oscillation of the pointer is minimized by electromagnetic damping. The moving coil is wound around a thin aluminum bobbin, or frame, and as this frame moves back and forth in the concentrated magnetic field, eddy currents are generated within the bobbin that produce their own fields which oppose the movement.

In order to use the basic D'Arsonval meter movement to measure the different variables, we must know some of its characteristics.

Full-scale current is the amount of current that must flow through the meter coil to deflect the pointer over the full calibrated scale.

Ohms-per-volt sensitivity is a measurement of meter sensitivity that is the reciprocal of the full-scale current and is the total amount of resistance for each volt of pressure needed to produce full-scale current. A meter that requires one milliamp

(1/1,000 amp) of current for full-scale deflection would require one thousand ohms in the meter circuit to limit the current through the meter to one milliamp. This meter is said to have a sensitivity of one thousand ohms per volt.

Many multimeters have a sensitivity of 20,000 ohms per volt and these meters require 1/20,000 amp, or 50 microamps, of current to move the pointer full scale.

Meter resistance is the total resistance of the meter that must be considered when making any computations regarding the current through the meter. Both the moving coil and the hairsprings have resistance, and in some meters there is a temperature compensating resistor in series with the coil. This resistor is made of a material whose resistance decreases with an increase in temperature which is opposite to the change in resistance of the coil. As a result, the meter resistance remains constant as the temperature changes.

Since current flow through the meter coil is extremely small but is proportional to amount of meter movement, this meter can be wired in parallel across a circuit to be measured, and will have a meter deflection proportional to voltage. So, like many other instruments in aircraft, it is "fibbing" to the pilot, in that the meter measures current but indicates values of voltage.

The **voltmeter** is used in place of the ammeter in some modern aircraft and automobiles to indicate

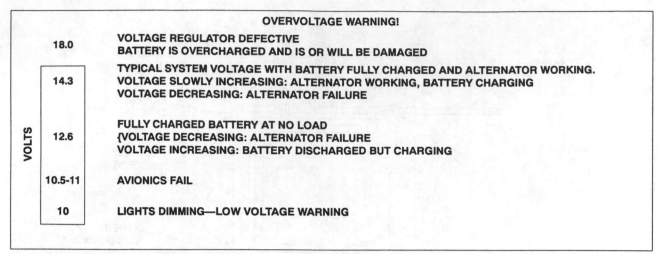

VOLTS		
		OVERVOLTAGE WARNING!
	18.0	**VOLTAGE REGULATOR DEFECTIVE** **BATTERY IS OVERCHARGED AND IS OR WILL BE DAMAGED**
	14.3	**TYPICAL SYSTEM VOLTAGE WITH BATTERY FULLY CHARGED AND ALTERNATOR WORKING.** **VOLTAGE SLOWLY INCREASING: ALTERNATOR WORKING, BATTERY CHARGING** **VOLTAGE DECREASING: ALTERNATOR FAILURE**
	12.6	**FULLY CHARGED BATTERY AT NO LOAD** **{VOLTAGE DECREASING: ALTERNATOR FAILURE** **VOLTAGE INCREASING: BATTERY DISCHARGED BUT CHARGING**
	10.5-11	**AVIONICS FAIL**
	10	**LIGHTS DIMMING—LOW VOLTAGE WARNING**

Figure 12-6. Expected operational voltages of a 12-volt electrical system, engine operating above 1500 RPM.

electrical system condition. This is possible because of the characteristics of modern storage batteries. Figure 12-6 indicates voltages expected in a 12-volt system under various conditions of battery charge and alternator operation. For 24-volt systems, double the values in figure 12-6.

Ammeters

The **ammeter** measures current flow. In order to measure large current flows without imposing a large current loss on the system, the D'Arsonval meter is used in parallel with a **shunt** which is a conductor whose size and therefore resistance is precisely established so that voltage drop will be proportional to the large amount of current flowing through it. The D'Arsonval meter is installed in parallel with the shunt so it measures the voltage drop across the shunt.

In the illustration of figure 12-7, the shunt has a 50-millivolt voltage drop across it when carrying 60 amps of current. A D'Arsonval meter is attached in parallel with the shunt in order to measure the voltage drop across the shunt. A meter is chosen that will produce a full scale deflection when measuring a voltage difference of 50 millivolts, so the meter will fully deflect when the shunt current is 60 amps. The meter in the illustration is partially deflected by a current of 25 amps flowing in the shunt. Note the direction of current flow and meter deflection. The battery is charging when electrons are flowing *from* the positive terminal, right? If this is confusing, think about the battery discharging,

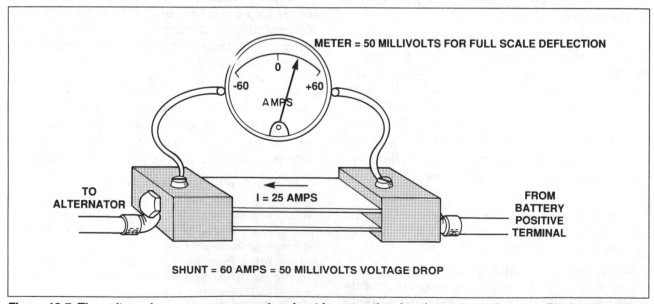

Figure 12-7. The voltage drop across an ammeter shunt is proportional to the amount of current flowing through it.

where electrons flow *out* of the negative terminal (region of excess electrons) and *into* the positive terminal (region of electron deficiency).

Loadmeters are ammeters installed in a circuit in such a way that they measure the load (current flow) in an alternator, generator or busbar. Generally, current flows only in one direction through a loadmeter, so an ammeter that deflects in only one direction from zero amps is a loadmeter. Placement of ammeters and loadmeters in the electrical system changes the information they provide to the pilot, so it is important that the pilot understand the placement and purpose of these instruments, which varies considerably in airplanes. More on this subject later in this chapter.

The Aircraft Electrical System

In over 12 years of teaching aircraft electrical systems to pilots, I have found it easiest for most pilots if they can fully understand one typical or generic electrical system. Then use that strong, basic understanding of the generic system to help themselves be able to understand the system in the aircraft that they fly. So, in the following pages, a generic system will be presented, bit-by-bit until it is complete. The reader is urged to put maximum effort into understanding this system, to the extent that the components of figure 12-17 can be correctly connected, not by memory but by understanding the purpose and function of each part of the system. Before marking on figure 12-17, you may wish to make copies of it for practice purposes.

Once the generic aircraft electrical system is understood, compare it to the electrical diagram of the aircraft you fly. It is such a good feeling when one finally understands something that before was a mystery!

The Source

A 12-volt system is the most common aircraft electrical system, so that will be the operating voltage of our generic study system. A lead-acid battery produces nearly 2.1 volts per cell, so six cells will be needed.

Typically, most systems are **negative ground** systems, meaning the negative terminal of the battery is connected to the airframe. See figure 12-8. It could be designed with the positive terminal grounded, and would work for lights and other purely resistance loads, but semiconductors (diodes, transistors, etc.) are totally sensitive as to which direction electrons move through their circuits, thus there is a need to standardize.

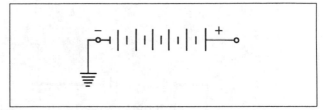

Figure 12-8. Symbol for a six-cell, 12-volt battery with the negative terminal connected to the ground (airframe).

Master Solenoid And Master Switch

14 CFR Part 23 of the The Federal Air Regulations deals with certification of normal and utility category aircraft. It states that there must be a way to electrically isolate the battery from the rest of the aircraft electrical system. In other words, the pilot must be able to shut off electrical power to the aircraft.

This is done with a solenoid (remember the difference between a relay and a solenoid?) switch called the **master solenoid**. This is a switch capable of carrying very large currents but is controlled by a very small current flowing through a small switch called a **master switch**. See figure 12-9.

When the master switch is closed, a circuit is completed which permits electrons to flow from the source (the negative terminal of the battery) into the ground, then up through the ground at the master switch, then through the master switch, up through the master solenoid coil, the internal connectors and back to the source at the battery positive terminal. While electrons flow along this path and through the coil, a magnetic field is generated by the coil which attracts the **contactor** downward against the force of the small spring which normally holds the contactor up. In the down position, the contactor makes electrical contact, allowing large amounts of current to flow from the battery positive terminal into the rest of the electrical system. Current flow through the master switch is kept small by the large resistance of the coil.

Can you trace the path of electrons flowing in the master switch circuit?

The Starter Circuit

The largest current user in the system is the (engine) starter motor. In a 12-volt system, as much as 400 amps is needed to start turning a cold engine. As the engine turns easier, current decreases to values typically about 30% of maximum, but that is still far more current than is used anywhere else in the system. When large current

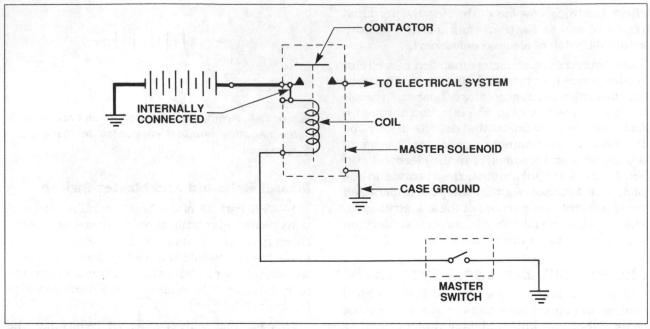

Figure 12-9. Master solenoid and master switch.

flows, large diameter cable is needed and the cable length is kept as short as possible to minimize voltage drop due to resistance in the cable. So there is an advantage to locating the battery near the starter motor. The battery is often mounted on the firewall for that reason, unless a CG problem requires mounting the battery aft of the cabin. On twins, the engine to be started first is often the one with the shortest cable run from battery to engine (after the first engine is started, system voltage goes from 12.0-12.6 to 14+ volts because one generator/alternator comes on-line). More current is available also, since both the battery and alter-

nator can contribute electron flow to start the second engine.

For the above reason, the large cable to the starter motor is connected as directly as possible to the battery (directly to the master solenoid).

There are two separate circuits in the starter circuitry of figure 12-10, which are the starter motor circuit and the control circuit. The control circuit allows the pilot to start and stop the starter motor by closing and opening the **starter switch**.

The starter switch may be a depressible button or toggle that is spring-loaded in the normally open

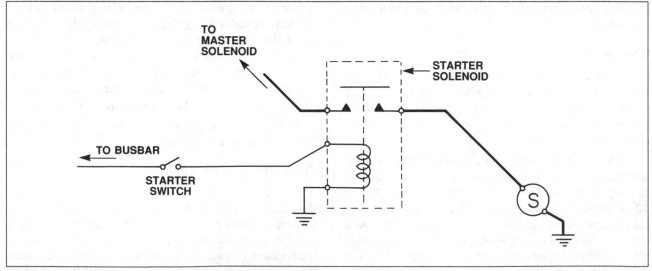

Figure 12-10. The starter motor and its control circuit.

position or it may be incorporated into the ignition switch in which case the starter switch is completely isolated (electrically) from the ignition system circuits.

When the starter switch is closed, a circuit is completed so electrons can flow from the battery into the battery ground, then through the airframe and up into the ground into the coil of the starter solenoid, then through the coil where the work is done, generating a magnetic field which pulls the contactor down to make a complete circuit through the starter motor. The electrons that flow through the coil continue on through the starter switch, its associated circuit breaker, the busbar, then back to the positive terminal of the battery. This flow continues until the starter switch is opened by the pilot.

When the starter solenoid contactor is pulled down, a circuit is completed to allow a large current to flow from the battery into the airframe, then through the starter motor (where the work is done), then through the starter solenoid contactor, the master solenoid contactor back to the source (the battery positive terminal). The starter motor circuit is the only circuit in the airplane that is not protected by a circuit breaker or fuse. This circuit is not protected because (1) it is not required by 14 CFR Part 23, (2) it would require a very large and expensive fuse or circuit breaker, and (3) years of experience indicate it is not needed.

Hint: *To follow any complete circuit in the same direction that electrons will flow, start first by identifying the source which will be the battery if the alternator is inoperative or the alternator if it is working. Electrons then flow out the negative or ground terminal of the source, through the airframe to the load (where the work is done), to the switch (that controls whether electrons flow or not), to the circuit protection (circuit breaker or fuse), then usually to the busbar, then back to the original source (the alternator or through the master solenoid contactor to the battery positive terminal. You may find this hint helpful when tracing a circuit until you get the knack of it.*

The Busbar

A **busbar** is a physically convenient place to terminate many wires neatly and safely. It is insulated from the airframe and, when the master switch is on, is connected to the positive terminal of the battery, so it is at essentially the same electrical potential (voltage). Actually, when electrons are flowing, the busbar voltage will be very slightly less than the voltage at the positive terminal due to a small voltage drop from the small resistance in the wire.

Aircraft certification requirements indicate that it is not acceptable to make wire connections except at a rigid point designed to accept wire terminals, so one cannot simply bare a wire and wrap another wire end around it to make a connection, then wrap it with electrical tape, as might be done in your car. If you see such a connection in an airplane, have a technician fix it. If left as is, it renders the aircraft unairworthy and likely voids the aircraft insurance. *This also means that you may not make electrical connections "in space" on the diagrams you do for this chapter. Terminate all connections at an appropriate terminal.* See figure 12-11.

Circuit breakers are often mounted directly on a busbar. See figures 12-12 and 12-13.

There may be more than one busbar in an aircraft. For example, if an avionics master switch is installed, that switch connects an avionics busbar to the ship's electrical system. See figure 12-13.

There may be several busbars in larger aircraft, for example, a pilot's side (L) busbar, a copilot's side (R) busbar and an emergency power busbar.

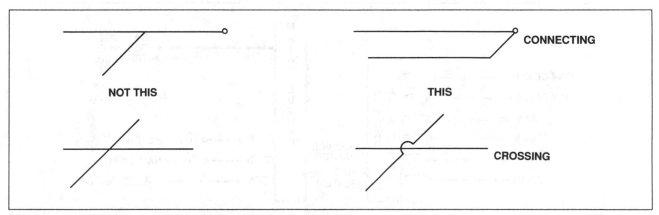

Figure 12-11. Do's and don'ts when drawing wiring diagrams.

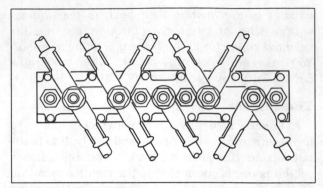

Figure 12-12. A busbar.

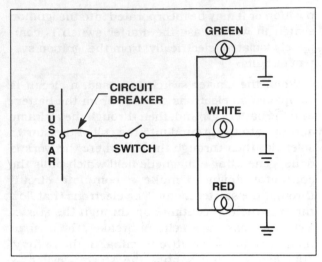

Figure 12-14. Navigation lights.

In the event of failure of both generators on a twin, all electrical power must come from the battery, so the pilot disconnects the main busbars. All items that are absolutely necessary for reduced-load operation are connected to the emergency power busbar, so all the work needed to shut off everything that is not absolutely needed in the aircraft is done by opening two switches to disconnect the L and R main busbars.

Individual Working Circuits

There are many circuits in the typical general aviation airplane, all of which provide some useful function. Nearly all are connected through the busbar, protected by a circuit breaker and controlled by the pilot with a switch.

An example of such a circuit is shown in figure 12-14. Current flows from the battery or alternator to ground, up through the ground to each individual light, then through the switch and circuit breaker to the busbar, then back to the source.

The Alternator

An **alternator** or **generator** can be thought of as an electron pump. Electrons aren't created or manufactured. They are already there, in the conductor, ready to move. In order for electrons to move, there must be a complete circuit and some pressure (voltage) to cause movement. The alternator provides the pressure by effectively sucking electrons in through the A terminal and pumping them out through the G terminal (see figure 12-15). A voltage is provided at the F terminal by the regulator that controls the output pressure (volts) and thus, the volume (amps) of the pump. How this is accomplished was described in detail in Chapters 10 and 11.

When the pilot engages the alternator switch, a circuit is completed to the regulator through the regulator ground. The regulator starts its function which is to sample the voltage in the aircraft system and control the output of the "pump" (alternator) to maintain adequate voltage.

The regulator samples system voltage by comparing the voltages at its terminals A and G. If the

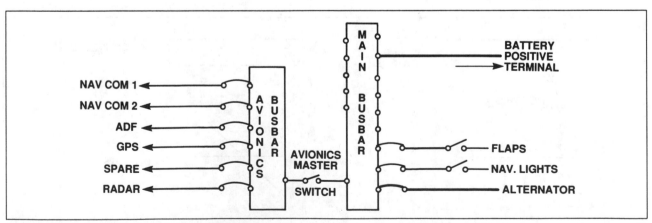

Figure 12-13. Circuit breakers installed on busbars.

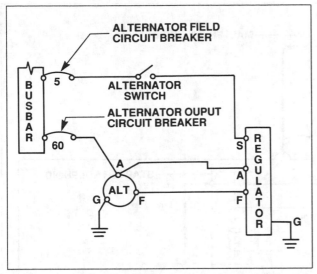

Figure 12-15. The alternator and regulator circuits.

voltage is low, the regulator increases voltage at its terminal F causing the alternator to take more power from the engine and use it to pump more electrons. If the system voltage is high (at or above 14.3 volts), the field voltage is decreased by the regulator to limit alternator output.

Pilots should remember that power is taken from the engine to drive the alternator only when the alternator is pumping electrons. The power taken from the engine is proportional to the current being produced by the alternator. Power taken from the engine is subtracted from the power available at the propeller. So, if someday you have to depart a short field with an obstacle with a heavy airplane on a hot, humid day with high density altitude, perhaps it would make sense to pull the alternator field circuit breaker, or open the alternator switch, and let the battery handle the electrical system until you have cleared the obstacle! Then, to avoid momentary over-voltage damage to avionics, shut off the avionics master switch, start the alternator, then turn the avionics back on.

Putting It All Together

Figure 12-16 shows all of the circuits we have just discussed put together as a complete generic system. Note that each circuit indicates whether heavy, medium or light amounts of current are carried.

How do the electrons know where to go? Many pilots ask this question! If a few principles are remembered, most of the confusion disappears. Keep in mind:

1. A complete pathway from the source through the load and back to the source is necessary if electrons are to flow.

2. A pressure, supplied by the source, is needed to cause flow.

3. Electrons will always follow the path of least resistance, to get back to the source.

Now, go back and review the *hint* located before the section on busbars, then, referring to the discussion of each individual component, see if you can trace the circuit (path of electron flow) through each part of the system. Start with the question: What is the source? If the alternator is operating, it is the source since it provides higher voltage than the battery. To complete the circuit, an electron must leave the source (negative terminal) and return to the source (positive terminal).

The dashed line through the master and alternator switches indicate that they are (in some aircraft) physically connected to each other in such a way that the master switch may be turned on with the alternator switch left off but the alternator switch *may not* be turned on while the master switch is off. If the master switch is turned off with the alternator on, the alternator would continue to energize the system but the battery would be unable to absorb and even out any voltage excursions (over-voltage peaks) produced by the alternator and caused by slow reaction time of the regulator. For this reason, the battery should not ever be taken out of the circuit while the alternator is operating. Even when fully charged, the battery acts as a very effective large capacitor or "voltage sink."

Ammeters And Loadmeter In The Circuit

Figure 12-16 has an ammeter installed between the master solenoid and the busbar (typical of Cessna aircraft circuits). In that position, an ammeter can sense the amount and direction of current flow in the battery circuit. A little study and circuit tracing will show that, in that position, the ammeter will *not* indicate current flowing in the starter motor circuit but will show current flowing in the starter motor control circuit.

Can you see the logic in the last statement?

In that position, the ammeter senses *only* current that is charging (flowing to the left through the ammeter) or discharging (flowing to the right) the battery. The one exception is current flowing through the master solenoid coil: if the battery is the source, the current flowing through the master solenoid coil (which is discharging the battery) is not sensed by the ammeter because it doesn't flow through it. When the alternator is the source, the

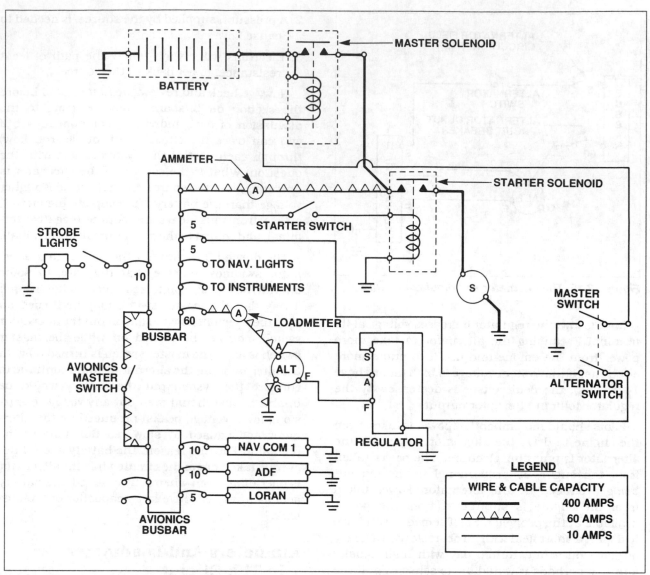

Figure 12-16. A typical "generic" aircraft electrical system.

ammeter indicates the amount of current that is charging the battery *plus* the amount of current flowing through the coil. When the battery is fully charged, no current flows into the battery because the voltage across its terminals equals the voltage across the alternator's A and G terminals. In this condition, the ammeter will show a very slight charge, equal in amount to the current flowing in the master solenoid coil.

Take a moment to trace the circuits and verify the statements of the previous paragraph.

An ammeter is also shown installed between the A terminal of the alternator and the busbar (typical of Piper aircraft systems). In this position, it obviously senses the current flow, or output, of the alternator. Since current can flow in only one direction through an alternator (because of the

diodes in the alternator's rectifier circuit), this ammeter qualifies as a loadmeter, indicating only the load (current) on the alternator. Operationally, a greater load will be indicated while the battery is charging which will decrease as the battery reaches full charge. When the battery is fully charged, the load of the electrical system will be indicated.

If the pilot turns on the landing lights, no indication will be seen on the ammeter (other than a momentary flicker indicating that the battery momentarily contributed until the regulator caused the alternator to react to the increased demand) while the loadmeter indication will increase to show the increased demand on the alternator caused by the landing lights. Both the ammeter and loadmeter give good indications of the "health" of the electrical system but they must be interpreted differently. A voltmeter is also a good indicator of system function. See the earlier discussion of voltmeters.

Electrical System Installation

Almost all aircraft currently produced in the United States are built under the provisions of FAR Part 23, which covers Normal, Utility, and Acrobatic category aircraft, or FAR Part 25, which covers transport category aircraft. The electrical systems required for transport category aircraft are so complex that we will make no attempt to cover their requirements here; but the type of aircraft normally used in general aviation meet the requirements of FAR Part 23, which we will discuss.

According to Part 23: An analysis must be made of the electrical load to assure that the power source has adequate capacity to carry all of the load, and that all of the wiring, cabling, and circuit protection devices are adequate for the load. The installation must be made in such a way that the risk of electrical shock to the crew, passengers, or ground personnel is reduced to a minimum.

Electrical power sources of multiengine aircraft must function either separately or in combination, and the failure of any component in one system must not impair the ability of any part of the other system to function properly.

There must be at least one generator to supply all of the electrical energy that is required for safe operation of the aircraft. This generator must be capable of continuously supplying its rated power, and it must be provided with adequate voltage control and provisions for preventing reverse current from draining the battery after shutdown. Suitable indicators must be provided to inform the pilot or crew of failure of the power source, and there must be a way for the pilot or crew to know the amount of power each source is producing.

Storage batteries must be installed in such a way that they will not overheat during a charge at maximum regulated voltage, and they must be so vented that any explosive or toxic gases will not cause a hazard, and no corrosive fluids can cause damage to any of the structure or adjacent essential equipment.

Fuses or circuit breakers must be incorporated in any circuit which could cause a hazard, except for the main circuit of the starter. No more than one essential circuit can be protected by a single circuit breaker or fuse, and if a circuit breaker is used, it must be of trip-free design, meaning that it cannot be closed into a fault, regardless of the position of its control. Automatic resetting circuit breakers generally are not permitted. If fuses are used, provisions must be made to carry one spare fuse of each rating, or 50% spares for each rating, whichever is greater.

A master switch is required that will disconnect the electrical power sources from the main bus, and the point of disconnection should be as near the source as practical.

The electrical wiring must be of sufficient size that it will carry the load without overheating from the current, nor should it have an excessive voltage drop.

All of the switches must carry the load for which they are intended, and all of them must be labeled so they are easily identified in flight.

These requirements for an electrical installation are merely excerpts from FAR Part 23, and before any alteration is made to an aircraft, the requirements should be studied in detail. Also to be studied are such methods as have been used and approved and are shown in Advisory Circular 43.13-1A and 2: Acceptable Methods, Techniques, and Practices for Aircraft Inspection and Repair, and for Aircraft Alteration.

With this background, it's time for a solo flight! See if you can wire the components of figure 12-17 (better make some copies of figure 12-17 before you write on it for further practice sessions) while following the admonitions at the bottom of the figure. Have a good flight!

Ignition Systems

Battery Ignition System

One of the earliest approaches to an ignition system was the make-and-break system in which an igniter, consisting of a pivoted, insulated contact, was installed inside the cylinder and connected through a kick coil to a battery (figure 12-18).

This insulated point made contact with a grounded point and when they were together, current flowed through them to ground. A cam, operated from the crankshaft, opened these points whenever ignition was required. As the points opened, an arc was drawn across them similar to that produced in arc welding.

While current was flowing through the contacts, a magnetic field was built up around the kick coil, and as the points opened, the current tried to

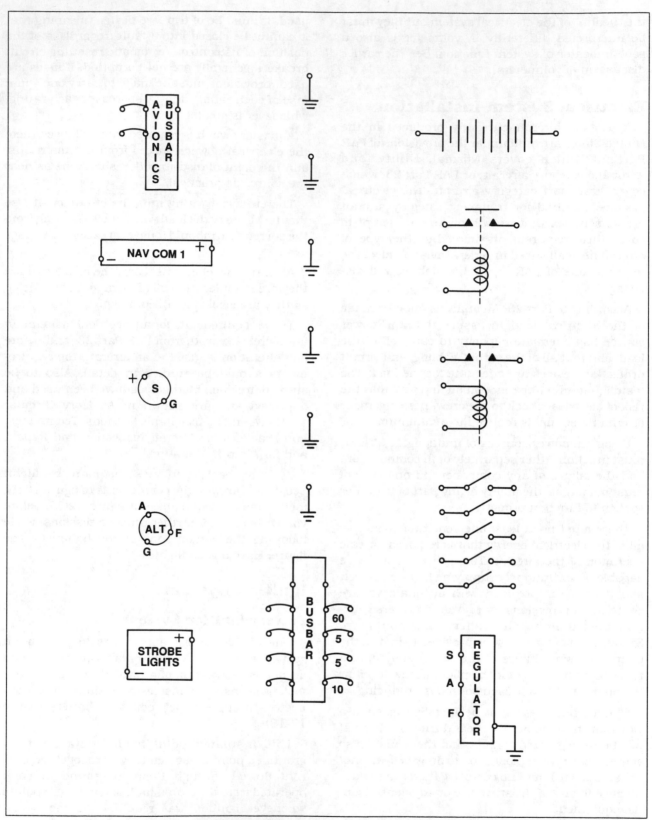

Figure 12-17. Practice hooking up these components into a generic electrical system. Don't use rote memory, use logical thinking in order to see if you understand. Please - connections only at terminals (see figure 12-11). Label all switches you use. Colors will make it easier to trace wires. Try not to make any SFA (smoke, fire and anxiety)! If you need more grounds or switches, draw them in and label the switches.

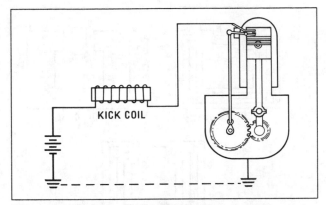

Figure 12-18. Make-and-break ignition system.

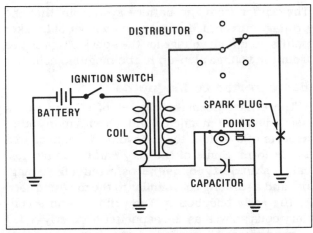

Figure 12-20. Schematic of a battery ignition system.

decrease. But the energy in the field tried to sustain it, so there was a sustained arc across the points which provided ignition. This system had some problems, so better things had to be developed.

Aircraft engines now almost universally use magneto ignition, but a few of the early engines used a battery system for starting. Since the principles of the two systems are similar, let's first analyze a battery ignition system (figure 12-19).

When the ignition switch is closed, current flows from the battery through the primary winding of the coil and the breaker points to ground. As this current flows, lines of flux build up around each of the turns of the primary and cut across the secondary turns, but since this buildup is rather gradual there will be no appreciable voltage induced into the secondary. When the cam opens the breaker points, the current in the primary abruptly stops flowing, and the magnetic field which has been held out by this current suddenly collapses.

The primary coil is made up of several hundred turns of relatively heavy wire, while the secondary

winding has thousands of turns of very small wire. The rapid collapse of the flux from the turns in the primary cutting across the many turns of the secondary induces a high voltage in the secondary. The distributor rotor is so timed that when the points break, it is aligned with the lead to the proper spark plug, and the high voltage will cause an arc across the spark plug electrodes, igniting the fuel-air mixture in the engine cylinder.

Primary current flows during the time the breaker points are opening. If provisions are not made to quench it, an arc will be drawn across the points, and they will soon burn and weld themselves together. A capacitor is connected in parallel with the points, so that as they begin to open and the resistance starts to build up, electrons follow what appears to be an easier path to ground, through the capacitor, and current immediately stops flowing at the points.

By the time the capacitor is fully charged, the points are opened wide enough that no current can flow, and so the primary field collapses almost instantly.

Magneto Ignition Systems

Battery ignition systems have an advantage over magneto systems for starting, because at slow speeds the primary current has plenty of time to build up to maximum and create a hot spark for starting the engine. But there is a corresponding disadvantage: as the engine speed increases, there is insufficient time for a complete buildup of the primary magnetic field, so the intensity of the spark decreases with engine speed.

Magnetos have been developed to provide a simple, dependable ignition system, *completely independent of the electrical system of the airplane.*

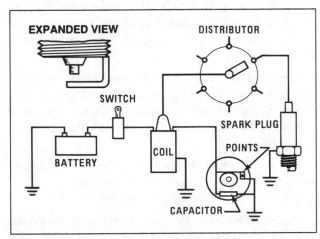

Figure 12-19. Battery ignition system.

They differ from the battery system in that the primary current is interrupted by a set of breaker points, and high voltage for the spark plugs comes from the voltage step-up in the magneto coil.

Basic Forms Of Magnetos

There have been three types of AC generators used in aircraft magnetos, but, as with many other aspects of aviation, these have been reduced to a single form. Some of the very earliest magnetos, called **shuttle-type magnetos**, used a fixed magnet and a rotating coil, similar to the magneto used in the early telephones. The primary and secondary coils as well as the capacitor (formerly called a condenser) all rotated, and a set of contacts opened at the proper time to interrupt the flow of primary current and induce a high voltage into the secondary winding.

The fact that the coils and capacitor were part of the rotating element brought problems, and so magnetos with fixed coils were developed.

Polar inductor magnetos, such as those made by the American Bosch Co., use fixed permanent magnets and fixed coils. But they obtain the reversal of flux by use of a soft iron rotor, which alternates the route the flux takes through the coil core (figure 12-21).

Figure 12-21(A): A soft iron rotor completes the magneto circuit so lines of flux can pass through the coil core from right to left.

Figure 12-21(B): Ninety degrees of magneto rotation later, the soft iron rotor completes the magnetic circuit again, but this time the flux lines pass through the coil core from left to right.

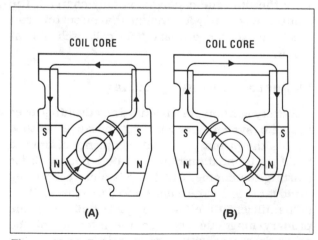

Figure 12-21. Polar inductor magneto.

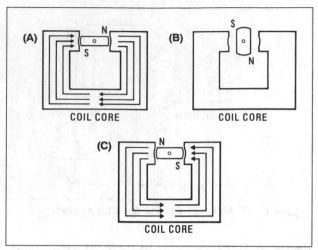

Figure 12-22. Magnetic circuit of a rotating magnet magneto.

Rotating Magnet Magneto

By far the most popular magnetos in use today have fixed coils and capacitors and rotating permanent magnets. This configuration was pioneered by the Scintilla Corporation, now the Electrical Components Division of the Bendix Corporation.

Figure 12-22 shows the magnetic circuit of a typical four- or six-cylinder, high-tension magneto.

In figure 12-22(A), the two-pole magnet is in its full register position with the north pole on the left side. Maximum flux now flows through the soft iron frame and coil core in a clockwise direction. As the magnet rotates, the amount of flux decreases until at the neutral position (figure 12-22(B)). There is no flux in the frame or core, but continued rotation brings the magnet again to full register (figure 12-22(C)), this time with the opposite polarity. The flux is again maximum in the frame and coil core, but is now in the opposite direction.

A plot of the static flux relative to the magnet position is shown in figure 12-23(A). Flux is maximum at full register, minimum at neutral, and again maximum, but in the opposite direction, at full register.

Figure 12-23(A). Static flux: The rotating magnet would produce a smooth curve similar to this if there were no windings around the coil core.

Figure 12-23(B). Primary current: Caused to flow by the changing flux in the coil core. The current stops abruptly when the points open.

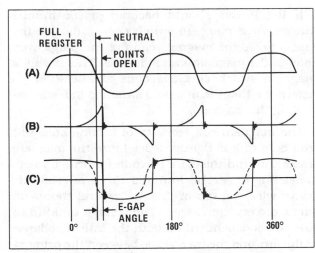

Figure 12-23. Flux curves in an aircraft magneto.

Figure 12-23(C). Resultant flux: The flux caused by the primary current delays the change in the flux from the magnet until the points open. The E-gap angle is the number of degrees the magnet rotates beyond the neutral position before the breaker points open.

Current generated in an electromagnetic generator, of which the magneto is a form, is determined by the *rate* at which the lines of flux change or cut across the windings of the coil. The primary coil is wound over the soft iron core, and in the full-register position of the magnet, the flux is maximum. But the *change* is minimum, so no current is being generated in the primary winding. As the magnet leaves the full-register position, there is no flux, but since it immediately starts to build in the opposite direction, the *change* in flux is maximum. And it is at the neutral position that the maximum amount of primary current *should* be caused to flow.

Resultant Flux

The condition which at first glance should exist actually does not, and for a good reason. As the flux begins to change, current starts to flow in the primary winding, and this current also causes a magnetic flux which, according to Lenz's law, *opposes* the flux which caused the current. The result of the primary flux reacting with the static flux is seen in the curve, figure 12-23, the resultant flux. The result is a delay in the buildup of the primary current. So instead of the maximum current flowing at the neutral position of the magnet, it flows several degrees after neutral.

The desired end result of a magneto is to have the maximum amount of voltage generated in the secondary winding of the coil. This voltage is deter-

mined by the rate of collapse of the magnetic field of the primary. In order to achieve the maximum rate of collapse, the primary current must be stopped *instantaneously* at the point of its maximum flow.

Electrical Circuit Of The Magneto

Primary Circuit

In figure 12-24 we see a portion of the primary circuit of a magneto.

Figure 12-24(A): When the points are closed, the current generated by the rotating magnet flows to ground.

Figure 12-24(B): As the points begin to open, current flows into the capacitor which appears to be a path to ground.

Figure 12-24(C): By the time the capacitor has charged, the points have opened far enough that current cannot flow through them, and no arcing will occur.

Current is generated in the primary coil and flows through the breaker points to ground, until the cam begins to open the points. The electrical inertia of the primary current tries to keep it flowing as the resistance across the points increases. Remember the effect of inductance?

When the current starts to decrease, the magnetic field around the coil begins to collapse. The collapse, or change, generates a current which *opposes the decrease* or attempts to keep the current flowing. If the reader will look back to figure 12-18, it can be seen that this circuit is similar to the kick coil in the make-and-break igniter of the early system. We wanted an arc then,

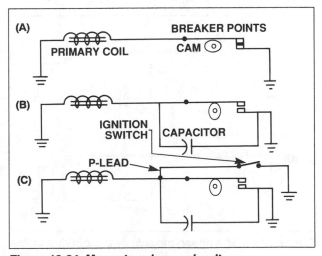

Figure 12-24. Magneto primary circuit.

but in this magneto, an arc is exactly what is *not* wanted.

To prevent an arc at the points, a condenser, more accurately called a capacitor, is placed in parallel with them (figure 12-24). As resistance across the points increases, electrons flow into the capacitor, since it *appears* to be a lower resistance path to ground. By the time the capacitor is charged, the points have opened far enough that electrons cannot flow across. The primary current has effectively been stopped instantaneously.

The capacitor not only prevents arcing of the points, but helps get a good flow of primary current started. In figure 12-25(A), let's assume that electrons have flowed into the capacitor as shown by the arrow. The capacitor charges up and then seeks to discharge through the points. But, as we have seen, the points are open by this time, and there is no path to ground through them. But the next half-cycle of alternating current has started to build up (figure 12-25(B)). The capacitor can now discharge through the primary coil and, in so doing, help get the flow of primary current going in the right direction. The magneto produces alternating current, so consecutive sparks have alternate polarity.

In order to stop the magneto from sparking the spark plug, a path to ground must exist in the primary circuit when the points open. This path to ground is provided from a point between the primary coil and the capacitor with a shielded wire called a **P-lead** to the ignition (magneto) switch. See figure 12-24(C). When the magneto switch is closed (OFF), a path to ground exists and the magneto will not spark the plug. If the magneto switch is open, it is in the ON position, enabling the magneto.

If the P-lead should become discontinuous (broken), the magneto switch cannot disable the magneto, so the magneto can fire the plugs even though the magneto switch is OFF. This is why a magneto switch check is done at idle RPM, to determine that the magneto switch is indeed controlling the magneto.

The magneto coil consists of a laminated soft iron core that is tightly wedged into the magneto frame. Around this core is wound a primary coil, consisting of several hundred turns of relatively heavy wire. On top of this are several thousand turns of very small wire. The voltage buildup in this coil is dependent on both the rate of collapse of the flux and the turns ratio between the primary and the secondary winding.

Secondary Circuit

Secondary voltage is taken from the high-voltage terminal of the coil through a carbon bush, to the rotor of the distributor, and then across an air gap to the high-voltage lead, which attaches to the spark plug. The distributor gear has teeth marked to align with a chamfered tooth on the magneto drive gear.

Special Forms Of Magnetos

In the interest of both safety and better combustion characteristics in the cylinder, all certificated aircraft engines have dual ignition. This requires two separate and independent ignition systems. Some early engines used a combination of battery ignition and a high-tension magneto; the battery system for starting, and then both systems for normal operation. Two separate magnetos are used on most modern engines, but with the need for more accessories and the limited number of drive pads available, the **dual magneto** is becoming popular.

This concept is not new, as dual magnetos were used on radial engines in World War II, including the Pratt and Whitney R-4360, a twenty-eight cylinder, four-row radial engine, which used seven dual, four-cylinder magnetos mounted around its nose section.

Dual magnetos may be considered as two separate ignition systems, as only the housing, rotating magnet, and cam are common to both systems. There are two sets of breaker points, two coils, two capacitors, and two distributors. The magnetos are similar in principle and operation to single magnetos. See figure 12-26.

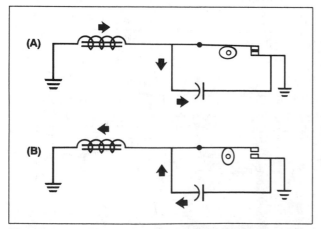

Figure 12-25. The capacitor aids in the buildup of the primary current.

Figure 12-26. The Bendix D-3000 dual magneto is actually two separate ignition systems, having only the rotating magnet, cams, and housing in common.

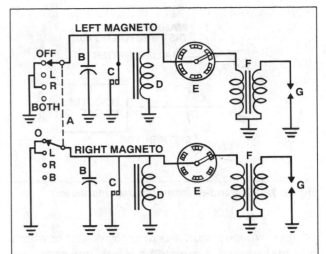

A - IGNITION SWITCH: ONE SWITCH FOR THE COMPLETE SYSTEM.

B - CAPACITORS: ONE IN EACH MAGNETO.

C - BREAKER POINTS: ONE SET IN EACH MAGNETO.

D - PRIMARY COILS: ONE IN EACH MAGNETO.

E - CARBON BRUSH DISTRIBUTOR: ONE IN EACH MAGNETO.

F - HIGH-TENSION TRANSFORMER: ONE FOR EACH SPARK PLUG. THESE ARE USUALLY LOCATED ON THE CYLINDER HEAD, NEAR THE SPARK PLUG.

G - SPARK PLUG: TWO IN EACH CYLINDER.

Figure 12-27. Low-tension ignition system.

The majority of the magnetos used on general aviation engines are of the rotating-magnet, high-tension, single-magneto configuration, but there are others.

One of the problems inherent with **high-tension magnetos** has been the inability of the harness and distributor to contain high voltage in the thin air of high altitudes. As the air becomes less dense, its insulating capability decreases, and sparks jump from the distributor finger to the housing, or an electrical breakdown occurs within the harness itself. Steps have been taken to prevent this, such as making the distributor physically larger, so there will be a greater distance for the spark to travel. Magnetos, distributors, and harnesses have been pressurized with compressed air to increase their resistance to breakdown. But one of the more practical developments, which has allowed magneto ignition to function at high altitude, has been the **low-tension magneto**.

Similar to the high-tension magneto in the generation of primary current, the low-tension magneto has a rotating magnet, cam-operated breaker points, and a capacitor. But it differs in the production of the spark in the secondary.

The coil in the low-tension magneto has only the primary winding, with its output going to a carbon-brush distributor. Primary current is carried to individual coils or transformers, one for each spark plug, mounted on the cylinder near the plug. When the breaker points open, the primary current through a specific coil is interrupted, its field collapses and generates a high voltage in the secondary winding, which is taken to the spark plug by a very short lead. See figure 12-27.

Aids To Starting

Aircraft magnetos provide a good, hot spark at idle, at cruise, and at high speeds, but some provision must be made for getting a hot spark when the engine first turns over slowly while starting. This spark must not only be hot, independent of the rotational speed of the engine, but it must come later than the normal spark. This allows the piston to be beyond top center when the pressure from the hot, expanding gases exerts its push on it.

Some early engines used a regular high-tension magneto, hand-cranked by the pilot, to provide a hot spark while the engine was being cranked (figure 12-28). The output of this **booster magneto**, as it was called, went directly to a trailing finger on the rotor of one of the distributors. This allowed the high-voltage booster output to fire the

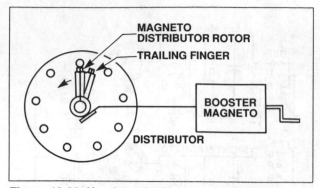

Figure 12-28. Hand-cranked booster magneto.

cylinder whose piston was near top center or which had just passed it, providing a hot, late spark *for* starting, and preventing kick-back *while* starting.

As smaller engines became popular in general aviation, starting procedures were simplified, and the booster magneto was replaced with the **impulse coupling** (figure 12-29).

A cam plate with two flyweights attached is keyed to the magnet shaft. The impulse coupling body rides over the cam and flyweight assembly and is the part driven by the engine.

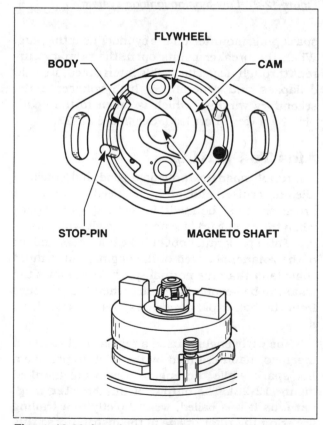

Figure 12-29. Impulse coupling.

The body and cam plate are connected through a heavy-duty, clock-type spring. Two stop pins are placed in the magneto housing, in such a position that as the starter rotates the engine, the flyweights move out and contact the stop pins.

This holds the magnet and cam plate still as the engine continues to turn, winding the spring. After a predetermined amount of rotation, projections on the coupling body contact the flyweights and release them from the stop pins. The spring now spins the magnet, producing a hot, late spark. As soon as the engine starts to fire and pick up speed, centrifugal force on the flyweights hold them away from the stop pins.

They lock the body and cam plate together so that the magneto operates as though it had a solid coupling. Magnetos with impulse couplings may be identified by an audible snapping, when the engine is pulled through by hand.

Impulse couplings provide good hot, late sparks for starting small engines. But as engine size increases and operational conditions become more severe, better starting systems are required.

The **vibrator starting system** is an improvement over the booster magneto, and has become pretty well standard. The induction vibrator produces pulsating direct current from pure DC (figure 12-30(A)). Pulsating DC from the induction vibrator is directed into the magneto to produce a high voltage in the magneto coil (figure 12-30(B)). Since this system does not put out a single spark, but instead a

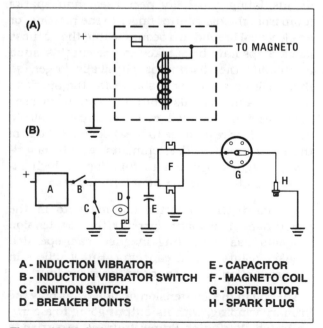

A - INDUCTION VIBRATOR
B - INDUCTION VIBRATOR SWITCH
C - IGNITION SWITCH
D - BREAKER POINTS
E - CAPACITOR
F - MAGNETO COIL
G - DISTRIBUTOR
H - SPARK PLUG

Figure 12-30. Induction vibrator system.

stream of sparks, it is called by the Bendix Corporation the **"Shower of Sparks" system**.

When voltage is applied to the coil, current flows and energizes it, making the core an electromagnet. The lower point is attracted to the core and opens the circuit, de-energizing the coil so the points will close and start the cycle over. The current output is in the form of pulsating direct current with about 200 pulses per second, and when it is fed into the primary of a magneto coil it is transformed into high voltage in the secondary.

The complete primary circuit (refer again to figure 12-30) includes not only the primary coil but the cam-operated breaker points, the capacitor and the ignition switch. With the ignition switch closed there is still the rise and fall of primary current, but there is not the sudden collapse needed to induce a secondary voltage.

For an aircraft engine to start, there must be a hot and *late* spark, and since aircraft magnetos have fixed timing, the normal spark cannot be retarded. To prevent the engine attempting to kick back from a spark occurring at its normal advance position, the Shower of Sparks must be timed so it will not fire the spark plug until the piston is at or near top dead center (figure 12-31).

If the output of the vibrator is fed into the primary of the magneto coil, voltage will be induced into the secondary, but when the breaker points short the pulsating DC to ground, no voltage will

be induced in the secondary. As soon as the points open, though, the current will go to ground through the primary and sparks will be produced as long as the points remain open. This produces the required hot spark, but does not help in getting the late spark. To accomplish this, a second set of points, the retard points, are installed in one of the magnetos, usually the left one. When the engine is being started, pulsating direct current goes to ground through both the run and the retard points which are in parallel.

The run points open first, at the normal advance position, but since the retard points are still closed, there is no spark at the secondary. At the proper position for the starting spark to occur, the retard points open, and the only path for the pulsating direct current is through the primary of the coil to ground. This provides a continual spark until the run points close. During this time, the distributor rotor is aligned with the electrode for one of the ignition leads.

The basic function of the Shower of Sparks system is rather simple (figure 12-32). But there are other things involved in starting an aircraft engine, besides providing the spark and to simplify starting. Some of these functions are incorporated in the ignition switch circuit, so when the switch is placed in the "Start" position, the following things happen:

1. The right magneto is grounded to prevent a spark at the advanced position, which could cause a kick-back and possibly damage the starter.

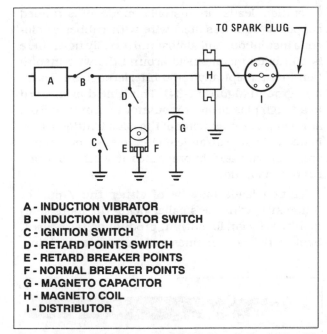

A - INDUCTION VIBRATOR
B - INDUCTION VIBRATOR SWITCH
C - IGNITION SWITCH
D - RETARD POINTS SWITCH
E - RETARD BREAKER POINTS
F - NORMAL BREAKER POINTS
G - MAGNETO CAPACITOR
H - MAGNETO COIL
I - DISTRIBUTOR

Figure 12-31. Provisions for obtaining a late spark with an induction vibrator system.

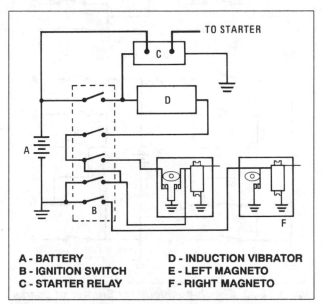

A - BATTERY   D - INDUCTION VIBRATOR
B - IGNITION SWITCH   E - LEFT MAGNETO
C - STARTER RELAY   F - RIGHT MAGNETO

Figure 12-32. Basic "Shower of Sparks" system.

2. The retard points circuit is completed so they become operational.

3. The vibrator is energized, sending pulsating direct current into the magneto.

4. The starter solenoid is energized, cranking the engine.

The ignition switch for a Shower of Sparks system is naturally more complicated than one for just the magnetos alone. Figure 12-32 shows the entire circuit. The switch is shown as a series of simple switches. But a rotary-type switch is normally used, one with an Off position fully counterclockwise, then Right, Left, Both, and a spring-loaded Start position.

Figure 12-33 shows the switch conditions for the various positions.

A—In the Start position, the starter solenoid and vibrator are energized, and the retard points are connected to the vibrator. The right magneto is grounded, and the left is operational.

B—In the Both position, all of the switches are open. This allows both magnetos to be operational, with the retard points out of the circuit of the left magneto.

C—In the Left position, all of the switches are open except the one which grounds the right magneto.

D—In the Right position, all of the switches are open except the one which grounds the left magneto.

E—In the Off position, both magnetos are grounded, and the switches to the starter solenoid and vibrator and those in the retard circuit are open.

Ignition Lead

Enough high voltage may be generated in a magneto to provide sufficient spark for igniting the fuel-air mixture, but if it is not carried to the spark plug without losses, the engine performance will deteriorate.

Modern aircraft carry a considerable amount of electronic equipment, and the communications and navigation systems must be able to receive weak signals from ground stations without interference from extraneous sources. Since high-voltage ignition systems constitute very effective radio transmitters, the energy radiated from the spark must be contained within the harness and grounded. Unless this is done, there will be enough interference to impair radio reception.

Ignition leads are usually made of stranded copper or stainless steel wire with rubber or silicone insulation, and almost universally now, there is a braided metal shield around the wire insulation to intercept any radio interference and carry it to ground (figure 12-34). This shield is encased in a tough plastic outer insulator to protect it from abrasion. The conductor of the slick ignition leads, instead of using stranded wire, uses a continuous spiral of wire and impregnates it with a silicone rubber insulation.

Ignition leads may be of either the 7-mm or 5-mm size, with the smaller being the more common size, by far, for current production harnesses. Some of the older engines use a separate shielding,

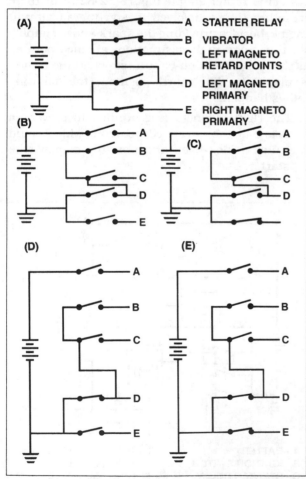

Figure 12-33. Switch positions for Shower of Sparks system: (A) Start, (B) Both, (C) Left, (D) Right, (E) Off.

Figure 12-34. Typical ignition lead.

complete with all of the elbows and nuts. The nuts may be loosened and unshielded wire pulled into the harness for lead replacement.

It is imperative when installing an ignition lead that there be no strain placed on the wire where it enters the terminal end of the spark plug. To prevent any strain, some manufacturers make elbows with several different angles, specifically 70°, 90°, 110° and 135°, while others make their harnesses with only one angle and depend on the flexibility of the wire to prevent any strain. One popular harness uses no elbow but when it is required to make a sharp bend, a bracket is used to hold the lead with a sufficient bend radius that it will not be damaged.

Spark plugs have either a 3/4-20 or 5/8-24 thread at the terminal end, and magneto harnesses are made to fit both types of spark plugs (figure 12-35).

Some of the older harnesses used a phenolic or ceramic tube with a coil spring at its end for the terminal connection in the spark plug; these are called **cigarettes**. The insulation was cut from the stranded conductor far enough back for the wires to stick through the small hole in the end of the

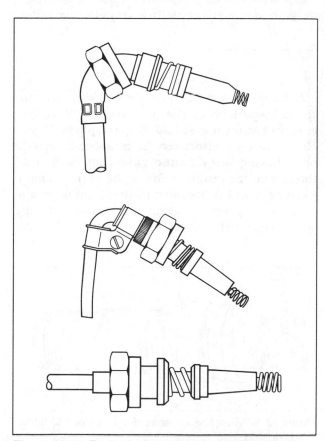

Figure 12-35. Typical spark plug lead terminals.

terminal, and the ends of the wires are fanned out to provide a good electrical contact and prevent the cigarette from slipping off of the wire. In some instances, a small aircraft nail or pin was slipped through the hole in the terminal into the strands of wire to provide a better connection.

The current practice is to use silicone rubber for the terminal connectors and to crimp the terminal to the wire rather than spreading the strands. The springs usually screw over the end of the terminal and may be replaced if they are damaged.

Replacement ignition leads have the spark plug end installed by the manufacturer, but the magneto end is assembled by the A&P after the lead is cut to length.

If a single spark plug lead becomes damaged, it can be replaced individually without having to replace the entire harness (figure 12-36). The various manufacturers have specific instructions which must be followed in detail for terminating an ignition lead, and their service manual must be available—and be adhered to!

Spark Plugs

Spark Plug Nomenclature And Identification

Since the days of the Wright Flyer, spark plugs have been one of the more critical parts of an aircraft engine, yet their seeming simplicity often precludes appreciation of their complexity.

The only function a spark plug has is to provide an insulated electrical terminal inside the combustion chamber of a reciprocating engine. To this terminal, a high voltage is applied in such a way that a spark will jump to a predetermined ground point and produce enough heat to ignite the fuel-air mixture. This spark plug must resist fouling by any of the contaminants in the combustion chamber. Its electrodes must undergo minimum erosion from either the heat in the cylinder or from the arcing action as the spark jumps between them.

As with all of the components in an airplane, spark plugs have progressed through the process of evolution. The very earliest spark plug was a make-and-break device in the cylinder, producing its spark by mechanically interrupting a flow of current. This was replaced by a spark plug similar to those used in automobiles and motorcycles, in which high voltage from an induction coil punched across an open gap. Then the utility of the airplane was increased by use of two-way radio communications, and **shielded ignition systems**

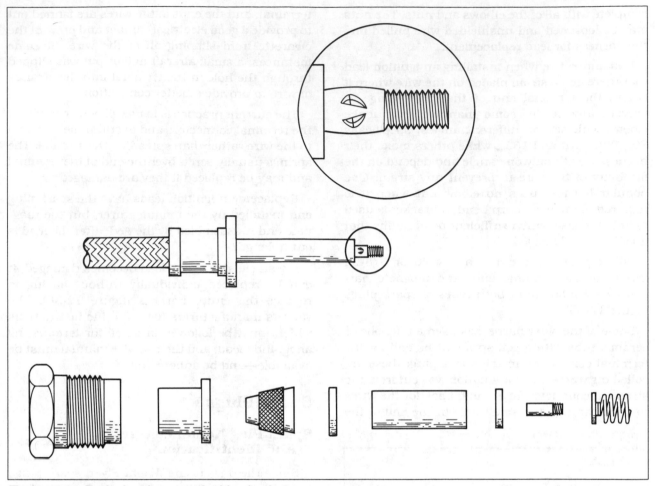

Figure 12-36. Typical replacement ignition lead.

were developed that contained the radiated energy in a wire braid around the spark plug leads and in a steel shell around the plug itself.

The single ground electrode used by automobiles was improved upon by the use of two, three, or four massive ground electrodes made of a special alloy of nickel. The center electrode evolved from a solid nickel rod to a nickel alloy sheath filled with copper for better heat conduction.

World War II brought out requirements for extended spark plug life, less tendency toward lead fouling with higher lead content fuels, and spark plugs that would resist ice bridging when attempting starts in cold, damp weather. The answer came in the form of fine-wire plugs, having their electrodes made of platinum wire.

Aircraft spark plugs are made with two sizes of shell threads, 14- and 18-mm; but, with very few exceptions, modern aircraft engines use the 18-mm spark plug.

Although a few of the older engines use un-shielded spark plugs, the great majority of engines in service today use **shielded spark plugs**. Figure 12-37 shows a **short-reach, massive-electrode** plug, having **low-altitude shielding**, or 5/8-24 threads on the terminal end of the barrel. A more watertight seal is provided in the terminal end of a spark plug by recessing the insulator in the spark plug shield and using a resilient seal on the

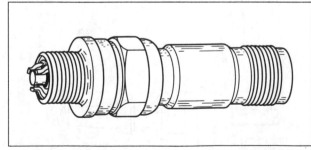

Figure 12-37. Short-reach, massive-electrode shielded spark plug.

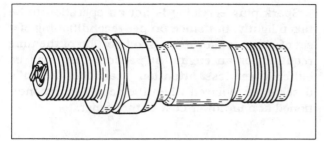

Figure 12-38. Long-reach, fine-wire electrode spark plug with "all-weather" shielding.

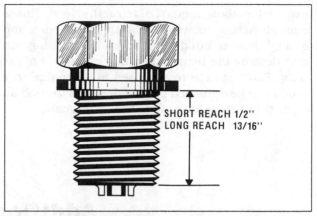

Figure 12-39. The length of the threads on the firing end of a spark plug determines its reach.

harness. This type of spark plug, figure 12-38, is called an all-weather or high-altitude spark plug and has a 3/4-20 thread on its terminal end.

The spark plug in figure 12-37 has a three-pronged insert in its firing end, permanently bonded in place, that forms the ground electrodes to and from which the spark jumps from the center electrode. As the spark jumps it erodes the electrode in much the same way, although on an infinitely smaller scale, as an electric arc welder. The multiple electrodes and large area provide a maximum amount of material, so that the interval between service of these plugs can be extended to the maximum. The center electrode of this type of spark plug as seen in figure 12-37 is a nickel sheath, completely filled with copper so there will be a maximum amount of heat transfer from the electrode, and hot spots will not build up in the nickel.

Fine-wire spark plugs, such as seen in figure 12-38, have a center electrode and two ground electrodes made of small-cross-section wires of either platinum or iridium. The small electrode cross section allows the spark plug to spark at a much lower voltage than the massive electrode, and at the same time the exotic materials of the electrodes resist erosion from both heat and sparking.

The length of the threads on the spark plug shell classifies it according to **reach** (figure 12-39). The shell is threaded for ½ inch on the short-reach spark plug, and for 13/16 inch on the long-reach.

The **heat range** of a spark plug refers to the ability of the insulator and center electrode to conduct heat away from its tip. Figure 12-40(A) shows a hot plug, one having a long path for the heat to travel to escape from the tip. Hot spark plugs are used in engines which have a relatively low amount of heat in their combustion chamber—low compression engines. The spark plug in figure 12-40(B) is a cold plug used in a hot-running, high-compression engine. This spark plug has a

relatively short path for the flow of heat from the insulator tip to the shell. The spark plug selected for each application must run hot enough to minimize fouling of the insulator tip, while at all times operating at temperature below that which could cause preignition.

Shielded ignition, while eliminating the problem of radio interference, causes another problem, that of accelerated electrode erosion. The shield acts as one plate of a capacitor and stores electrical energy. The instant the spark occurs, this capacitor is very rapidly discharged, resulting in high values of current. The discharge makes up what is known as the capacitive component of the spark. High-amplitude, rapidly changing current is responsible for a large portion of the radiated interference and electrode erosion, but has no effect on engine performance due to its very short duration. A **resistor** is built into the spark plug to limit the peak current allowed to flow, and thus minimizes electrode erosion.

In the course of one hundred hours of operation, a plug will spark about eight million times and be

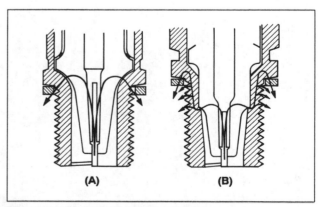

Figure 12-40. Hot and cold spark plugs.

exposed to about a quart of tetraethyl lead. This is enough arcing to wear away the electrodes and enough lead to build up deposits in the firing end and destroy the heat-conduction capability of the plug. Platinum electrodes used in some fine-wire plugs can be damaged by lead deposits to such an extent that the plug may not be salvageable.

Spark plug servicing is not an operation to be taken lightly, but since proper reconditioning of a set of plugs constitutes a good portion of the time required for an engine inspection, they may be either given less attention than they actually deserve, or replaced when they could be reconditioned at a far lower cost to the owner.

Additional Reading

1. Aircraft Ignition and Electrical Power Systems; EA-IGS; IAP, Inc., Publ; 1985.

2. A&P Technician General Textbook; EA-ITP-G2; IAP, Inc., Publ; 1991; Chapter 3.

3. Bygate, J.E.; Aircraft Electrical Systems; EA-357; IAP, Inc. Publ; 1990.

4. Airframe and Powerplant Mechanics General Handbook; EA-AC65-9A; IAP, Inc, Publ; 1976; Chapter 8 and 9.

5. Powell, J., Aircraft Radio Systems; EA-356; IAP, Inc, Publ; 1981.

6. Aircraft Batteries; EA-AB-1; IAP, Inc., Publ; 1985.

Study Questions And Problems

1. What characteristics of series and parallel circuits can be used to tell them apart?

2. How many horsepower will be taken away from the propeller by an alternator producing 25 amps at 14 volts? Hint: Look up the relationship between watts and horsepower. Assume the alternator is 74% efficient.

3. How much decrease in rate-of-climb performance will the conditions of Question #2 cause for a 2500 pound aircraft that was climbing at 480 FPM with the alternator switch off and the propeller operating at 82% efficiency? Hint: THP = BHP × prop efficiency and ROC = $\dfrac{\text{ETHP} \times 33{,}000}{\text{WT}}$.

4. How can a D'Arsonval meter, which is usually used as a voltmeter, be used as an ammeter?

5. In an aircraft with a 12-volt electrical system, what would a voltmeter reading of 12.4 volts indicate about the system if the reading was increasing slowly? Decreasing slowly?

6. What principal differences exist between an ammeter and a loadmeter?

7. How can a pilot tell if the instrument is an ammeter or a loadmeter by looking at the face of the instrument in the cockpit?

8. Compare the lead-acid and nickel-cadmium storage batteries by listing the advantages and disadvantages of each.

9. What device(s) electrically isolate the battery from the aircraft's electrical system?

10. What will happen to the aircraft engine if there is a total failure of the aircraft electrical system?

11. What is the only circuit in the aircraft electrical system that is not protected by a fuse or circuit breaker? Why is it not protected?

12. How does an alternator produce electricity?

13. How is an alternator's output controlled?

14. Do electrons that operate the strobe light of figure 12-16 pass through the ammeter if the alternator is working? If so, which way? If the alternator is inoperative? If so, which way?

15. Do Question #14 using the master solenoid coil as the active component.

16. Wire up figure 12-17, then check it, circuit-by-circuit to see if it will work properly. Then check it, wire-by-wire, with figure 12-16.

17. Using color coding, trace each circuit on the electrical system diagram of the aircraft you fly, with the same circuit on the generic diagram. Number each segment of each circuit (wire between two terminals) the same on both diagrams.

18. Does the spark from a battery ignition system become stronger or weaker as the engine speed increases?

19. Same question as #18 for magneto ignition systems.

20. Is the primary current in a magneto system pulsating direct current or alternating current?

21. On what kind of an airplane should a low-tension magneto be used? Why?

22. How does a dual magneto differ from a single magneto?

23. When a magneto switch is open is the magneto "ON" or "OFF?"

24. Explain how a broken P-lead can be dangerous and how to check for a broken P-lead.

25. What is the purpose of an impulse coupling and how does it work?

26. What is a "cigarette" in an aircraft ignition system?

Chapter XIII
Hydraulic Systems And Landing Gear

History Of Fluid Power Applications

Fluid power may be thought of as those systems in which work is accomplished by the movement of a fluid. This fluid may be either compressible, as a gas (**pneumatics**), or incompressible, a liquid (**hydraulics**).

Since our earliest recorded history, man has used fluid power; first, to move himself or objects from one place to another by floating down a stream. The tremendous power in fluids in motion was observed when man saw the havoc wreaked by a wind storm or the devastation from a river on a rampage. This power was first harnessed by the waterwheel as early as the first century BC, and later by the windmill.

The overshoot waterwheel which provided power for our early industrial plants has given way to the high-speed turbine waterwheels in modern hydroelectric generators, which provide much of the electrical energy we use today.

Waterwheels and windmills are examples of fluid power utilizing an open system. Aircraft fluid power systems use a **closed system**, in which the moving fluid is confined in such a way that its pressure may be increased. Thus more work can be done by less fluid.

Closed hydraulic systems are familiar to us in the form of the hydraulic jack used to lift automobiles or the adjustable barber or dental chair. In aircraft factories, hydropresses exert tons of hydraulic force to form many of the complex sheet metal parts.

Airplanes would be far less efficient if it were not for fluid power devices; hydraulic brakes allow the pilot control of the airplane on the ground, without requiring a complex mechanical linkage system. Hydraulic retraction systems pull the heavy landing gear into the wheel wells to decrease wind resistance. Hydraulic-boosted controls make flying high-speed jet aircraft possible. And air under pressure is used to break ice off the wing and tail surfaces.

Basic Laws Of Fluid Power

Fluid power, as any other branch of physics, conforms to certain definite and well defined laws.

Let's review some basic principles before proceeding with a discussion of hydraulics.

Force is energy exerted, or brought to bear, and is the cause of motion or change. In practical fluid power, it is expressed in pounds or in metric terms of newtons or dynes.

When a force is applied over a given area it is termed **pressure** and is expressed in pounds per square inch or grams per square centimeter. Other units of pressure may be encountered, but, for our purpose, these may all be converted back into basic units of force and area. So, Pressure = Force × Area.

Any time a force causes an object to move, **work** is done. Work is expressed in foot-pounds or in kilogram-meters.

Any container for a fluid has a certain volume which may be expressed in cubic units. **Volume** is usually considered to be the base of the container in square units times its height in the same units. Sometimes fluids are held in spherical containers whose volume may be found by the formula

$$V = \frac{4\pi R^3}{3}$$

in which the volume will be given in cubic units.

Work is simply the product of force and distance, but the amount of **power** required to accomplish a given amount of work must take into consideration the **time** required. Power is therefore force times distance, divided by time.

$$Power = \frac{Force \times Distance}{Time}$$

A standard unit of power is the horsepower. This is 33,000 foot-pounds of work done in one minute; or, if a shorter period of time is desired, 550 foot-pounds of work done in one second.

In the metric system, a metric horsepower is 4500 kilogram-meters per minute, or 75 kilogram-meters in one second. One metric horsepower is equal to 0.986 horsepower.

In hydraulic systems, power may be computed by considering the flow rate in gallons per minute (one gallon = 231 cubic inches) and the pressure in pounds per square inch, to get force-distance-time relationship. One gallon per minute of flow under a pressure of one pound per square inch will produce 0.000583 horsepower.

Horsepower = Gallons per minute × pounds per square inch × 0.000583

Fluid Statics And Dynamics

The Law Of Conservation Of Energy

Perhaps the most basic law of physics deals with our relationship to energy. Man has not been given the prerogative to create or destroy energy. We have energy at our disposal and are able to change its form, but in the final analysis we end up with exactly the same amount as we started with. In almost any type of mechanical device, we seem to lose energy because of its inefficiency; what actually happens, however, is that some of the energy has been transformed into an unusable form, such as heat from friction.

Basically, energy is found in one of two forms: **kinetic** which exists in an object due to its motion; and **potential**, which exists in an object because of its position or the arrangement of its parts. In fluid power, a practical way to look at this is to consider potential energy expressed in a fluid as its **pressure** and kinetic energy expressed as its **velocity**.

A **paradox** is a true statement that does not readily appear to be true. In this case, as illustrated in figure 13-1, the static pressure (head) exerted by a column of fluid is proportional to the height of the top of the fluid and is not affected by its volume. What this means is: if we have a container of liquid in any of the shapes of figure 13-1, the pressure indicated by the gauge at the bottom of the container will depend on the height of the top of the fluid and will not be affected by the shape of the container or by the amount of fluid. This is sometimes referred to as the **hydrostatic paradox**.

Power transmission in a closed hydraulic or pneumatic system is best explained by **Pascal's law**. Stated in a simple form, it says, "Pressure in an enclosed container is transmitted equally and undiminished to all parts of the container and acts at right angles to the enclosing walls."

In figure 13-2, we see a closed container filled with fluid. Pressure gauges are arranged around the container to measure the pressure created by force F pushing down on the piston. The pressure will be F ÷ A pounds per square inch and will be the same on every gauge regardless of the position in the system, or the shape of the container. This law is taken advantage of in an automobile brake system. When the brake pedal is depressed, pressure is transmitted equally and undiminished to all of the wheels, regardless of the distance between the wheel and the brake

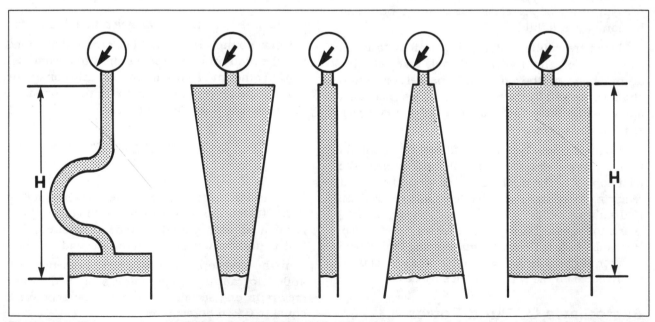

Figure 13-1. The pressure exerted by a column of liquid is dependent on its height and density, but is independent of its volume.

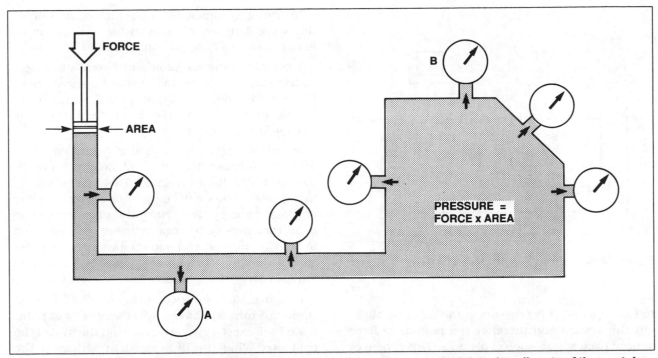

Figure 13-2. Pressure in an enclosed container is transmitted equally and undiminished to all parts of the container and acts at right angles to the enclosing walls.

pedal. The same concept applies to fluid power systems in aircraft.

The alert reader may ask, "Isn't the pressure at gauge A in figure 13-2 slightly greater than the pressure at gauge B because of a difference in head?" Yes it is, but the difference is indeed slight compared to the operating pressures of a closed system.

An application of Pascal's law allows us to see the **mechanical advantage** there is in a hydraulic system. To briefly review the principle of mechanical advantage, look at the balance in figure 13-3. The product of weight 1 and length 1 makes up moment 1. This force tries to rotate the board counterclockwise and is opposed by moment 2, the product of force 2 and length 2. In a system such as this, it is possible for a small force to lift a large weight. Since we do not get something for nothing, the price is paid in distance moved. The work (force × distance) done by one side of the balance is exactly the same as that done on the other side. For instance, if L_1 is 40 inches and L_2 is 20 inches, W_2 of 100 pounds could be balanced by W_1 of 50 pounds. If W_2 is to be lifted a distance of one foot, W_1 must move two feet. the work done on side 1 is two feet times 50 pounds, or 100 foot-pounds, which is the same as side 2, one foot times 100 pounds, 100 foot-pounds.

The same mechanical advantage may be obtained with a hydraulic system, according to Pascal's law. Figure 13-4 illustrates a simple hydraulic jack. The piston on side 1 has an area of one square inch, and on side 2, an area of ten square inches. If a force of ten pounds is applied to side 1, there will be a pressure generated in the system of ten pounds per square inch. According

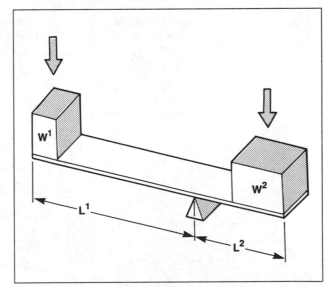

Figure 13-3. Length 1 x Weight 1 = Length 2 x Weight 2.

237

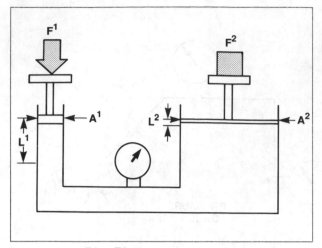

Figure 13-4. $\dfrac{F1}{A1} = \dfrac{F2}{A2}$ $F1 \times A2 = F2 \times A1.$

to Pascal's law, this pressure is the same throughout the system and therefore ten pounds of force acts on each square inch of piston 2. This produces a force of 100 pounds which will balance $F2$.

When piston 1 is moved down one inch, one cubic inch of fluid is pushed from side 1 into side 2. This spreads out over the entire area of piston 2 and lifts it only one tenth of an inch. Piston 1 would have to move down ten inches to lift piston 2 one inch.

The work done by side 1 is ten inches times ten pounds or 100 inch-pounds, and that by side 2 is one inch times 100 pounds, also 100 inch-pounds.

Hydraulic systems are quite efficient, and practically speaking we will not consider system losses in our basic study of system operation.

The hydrostatic paradox and Pascal's law deal with **static** conditions; that is, fluid not in motion. When fluid is put into motion, other things begin to happen, which are best explained by **Bernoulli's principle**.

In considering fluid in motion, we must begin with the premise that we will hold the energy content of the fluid constant. This means simply that we will neither add energy, nor will we consider any loss of energy. The principle then becomes one of an exchange between kinetic energy expressed in fluid as velocity, and potential energy, the pressure of the fluid. **Total Energy**, E_T, is the sum of kinetic energy, E_K, and potential energy. E_P.

In figure 13-5, an incompressible fluid flowing through a tube will have a specific velocity at point A, and will exert a given pressure on the wall of the container. When this fluid meets a restrictor in the line, such as point B, if no energy is added and none is lost, the velocity of the fluid must increase to get through the smaller area. The kinetic energy increases at the expense of the potential energy so the pressure at point B drops. At point C the area is back to the original, and velocity and pressure are the same as at A. It may be safely said that as long as the Total Energy in a flow of fluid remains constant, any increase in its velocity will result in a decrease in its pressure.

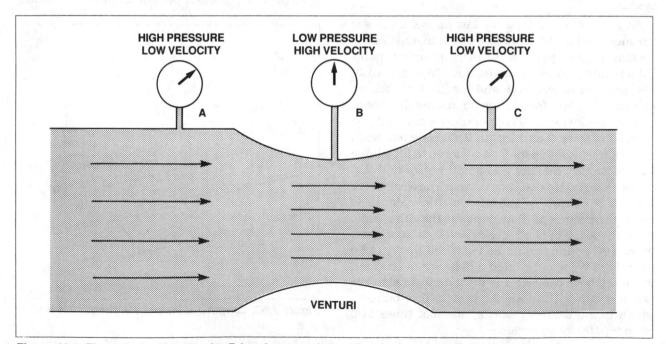

Figure 13-5. The velocity (E_K) at point B has increased; therefore its pressure (E_P) has decreased.

Hydraulic Fluids

In the open hydraulic system of rivers and waterways, water is the hydraulic fluid. Aircraft engines use hydraulic lifters to open the valves; these lifters use engine oil as the fluid. Turbine engines vary inlet guide vanes and open bleed valves, using fuel as a fluid. A pneumatic system uses air as its fluid. We can almost say any fluid can be used to transmit a force, but in an aircraft hydraulic system we have requirements other than just the transmission of a force.

An aircraft hydraulic fluid must be as incompressible as practical, and it must flow through the lines with a minimum of friction. It must have good lubricating properties so the pump and system components will not wear excessively, and it must not foam in operation. It must also be compatible with the metal in the components and with the elastic material of the seals.

Types Of Fluids

Vegetable Base Fluid

MIL-H-7644 fluid has been used in the past when hydraulic system requirements were not so severe as they are today. This fluid was essentially castor oil and alcohol and was dyed blue for identification. Natural rubber seals may be used with this fluid, and the system may be flushed with alcohol. It is quite unlikely that vegetable base fluid will be encountered in servicing modern aircraft.

Mineral Base Fluid

MIL-H-5606 is the most commonly used hydraulic fluid today. It is basically a kerosene-type petroleum product, having good lubricating characteristics and additives to inhibit foaming and prevent corrosion formation. It is quite stable chemically and has very little viscosity change with temperature. 5606 fluid is dyed red for identification, and systems using this fluid may be flushed with naphtha, varsol, or Stoddard solvent. Neoprene seals and hose may be used with 5606 fluid.

Synthetic Fluid

Mineral base hydraulic fluid has the limitation of flammability. A leak under high pressure will release a combustible spray of oil into the airplane and cause a fire hazard. To overcome this restriction and at the same time have a fluid capable of standing up under the higher pressures and temperatures required by the modern high-speed jet aircraft, a synthetic phosphate ester fluid has been developed. The most commonly used fluid of this type is MIL-H-8446, commonly known as Skydrol®-500A. This fluid is colored light purple (other grades of Skydrol are green or amber), is very slightly heavier than water, and has a wide range of operating temperatures—from somewhere around –65 °F to over 225 °F for sustained operation.

Skydrol is not without its problems for the pilot and technician, however, as it is quite susceptible to contamination by water from the atmosphere. It must be kept tightly sealed. It will attack polyvinyl chloride and must not be allowed to drip onto electrical wiring, as it will break down the insulation. It will also lift the finish from an aircraft if it is not epoxy or polyurethane. Butyl, silicone rubber, or Teflon® seals may be used with Skydrol.

Contamination Detection And Protection

Hydraulic systems operate with high pressures, and the components have such close fitting parts that any contamination will cause their failure.

When servicing a hydraulic system, be sure that only the proper fluid is used. The service manual of the airplane specifies the fluid, and the reservoir should also be marked with the type of fluid required.

Systems using 5606 fluid may be kept free from contamination by keeping the reservoir tightly closed. Old 5606 fluid has a somewhat sour smell and is darker in color than fresh fluid. Any fluid suspected of contamination or aging should be drained and the system flushed and filled with fresh fluid.

The Hydraulic Power Pack

To simplify hydraulic systems, the hydraulic reservoir, control valves, electric-motor-driven hydraulic pump and other essential parts are incorporated into one assembly, normally called a **power pack**. Figures 13-6A and B show the hydraulic system of a typical light twin-engine airplane using a power pack. This system utilizes a reversible pump driven by a reversible DC (direct current) electric motor.

Figure 13-6A shows the motor driving the pump in the direction which lowers the landing gear. When the gear-type pump turns in the direction indicated, the fluid comes from the reservoir down through the right check valve, around the outside of the gears in the pump, where it is pressurized to the pressure permitted by the low-pressure control valve which provides enough pressure to move the gear to the down-and-locked position. From the pump, fluid moves down to the shuttle valve, pushing its piston to the left, compressing its spring and opening the passage to the down-side of the three landing-gear cylinders. This same

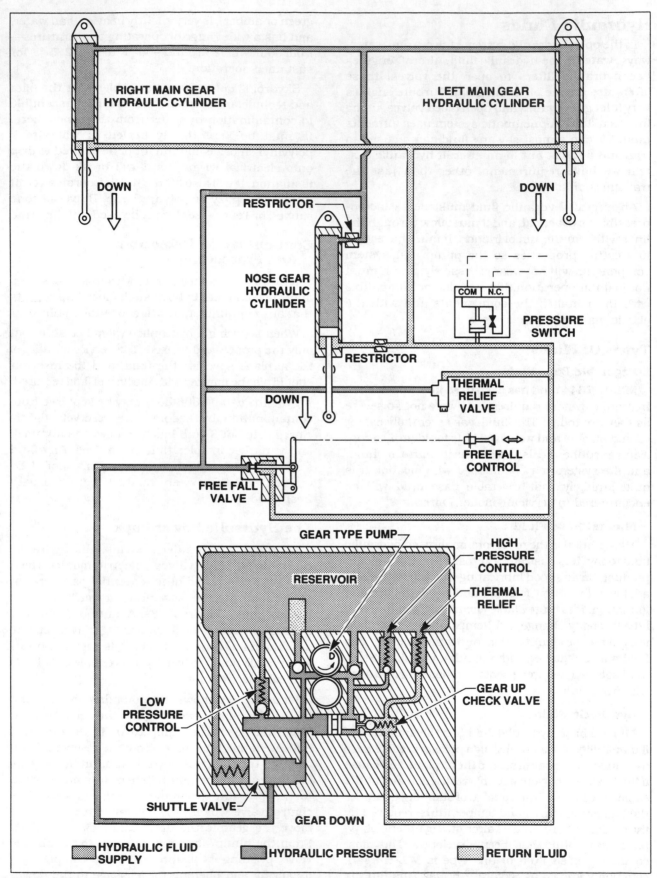

Figure 13-6A. Power-pack-type hydraulic system. In this condition, the landing gear is being lowered.

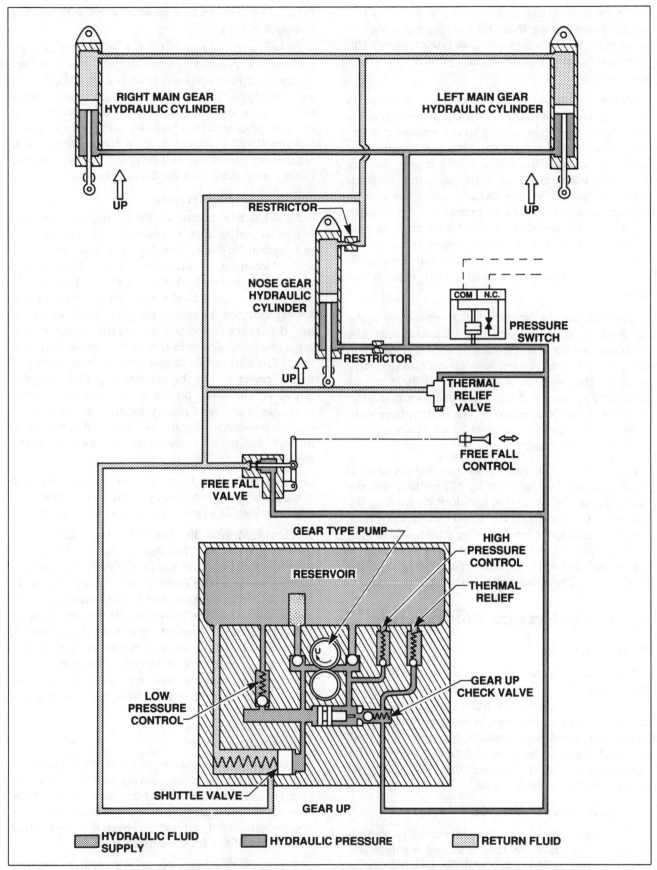

Figure 13-6B. Power-pack-type hydraulic system. In this condition, the landing gear is being retracted.

pressure moves the gear-up check valve to the right, opening it so that fluid returning from the up-side of the cylinders can flow back through the pump to the down-side of the cylinders or to the reservoir.

When the landing gear position selector is moved by the pilot to the GEAR UP position (Figure 13-6B), the pump is turned in the opposite direction. Fluid is drawn from the reservoir through the filter and around the gears in the pump, moving fluid from left to right through the pump, through the gear-up check valve (which holds the gear in the up position when the pump is off in cruise flight), to the up-side of the three landing-gear cylinders. The return fluid passes through the shuttle valve to the reservoir. Note that the shuttle valve piston has been moved to the right by its spring.

When all three gear cylinders have moved to the up position, pressure will build up and open the pressure switch that shuts off electrical power to the pump motor. There are no mechanical up-locks in this system — the gear is held up by hydraulic pressure. If the pressure bleeds off, the switch will start the pump motor, causing pressure to be restored before the gear has a chance to fall out of the wheel wells.

Emergency gear extension for this system is accomplished by the pilot by (1) pulling out the landing gear motor circuit breaker, (2) putting the gear selector handle in the GEAR DOWN position, and (3) opening the free fall valve which allows fluid to flow from the up side of the three cylinders to the down side in response to the weight of the landing gear "free-falling" to the down position.

Hydraulic System Components

Reservoirs

The **hydraulic reservoir** not only stores fluid for the system, but serves as an expansion chamber and a point at which the fluid can purge itself of any air it has accumulated in its operational cycle. Reservoirs must have enough capacity to hold all the fluid that can be returned to the system with any configuration of the gear, flaps, and all other hydraulically-actuated units.

Nonpressurized Reservoirs

The fluid return into the reservoir is usually directed in such a way that foaming is minimized and any air in the fluid will be swirled out or extracted. Some reservoirs have filters built into

them at the return line so all of the fluid entering the tank is strained.

Reservoirs for most of the medium-size hydraulic systems have two outlets; one is either located partially up the side or connected through a **standpipe**. This outlet feeds the engine-driven pump so in the event of a break in a system line or any type of leak that loses all of the pump's fluid, there will still be some fluid in the reservoir. The emergency pump outlet is near the bottom and the hand pump draws its fluid from there.

Pressurized Reservoirs

As airplanes began to fly at higher altitudes where the outside air pressure is low, the returning hydraulic fluid developed a bad tendency toward foaming. To minimize this condition, reservoirs were pressurized. One of the early methods was to inject air into the returning fluid through an aspirator or a venturi tee. The fluid returning into the reservoir flowed through the venturi creating a low pressure, and the air was pulled into the fluid. The aerated fluid was swirled into the top of the reservoir where the air was expelled from the liquid. A relief valve on the reservoir maintained a pressure of about 12 psi on the fluid. Some turbine engine-powered aircraft use a small amount of filtered compressor bleed air to pressurize the reservoir.

Jet aircraft that fly at very high altitudes and systems which place heavy demands on the fluid require reservoirs pressurized to a higher pressure.

When hydraulic fluid is used to perform work, heat is generated. This heat must be dissipated from the fluid, and, in the larger systems, such as the one used in the Boeing 727, heat exchangers are installed in the fuel tanks. The hydraulic fluid, on the way back into the reservoir, passes through these coils and its heat is given up to the fuel in the tank. Restrictions are made regarding the operation of any of these hydraulic systems on the ground when there is less than a certain amount of fuel in the tank in which these heat exchangers are located.

Hydraulic Pumps

Fluid power is available in an aircraft hydraulic system when fluid is moved under pressure. Pumps used in these systems are simply fluid movers, rather than pressure generators. Pressure can be generated only when there is a restriction to the flow of the fluid being moved.

There are two basic types of hydraulic pumps: those operated by hand, and those driven by some source of power, such as by an electric motor or an aircraft engine.

Hand Pumps

Single-action pumps move fluid only on one stroke of the piston, while **double-action** pumps move fluid with both strokes. Double-action pumps are the only ones commonly used in aircraft hydraulic systems because of their greater efficiency. Figure 13-7 is a diagram of a piston rod displacement hand pump.

On the stroke during which the piston is pulled out of the cylinder, fluid is drawn in through the inlet check valve and the fluid on the back side of the piston is forced out the pump outlet. When the piston is forced into the cylinder, the rod displaces part of the fluid and some of it is again forced out the discharge.

Let's assume some values: the large end of the piston has an area of two square inches, the rod displaces one square inch, and the piston moves one inch. When the piston moves out of the cylinder, two cubic inches of fluid is drawn in. Now when the piston is moved into the cylinder, the two cubic inches of fluid is forced out, but the space behind the piston has only one cubic inch of volume; so one cubic inch of fluid must be forced out the pump discharge port. When the piston is again pulled out of the cylinder, the remaining one cubic inch is forced out of the pump. Every time the piston is moved out of the cylinder, two cubic inches of fluid is taken in and one cubic inch is discharged. Each time the piston moves into the cylinder, one cubic inch is discharged but no fluid is taken into the pump.

If a force of 500 pounds is exerted on the piston as it is pulled out of the cylinder, a pressure of 500 psi will be built up. On the return stroke, however, since there is a working piston area of two square inches, the same 500 pounds of force will generate only 250 psi pressure.

Power Pumps

These may be classified as either constant or variable displacement pumps. A **constant displacement pump** is one that moves a given amount of fluid each time it rotates. A pump of this type must have some sort of unloading device or regulator to prevent its building up so much pressure that is will rupture a line or perhaps damage itself.

Variable displacement pumps move a volume of fluid proportional to the demands of the system.

These pumps are quite often of the piston type, and their output volume is varied by changing the stroke of the piston.

One of the simpler constant displacement pumps used to move a rather large volume of fluid under a relatively low pressure is the vane pump, figure 13-8.

The vanes are free-floating in the rotor and are held against the wall of the steel sleeve by a spacer. As the rotor turns in the direction indicated by the arrow, the volume between the vanes on the inlet side increases and the volume between the vanes on the discharge side decreases. This draws fluid into the pump and forces it out the other side. This type of pump finds wide use in the aircraft for moving fuel for piston engines, and also air for gyro instruments and pneumatic deicer boots. It is used also to a somewhat lesser degree for hydraulic systems.

To move the medium volumes of fluid under medium pressures required by some hydraulic systems, gear pumps are used. The two types commonly used for aircraft hydraulic systems are the spur gear and the gerotor.

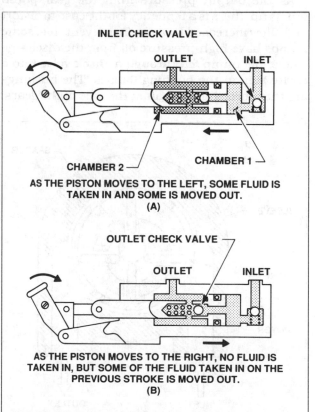

AS THE PISTON MOVES TO THE LEFT, SOME FLUID IS TAKEN IN AND SOME IS MOVED OUT.
(A)

AS THE PISTON MOVES TO THE RIGHT, NO FLUID IS TAKEN IN, BUT SOME OF THE FLUID TAKEN IN ON THE PREVIOUS STROKE IS MOVED OUT.
(B)

Figure 13-7. The piston rod displacement hand pump is a double-acting pump. Fluid is pushed out of the pump on every stroke, but taken in only on every other stroke.

The simple **spur gear pump**, figure 13-9, uses two meshing gears closely fitted into a housing. One of the gears is driven by the engine accessory drive or an electric motor, and this gear drives the other. As the gears rotate in the direction shown by the arrows, the space between the teeth on the inlet side becomes larger. Fluid is pulled into this space, trapped between the teeth and the housing and carried around to the discharge side of the pump. Here, the teeth of the two gears come into mesh, providing a barrier which forces the fluid out the pump discharge. A small amount of fluid is allowed to leak past the gears and around the shaft for lubrication, cooling and sealing. This fluid drains into the hollow shafts of the gears and is picked up by the low pressure at the inlet side of the pump. A weak relief valve holds the oil in the hollow shafts until it builds up a pressure of about 15 psi. This so-called **case pressure** is maintained so that in the event the shaft or seal becomes scored, fluid will be forced out rather than air being drawn into the pump. Air would otherwise displace the fluid needed for lubrication and the pump would be damaged.

As the output pressure from the gear pump builds up, there is a tendency for the case to distort and allow increased leakage. To prevent this, some pumps have high-pressure oil from the discharge side of the pump fed through a check valve into a cavity behind the bushing flanges. The bushings are thus forced tight against the side of the gears, decreasing the side clearance and minimizing leakage, also compensating for bushing wear.

The **gerotor pump** is sort of a combination internal-external gear pump. Figure 13-10 shows its operation. Its four-tooth spur gear is driven by an accessory drive from the engine, and as it turns it rotates the five-tooth internal gear rotor. Looking at the relationship between the two gears, one will see that as the spur gear rotates and turns the internal gear, the space between the teeth gets larger on one side, smaller on the other. Covering these gears is a plate with a crescent-shaped opening above each side of the gears. The opening above the space which is getting larger is the inlet side of the pump, and the opening above the side having the gears coming into mesh is the outlet.

Hydraulic Valves

Flow Control Valves

One of the more common flow control valves is the **selector valve**, which determines the direction of flow of fluid to retract or extend the landing gear or to select the position of the flaps. There are two common types of selector valves: the **open center valve**, which directs fluid through the center of the valve back to the reservoir when a unit is not being actuated, and the **closed center valve**, which stops the flow of fluid when it is in the neutral position. Both valves direct fluid under pressure to one side of the actuator and vent the opposite side to the reservoir.

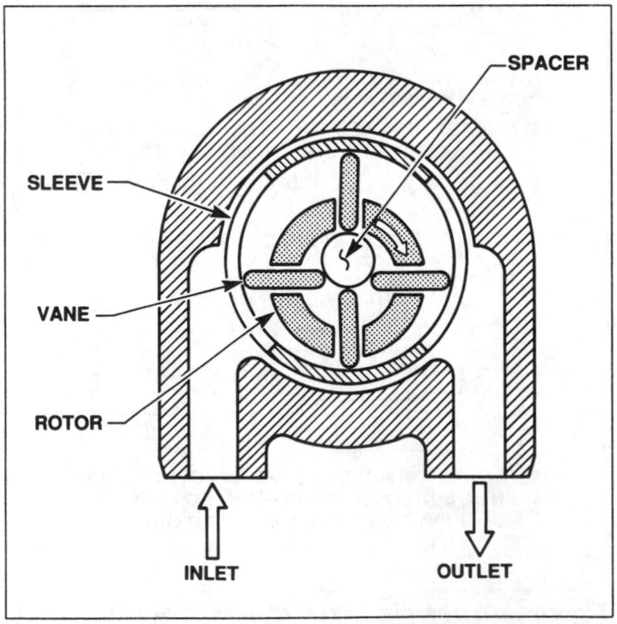

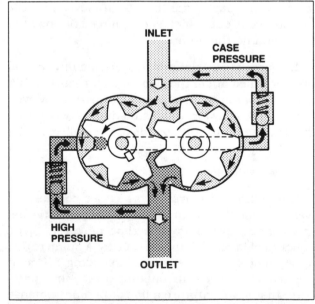

Figure 13-8. The vane pump is constant displacement and moves a relatively large volume of fluid under relatively low pressures.

Figure 13-9. A spur gear pump moves a medium volume of fluid under pressures up to about 1500 psi.

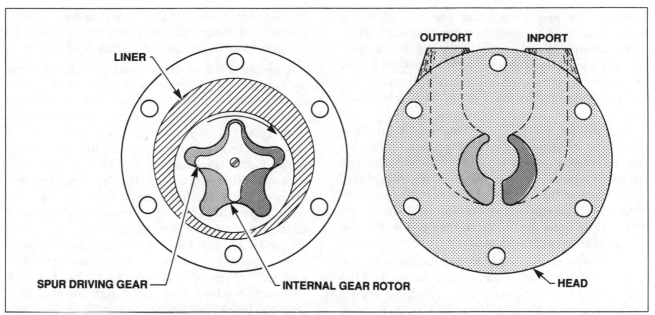

Figure 13-10. *The gerotor pump is a special form of gear pump, producing up to about 1500 psi pressure with a moderate flow.*

For systems using relatively low pressure for actuation, a simple plug-type selector valve, figure 13-11, is often used. In one position, the pressure port and actuator port 1 are connected. Actuator port 2 is connected to the return line. When the selector handle is turned ninety degrees, the actuator ports are reversed to the pressure and return lines.

Higher pressure systems require a more positive shutoff of fluid flow and a poppet-type selector is often used. In figure 13-12(A) the upper right and lower left poppets are off their seat, and the fluid is flowing such a way that the pressure is supplied to actuator line 2 and line 1 is connected to the return line. When the handle is in the neutral position, all of the poppets are closed, and no fluid flows. When the handle is rotated the opposite position, figure 13-12(B), the pressure and return lines feed the opposite sides of the actuator.

Modern aircraft with retractable landing gear often have doors that close in flight to cover the wheel well and make the airplane more streamlined. To be sure the landing gear does not extend while the doors are closed, **sequence valves** are used. These are actually check valves which allow a flow in one direction but may be opened manually so fluid can flow freely in both directions, figure 13-13.

Figure 13-14 shows the position in a landing gear actuation system in which these valves would be installed. The wheel well doors must be fully

open before the sequence valve will allow fluid to flow into the main landing gear cylinder. The return fluid flows unrestricted through the sequence valve on its way back into the reservoir.

Priority valves, figure 13-15, are similar to sequence valves except that they are opened by hydraulic pressure rather than by mechanical contact. They are called priority valves because such devices as wheel well doors, which must

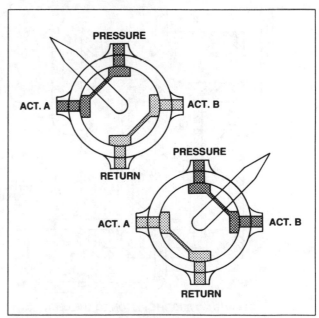

Figure 13-11. *A plug-type rotary selector valve used in closed center hydraulic systems.*

245

operate first, require a lower pressure than the main landing gear, and the valve will shut off all the flow to the main gear until the doors have actuated and the pressure builds up at the end of the actuator stroke. When this buildup occurs, the priority valve opens and fluid can flow to the main gear.

Modern jet aircraft are dependent on their hydraulic systems, not only for raising and lowering the landing gear, but for control system boosts, thrust reversers, flaps, brakes and many auxiliary systems. For this reason most aircraft use more than one independent system; and in these systems, provisions are made to fuse or block a line if a serious leak should occur.

Of the two basic types of **hydraulic fuses** in use, one operates in such a way that it will shut off the flow of fluid if sufficient pressure drop occurs across the fuse. Fluid flows from A to B in figure 13-16. The spring keeps the passage open and fluid flows into the system for normal operation. If a break should occur in the line, the pressure at side B will drop, and fluid pressure will force

the slide over to close the valve and prevent further loss of fluid, figure 13-16(B). Reverse flow is in no way hindered or controlled by this type of fuse.

A second type of fuse, figure 13-17, does not operate on the principle of pressure drop, but it will shut off the flow after a given amount of fluid has passed through the line. In the static condition, figure 13-17(A), all of the passages are closed off. When fluid begins to flow in the direction of normal operation, figure 13-17(B), the sleeve moves over, compressing the large spring and opening the valve for normal flow. At this time some fluid enters the small orifice and begins to drift the piston over. Normal operation of the unit protected by this fuse does not require enough flow to allow the piston to drift completely over and seal off the line. If there is a leak, however, sufficient fluid will flow that the piston will move over and block the line, as seen in figure 13-17(C). Reverse flow is provided for as the fluid acts on the small piston, compressing its spring and opening the passages for the return fluid, figure 13-17(D).

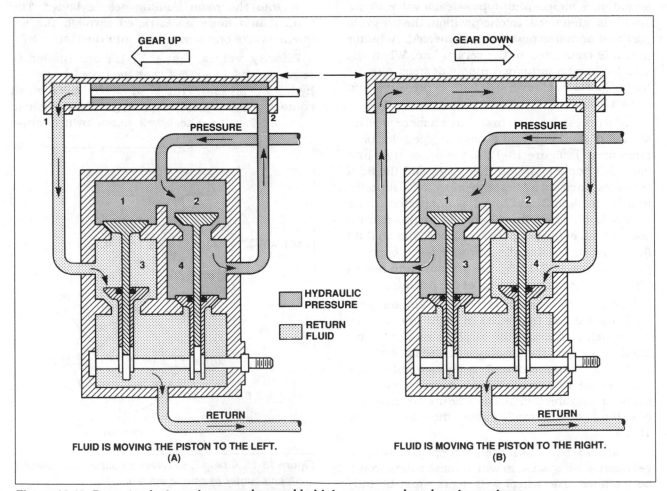

FLUID IS MOVING THE PISTON TO THE LEFT.
(A)

FLUID IS MOVING THE PISTON TO THE RIGHT.
(B)

HYDRAULIC PRESSURE

RETURN FLUID

Figure 13-12. Poppet selector valves may be used in high-pressure closed center systems.

246

There are many instances in an aircraft hydraulic system where it is desired to allow flow in one direction and prevent it in the opposite direction. This is accomplished by the use of **check valves**. There are several types of these valves; the ball, cone and swing valves, figure 13-18(A), (B), and (C) are the most common.

Certain applications require full flow one way and restricted flow in the opposite direction. An example of this is in a landing gear system where the weight of the gear and air loads causes the extension to be excessively fast, and the weight of the gear against the air loads requires every bit of pressure possible to get the gear up. An **orifice check valve** is installed in such a way that fluid flowing into the gear-up lines finds no restriction, and fluid leaving the gear-up side is restricted by the orifice in the check valve, figure 13-19.

Pressure Control Valves

The simplest type of pressure control valve is the **relief valve**. It is used primarily as a backup rather than a pressure control device because of the heat generated and power dissipated when it relieves pressure. System pressure relief valves are set to relieve at a pressure above that maintained by the system pressure regulator, and only in the event of a malfunction of the regulator will the relief valve be called into service. In systems where fluid may be trapped in a line between the actuator and its selector valve, there is the problem of pressure buildup by thermal expansion of the fluid. Thermal relief valves are installed in these lines to prevent damage by relieving a small amount of fluid back into the return line. A simple relief valve is seen in figure 13-20.

Closed-center hydraulic systems require a regulator to maintain the pressure within a specified range and to keep the pump unloaded when no unit is being actuated. The simplest **pressure regulator** is the balanced type whose principle of operation is shown in figure 13-21.

Starting with a discharged, or flat, system, the pump pushes fluid through the check valve into the system and the accumulator. When no fluid is required for actuation, the accumulator fills and pressure builds up. This pressure pushes up on the piston and also down on the ball. A condition is reached where there is a balance of forces: both the fluid pressure on the ball and the spring act downward and the hydraulic pressure on the piston pushes up. At the condition of balance, when the pressure is 1500 psi, there will be a force of 1500 pounds pushing up on the piston. There will

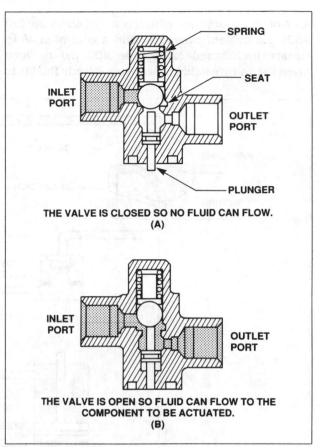

Figure 13-13. A sequence valve acts as a check valve until its plunger is mechanically pushed up, opening the return path.

be a total downward force of 1000 pounds applied by the spring, and 1/3 of 1500, or 500 pounds of hydraulic force pushing down on the ball. Now if the pressure rises above this value, since the spring force is constant and not affected by the hydraulic pressure, the piston will move up and unseat the ball. When the ball unseats, the fluid returns to the reservoir and the pump pressure drops to essentially zero. The check valve seats and holds pressure trapped in the accumulator and the system. This unloaded condition will continue until the pressure on the system drops to 1000 psi, at which point the spring can force the piston down and allow the ball to seat. When the ball seats. the pressure again rises to the unloaded pressure.

If it is desired to reduce the pressure in some branches of a hydraulic system, a simple **pressure reducing valve** may be used. The valve shown in figure 13-22 reduces the pressure by the action of a balance of hydraulic and spring forces.

Let's assume that a piston having an area of one square inch is held against its seat by a spring with 100 pounds of force. The piston has a shoulder area of ½ square inch which is acted on by the full 1500 psi system pressure, and a seat area of ½ square inch acted on by the 200 psi reduced pressure. A tiny hole in the piston bleeds fluid into the chamber behind the piston and the relief valve maintains this pressure at 750 psi. This relief action is determined by the pressure inside the piston cavity acting on one side of the relief ball

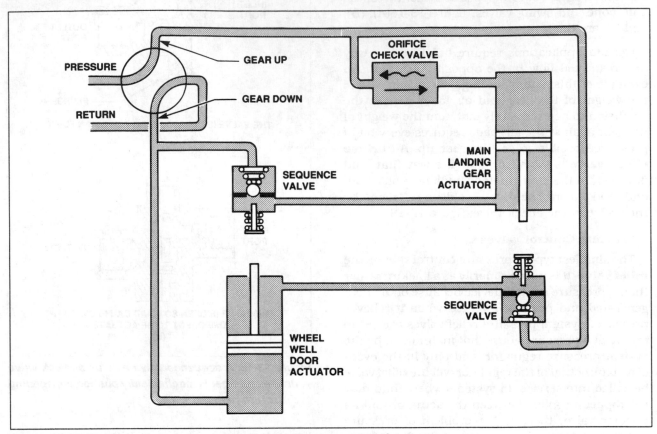

Figure 13-14. Position of sequence valves in a landing gear retraction system.

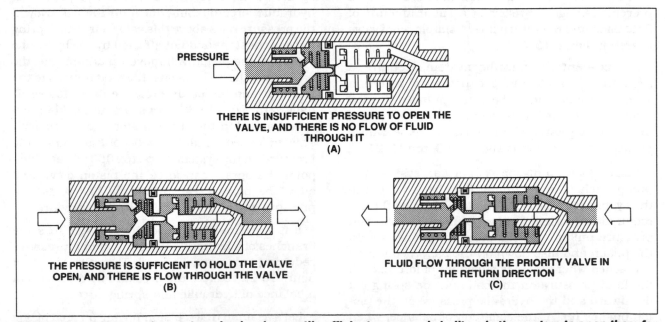

THERE IS INSUFFICIENT PRESSURE TO OPEN THE VALVE, AND THERE IS NO FLOW OF FLUID THROUGH IT
(A)

THE PRESSURE IS SUFFICIENT TO HOLD THE VALVE OPEN, AND THERE IS FLOW THROUGH THE VALVE
(B)

FLUID FLOW THROUGH THE PRIORITY VALVE IN THE RETURN DIRECTION
(C)

Figure 13-15. Priority valves act as check valves until sufficient pressure is built up in the system to open them for full flow.

and the spring and reduced pressure (200 psi) acting on the opposite side. When the reduced pressure drops, the hydraulic force on the ball drops, allowing it to unseat. This decreases the hydraulic force on the piston and allows it to move up. Fluid now flows into the reduced pressure line and restores the 200 psi. This increased pressure closes the relief valve so that the pressure behind the piston can again come up to 750 psi and seat the valve. Rather than the piston chattering, the tiny bleed hole causes it to have a relatively smooth action, and it remains off its seat just enough to maintain the reduced pressure as it is used.

Accumulators

Hydraulic fluid is noncompressible, and in order to store fluid under pressure, a compressible fluid must be utilized. These conflicts are resolved by the use of an accumulator. There are three basic types of accumulators; two are hollow steel spheres, divided into two compartments by either a diaphragm or a bladder; the other is a steel cylinder with a floating piston forming the two compartments. Figure 13-23 shows each type.

The accumulator is charged with compressed air or nitrogen to a pressure of approximately one-third the system pressure. As the pump forces hydraulic fluid into the accumulator, the air is further compressed and exerts a force on the

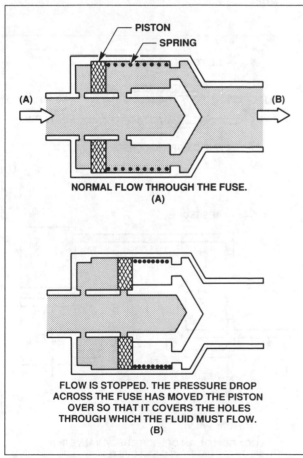

Figure 13-16(A)—Flow rate is low enough that the piston is not moved over against its spring. (B)—Flow rate is high enough that the pressure drop across the orifice moves the piston over and stops the flow.

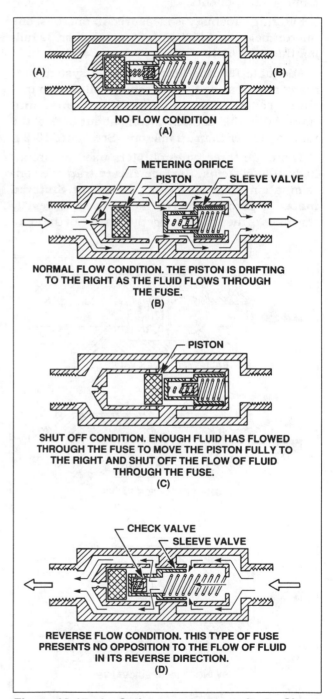

Figure 13-17. 1—Outlet to actuator unit; 2—Sleeve valve; 3—Check valve; 4—Piston; 5—Metering orifice; 6—Inlet from selector valve.

249

hydraulic fluid, holding it under pressure after the system pressure regulator has unloaded the pump.

Filters

One of the more important requirements for hydraulic fluid is its cleanness. Solid particle contaminants must all be removed as they can damage components.

Filtering capability is measured in microns, one micron being one millionth of a meter, or 39 millionths of an inch.

Adequate filtering for a hydraulic system normally requires the filter to remove all contamination greater than about 25 microns, and **nominally** (this actually means about 95% of the time) all larger than 10 microns. See figure 13-24.

There are three types of filters used in aircraft hydraulic systems. **Surface filters** trap the contamination on the surface of the element. **Sintered metal**, a porous material made up of extremely tiny balls of metal fused together, is one of the more

popular types of surface filtration. These filters usually have a bypass valve which opens to allow the fluid to bypass the element if it should clog.

Micronic filters are made up of specially treated cellulose paper elements pleated to provide more area. Filters of this type are often installed in the return line into the reservoir where the pressure drop is low. See figure 13-25.

A filter similar to the one with the paper element has a stainless steel wire mesh, such as shown in figure 13-26. This wire will retain about 95% of all of the contamination larger than 5 to 10 microns.

Edge filters, often called Cuno filters, are composed of stacks of thin metal discs with scrapers between them. All of the fluid flows between the discs and contaminants are stopped on the edge. See figure 13-27. These filters are cleaned by turning the shaft which rotates the discs and scrapes the contamination from between them into the outer housing where it can be removed by draining.

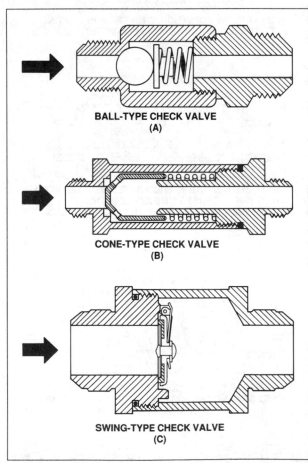

Figure 13-18(A)—Ball check valve. (B)—Cone check valve. (C)—Swing check valve.

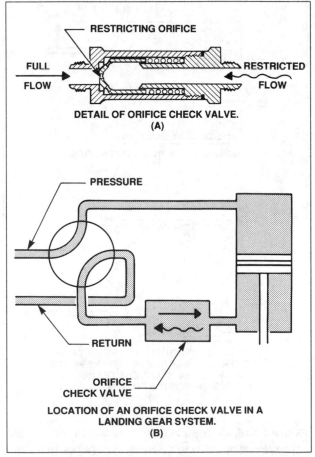

Figure 13-19(A)—Orifice check valve. (B)—Orifice check valve, installed in a landing gear system.

Fluid Lines

Most **rigid lines** for hydraulic or pneumatic systems are made of 5052-0 aluminum alloy. This metal is easy to form and has sufficient strength for almost all aircraft hydraulic system installations. If additional strength is needed, for instance, for higher pressure systems, or for oxygen installation, stainless steel tubing may be used.

Low-pressure flexible fluid lines are seldom used for hydraulic systems, but in the process of considering fluid lines, they should be discussed. This type of hose, MIL-H-5593, has a seamless synthetic rubber inner liner and a single cotton braid reinforcement. All of this is covered with either smooth or ribbed synthetic rubber. Maximum pressure for this hose runs from 150 psi for ⅜″ ID to 300 psi for ⅛″ ID. See figure 13-28.

Medium-pressure hose, MIL-H-8794 hose, figure 13-29, has a smooth synthetic rubber inner liner, covered with a cotton braid. This, in turn, is covered with a single layer of steel wire braid, and over this is a rough, oil-resistant cotton braid.

The operating pressure for a hose varies with its size; the smaller the hose, the higher the allowable operating pressure. Generally, MIL-H-8794 hose is used in systems operating at about 1500 psi. All flexible hoses used in aircraft fluid power systems have a lay-line, a yellow painted stripe, along its length to allow the technician to see at a glance whether or not the hose is twisted. The lay-line should not twist around the hose when installed.

High-pressure hose, MIL-H-8788 hose, has a smooth synthetic rubber inner liner, two high-tensile carbon steel braid reinforcements, a fabric braid, and a smooth black synthetic rubber outer cover, figure 13-30.

Another high-pressure hose, similar to MIL-H-8788, has a butyl inner liner and a smooth synthetic rubber outer cover colored green instead of black. The lay and marking are white instead of yellow. This hose is to be used only with Skydrol and is suitable for pressures up to 3000 psi, as is the case with MIL-H-8788.

Teflon hose has a liner made of tetrafluorethylene, or Teflon resin, and is covered with stainless steel braid. Medium-pressure hose, figure 13-31, is covered with one stainless steel braid, and high-pressure hose has two. Teflon has very desirable operating characteristics and may be used in fuel, lubricant, hydraulic, and pneumatic systems in modern aircraft. This hose has one characteristic, though, that the A&P must

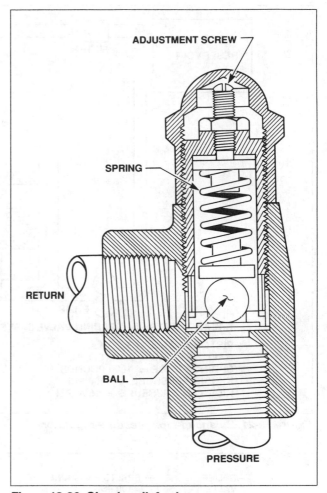

Figure 13-20. Simple relief valve.

be aware of to get the best service from it. The inner liner for this hose is extruded and will pre-form or "take a set" after it has been used with high-temperature or high-pressure fluid. After hose of Teflon has been used, it should not be bent or have any of its bends straightened out. When this tubing is removed from the airplane, it should be supported in the shape it had at the time it was installed if it is to be used again.

High-pressure Seals

Seals are used throughout hydraulic and pneumatic systems to minimize internal leakage and the loss of system pressure. There are two types of seals in use: **gaskets**, where there is no relative motion between the surfaces, and **packings**, where relative motion does exist.

There are many different kinds of seals used in aircraft applications, ranging from flat paper gaskets up through complex, multi-component packings. **V-ring packings** or **chevron seals**, figure 13-32, have found extensive use in the past.

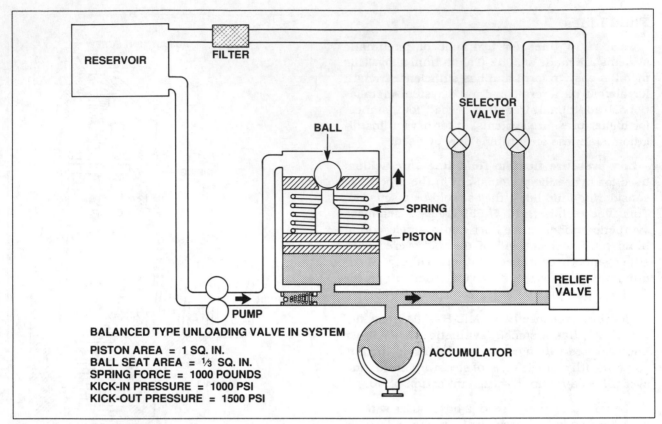

Figure 13-21. Balanced-type pressure regulator.

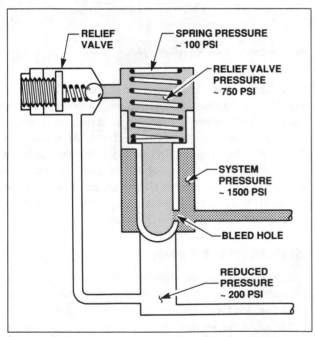

Figure 13-22. Pressure reducing valve.

Figure 13-32(A) shows a cross section of a chevron seal, a single direction seal with the pressure on the side of the lip. Chevrons, a type of compression seal, are usually installed either in pairs or in larger stacks and require a metal backup ring and a spreader. The amount of spread of the seal is determined by the tightness of the adjusting nuts, figure 13-32(B).

Most modern hydraulic and pneumatic systems use **O-rings** for both packings and gaskets. O-rings fit into grooves in one of the surfaces being sealed. The groove should be about 10% wider than the width of the seal, and deep enough that the distance between the bottom of the groove and the other mating surface will be a little less than the width of the O-ring, figure 13-33(A). This provides the squeeze necessary to seal under conditions of zero pressure. Figure 13-33(B) shows the proper squeeze of an O-ring.

Figure 13-33(C) illustrates the leakage that may be expected when there is no squeeze. As the pressure of the fluid increases, as in figure 13-33(D), the ring tends to wedge in tight between the wall of the groove and the other mating surface.

An O-ring of the appropriate size can withstand pressures up to about 1500 psi without distortion, but beyond this, there is a tendency for the ring to extrude into the groove between the two mating surfaces. To prevent this, an anti-extrusion or backup ring is used, usually made of Teflon.

Actuators

Linear

The ultimate function of a hydraulic or pneumatic system is to convert the pressure in the fluid into work. In order to do this, there must be some movement. **Linear actuators** consist of a cylinder and piston. The cylinder is usually attached to the aircraft structure and the piston to the component being moved (figure 13-34). In this

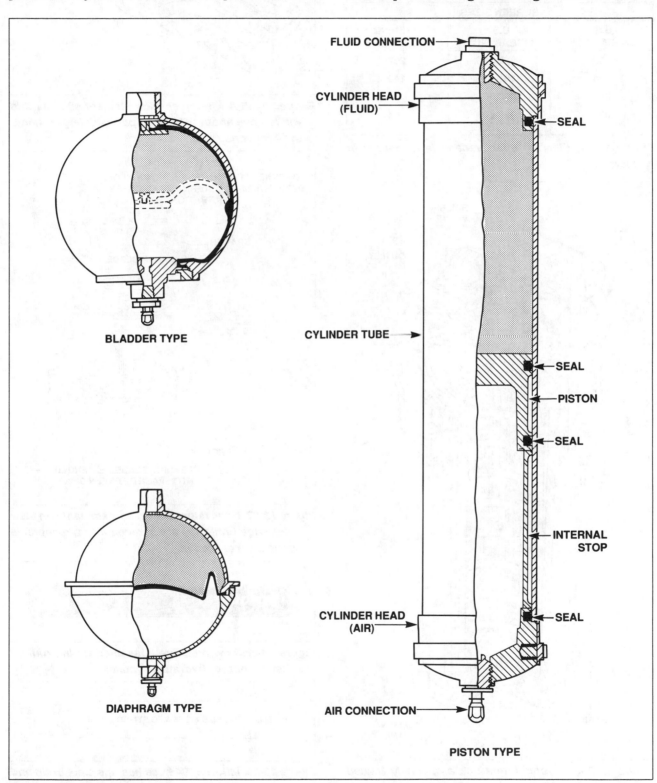

Figure 13-23. Accumulators.

RELATIVE SIZES	
GRAIN OF SALT	70 MICRONS
LOWER LIMIT OF VISIBILITY (NAKED EYE)	40 MICRONS
WHITE BLOOD CELLS	25 MICRONS

SCREEN SIZES		
U.S. SIEVE NO.	U.S. LINEAR INCH	OPENING IN MICRONS
50	52.36	297
100	101.01	149
200	200.00	74
325	323.00	44

Figure 13-24. Filtering effectiveness is measured in microns.

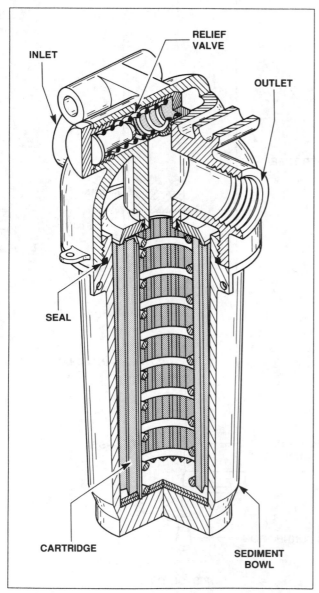

Figure 13-25. Micronic filters use a specially treated pleated paper filter element.

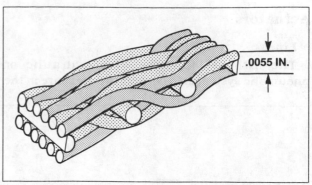

Figure 13-26. A wire mesh filter element such as this will remove about 95% of all particles larger than 5 to 10 microns.

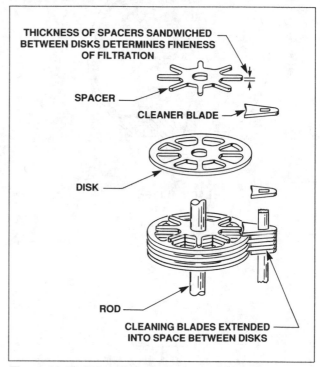

Figure 13-27. Edge filters collect contaminants on their outer edge until they are scraped into a receptacle by the cleaner blades.

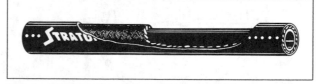

Figure 13-28. Low-pressure hose used for instrument lines, but not for hydraulic systems.

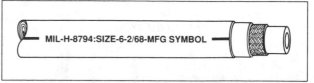

Figure 13-29. MIL-H-8794 hose has one steel braid and an outer cover of rough cotton braid.

case, the cylinder attaches to the main wing spar through a trunion fitting.

There are three basic types of linear actuators. The single-acting actuator has the piston moved in one direction by hydraulic force and returned by a spring, figure 13-35(A), while the double-acting actuator may be either unbalanced, figure 13-35(B), or balanced, figure 13-35(C).

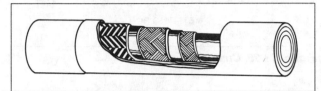

Figure 13-30. MIL-H-8788 has two steel wire braids for reinforcement and is covered with a smooth outer cover.

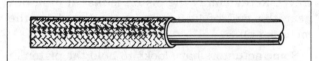

Figure 13-31. Medium-pressure hose of Teflon has a Teflon liner and a single stainless steel braid for reinforcement.

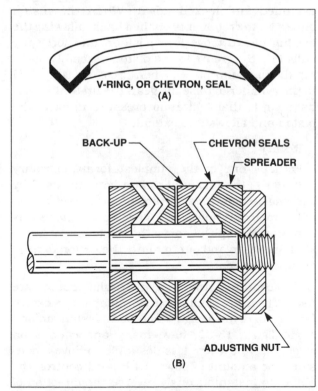

V-RING, OR CHEVRON, SEAL
(A)

BACK-UP — — CHEVRON SEALS
— SPREADER

ADJUSTING NUT —
(B)

Figure 13-32. A chevron seal is a single direction seal held against the inside bore of the actuator by a spreader and an adjusting nut.

Unbalanced actuators have more area on one side of the piston than on the other because of the piston rod. An example is that of figure 13-35(B). To raise the gear, as much force as possible is required so that the fluid pushes against the full area of the piston. The weight of the gear plus the dynamic pressure of the air helps get the gear down, so a smaller amount of force is required to lower and lock the gear. The fluid is put into the end of the actuator with the rod and it pushes on that portion of the piston not taken up by the shaft.

A balanced actuator has a shaft on both sides of the piston so the area is the same on each side and the same force is developed in each direction. Balanced actuators are commonly used for hydraulic automatic pilot servo actuators.

Linear actuators may have features that adapt them to special jobs. Figure 13-36 illustrates a cushioned actuator that allows the piston at the

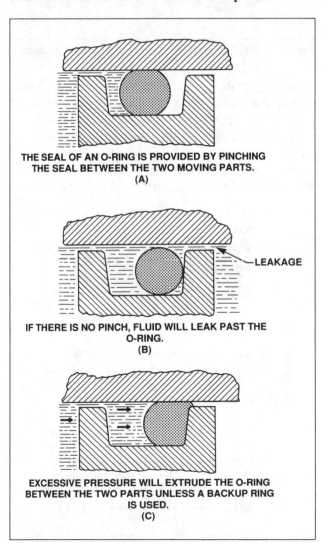

THE SEAL OF AN O-RING IS PROVIDED BY PINCHING THE SEAL BETWEEN THE TWO MOVING PARTS.
(A)

—LEAKAGE

IF THERE IS NO PINCH, FLUID WILL LEAK PAST THE O-RING.
(B)

EXCESSIVE PRESSURE WILL EXTRUDE THE O-RING BETWEEN THE TWO PARTS UNLESS A BACKUP RING IS USED.
(C)

Figure 13-33. The O-ring as a hydraulic seal.

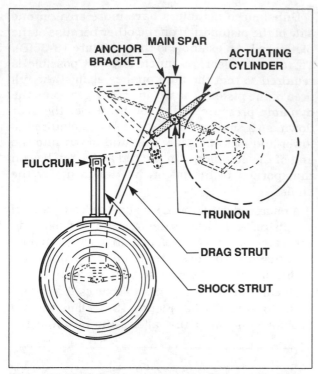

Figure 13-34. Typical retractable landing gear actuator.

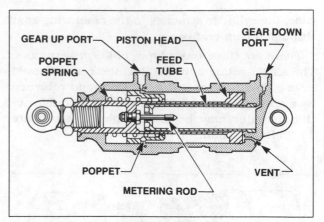

Figure 13-36. Cushioned actuator.

beginning of its stroke to move slowly, accelerate during the middle part of its stroke, and snub or decrease its speed at the end of the stroke.

Fluid flows through the feed tube, around the metering pin, into the hollow piston shaft. The restriction of the fluid allows the piston to move out slowly. When the metering pin pulls all the way

out of the orifice, the piston extends faster until the piston head contacts the poppet. The movement of the piston is slowed by the poppet spring, and it comes to a smooth stop at the end of its travel. Retraction of the piston is fast at first, but near the end of its stroke it is slowed by the metering pin.

Some actuators have locks to hold the piston in a retracted position until the hydraulic pressure releases them. Figure 13-37 shows a linear actuator used for the main landing gear. In figure 13-37(A), the piston is retracted and the gear extended and locked down. When the piston moves in, the retainer is pushed back, allowing the locking pin to move into position and force the balls into the groove in the piston assembly, locking the piston in place. When pressure is applied to the cylinder to raise the gear, the first thing that happens is that hydraulic pressure unlocks the piston and allows it to extend.

Rotary

Perhaps one of the simplest forms of rotary actuator is the rack and pinion actuator used by the single-engine Cessna aircraft for the retraction of the main landing gear. Figure 13-38 shows this actuator: a simple piston with rack teeth cut in the shaft moves in and out to rotate the pinion to raise or lower the gear.

For continual rotation, hydraulic motors are used. These are similar to hydraulic pumps except for certain detail design differences. Piston motors, as in figure 13-39, have many applications on larger aircraft where it is desirable to have a considerable amount of power with good control, the ability to instantaneously reverse the direction of rotation, and no fire hazard if the motor is stalled.

Vane-type motors are also used, but instead of these being as simple as a pump, they require

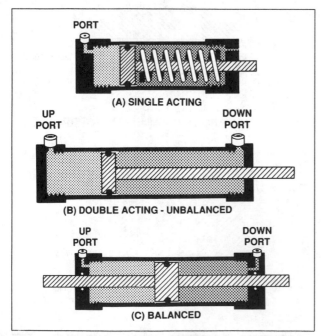

Figure 13-35. Typical linear actuators.

provision to balance the load on the shaft. This is done by directing some of the pressure to both sides of the motor, figure 13-40.

Evolution Of The Aircraft Hydraulic System

When flying was less complex, there was very little need for hydraulic systems. The airplanes flew so slow that drag was of no great concern, so the landing gear could hang down in the wind. Landing speeds were so low there was no need for flaps—once on the ground, the tail skid served as a very effective brake. Paved runways, however, brought out the need for brakes, and the simple hydraulic system came into being.

Sealed Brake System

A simple diaphragm-type master cylinder and expander tube brake, as in figure 13-41, is a complete hydraulic system in its most simple form.

The entire system is sealed, and when the pedal is depressed, fluid is simply moved into the expander tube, which expands and pushes the brake blocks against the drum.

Vented Brake System

The need to vent the brake system so heat expanding the fluid will not cause the brakes to drag brought about the piston-type master cylinder to replace the sealed diaphragm unit. See figure 13-42.

Single-Acting Actuator System

Landing speeds began to increase, so flaps were added. To lower the flaps hydraulically requires more fluid than one stroke of the pump can supply, so a simple pump and selector valve are added to the single-acting cylinder of figure 13-43. When the selector valve is rotated to the flaps-down position and the hand pump worked, fluid forces the piston out and the flaps down. Air loads on the flap and a spring in the cylinder raise the flaps when the selector valve is rotated to the flaps-up position. Two check valves in the pump allow as much fluid to be pumped as the pilot needs.

Powered Pump Systems

Manual Pump Control Valve

Faster aircraft with retractable landing gear resulted in a more complex system, such as diagrammed in figure 13-44.

In this basic system, an engine-driven pump receives its fluid from the reservoir, and, when no

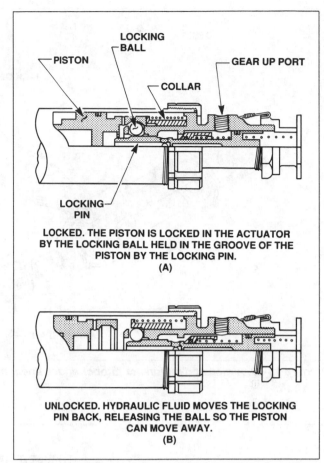

Figure 13-37(A)—Landing gear down-lock in locked position. (B)—Landing gear down-lock in unlocked position.

unit is being actuated, pumps the fluid through the pump control valve back into the reservoir through the filter. Fluid circulates all the time, but since there is very little restriction in the system, almost no power is taken from the engine to drive the pump. When the pilot wants to lower the landing gear, the selector valve is placed in the gear-down position and the pump control valve closed, shutting off the return to the reservoir. The pump now forces fluid into the landing gear cylinders and lowers the gear. When the gear is completely down, the pressure builds up but cannot damage the system because it is relieved by the system pressure relief valve until the pilot opens the pump control valve and unloads the pump. Some pump control valves are automatic and will open when the unit is fully actuated. A hand pump draws its fluid from the bottom of the reservoir, and may be used to pressurize or actuate the system when the engine is not in operation, or in case of a pump or engine failure.

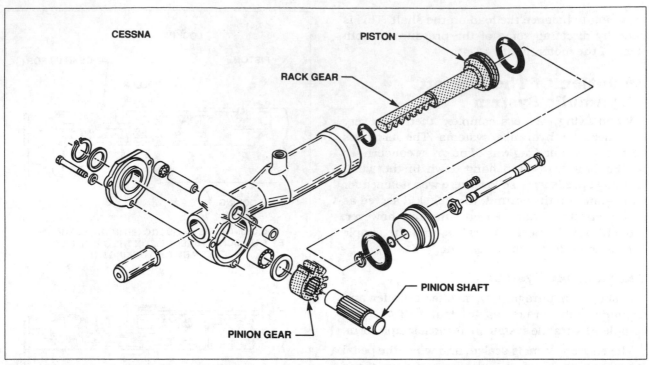

Figure 13-38. A rack and pinion actuator changes linear movement of the piston into rotary movement of the pinion shaft.

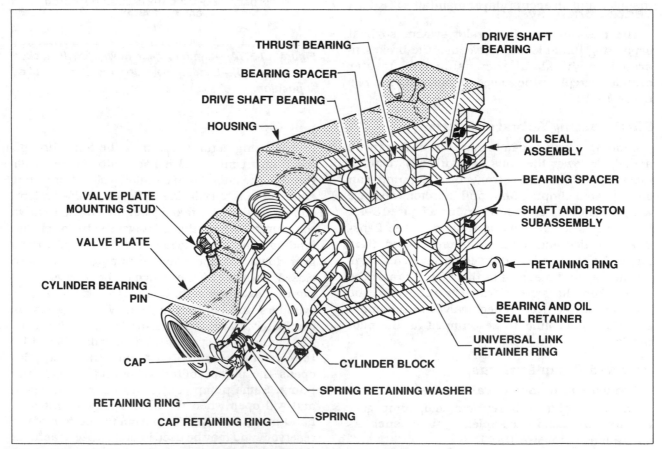

Figure 13-39. A piston-type hydraulic motor is similar to a piston pump.

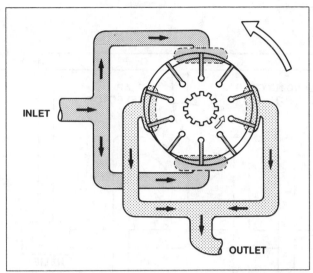

Figure 13-40. A balanced vane-type hydraulic motor directs pressure to vanes on both sides of the motor at the same time.

Automatic Unloading Valve

The final stage in the evolution of the basic hydraulic system comes when an automatic pump control valve, called an unloading valve, is installed and an **accumulator** maintains pressure on the system at all times. The basic system is seen in figure 13-45.

The engine-driven pump is fed from the reservoir and moves fluid through the unloading (pressure relief) valve into the system pressure manifold. This charges the accumulator and holds pressure on the hydraulic fluid in the system. When the pressure reaches a predetermined value, the un-loading valve traps it in the system and directs the output of the pump back into the reservoir through the filter. When either the landing gear or flaps is actuated, the pressure in the system drops and

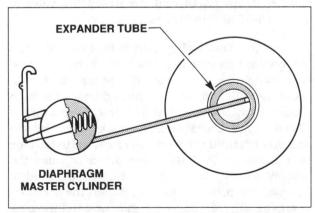

Figure 13-41. The expander tube brake and diaphragm master cylinder is a hydraulic system of the most simple sort.

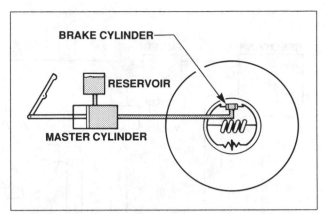

Figure 13-42. The vented master cylinder provides for expansion of fluid and replenishing lost fluid.

will direct the pump output back into the system to supply fluid for the actuation. The hand pump and system pressure relief valve serve the same function as on the simpler systems.

Open Center System

In an effort to minimize the complexity of a hydraulic system for small and medium size airplanes, an **open center system** may be used. This system, figure 13-46, uses special selector valves in series with each other. When no unit is being actuated, the fluid flows from the reservoir through the open center of the valves and back into the reservoir. When actuation is called for, the selector valve directs fluid into one side of the actuator; fluid from the opposite side returns to the reservoir through the other side of the valve. When the actuation is complete, the pressure will rise and automatically shift the valve into its open center position, allowing fluid to circulate through the system with almost no load on the pump.

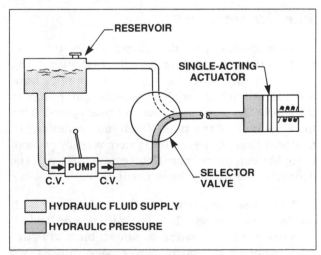

Figure 13-43. A simple single-acting pump allows more than one stroke to be used for actuation.

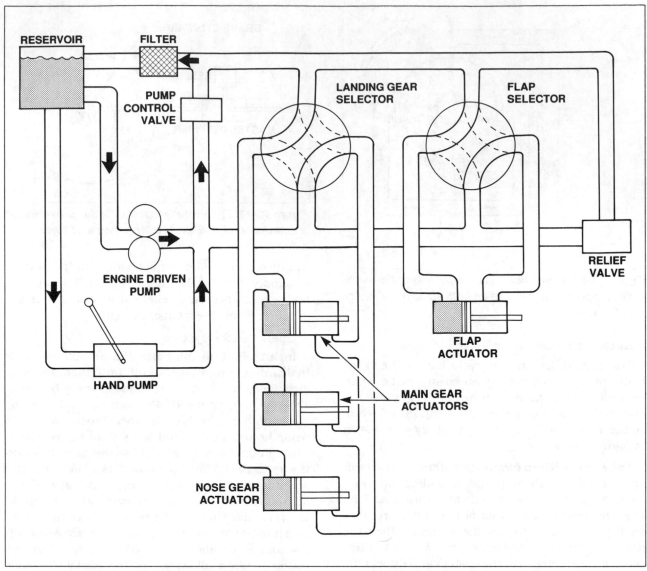

RESERVOIR

FILTER

PUMP
CONTROL
VALVE

LANDING GEAR
SELECTOR

FLAP
SELECTOR

ENGINE DRIVEN
PUMP

RELIEF
VALVE

HAND PUMP

FLAP
ACTUATOR

MAIN GEAR
ACTUATORS

NOSE GEAR
ACTUATOR

Figure 13-44. A manual pump control valve keeps the pump unloaded when no unit is being actuated.

The Power Pack

Engine driven hydraulic pump systems, unless an electric clutch is fitted, require the hydraulic pump to operate any time the engine is running. Continuous operation requires a well-built pump to withstand the wear and tear and takes power from the engine all of the time. When one contemplates how little time the hydraulic system is really needed, it would seem much more appropriate to operate the hydraulic pump only when needed.

The **power pack** drives the hydraulic pump with an electric motor which is easily turned on and off with one pressure switch as shown back in figure 13-6. Another design of power pack utilizes two pressure-operated electric switches to turn the electric motor on when pressure is low and off

when pressure is high. Thus, the pump control valve of figure 13-44 and the unloading valve of figure 13-45 are not needed.

When the electrical system is first turned on, if the system pressure is below that of the low pressure switch (i.e., 800 psi), the pump runs until pressure builds to a value that activates the high pressure switch (i.e., 1800 psi) which shuts off the electric motor. Unless there are leaks, pressure will remain high until the pilot calls for work to be done by the system (put the flaps down or raise the landing gear) by opening a valve, causing fluid to flow and the pressure to drop. When the pressure lowers to 800 psi, the pump will run until the gear or flaps have completed movement. Fluid movement stops, pressure rises to 1800 psi and the pump motor is shut off, thus completing the cycle.

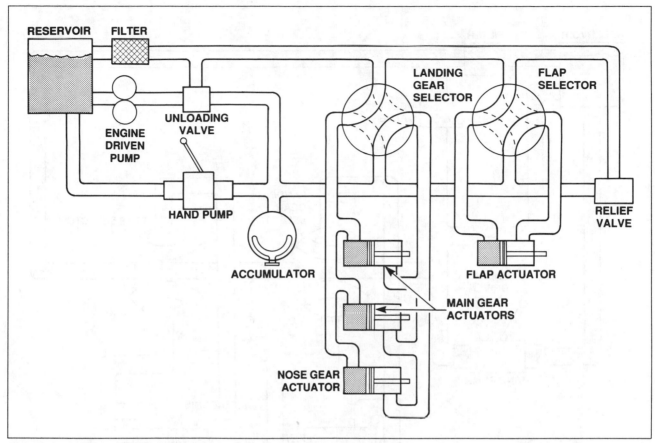

Figure 13-45. The accumulator holds pressure on the system, and the automatic unloading valve keeps the pump unloaded as long as the pressure is maintained.

Often, there will be a light on the panel, indicating when the pump motor is running so the pilot can monitor pump operation. The pump should operate only during hydraulic system use. If it continues to operate when the system is not supposed to be moving gear or flaps, a fault is indicated (hydraulic leak or pressure switch malfunction). The pilot should stop the pump motor by pulling the hydraulic system circuit breaker. The circuit breaker can be re-engaged when the pilot wishes to lower flaps or landing gear, then pulled again when hydraulics are no longer needed.

If a significant hydraulic leak exists, the gear may not extend fully (look for evidence of hydraulic fluid on the after parts of the aircraft, such as the horizontal stabilizer). Too much fluid has been pumped out of the reservoir—the pump can't access more fluid from the reservoir to build pressure. But the hand pump can because its reservoir tap is at a lower point on the reservoir, giving it access to an emergency supply of fluid that can't be pumped overboard by the powered pump.

As can be seen, there are several kinds of hydraulic systems. Some power only the gear—

others include flaps. A few airplanes power gear, flaps *and brakes* from a single hydraulic system. Imagine the problems facing the pilot landing on a short field if the system quietly fails after gear and flaps are down on final approach—no brakes!

Obviously, there is a great need for the pilot to know and understand how the hydraulic system works, how to check fluid levels and find evidence of leaks, as well as know the emergency procedures in case of failures.

Aircraft Landing Gear

There is perhaps no other single part of an airplane structure that takes the beating the landing gear is subjected to. A single hard landing can apply forces many times the weight of the airplane to the tires, wheels and shock absorbing system.

Shock Absorbers

Not all airplanes use shock absorbers—the popular Cessna single-engine series of airplanes does not use a shock absorber for its main gear, figure 13-47. Instead, either a steel leaf or tubular spring gear accepts the energy of the landing

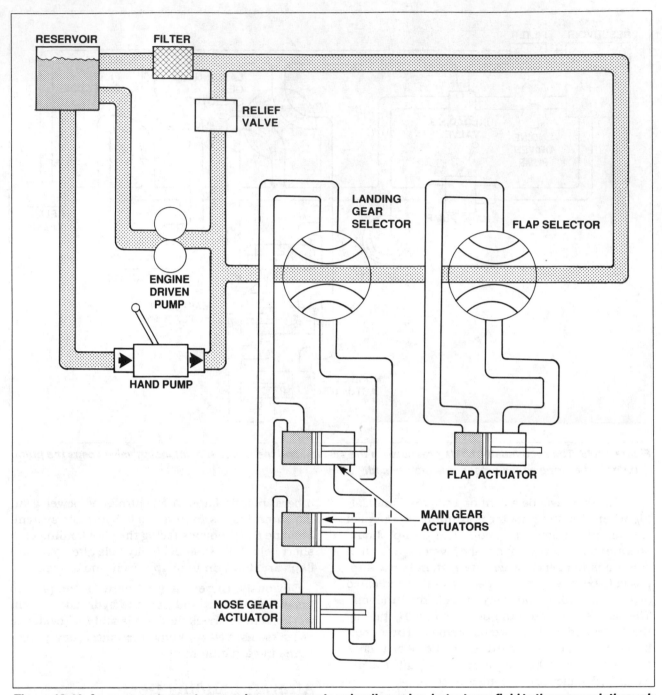

Figure 13-46. An open center system requires no separate unloading valve, but returns fluid to the reservoir through the open center of the selector valves when no unit is being actuated.

impact *and returns it to the airplane.* In a properly conducted landing, energy is returned in such a way that no rebound is caused. Another type of landing gear with a shock absorber is the **bungee** shock cord gear used on the early models of Piper aircraft, figure 13-48. Elastic shock cord composed of many small strands of rubber encased in a loose weave cotton braid stretches with the landing impact *and returns the energy to the airframe.*

To absorb shock, the energy of the landing loads must be converted into some other form of energy. This is done on most modern airplanes by an **oleo (oil) shock strut**, figure 13-49, although only a part of the energy is converted. The rest is passed to the airframe.

The wheel is attached to the piston of the oleo strut which is held in the cylinder by torsion links or scissors. These allow the piston to move in and out but not to turn. The cylinder is attached to the

Figure 13-47. The spring steel landing gear accepts the shock but does not actually absorb it.

Figure 13-48. Bungee shock cord rings accept the shock but do not absorb it.

structure of the airplane. Figure 13-50 shows the inside of the strut composed of two chambers separated by an orifice, with a tapered metering pin moving in the orifice. The strut is completely collapsed and filled with aircraft hydraulic fluid; then compressed air or nitrogen is introduced into the strut to extend it to a specified height with the weight of the airplane on it. As the strut extends, the oil drains into the lower compartment. On touchdown, the piston is forced into the cylinder, and oil passes from the lower chamber into the upper through the metering orifice. The orifice restricts the flow, thus generating heat as the oil is forced into the upper compartment. This heat is from the energy of the landing impact. Notice in figure 13-50 that the metering pin is tapered. As the piston is pushed farther up into the cylinder, the size of the orifice is decreased and the passage of the oil is more restricted. This provides a progressively stiffer shock strut and smoother shock absorbing action. Notice, also, that there is an enlarged area or knob at the end of the metering pin. This prevents rebound. If the airplane should bounce on landing, the strut will attempt to extend fully; but when the knob is reached, it will snub or slow down the extension.

Wheel Alignment

It is important for the wheels of an airplane to be in proper alignment with the airframe. Airplanes using oleo shock struts have their wheels aligned by inserting shims between the arms of the torque links, figure 13-51.

Airplanes using spring steel landing gear have their wheel alignment changed by adding or removing shims from between the axle and the landing gear strut, figure 13-52.

Nose Wheel Steering And Shimmy Dampers

Almost all tricycle gear airplanes have provisions for steering the airplane by controlling the nose wheel, but some have a castering nose wheel and steering is done by independent braking of the main wheels. Some light airplanes with steerable nose wheels have a direct linkage between the rudder pedals and the nose gear; others have their nose gear steerable through a specific range, after which it breaks out of steering and is free to caster up to its limits of travel.

Figure 13-49. The oleo strut piston is prevented from turning in its cylinder by torsion links which allow in and out movement, but no rotation.

263

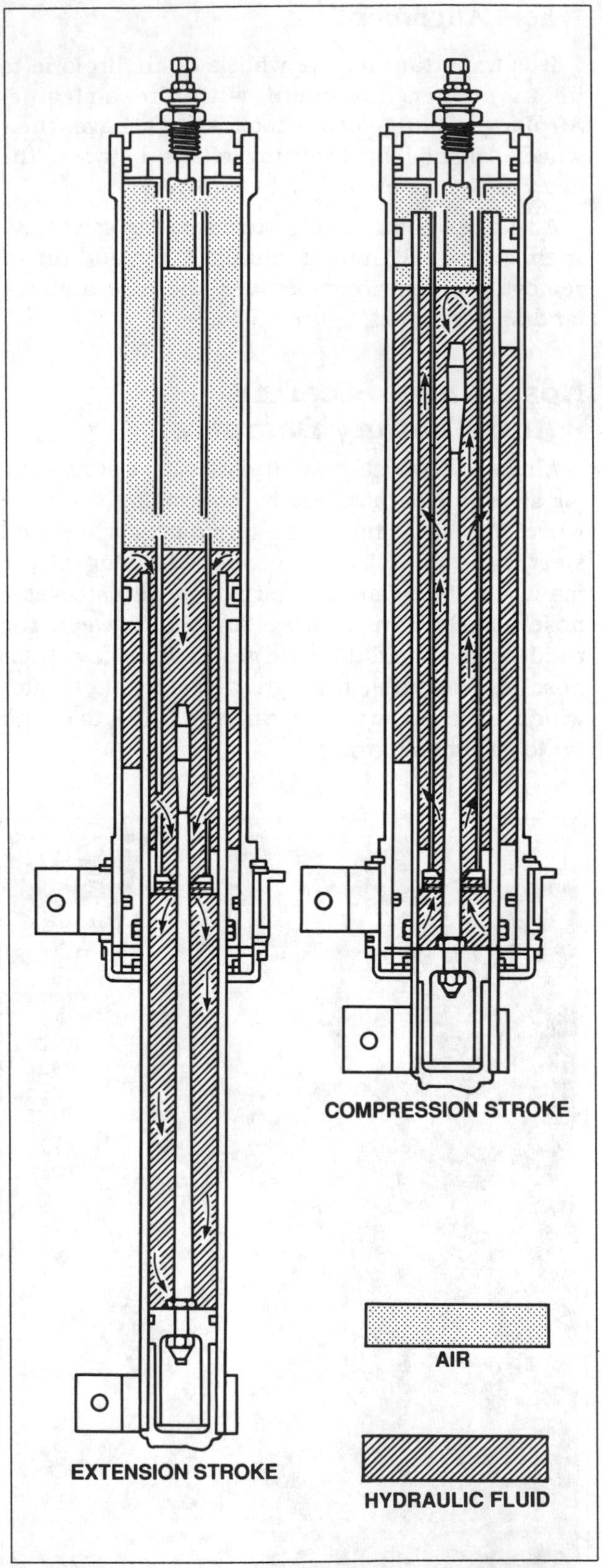

Because of the geometry of the nose wheel, it is possible for it to shimmy at certain speeds. To prevent this, a shimmy damper, figure 13-53, is installed between the piston and cylinder of the nose gear oleo strut. These dampers range from

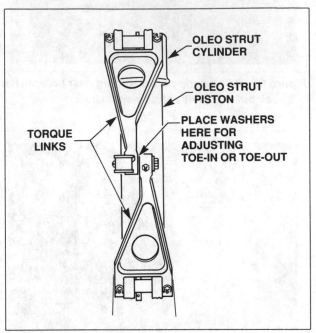

OLEO STRUT
CYLINDER

OLEO STRUT
PISTON

PLACE WASHERS
HERE FOR
ADJUSTING
TOE-IN OR TOE-OUT

TORQUE
LINKS

Figure 13-51. Wheel alignment on landing gear equipped with oleo shock struts is done with shims between the arms of the torque links.

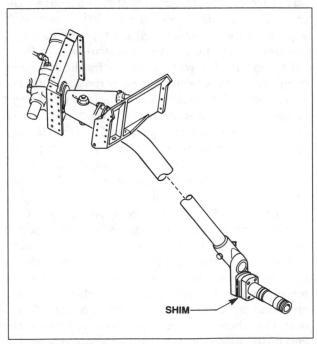

SHIM

Figure 13-52. Wheel alignment on a spring steel landing gear is done with shims between the axle and the gear leg.

COMPRESSION STROKE

EXTENSION STROKE

AIR

HYDRAULIC FLUID

Figure 13-50. An oleo shock strut converts some of the mechanical energy of the impact into heat in the fluid as the fluid is forced through the small metering orifice.

Figure 13-53. A typical nose wheel shimmy damper which can be serviced.

the simplest sealed unit of figure 13-54, through the complex rotary dampers of figure 13-55. As the nose wheel fork rotates, hydraulic fluid is forced from one compartment into the other through a small orifice. This restricts rapid movement of the gear in a shimmy but has no effect on normal steering.

Large aircraft normally have the nose gear steered by hydraulic action. This is done by directing hydraulic fluid into one of the steering cylinders, figure 13-56. Fluid from the main hydraulic system flows into the steering control valve, figure 13-57. A control wheel operated by the pilot directs pressure to one side of the nose wheel steering pistons; fluid from the opposite side is vented back into the reservoir through a pressure relief valve that holds a constant pressure on the system to snub shimmying. An accumulator in the line to the relief valve holds pressure on the system when the steering control valve is in the neutral position.

SHIMMY DAMPER

Figure 13-54. A typical sealed nose wheel shimmy damper.

Landing Gear Retraction Systems

As the speed of aircraft becomes high enough that the parasite drag of the landing gear hanging out the airstream is greater than the induced drag caused by the added weight of a retracting system, the gear should be designed to be retracted into the structure. Small aircraft use simple mechanical retraction systems, some use a hand crank to drive the retracting mechanism through a roller chain, but the simplest of all uses a direct hand lever, push-pull rods and bellcranks to raise and lower the wheels. Many aircraft use electric motors to drive the gear retracting mechanism and some European-built aircraft use pneumatics.

Electrical Landing Gear Retract Systems

Retractable landing gear with their warning and indication systems give a real workout in electrical circuit reading and troubleshooting. In figure 13-58 the landing gear circuit for a typical twin-engine general aviation airplane is shown. The landing gear is hydraulically operated by a reversible DC electric motor driving a reversible hydraulic pump, or is electrically operated by a reversible DC motor driving a mechanical linkage. The motor turns in one direction to lower the gear and in the opposite direction to raise it.

Let's follow the system as the gear is operated. Consider the airplane to be in the air with the landing gear down and locked, but with the landing gear

Figure 13-55. A vane-type shimmy damper used for nose wheel steering as well as damping the tendency to shimmy.

selector switch having just been placed in the GEAR UP position. This is seen in figure 13-58(A). The down-limit switches in each of the three landing gears are in the DOWN position and the up-limit switches are all in their NOT UP position. Current flows through the NOT UP side of the up-limit switches through the FLIGHT side of the squat switch (a switch on one of the struts that is in one position when the weight of the airplane is on the landing gear, and in the other position when there is no weight on the wheels) and through the hydraulic pressure switch. This switch is closed when the hydraulic pressure is low, but opens when the hydraulic pressure rises to a preset value.

From the pressure switch the current goes through the coil of the LANDING GEAR UP relay and to ground through the landing gear selector switch. This flow of current creates a magnetic field in the LANDING GEAR UP relay and closes it, so that current can flow from the main bus through the relay contacts and the windings that turn the motor in the direction to raise the landing gear. As soon as the landing gear is fully up, the up-lock switches in each gear move to the UP position and current is shut off to the landing gear motor relay and the motor stops. The down-limit switches have moved to the NOT DOWN position, and the three down-and-locked lights go out. If the throttle is closed when the landing gear is not down and locked, the warning horn will sound.

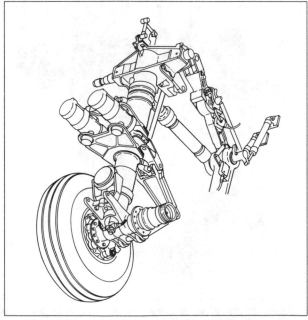

Figure 13-56. This large transport nosewheel steering is done by two steering cylinders.

The landing gear may be lowered in flight by moving the landing gear selector switch to the GEAR DOWN position. Follow this in figure 13-58(B). Current flows through the NOT DOWN side of the down limit switches, through the coil of the LANDING GEAR DOWN relay, and to ground through the landing gear selector switch. The motor turns in the direction needed to produce hydraulic pressure to lower the landing gear. when the landing gear is down and locked, figure 13-58(C), current flows through the DOWN sides of the down-limit switches, and the green GEAR DOWN AND LOCKED lights come on. In the daytime, current from these lights goes directly to ground through the closed contacts of the light dimming relay, but at night when the navigation lights are on, current from this light circuit energizes the relay, and current from the indicator lights must go to ground through the resistor. This makes the lights burn dimly enough that they will not be distracting at night.

If any of the limit switches are in a NOT UP or a NOT DOWN position, a red UNSAFE light will light up. But when the landing gear selector switch is in the LANDING GEAR DOWN position and all three gears are down and locked, the light will be out. If the selector switch is in the LANDING GEAR UP position and all three gears are in the up position, the light will also be out.

If the airplane is on the ground with the squat switch in the GROUND position, and the landing gear selector switch is moved to the LANDING GEAR UP position, the warning horn will sound but the landing gear pump motor will not run. See figure 13-58(C).

Hydraulic Systems

A simple hydraulic landing gear system uses a hydraulic power pack containing the reservoir, a reversible electric motor-driven pump, selector valve, and, in some instances, an emergency hand pump and any special valves required. Figure 13-6A is a schematic of a system such as this.

To raise the landing gear, the gear selector handle is placed in the gear-up position. This starts the hydraulic pump, forcing fluid into the gear-up side of the actuating cylinders, raising the gear. The initial movement of the piston releases the landing gear down-lock, so the gear can retract. When all three gears are completely retracted, up-limit switches stop the pump. There are no mechanical up-locks, so the gear is held up by hydraulic pressure. A pressure switch starts the pump and restores pressure if it drops to a predetermined

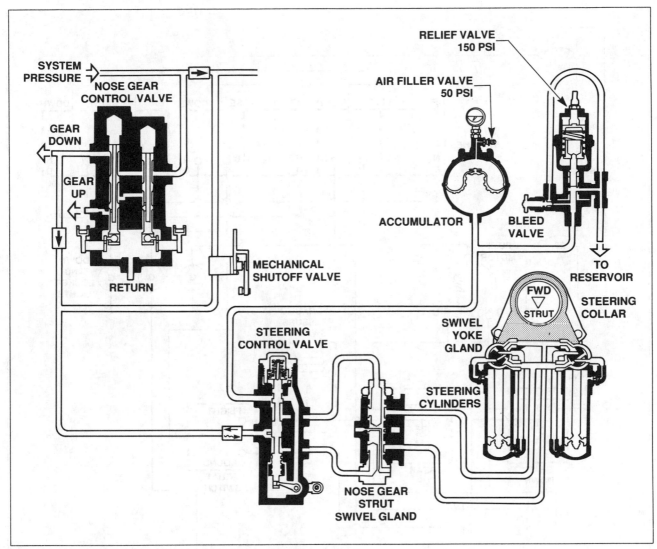

Figure 13-57. Nosewheel steering system in a jet transport.

level and any one of the wheels drops away from its up-limit switch.

To lower the gear, the selector is placed in the gear-down position, releasing the pressure on the up side of the cylinders through the power pack. The shuttle valve moves over and the gear falls down and locks. When all three gears are down and locked, the limit switches shut the pump motor off.

All retractable landing gear systems must have some means to lower the gear in the event the main extension system should fail. This simple system depends on the gear free-falling and locking into position. To actuate the emergency extension, a control on the instrument panel opens a valve between the gear-up and gear-down lines that

dumps fluid from one side of the actuator to the other and allows the gear to fall and lock in place.

More complex landing gear systems use compressed air or nitrogen to provide the pressure for emergency extension of the gear. In systems using this type of emergency extension, a shuttle valve is installed in the actuator where the main hydraulic pressure and the emergency air pressure meet, figure 13-59. For normal operation, fluid enters the actuator through one side of the shuttle. In the event of failure of the hydraulic system, the gear handle may be placed in the gear-down position and the emergency air supply released into the system. The shuttle valve moves over, directing compressed air into the actuator and sealing off the line to the normal hydraulic system.

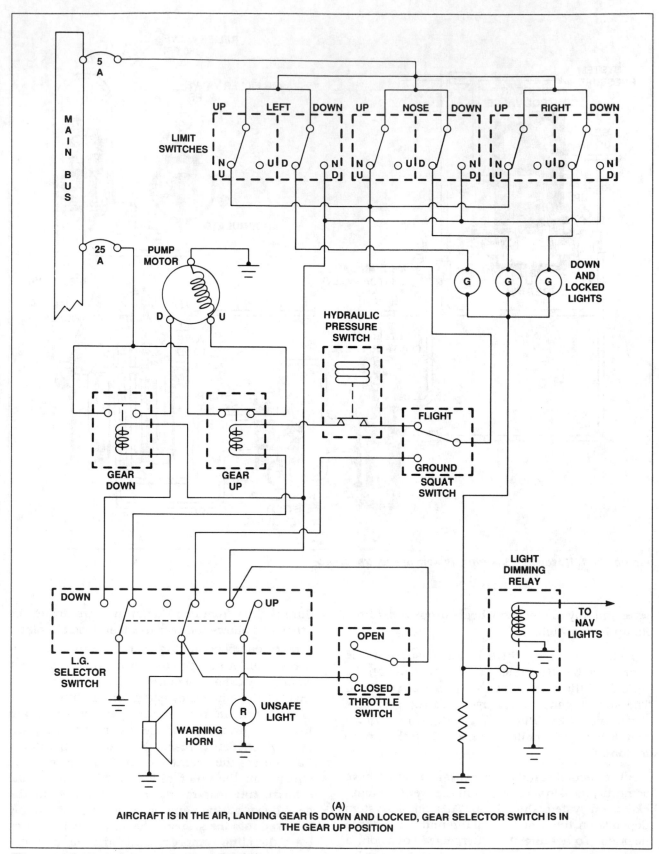

(A)
AIRCRAFT IS IN THE AIR, LANDING GEAR IS DOWN AND LOCKED, GEAR SELECTOR SWITCH IS IN THE GEAR UP POSITION

Figure 13-58(A). Landing gear circuit (1 of 3).

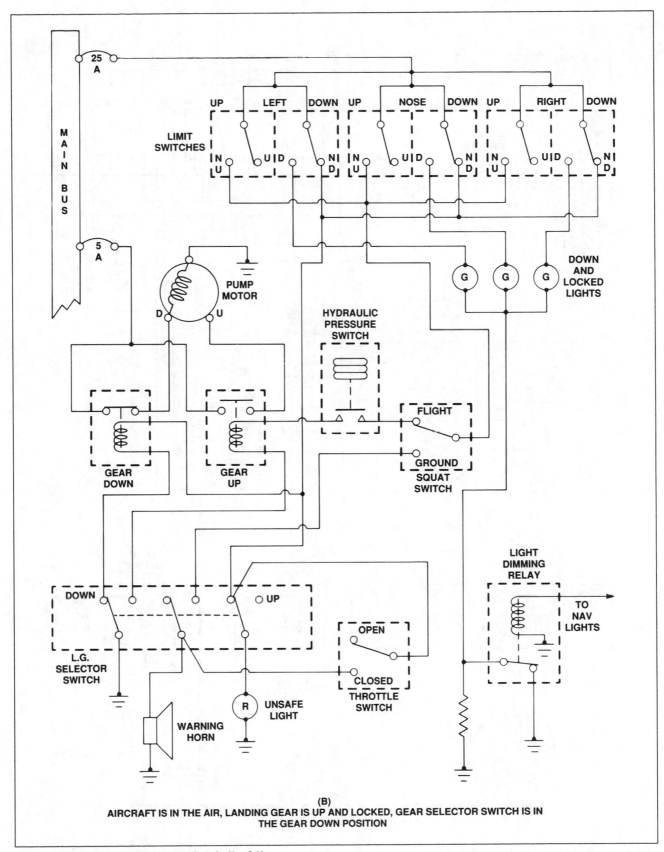

(B)
AIRCRAFT IS IN THE AIR, LANDING GEAR IS UP AND LOCKED, GEAR SELECTOR SWITCH IS IN
THE GEAR DOWN POSITION

Figure 13-58(B). Landing gear circuit (2 of 3).

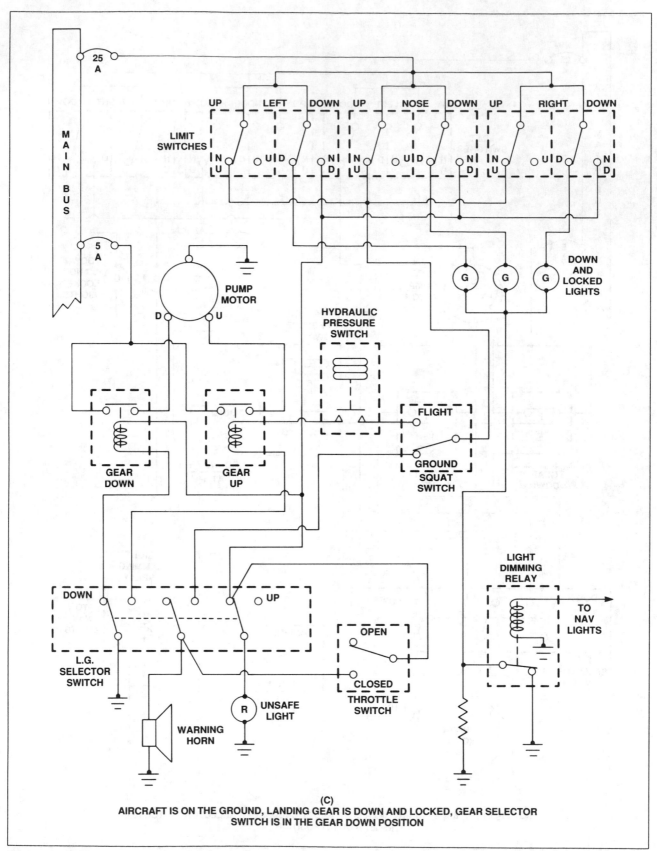

(C)
AIRCRAFT IS ON THE GROUND, LANDING GEAR IS DOWN AND LOCKED, GEAR SELECTOR SWITCH IS IN THE GEAR DOWN POSITION

Figure 13-58(C). Landing gear circuit (3 of 3).

If the pneumatic emergency extension system is used, the gas must be bled from the hydraulic system before normal operation may resume.

Landing Gear Retraction— An Operational Philosophy

Landing gear emergencies (not able to get the gear down or unable to get a "gear down and safe" indicator light) happen once in awhile but it is a "nice" emergency in that the pilot has plenty of time to review the manufacturer's emergency procedures, assess the situation, ask for advice, fly by the tower for a visual inspection, and assemble whatever emergency equipment is needed before committing to a landing. A good understanding of the landing gear system in the aircraft being flown will assist the pilot in the assessment phase of the emergency.

It is in the realm of landing gear retraction, rather than extension, that pilots' sometime create problems for themselves needlessly. *When* should the landing gear be retracted? Pilots I have flown checkrides with seem to have many answers to that question. I have heard answers such as: "At 400 feet AGL", "When there is no more runway available to land on", "When the departure end of the runway passes under the nose." When asked "Why?", most responded: "Because that's what my flight instructor taught me to do."

A logical look at this question indicates that there is no one good answer to this question, so I won't attempt to provide one, but rather, present the reader with a logical decision loop which, with some thought and discussion, may be improved upon. See figure 13-60.

In thinking through the "what if's" associated with a complete engine failure on takeoff, I have come to believe the following to be self-evident:

1. I would rather land with *gear down* on the runway even if it is on the last foot of runway because (a) wheels down allows braking—the energy available to injure is a function of the square of the velocity—decreasing speed by half decreases energy by 75%, (b) gear down provides directional control even if only momentarily, and (c) shearing the landing gear absorbs energy and slows the aircraft.

2. With most aircraft, landing gear *in transit* produces more drag than gear down and may decrease lift due to the change in wing geometry, so generally (on single-engine aircraft) landing gear is left down until obstacles are cleared. There may be exceptions

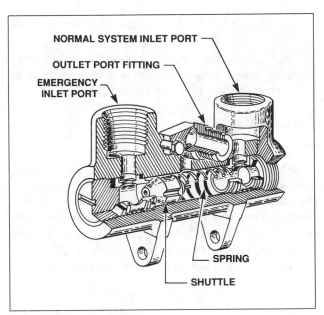

Figure 13-59. Shuttle valve.

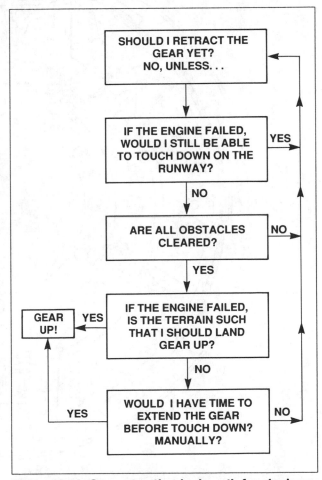

Figure 13-60. Gear retraction logic path for single-engine aircraft.

to this—check the operating procedures manual for the aircraft to be flown.

3. Gear up landings generally are made on water, deep snow, deep mud and any other very soft, deep material, to avoid pitch upset.

Aircraft Brakes

Wheel Units

Aircraft brake systems slow the airplane down by exchanging kinetic energy from the motion of the airplane into heat energy generated by the friction between the linings and the brake drum or disc. There are two basic types of brakes in use: the **energizing brake**, in which the weight of the airplane is used to apply the brakes, and the **nonenergizing brake** where the weight of the airplane does not enter into the stopping action.

Energizing brake systems are typified by the drum-type brake, used on automobiles, the duo-servo type; that is, the weight of the vehicle aids in the application of the brake in both the forward and reward direction. Some smaller airplanes using energizing brakes do not have duo-servo action, as only the forward motion is used to help apply the brake. Servo-action, or energizing, brakes have their shoes or linings attached to a floating back plate, figure 13-61. When the brakes are applied, the pistons in the brake cylinder move out, pushing the linings against the rotating drum. Friction attempts to rotate the linings, but they are restrained at their back end so the rotation wedges

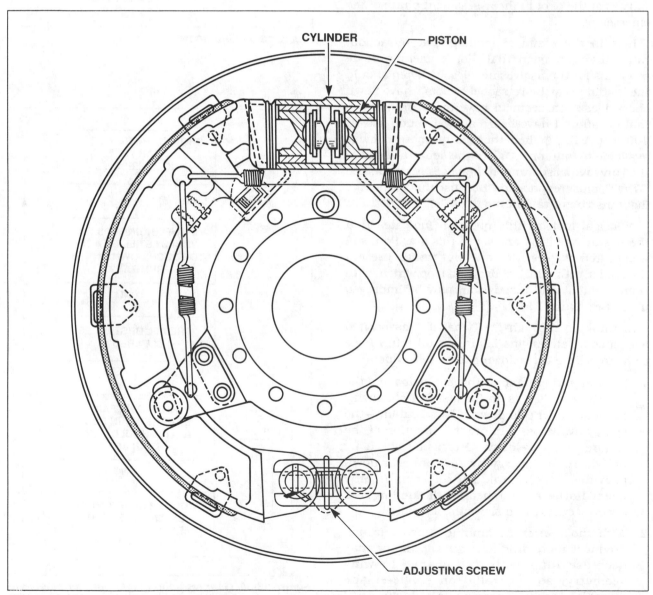

Figure 13-61 Energizing-type drum brake.

the lining against the drum. When the hydraulic pressure is released, the retracting spring pulls the linings away from the drum and releases the brakes.

Nonenergizing brakes include **expander tube brakes** and **disc brakes**. In this brake, hydraulic fluid from the master cylinder is directed into the synthetic rubber tube around the axle assembly, figure 13-62. When this tube is expanded by hydraulic fluid, it pushes the lining blocks against the drum and slows the airplane. The heat generated in the lining is prevented from damaging the expander tube by thin stainless steel heat shields between each of the lining blocks.

The most popular brake for modern light aircraft is the **single disc**. This brake has a caliper which applies a squeezing action between the fixed linings and the rotating disc. There are two types of single disc brakes, one having the disc keyed into the wheel and free to move in and out as the brake is applied; the other has the disc rigidly attached to the wheel, and the caliper moves in and out on a couple of anchor bolts.

Figure 13-63 is typical for the Goodyear single-disc brake. The disc is keyed to rotate with the wheel by hardened steel drive keys but is free to move in and out on the keys. It is prevented from rattling by disc clips. The housing is bolted to the axle and holds the two linings, one fixed in a recess in the outer portion of the housing, and the other riding in the inner side. The disc rotates between the two linings and is clamped when hydraulic fluid under pressure forces the inner lining against the disc. A piston sealed with an O-ring actually applies the push.

The Cleveland brake uses a disc solidly bolted to the inner wheel half, figure 13-64. The brake assembly, figure 13-65, consists of a torque plate bolted to the axle with tow anchor bolts holding the brake cylinder, linings, and pressure plates. When the brakes are applied, hydraulic fluid forces the pistons out and squeezes the disc between the linings.

Multiple disc brakes are utilized when greater braking force is needed. It is a matter of simple physics that determines the size brake required for a given airplane. The gross weight of the airplane, the speed at the time of brake application, and the density altitude all determine the amount of heat generated and dissipated in the brake. As airplane size and weight have gone up, the need for larger braking surface has increased also. This brought out, first, the **thin disc multiple disc brake**, and,

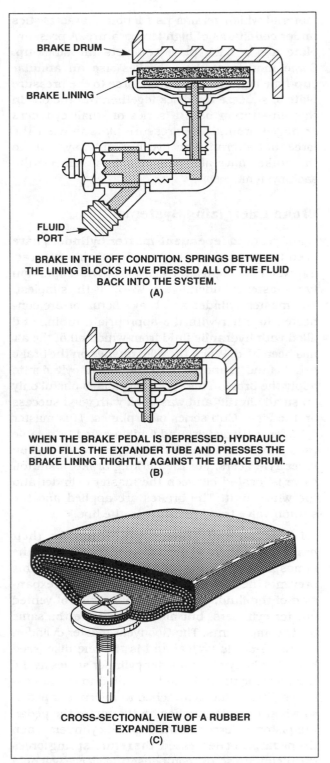

BRAKE IN THE OFF CONDITION. SPRINGS BETWEEN THE LINING BLOCKS HAVE PRESSED ALL OF THE FLUID BACK INTO THE SYSTEM.
(A)

WHEN THE BRAKE PEDAL IS DEPRESSED, HYDRAULIC FLUID FILLS THE EXPANDER TUBE AND PRESSES THE BRAKE LINING THIGHTLY AGAINST THE BRAKE DRUM.
(B)

CROSS-SECTIONAL VIEW OF A RUBBER EXPANDER TUBE
(C)

Figure 13-62. An expander tube brake.

more recently, with the advent of the jet aircraft, the **segmented rotor multiple disc brake**, figure 13-66. In the brake shown here, there are five rotating discs, keyed into the wheel, and between each disc there is a stator plate. Riveted to each side of these stators are wear pads made of a

273

material which retains its friction characteristics under conditions of high temperature. A pressure plate and a back plate complete the stack-up. Some models of these brakes use an annular cup-type actuator to apply a force to the pressure plate to squeeze the discs together, but the one in this illustration uses a series of small cylinders arranged around the pressure plate to exert the force. Jet aircraft with two hydraulic systems to the brakes have alternate cylinders connected to each system.

Brake Energizing Systems

For years **independent master cylinders** have been the most commonly used pressure generating system for light aircraft brakes. The diaphragm type master cylinder, figure 13-67, is the simplest. The master cylinder and brake actuator are connected together with the appropriate tubing and filled with hydraulic fluid from which all of the air has been bled. When the pilot pushes on the brake pedal, fluid is moved into the wheel cylinder to apply the brake. This type of system is useful only on small aircraft and was used with good success on the Piper Cub series of airplanes. This master cylinder is turned around and operated by a cable from a pull handle under the instrument panel on Piper Tri-Pacers. For the parking brake, a shutoff valve is located between the master cylinder and the wheel unit. The brakes are applied and the shutoff valve traps pressure in the line.

Larger aircraft require more fluid for their brakes, and there is need to vent this fluid to the atmosphere when the brakes are not applied. This prevents the brakes dragging from thermal expansion of the fluid. There are many types of vented master cylinders, but all of them have the same basic components. The Goodyear master cylinder, figure 13-68, is typical and is the one discussed here. The body of the master cylinder serves as the reservoir for the fluid and is vented to the atmosphere. The piston is attached to the rudder pedal, so when the pilot pushes on the top of the pedal, the piston is forced down into the cylinder. When the pedal is not depressed, the return spring forces the piston up so the compensating sleeve will hold the compensator valve open. Fluid from the line to the wheel unit is vented to the atmosphere. When the pedal is depressed, the piston is pushed away from the compensating sleeve, and the special O-ring and washer, the Lock-O-Seal, seals fluid in the line to the brake. The amount of pressure applied to the brake is proportional to the amount the pilot pushes. When the pedal is released, the

compensator opens and vents the brake line into the reservoir.

The parking brake for this type of master cylinder is a simple ratchet mechanism that holds the piston down in the cylinder. To apply the parking brake, the pedal is depressed and the handle pulled. This locks the piston. To release the

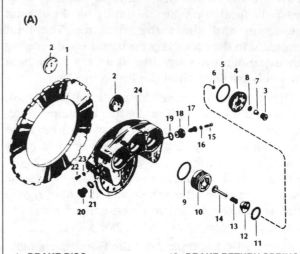

(A)

1. BRAKE DISC
2. LINING PUCK
3. ADJUSTING PIN NUT
4. CYLINDER HEAD
5. O-RING GASKET
6. O-RING PACKING
7. ADJUSTING PIN GRIP
8. WASHER
9. O-RING PACKING
10. PISTON
11. INTERNAL RETAINER RING
12. SPRING GUIDE
13. BRAKE RETURN SPRING
14. ADJUSTING PIN
15. BLEEDER SCREW
16. WASHER
17. BLEEDER VALVE
18. BLEEDER ADAPTER
19. GASKET
20. FLUID INLET BUSHING
21. GASKET
22. SCREW
23. WASHER
24. BRAKE HOUSING

(B)

Figure 13-63(A)—Goodyear single-disc brake. (B)— Single-disk brake used on a small general aviation aircraft.

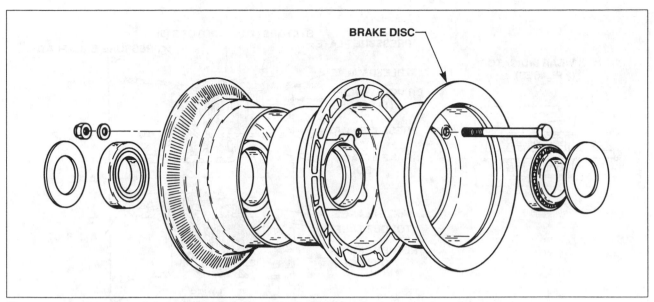

Figure 13-64. The disc of the Cleveland brake bolts to the wheel.

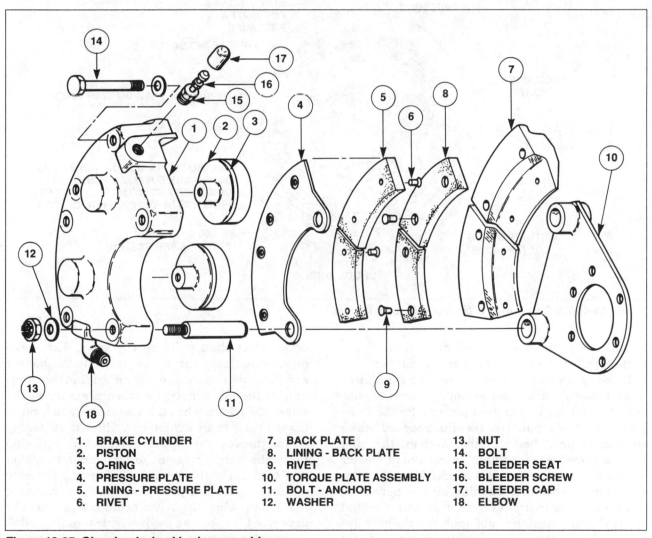

1. BRAKE CYLINDER	7. BACK PLATE	13. NUT
2. PISTON	8. LINING - BACK PLATE	14. BOLT
3. O-RING	9. RIVET	15. BLEEDER SEAT
4. PRESSURE PLATE	10. TORQUE PLATE ASSEMBLY	16. BLEEDER SCREW
5. LINING - PRESSURE PLATE	11. BOLT - ANCHOR	17. BLEEDER CAP
6. RIVET	12. WASHER	18. ELBOW

Figure 13-65. Cleveland wheel brake assembly.

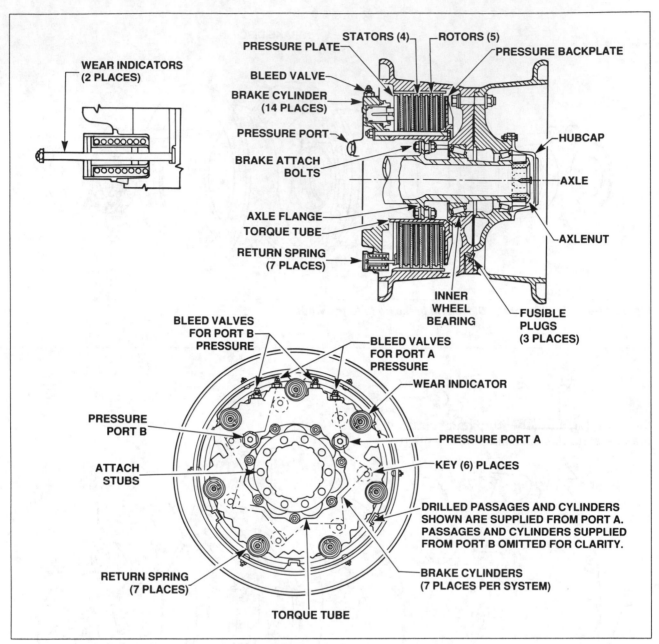

Figure 13-66(A). Segmented rotor brake for a jet airliner.

brake, the pedal is depressed more than at the initial application so the ratchet can release.

There is an airplane of a size which requires more braking force than an independent master cylinder can apply, yet does not require the complex system of a power brake; the **boosted brake system** is used here. In this system, the pilot applies pressure as with any independent master cylinder. If more pressure is needed than the pilot can apply, continued pushing on the pedal will introduce some hydraulic system pressure behind the piston and help the pilot apply force. This valve is installed on the brake pedal in such a way that

brake application pulls on the valve. The initial movement closes the space between the poppet and the piston so fluid can be forced into the wheel unit. If there is a need for more pressure at the wheel, the pilot pushes harder on the pedal, causing the valve to straighten out. This straightening action moves the spool valve over, allowing hydraulic system pressure to get behind the piston and help apply the brakes. The spring provides regulator action and prevents pressure continuing to build up when the brake pedal is held partially depressed. As soon as the pilot releases the pedal, the spool valve moves back and relieves the system

Figure 13-66(B). Multiple-disk brake used in the wheel of a large jet-transport aircraft.

pressure back into the return manifold; the brake valve then acts as an independent master cylinder. See figure 13-69.

Almost all large aircraft use **power brakes** operated by pressure from the main hydraulic system. This is not done by simply valving part of the pressure into the actuating units, since a brake system has special requirements that must be met. The brake application must be proportional to the force the pilot exerts on the pedals. The pilot must be able to hold the brakes partially applied without there being a buildup of pressure in the brake lines. Since these brakes are used on airplanes so large the pilot has no way of knowing when one of the wheels is locking up, there must be provisions to prevent any wheel skidding. The pressure actually supplied to the wheel unit must

be lower than the pressure of the main hydraulic system, so a system of lowering or deboosting the pressure may be incorporated in the system. Since the wheels are susceptible to damage, provisions should also be made to lock off the fluid from a wheel in the event a hydraulic line is broken. Finally, there must be an emergency brake system that can actuate the wheel units in the event of a failure of the hydraulic system.

Figure 13-70 is a simplified schematic of a typical power brake system in a large jet aircraft. The brakes get their fluid from the main hydraulic system and a check valve and accumulator hold pressure for the brakes in the event of a hydraulic system failure. The pilot and copilot operate the power brake control valves through the appropriate linkages. These valves are actually regulators which provide an amount of pressure to the brake system proportional to the force the pilot applies to the pedals. Once the pressure is

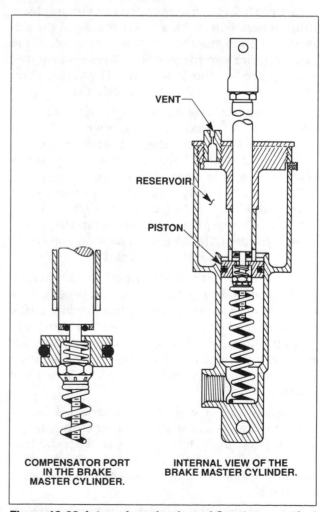

COMPENSATOR PORT IN THE BRAKE MASTER CYLINDER.

INTERNAL VIEW OF THE BRAKE MASTER CYLINDER.

Figure 13-68. Internal mechanism of Goodyear vertical master cylinder.

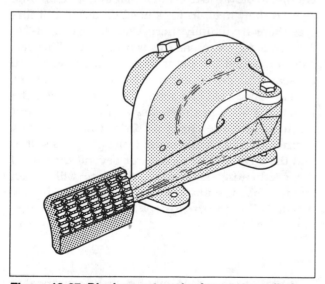

Figure 13-67. Diaphragm-type brake master cylinder.

reached, the valve holds it as long as that amount of force is held on the pedals. In large aircraft, the pilot does not have a feel for every wheel, so an anti-skid system is installed to sense the rate of deceleration of each wheel and compare it with a maximum allowable deceleration rate. If the wheel attempts to slow down too fast, as it does at the onset of a skid, the anti-skid valve will release the pressure from that wheel back into the system return manifold.

The pressure applied by the brake control valve is too high for proper brake application, so a debooster is installed in the line between the anti-skid valve and the brake. This lowers the pressure and increases the volume of fluid supplied to the wheel units.

Aircraft Wheels, Tires And Tubes

Wheels

Figure 13-71 is an illustration of a **single-piece, drop center wheel**. This has, because of the difficulty in changing the tire, been replaced by the more popular **two-piece wheel**. Tires are removed and replaced on this type of wheel by prying them over the rim, as with an automobile tire.

Some wheels have such a high flange and the tires arc so stiff that single-piece wheels are impractical; so the removable rim wheel came into being. In this wheel, figure 13-72, the outer rim is removable and is held in place when the tire is inflated, with a snap ring. Great care must be exercised when inflating a tire on a wheel with a removable flange, because if the snap ring is not properly seated, the flange can blow off and create a hazard to anyone standing nearby.

By far the most popular wheel in all sizes on modern airplanes is the **two-piece wheel**. It is either cast or forged of aluminum or magnesium alloy and the halves are bolted together with high-strength bolts. Hardened steel bearing races are shrunk into the hub and either the brake disc, drum, or disc retainers are shrunk or bolted into place.

When two-piece wheels are used with tubeless tires, the halves are sealed with an O-ring, figure 13-73. This ring is lubricated with a special low-temperature grease and must fit into the groove in the wheel half with no twisting or any indication of damage.

Large jet aircraft have such high landing speeds and so much weight that tremendous amounts of energy are involved in any heavy braking. This

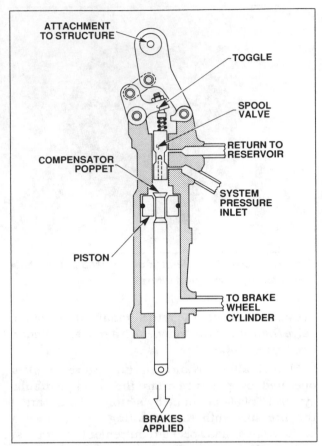

Figure 13-69. Boosted brake master cylinder. A—Pull rod; B—Port to brake cylinder; C—Piston; D—Compensator poppet; E—System pressure inlet port; F—Port to system reservoir; G—Spool valve; H—Adjustment level; I—Attachment pivot.

energy is dissipated in the form of heat and may cause the air pressure in the tire to rise high enough to blow the tire and inflict injury on anyone around the wheel. In tires used on aircraft of this type there are usually thermal fuses, figure 13-74. A thermal fuse such as this is a special hollow bolt with a low-melting-point plug in its center as an air seal. This particular fuse indicates whether or not excessive heat has been encountered. After the fuse is installed in the hollow bolt the stem is cut off flush with the cap nut. If the wheel becomes overheated from brake action, the plug will soften and the stem will be forced out beyond the nut. If the overheating is serious, the fuse will soften enough that the air will blow it out of the hollow bolt and allow the tire to deflate without the danger of its blowing up.

Tires

Aircraft tires, while similar in appearance to those used on ground vehicles, are quite different in their construction and operational requirements. They are subject to extreme loads when the airplane makes contact with the ground, and the stresses imposed on the carcass are high when the wheel must suddenly accelerate from zero velocity to the touchdown speed of the airplane. An airplane tire does not have the sustained high temperatures an automobile tire faces on hot sum-

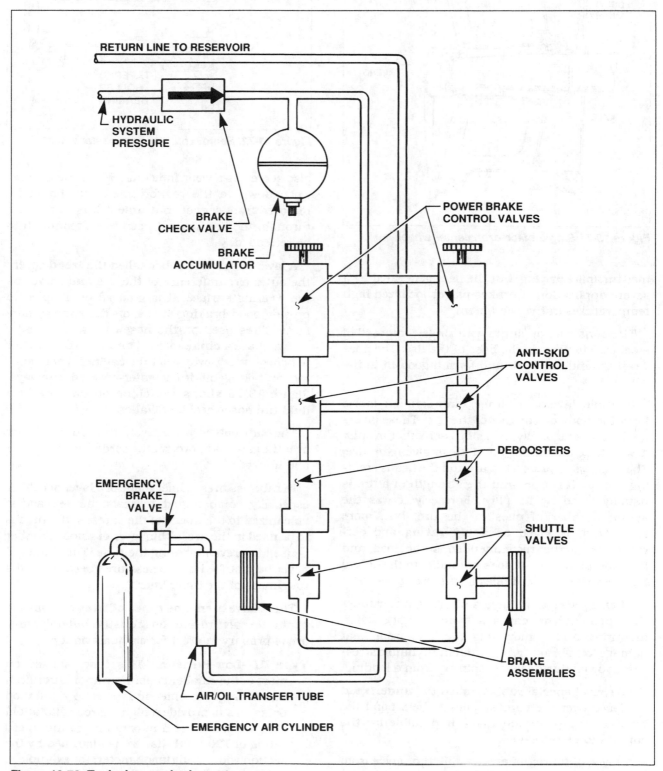

Figure 13-70. Typical power brake system.

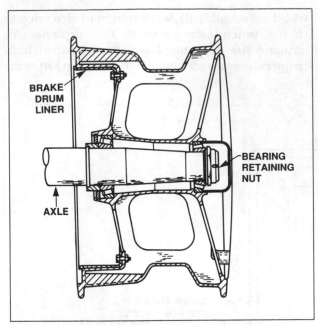

Figure 13-71 Single-piece drop center wheel.

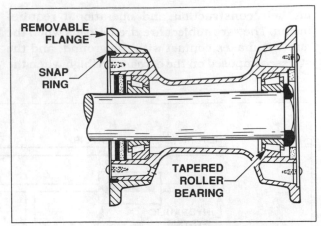

Figure 13-72. Removable rim drop center wheel.

mer turnpike driving, but the heat generated by a severe application of brakes places localized high temperatures in the bead area.

Tire construction starts with rings composed of steel wire to provide the base of the tire. The plies wrap around the **bead** and form a good fit to the wheel.

Multiple layers of rubber-impregnated fabric form the body or **carcass** of the tire. These layers are laid up diagonally, crossing each other in such a way that maximum strength is given the tire. These layers or **plies** wrap around the beads creating the ply turnups, and the strength of a tire is usually rated by its plies. Formerly it was the actual number of plies in the tire, but, more recently, it is given as a **ply rating** and corresponds to the recommended static load and inflation pressure—it does not indicate the actual number of fabric cord plies in the tire carcass.

Chafing strips are layers of fabric and rubber that protect the carcass from damage when mounting and demounting the tire, from heat damage from braking, and from the dynamic stresses involved during wheel spin-up from a landing.

A layer of special rubber called the **undertread** is placed over the carcass plies to help bond the tread to the plies and make it possible for the carcass to be retreaded.

A layer of fabric, often of a different color from that of the main plies, is laid over the undertread to stabilize the tread for high-speed operation. It

also serves as a wear indicator. When the tread is worn down to the reinforcement, it should be removed from service; but unless it is worn clear through, the tire has not been worn too much to be retreaded.

A layer of special rubber called the **tread** covers the outer circumference of the tire and serves as the wearing surface. It is grooved or dimpled to provide good braking action on the runway surfaces. Tires used on the nosewheels of some jet aircraft have a **chine** molded into the tread rubber to deflect water away from the engines when landing or taking off from water-covered runways. Figure 13-76 shows the chine on one tire of a dual-tire nosewheel installation.

The **sidewall** is a relatively thin rubber cover over the carcass to protect the cords form exposure and injury.

In tubeless tires, a liner made of a layer of rubber especially compounded to resist air leakage is vulcanized to the inside of the carcass. If a tube is to be used in the tire, a thin layer of smooth rubber is used to prevent chafing the tube. Tubes should never be installed in tubeless tires because of the roughness of the inner liner.

There have been nine **types of tires** established by the aircraft tire and rim industry, but only three are of primary interest for civilian aviation:

Type III—Low-Pressure. This type is used on most of the smaller general aviation aircraft. It has a large volume and uses low inflation pressures to provide a high degree of flotation. Type III tires have a maximum ground speed rating of 120 MPH. Its size is identified by the section width and rim diameter; for example, a 7.00-6 tire has a width of seven inches, and fits a six-inch diameter rim.

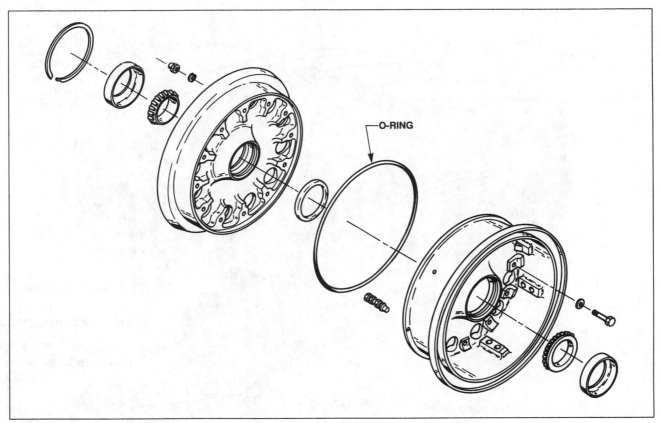

Figure 13-73. Two-piece wheel construction.

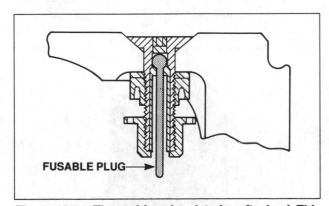

FUSABLE PLUG

Figure 13-74. Thermal fuse in a jet aircraft wheel. This melts and blows out, allowing an overheated tire to deflate rather than blow out.

Type VII—Extra High Pressure. These tires are used on military and civilian jet aircraft and have ground speed ratings up to 250 MPH; inflation pressures for some of them are as high as 315 psi. The size of these tires is identified by their outside diameter and section width; for example, a 34 × 11 tire has a 34 inch outside diameter and a section width of eleven inches.

New Design. All of the new designs of aircraft tires are being dimensioned according to this category, rather than by the methods of the older types: A tire identified as 15 × 6.06-6 has an outside diameter of 15 inches, a section width of six inches, and a bead seat diameter of six inches.

Tire Inspection And Repair

Tread wear should be a part of every preflight inspection. Aircraft tires take a pounding on every landing and the tread is rapidly worn off as the rough surface of the runway is used to accelerate the wheel from zero to touchdown velocity. Most aircraft tires have their tread used up long before the carcass is worn out, so it is possible to retread them and thus extend their useful life. A careful inspection and attention to any small defect and wear indication will help get the maximum use of a tire.

Figure 13-77(A) is the cross section of a tire showing normal wear on the tread. This tire has been properly inflated, as is indicated by the even wear of all sections of the tread.

Figure 13-77(B) shows the way a tread will wear if the tire has been operated in an over-inflated condition. The center of the tire wears first and the tread ribs at the edges show less wear.

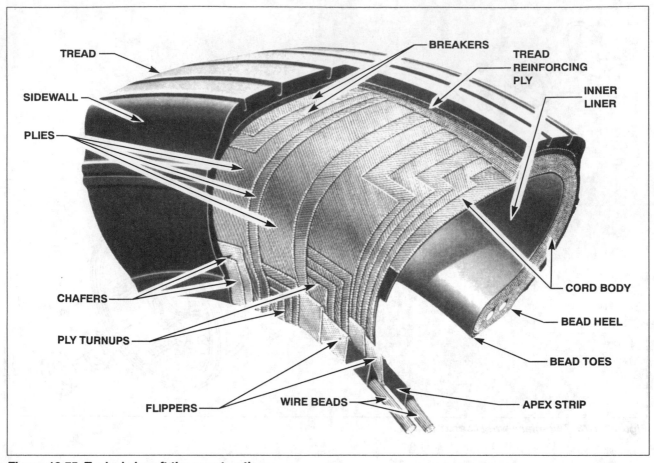

Figure 13-75. Typical aircraft tire construction.

Operating a tire with too little air pressure will cause the tread ribs on the sides to wear first and the center ribs to show less wear, figure 13-77(C).

A tire should be removed from service by the time the tread reinforcement shows through. Figure 13-77(D) shows a tread section that has worn down to the breaker plies and is too far worn to be retreaded.

Retreading can be done and effect quite a savings in the operating budget if certain precautions are observed. There are several types of retreading used on aircraft tires. Top-capping may be used when there is little shoulder wear; only the tread is removed and a new one applied.

Full capping is done by removing the old tread and applying new material which comes down several inches over the shoulder of the tire. A three-quarter retread replaces the full tread and one side of the sidewall. For this, the old sidewall is buffed down and new material is applied from the bead to the edge of the new tread.

Figure 13-76. Chine-type for nose wheels of jet aircraft having fuselage mounted engines. This throws water away from the engine inlet.

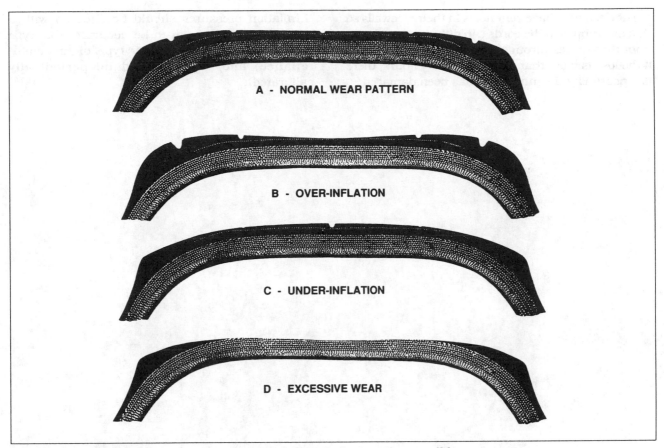

Figure 13-77. Tire tread wear is a good indicator of the tire's operating condition.

Tires may be retreaded if the carcass is sound and there are no flat spots which extend into the carcass plies.

Only approved repair stations may retread aircraft tires, and the retreaded tire must be identified by the letter "R" followed by a number indicating the number of times the tire has been retreaded, the month and year the retreading was done, and the name of the agency doing the work.

Since a retreaded tire is often larger than a new one, because of a greater amount of tread material used, it is important that the tire not foul in the wheel well of a retractable landing gear airplane. A retraction test should be performed after a retreaded tire is installed.

Proper **inflation** is the most important maintenance function to get maximum service from aircraft tires. The pressure should be checked with an *ACCURATE* pressure gauge. Pressure should be checked when the tires are cool, at least two or three hours after the airplane has been flown.

When tube-type tires are installed, air is usually trapped between the tube and the tire. This gives a false pressure indication until the trapped air seeps out around the beads or around the valve stem. Tubeless tires must be checked for proper inflation after allowing a waiting period of from 12 to 24 hours, because the nylon of which the plies are made will stretch and decrease the pressure.

When a tire is installed on the airplane, and weight is put on it, the volume of the air chamber is decreased so that the pressure will be about 4% greater than with no load applied.

If a tire consistently loses air pressure, it should be carefully checked before condemning it. Tubeless tires must be **fully** inflated and allowed to sit for 12 to 24 hours and rechecked. In this period of time they will lose approximately 10% of their pressure due to the nylon stretch. Another natural cause of pressure loss is temperature change. Tire pressure will change about one psi for every 4°F change in temperature.

Tubeless tires have vent holes in their sidewalls so that air trapped in the cords can diffuse. Air will seep from these holes throughout the life of the tire, but if the loss is more than about 5% in 24 hours, there is a possibility the inner liner has been damaged.

Inflation pressures should be checked with a gauge which is known to be accurate. Dial-type gauges are the best for any type of tire maintenance, and these should be periodically calibrated.

Additional Reading

1. Airframe and Powerplant Mechanics Airframe Handbook; EA-AC65-15A; IAP, Inc., Publ; 1976; Chapters 8 and 9.
2. A&P Technician Airframe Textbook; EA-ITP-A2; IAP, Inc., Publ; 1991; Chapters 9 and 12.
3. Aircraft Hydraulic Systems; EA-AH-1; IAP, Inc., Publ; 1985.
4. Aircraft Wheels, Brakes and Anti-Skid Systems; EA-AWB; IAP, Inc., Publ; 1985
5. Aircraft Tires and Tubes; EA-ATT-2; IAP, Inc., Publ; 1985

Study Questions And Problems

1. What are the principal technical differences between pneumatics and hydraulics?

2. What are the advantages of a closed hydraulic system, as compared to an open system?

3. In a 1500 psi hydraulic system, how much power is exerted when 2000 cubic centimeters of fluid flows to a lower pressure of 100 psi in 15 seconds, in order to retract the landing gear?

4. Name and state the two scientific laws which are the principle upon which all hydraulic systems work (static and dynamic flow).

5. Will the hydraulic fluid used in the aircraft you fly burn if ignited?

6. What type of braking system, as described in this chapter, are the brakes on the aircraft you fly?

7. What is the purpose of an accumulator in a hydraulic system?

8. What is the purpose of an accumulator in a propeller hydraulic system (explain in as much detail as you can, in your own words).

9. What is the most common seal used in today's hydraulic systems?

10. How might an unbalanced actuator be the best choice for a particular use? In your explanation, give an aircraft example other than landing gear.

11. How does a shimmy damper work?

12. Name and describe three types of landing gear emergency extension systems.

13. Are the brakes on the aircraft you fly of the energizing or nonenergizing type? How can you tell?

14. How many brake master cylinders does the airplane you fly have? Where are they located?

15. How do aircraft tires differ from automobile tires?

16. What factors affect the pressure in an aircraft tire, assuming it is not leaking, and to what extent?

17. If you are flying solo in a complex aircraft, when would you put the gear up?

Chapter XIV
Pneumatic And Deicing Systems

Pneumatic Systems

Modern aircraft use compressed air, or pneumatic, systems for a variety of purposes. Some use pneumatics rather than hydraulics for the operation of the landing gear, flaps, brakes, cargo doors and other forms of mechanical actuation. Other aircraft using hydraulics for these major functions may have a cylinder of compressed air or nitrogen as a backup source of power to lower the landing gear and apply the brakes in the event of a failure of the hydraulic power. Still other aircraft use pneumatics only for deicing and for the operation of various flight instruments and, finally, some aircraft use pneumatic systems only to provide a positive air pressure in the cabin for flight at high altitude, where pressurization is used to supply the passengers and crew with the needed oxygen. We will look at each of these types of systems in descending order of their complexity.

Low-pressure Pneumatic Systems

Compressed air under a low pressure is used to drive some of the instruments and inflate the pneumatic deicer boots. This air pressure is usually provided by an engine-driven vane-type air pump. The main use of these pumps is to drive vacuum-operated instruments, so these pumps are commonly called vacuum pumps although, in reality, they are air pumps.

A Full Pneumatic System

The majority of airplanes built in the United States use hydraulic or electric power for such heavy duty applications as the operation of the landing gear, flaps and brakes, but many of the European designed and built aircraft use compressed air for these functions. Nitrogen is desirable when an inexhaustable supply is not needed because dry nitrogen is chemically inert and without moisture so it won't corrode internal parts.

Some of the advantages of using compressed air over hydraulics or electrical systems are:

1. Air is universally available in an inexhaustible supply.

2. The units in a pneumatic system are reasonably simple and lightweight.

3. Compressed air, as a fluid, is lightweight and, since no return system is required, weight is saved.

4. The system is relatively free from temperature problems.

5. There is no fire hazard, and the danger of explosion is minimized by careful design and operation.

6. Installation of proper filters minimizes contamination as a problem.

Figure 14-1 shows a typical full pneumatic system as is used on a popular European-built twin-engine commuter transport airplane (F-27). Each of the two compressors is a four-stage piston-type pump driven from the accessory gearbox of the two turboprop engines. Air is taken into the first stage through an air duct and is compressed, then passed successively to the other three stages of the pump. The discharge air from the fourth stage is routed through an intercooler and a bleed valve to the unloading valve. The **bleed valve** is kept closed by engine oil pressure and, in the event of a loss of the engine lubricating oil, the valve will open and relieve the pump of any load.

The **unloading valve** maintains pressure in the system between 2,900 and 3,300 psi. When the pressure rises to 3,300 psi, a check valve traps it and dumps the output of the pump overboard. When the system pressure drops to 2,900 psi, the output of the pump is directed back into the system.

A **shuttle valve** in the line between the compressor and the main system makes it possible to charge the system from a ground source. When the pressure from the external source is higher than that of the compressor, as it is when the engine is not running, the shuttle slides over and isolates the compressor.

Moisture in a compressed air system will condense and freeze when the pressure of the air is dropped for actuation and, for this reason, every bit of water must be removed from the air. A **separator** collects the water that is in the air on a baffle and holds it until the system is shut down. When the inlet pressure to the separator drops below 450 psi, a drain valve opens and all of the

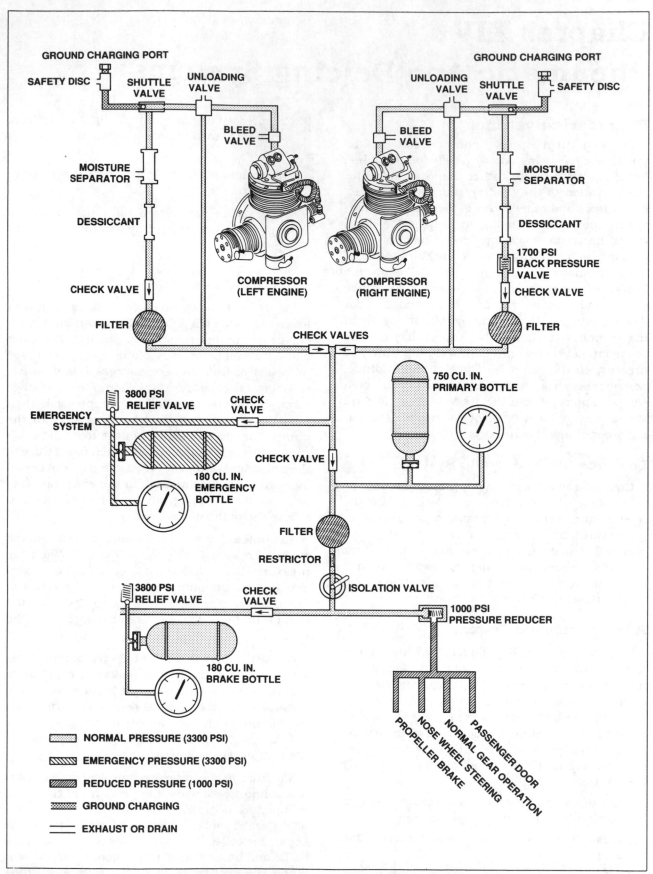

Figure 14-1. Full pneumatic system for a twin-engine turboprop airplane.

accumulated water is blown overboard. An electric heater prevents the water collected in the separator from freezing.

After the air leaves the moisture separator with about 98% of its water removed, it passes through a **desiccant**, or chemical dryer, to remove the last traces of moisture.

Before it enters the actual operating system, the air is filtered through a 10-micron sintered-metal filter, and when it is realized that the lower level or visibility with the naked eye is about 40 microns, it can be seen that this provides really clean air to the system.

A **back pressure valve** is installed in the right engine nacelle. This is essentially a pressure relief valve in the supply line that does not open until the pressure from the compressor or ground charging system is above 1,700 psi, and this assures that the moisture separator will operate most efficiently. If it is necessary to operate the system from an external source of less than 1,700 psi, it is connected into the left side where there is no back pressure valve.

There are three **air storage bottles** in this airplane; a 750-cubic-inch bottle for the main system, a 180-cubic-inch bottle for the normal brake operation, and a second 180-cubic-inch bottle for emergency operation of the landing gear and brakes.

A manually operated isolation valve allows a technician to close off the air supply so the system can be serviced without having to discharge the storage bottle.

The majority of the components in this system operate with a pressure of 1,000 psi, so a pressure reducing valve is installed between the isolation valve and the supply manifold for normal operation of the landing gear, passenger door, drag brake, propeller brake, and nose wheel steering. This valve not only reduces the pressure to 1,000 psi, but it also serves as a backup pressure relief valve.

The **emergency system** stores compressed air under the full system pressure of 3,300 psi and supplies it for landing gear emergency extension.

Emergency Backup Systems

All aircraft with retractable landing gear must have some method of assuring that the gear will move down and lock in the event of failure of the main extension system. One of the simple ways of lowering and locking a hydraulically actuated landing gear is by using compressed air or nitrogen stored in an emergency cylinder. The gear selector is placed in the gear down position to provide a path for the fluid to leave the actuator and return into the reservoir. Compressed air is then released from the emergency cylinder into the actuator through a **shuttle valve**. See figure 13-60. This valve is moved over by air pressure to close off the hydraulic system so little or no air can enter the rest of the hydraulic system. The air pressure is sufficient to lower and lock the landing gear against the flight loads.

Emergency operation of the brakes is also achieved in some airplanes by the use of compressed air. When the pilot is sure there is no hydraulic pressure to the brakes, the pneumatic brake handle is used. Clockwise rotation of this handle increases the brake pressure, and when the handle is held stationary, the pressure is constant. Nitrogen or air pressure released by this control handle forces hydraulic fluid in the transfer tube into the main wheel brakes through shuttle valves. When the brake handle is rotated counterclockwise, pressure is released and the gas is exhausted overboard.

Low-pressure Pneumatic Systems For Instruments

Many aircraft use air-driven gyro instruments as either the primary gyro instruments or as backup instruments when the primary gyros are electrically driven.

For many general aviation aircraft, all of the air-driven gyro instruments use an engine-driven vacuum pump to evacuate the instrument case, and filtered air is pulled into the instrument to spin the gyro. The reason for this is that it is much easier to filter air being pulled into the instrument than it is to filter the air after it has been pumped by an engine-driven pump lubricated by engine oil. The output of these pumps always contains some particles of oil.

Pressurized aircraft create extra problems for suction-operated instruments, and the latest generations of air-driven gyros for these aircraft use pressure. Turbine-powered aircraft bleed some of the pressure from the engine compressor, regulate and filter it, and then direct it over the gyros. Aircraft with reciprocating engines use engine-driven air pumps to provide the airflow for the gyros. This air is regulated and filtered before it reaches the instrument.

There are two types of air pumps used to provide instrument airflow, and both are vane-type pumps. See figure 14-2. Sliding vanes are rotated

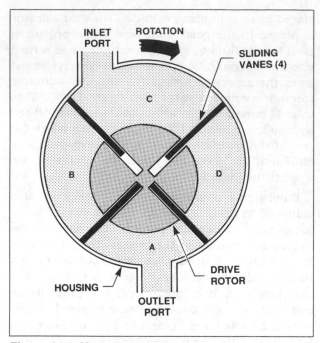

Figure 14-2. Vane-type air pump.

Figure 14-4(A). Dry-type air pump.

by the driveshaft and held out against the housing of the pump by centrifugal force. As the shaft turns, the chambers located at positions *A* and *B* become larger, while those at positions *C* and *D* decrease in size. Air is pulled into the pump at the

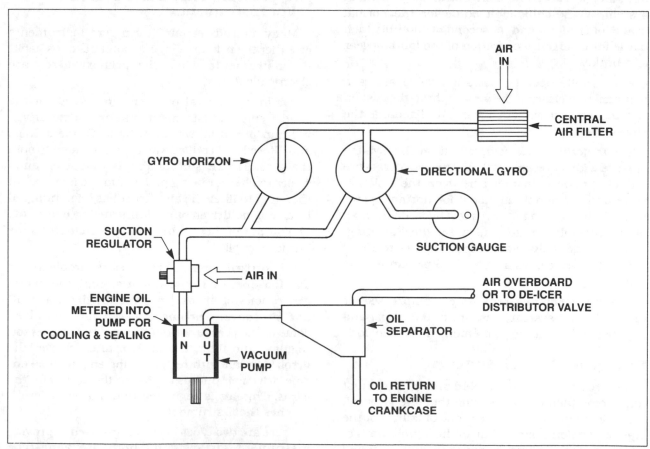

Figure 14-3. Vacuum system using a wet-type vacuum pump.

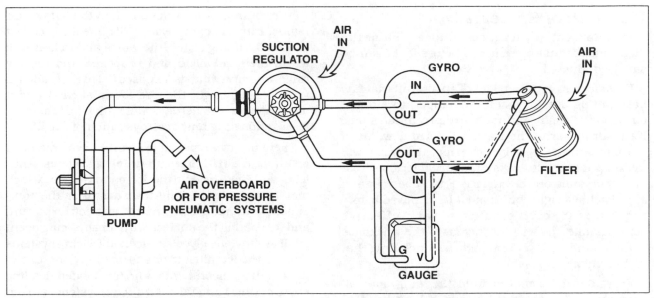

Figure 14-4(B). A vacuum system for the instruments of an aircraft using a dry-type air pump.

position where the chambers are large, and it is compressed as the chamber size decreases. "Wet" vacuum pumps use steel vanes moving in a cast-iron housing and are sealed and lubricated by engine oil metered into the inlet air port. This oil is discharged with the air and is removed by an oil separator before the air is either used for inflating deicer boots or is pumped overboard. See figure 14-3.

The more modern instrument air systems use "dry" pumps that have carbon vanes and rotors and require no external lubrication. These pumps may be used to drive the instruments by producing a vacuum and pulling air through them, as is seen in figure 14-4, or by using the output of the pump to force the air through the instruments (figure 14-5). Instrument systems are discussed in much more detail in Chapter 17.

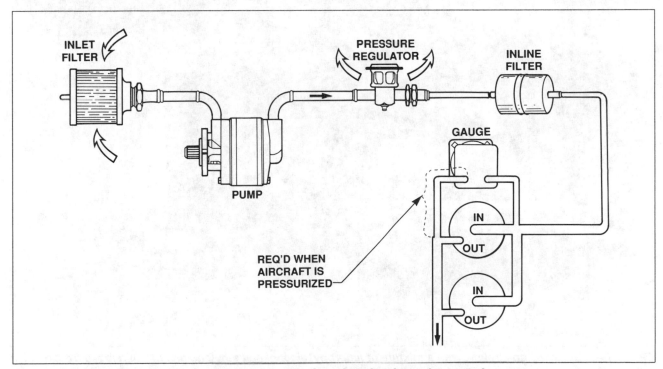

Figure 14-5. A pressure system for the instruments of an aircraft using a dry-type air pump.

Ice Control Systems

Ice in almost any form constitutes a hazard to flight and it must be removed before flight can be safely conducted.

Frost forms on the surface of an aircraft that has been sitting outside when the temperature of the air drops at night and moisture precipitates out. If the air is warm, the water will form dew, but if the temperature is below freezing, the water will freeze as it precipitates out and it will form as frost in tiny crystals on the surface. Frost does not add appreciable weight, but it *must* be removed before flight because it creates a very effective aerodynamic spoiler that increases the thickness of the boundary layer and adds so much drag that flight may be impossible.

Frost can be removed from the wing and tail surfaces by brushing it off with a long handled T-broom. Or, better yet, it can be prevented from forming on the surfaces by covering them when the airplane is secured for the night. Spraying the surfaces with a deicing solution of ethylene glycol and isopropyl alcohol just before flight will effectively remove all traces of the frost. See figure 14-6.

Aircraft that fly into clouds when the outside air temperature is near freezing will quite likely collect ice on all leading edges (the wings and tail, as well as on the windshield and propeller, and on any radio antenna that is exposed). This ice adds a great deal of weight the aircraft must carry and it also changes the aerodynamic shape of the surfaces, adds drag and destroys much of the lift.

There are two types of ice control systems used on aircraft: **anti-icing** and **deicing** systems. Anti-icing systems prevent the formation of ice, which may be done by heating the surface or the component with hot air, engine oil, or electric current, and/or coating the surface with an anti-icing agent such as ethylene glycol or alcohol. Deicing systems remove the ice after it has formed, by the use of pneumatic deicer boots and/or heated leading edges of wings and tails. Anti-icing systems are not usually capable of removing ice once it has formed.

Thermal anti-icing utilizes heated air that is directed through a specially designed heater duct in the leading edge of the wing and the tail surfaces to heat this portion of the airfoil and prevent the formation of ice. This air can be heated, in

Figure 14-6. Spraying an airplane with a mixture of isopropyl alcohol and ethylene glycol will remove frost and ice from the surface and prevent its refreezing.

292

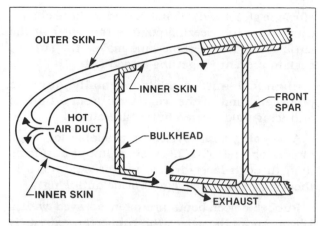

Figure 14-7. Thermal anti-icing is accomplished by flowing hot compressor bleed air from the engine through a duct in the leading edge of the wing.

Figure 14-8. Pitot heads such as this prevent ice plugging the entry hole by warming them with an electric heater.

reciprocating engine aircraft, by using combustion heaters or heater shrouds around the engine exhaust system.

Most aircraft that use the thermal anti-icing systems today are turbine powered, and it is a simple matter to use some of the heated compressor bleed air to heat the leading edges and prevent the formation of ice. See figure 14-7.

The Boeing 727 takes bleed air from the two outboard engines and directs it through the wing anti-icing control valves to a common manifold and then out into the wing leading edge ducts. The two inboard leading edge flaps and eight leading edge slats are protected with this hot air. These portions of the wing are protected from overheating by overheat sensor switches. If they sense an overheat condition, they turn on an overheat warning light and close the anti-icing valves, shutting off the flow of hot air into the ducts. When the duct temperature drops to an allowable range, the overheat light will go out, and hot air will again flow into the duct.

The center engine of the Boeing 727 has its air intake at the rear top of the fuselage and because of this, some of the anti-icing hot air is ducted to the upper VHF radio antenna to prevent ice forming on it and breaking off to be ingested into the center engine. Turbine engines are susceptible to ice damage if chunks of ice form on some of the exposed portions of the engine and break off and are sucked into the compressor. The engines on the Boeing 727 have hot compressor bleed air directed through the inlet guide vanes, the engine bullet nose, and through the oil cooler scoop for the constant speed drive, as well as for the inlet duct for the center engine.

While it is not exactly a part of the thermal anti-icing system, many transport airplanes keep the windows in the cabin area free of fog and frost by directing warm air between the panes of the windows.

Electric anti-icing has many applications. Pitot heads installed on almost all aircraft that may possibly encounter icing are electrically heated. These heaters are so powerful that they should not be operated on the ground because, without an adequate flow of air over them, there is a possibility that they will burn out. Always treat a pitot head like a hot burner on a stove—it can cause severe burns if touched while it is on. See figure 14-8. These heaters require enough current that the load meter will deflect noticeably when the heater is on.

Static ports and stall warning vanes on many aircraft are also electrically heated. The static port on some of the smaller aircraft are not heated, but if there is no provision for melting the ice off of this vital pressure pickup point, the aircraft should be equipped with an alternate static air source valve. This valve allows the pilot to reference the flight instruments to a static source inside the aircraft if the outside static port should become covered with ice or otherwise plugged.

Windshields and cockpit windows may be electrically heated to prevent ice obstructing the vision of the pilot and the copilot. There are two methods of heating these components. One method uses a conductive coating on the inside of the outer layer of glass in the laminated windshield, and the other method uses tiny resistance wire embedded inside the laminated windshield.

The windshield of a high-speed jet aircraft is a highly complex and costly component. For all of the transport category aircraft, these windshields must not only withstand the pressures caused by pressurization and normal abuse and flight loads, but they must also withstand, without penetration, the impact produced by a four-pound bird striking the windshield at a velocity equal to the airplane's design cruising speed. For a windshield to be this strong, it is built as a highly complex sandwich, with some of the business jet windshields about an inch and a half thick, made of three plies of tempered glass with layers of vinyl between them. The inner surface of the outer ply of glass is coated with a conductive material through which electric current flows to produce enough heat to melt off any ice that forms on the windshield. There are temperature sensors and an elaborate electronic control system to prevent these windshields becoming overheated. See figure 14-9.

The windshields are heated not only to prevent ice, but to strengthen them against bird strikes. When the windshield is heated, the vinyl layers are less brittle and will withstand an impact with much less chance of penetration than they will when they are cold.

Chemical anti-icing is accomplished by coating certain surfaces and components of an aircraft with either isopropyl alcohol, or a mixture of

ethylene glycol and alcohol. Either of these chemicals lowers the freezing point of the water at the surface and at the same time makes the surface slick to prevent ice getting a good grip.

Chemical anti-icing is normally done to propellers, and to the windshield from a tank of anti-icing fluid carried in the aircraft.

A **weeping wing** system is available for small aircraft as a retrofit. This system pumps anti-icing fluid through laser-drilled, microscopic holes in the wing's leading edge to provide anti-icing.

Rubber deicer boots are often sprayed with a silicon spray that gives the rubber an extremely smooth surface so ice does not adhere as well to it.

Deicing Systems

Anti-icing systems prevent the formation of ice on the protected component, but it has been found that for keeping the wings and tail surfaces of some of the slower airplanes free of ice, it is more effective to allow the ice to form on the surface and then crack it so the airflow over the surface will carry the ice away. This is more effective than melting the ice on the leading edge, because melting causes the water to flow back to an unheated portion of the surface and refreeze, forming a ridge that becomes an effective aerodynamic spoiler. Some of the first systems developed to combat ice formation were deicing systems.

Airline flying was hindered in the days of the Ford Trimotor and the Curtiss Condor by the exposed wires and struts on which ice could accumulate. Pilots did not dare fly into clouds where ice could exist. But with improved instruments and radio, and with the introduction of the Boeing 247 and the Douglas DC-2, flight into icing conditions did occur. To remove the ice, the B.F. Goodrich Company developed a rubber deicer boot that was installed on the leading edges of the wings and the empennage.

A **rubber boot** containing several longitudinal tubes is glued to the leading edge of the surface. Air from the discharge of the engine-driven vacuum pump is passed through a timer-operated distributor valve into the tubes of the boot in a sequential manner. Figure 14-11(A) shows the boot as it is installed on the leading edge of a wing with all of the tubes deflated. When they are deflated, suction from the suction side of the pump or from an ejector around the pump discharge line holds the tubes evacuated so air flowing over the boot will not cause the tube to distort the shape of

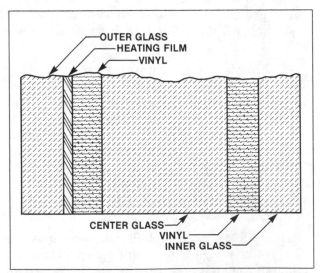

Figure 14-9. The windshield of a modern jet aircraft is made up of laminations of glass and vinyl. It is heated by electric current flowing through a conductive film on the inside of the outer layer of glass.

Figure 14-10. This airplane can fly into icing conditions because the wing and tail surfaces are deiced with pneumatic deicer boots and the propeller is protected by electric or chemical deicer anti-ice systems.

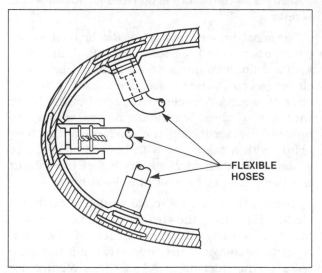

Figure 14-11(A). Pneumatic deicer boot with all tubes deflated.

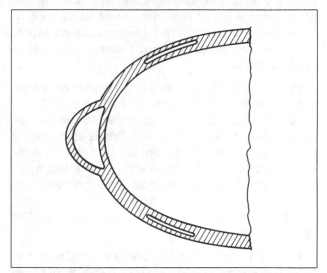

Figure 14-11(B). Pneumatic deicer boot with the center tube inflated and the outer tubes deflated.

the leading edge of the wing. In figure 14-11(B) the center tube is inflated and any ice that has formed over it will crack. The center tube now deflates and the outer tubes inflate and push up the cracked ice so air flowing over the wing will get under it and blow it off of the surface. All of the tubes now

deflate and are held tight against the boot by suction until the ice reforms, and then the cycle repeats itself.

The cycle of operation causes the tubes to inflate in a symmetrical manner so the disruption of lift during the inflation will be uniform and will not

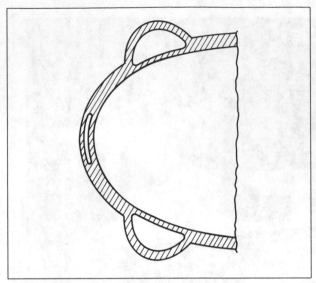

Figure 14-11(C). Pneumatic deicer boot with the center tube deflated and the two outer tubes inflated.

cause any flight control problems. The manufacturer of the aircraft has determined by flight tests the proper cycle time for the operation.

Larger aircraft that use this type of deicing system have an electric motor-driven timer to operate solenoid valves that, when the system is turned, will continually cycle the system through all of the tubes, and then provide the proper duration of rest time to allow the ice to form over the boots. Then the cycle is repeated. Any time the tubes are not inflated, suction is applied to them.

Smaller aircraft do not use the elaborate timer, but are turned on by the pilot when it is determined that an accumulation of ice on the leading edges should be broken off. When the deicing switch is turned on, the boots will cycle through one, two or three operating cycles, depending upon the design of the system, and then the tubes will be connected to the vacuum side of the air pump to hold them tight against the leading edge. See figure 14-12.

The air for inflating the boots normally comes from the engine-driven air pump. Some of these pumps are of the "wet" type which use engine oil taken into the pump through holes in the mounting flange to lubricate and seal the steel vanes. This oil must all be removed by an oil separator and sent back into the engine crankcase before the air can be used to inflate the deicer boots.

Dry-type pumps are used for some installations, and these pumps do not require an oil separator as they use carbon vanes which make the pump self-lubricating. Carbon dust is a product of these pumps, so a filter is used.

Some deicing systems that are used only occasionally inflate the boots from a cylinder of compressed air that is carried just for this purpose.

There are several configurations of deicer boots, but all accomplish their work in the same way. They allow the ice to form and then break it off as the tubes inflate. Figure 14-13 shows some of the more commonly used configurations. Some boots use spanwise tubes that inflate alternately, and some inflate simultaneously. Other configurations of boots have chordwise tubes that may inflate either alternately or simultaneously. The configuration of the tubes is determined by flight test and, naturally, only the specific boot that is approved for the aircraft should be used.

When rubber deicer boots were first developed, adhesives had not been developed to the state they are today, and these boots were installed on the leading edge of the surfaces with machine screws driven into Rivnuts installed in the skin. This type of installation can be identified by a narrow metal fairing strip that covers the screw heads at the edges of the boots. Almost all of the newer boot installations fasten the boots to the surface with adhesives so that there is no need for Rivnuts and screws.

The most important part of deicer boot maintenance is to keep the boots clean. Wash the boots with a mild soap and water solution, and if any cleaning compounds have been used with the aircraft, wash all traces of them off of the boots with lots of clean water. Oil or grease may be removed by scrubbing the surface of the boot lightly with a rag damp with benzoil or lead-free gasoline and then wiping it dry before the solvent has had a chance to soak into the rubber.

Repairs than can be made to deicer boots include refurbishing the surface of the boot, repairing scuff damage to the surface of the boot, repairing damage to the tube area, and repairing tears in the fillet area. All of these repairs are detailed in the manufacturer's service manuals, and these instructions must be followed in explicit detail.

Many modern propellers installed on both reciprocating and turboprop engines are deiced with an **electrothermal deicer system**.

Rubber boots with heater wires embedded in the rubber are bonded to the leading edges of the propeller blades, and electrical current is passed

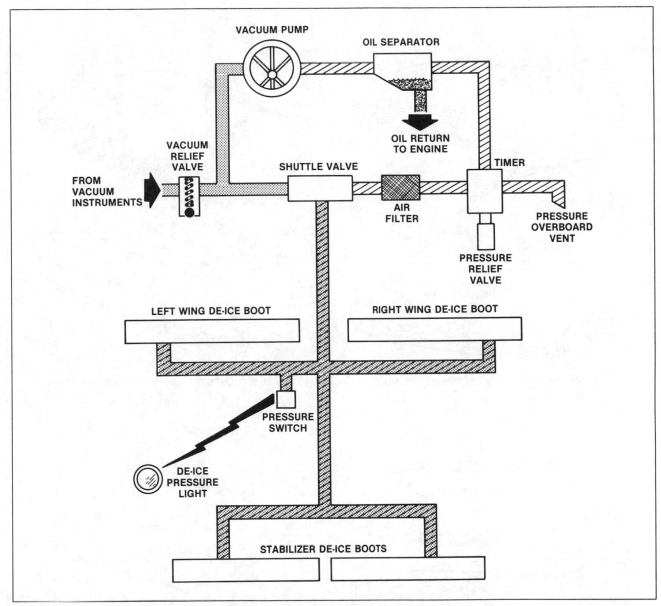

Figure 14-12. Pneumatic deicer boot system for a single-engine general aviation airplane.

through these wires to heat the rubber and melt any ice that has formed, so centrifugal force and wind can carry the ice away.

The boots in some installations are made in two sections on each blade. Current flows for about a half minute through the outboard section of all blades and then for the same time through the heaters on the inboard section of all of the blades. The time the current flows has been proven by flight tests to be sufficient to allow ice to form over the inactive section and long enough to loosen the ice from the section that is receiving the current. See figure 14-14.

The complete propeller deicer system consists of the following components:

1. Electrically heated deicers bonded to the propeller blades.

2. Slip-ring and brush block assemblies that carry the current to the rotating propeller.

3. Timer to control the heating time and sequence of the deicing cycle.

4. An ammeter to indicate the operation of the system.

5. All of the wiring, switches and circuit breakers necessary to conduct electrical power from the aircraft electrical system into the deicer system.

The slip-ring assembly is mounted on the propeller either through a specially adapted

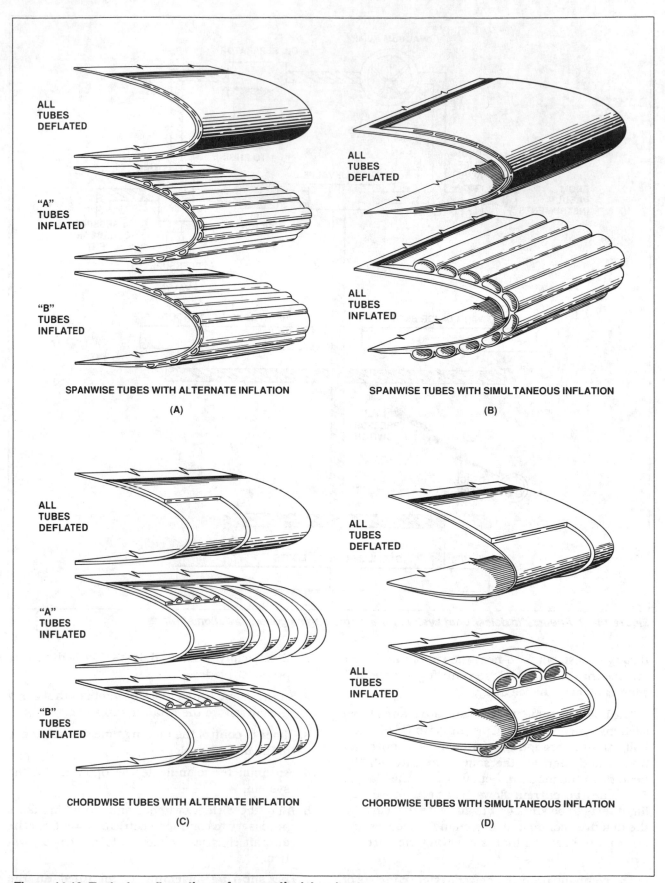

ALL
TUBES
DEFLATED

"A"
TUBES
INFLATED

"B"
TUBES
INFLATED

SPANWISE TUBES WITH ALTERNATE INFLATION

(A)

ALL
TUBES
DEFLATED

ALL
TUBES
INFLATED

SPANWISE TUBES WITH SIMULTANEOUS INFLATION

(B)

ALL
TUBES
DEFLATED

"A"
TUBES
INFLATED

"B"
TUBES
INFLATED

CHORDWISE TUBES WITH ALTERNATE INFLATION

(C)

ALL
TUBES
DEFLATED

ALL
TUBES
INFLATED

CHORDWISE TUBES WITH SIMULTANEOUS INFLATION

(D)

Figure 14-13. Typical configurations of pneumatic deicer boots.

Figure 14-14. An electrical deicing system is used to remove ice from this propeller.

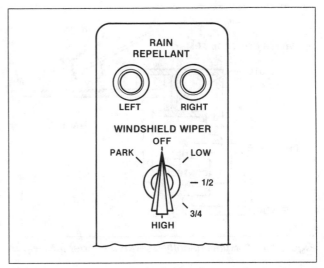

Figure 14-15. Rain control panel on a jet transport airplane.

starter gear, or attached to the spinner bulkhead or the crankshaft flange.

The brush block is mounted on the engine so the three brushes will ride squarely on the slip rings.

The timer controls the sequence of current to each of the deicers. The sequence of heating is important, to provide the best loosening of the ice so it can be carried away by the centrifugal force. And it is also important that the same portion of each blade be heated at the same time, to prevent an out-of-balance condition.

The ammeter shows the operation of the system and assures the pilot that each heater element is taking the required amount of current.

Rain Control Systems

Almost all of the small general aviation aircraft use transparent acrylic plastic windshields, and this soft material is so easy to scratch that windshield wipers are seldom installed. One way to minimize the effect rain has on visibility in flight is to keep the windshield waxed so water will not be able to spread out over the surface, but will bead up so the wind can blow it away.

Larger and faster aircraft that routinely operate in rain have rather elaborate rain control systems. There are three methods used to control the effects of rain and they may be used together. We will discuss each of these three methods separately. They are: mechanical windshield wipers, chemical rain repellant, and a high-velocity air blast.

Windshield wipers for aircraft are similar to those used on automobiles except they must be able to withstand the air loads caused by the high speeds of operation. Electrical windshield wipers are usually operated by a two-speed DC motor that drives a converter. This converter changes the rotary output of the motor into the reciprocating motion needed for the wiper blades. When the windshield wiper switch is turned Off, the control circuit is open, but the motor continues to run until the blades are driven to the Park position. The motor then stops, but the control circuit is armed so the motor will start when the windshield wiper switch is turned to either the Fast or Slow position. Some installations have a separate position on the speed selector switch that allows the pilot to drive the wiper blades to the Park position before putting the switch in the Off position. See figure 14-15.

Some aircraft use hydraulic windshield wipers that use pressure from the main hydraulic power system to drive the wiper blades (figure 14-16). Hydraulic fluid under pressure flows into the control unit, which periodically reverses the direction of the flow of fluid to the actuators. Inside the actuators are pistons which move a rack and pinion gear system. As the pistons move in one direction, the wiper will move one way, but when the flow is reversed, the piston and the wiper blades will move in the opposite direction. When the control valve is turned Off, the blades are driven to and held in the Park position. Speed control is accomplished by varying the flow rate through a variable orifice in the fluid line.

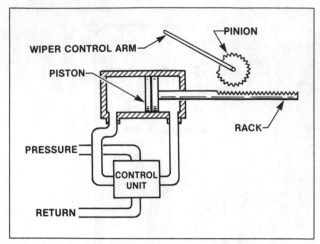

Figure 14-16. Hydraulically operated windshield wiper.

Windshield wipers must never be operated on a dry windshield, and the blades must be kept clean and free of any type of contaminants that could scratch the windshield. If the windshield wiper should ever have to be operated for maintenance or adjustment, the windshield must be flooded with ample quantities of fresh, clean water and kept wet while the wiper blades are moving across the glass.

Many of the jet transport aircraft have a **chemical rain repellant system** that uses a liquid chemical sprayed on the windshield to prevent the water reaching the surface of the glass (figure 14-17). Since it cannot wet the surface and spread out, the water will form beads and the wind can easily carry it away and leave the glass free of water so the pilot's visibility will not be distorted.

The repellent is a syrupy liquid that is carried in pressurized cans connected into the rain repellent system. When the aircraft is flying in rain so heavy that the windshield wipers cannot keep the windshield clear, the pilot can depress the Rain Repellent button and a single timed

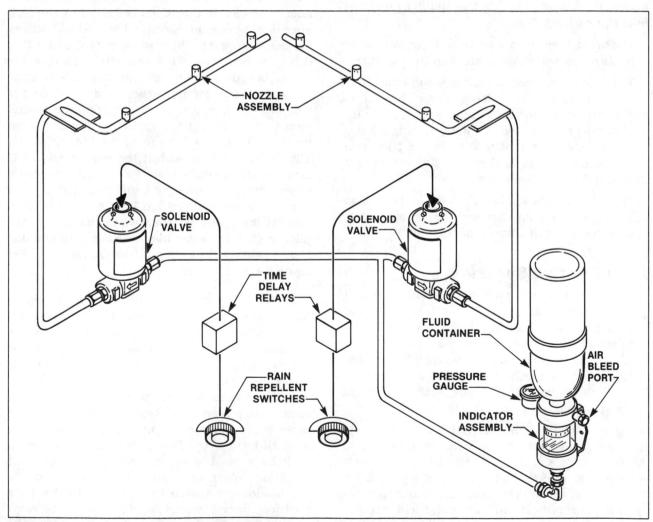

Figure 14-17. Chemical rain repellent system used on a jet transport airplane.

application of the liquid will then be sprayed out onto the windshield. The windshield wipers then spread the liquid out evenly over the wiped surface.

The liquid should never by sprayed onto the windshield unless the rain is sufficiently heavy, because too much repellent can smear on the windshield and be difficult to see through. The repellent is difficult to remove if it is spayed onto a dry windshield.

The operating system consists of two pressurized containers of repellent and two DC solenoid valves that, once actuated, are held open by a time-delay relay. When the Rain Repellent push-button switch is depressed, the fluid flows for the required period of time, which is less than a second, and then the valve closes until the push-button is again depressed. The number of times the button is depressed is determined by the intensity of the rain.

High-pressure compressed air may be ducted from the engine bleed air system into a plenum chamber and then up against the outside of the windshield in the form of a high-velocity sheet of warm air. This air blast effectively prevents the rain hitting the windshield surface and adhering to it, and provides anti-icing of the windshield, or deicing at lesser velocities.

Study Questions And Problems

1. Name and describe the function of each of the basic parts of a full pneumatic system.

2. What tasks are performed by air in the aircraft you fly?

3. What type of pneumatic pump drives what instruments in the aircraft you fly?

4. What is the difference between a deicing and an anti-icing system?

5. What is the source of warm air used by a thermal anti-icing system?

6. How is electrical power transmitted from the airframe to the electrothermal deicing boots on a spinning propeller?

7. What causes a deicing boot to inflate? Deflate?

8. How, other than windshield wipers, is rain kept off of aircraft windshields?

Chapter XV
Aircraft Structures And Flight Controls

Evolution Of Aircraft Structures

The early dreamers of flight had little concept of a practical structure for their machines. The Greeks had Daedalus and his son Icarus flying with wings made of feathers and wax, while other dreamers conjured up machines resembling birds. Even the genius Leonardo da Vinci conceived a flying machine which had flapping wings attached to a body patterned after that of a bird, DeLana and Cayley both felt that they had to incorporate characteristics of another means of transportation, as the machines in their drawings and the models they built had bodies resembling boats.

It was only with the discovery that lift could be produced by air flowing over a cambered surface that aerodynamics took a practical turn. The gliders of Lilienthal and Chanute proved that manned flight was possible, and by using the results of their experiments, the Wright brothers developed a biplane glider with which they solved the biggest problem of the time—the problem of control.

The early flying machines produced by the Wrights, Glenn Curtiss, Henri Farman, Alberto Santos-Dumont, and the Voisin brothers all had a common type of structure. The wings used ribs of bent wood and were covered with cloth fabric to form the lifting surfaces. The bodies were little more than open girder frameworks made of bamboo or strips of wood held together with piano wire (figure 15-1). Auxiliary surfaces similar in construction to the wings, but smaller, were attached to the body either ahead of or behind the wings in an attempt to provide stability and control.

Figure 15-1. The first airplanes used an open truss to hold the occupants and the engine and to provide an attachment for the lifting and control surfaces.

With the basic problems of flight and control solved, airplanes evolved into a more or less standard configuration. Up through World War I, most airplanes were built with a **truss-type structure** and had struts and wire-braced wings. The occupants sat in open cockpits in the fabric-covered body, or fuselage. Almost all of these airplanes had the engine up front and the auxiliary surfaces at the tail.

Increased knowledge of flight and the experience gained in building strong, lightweight structures allowed builders to turn their attention to the problem of decreasing the resistance these crude machines offered to the air through which they passed. This resistance, or friction, robbed the planes of much of their potential for speed.

To minimize wind resistance and yet retain the strength provided by the truss structure, the designers attached a superstructure of wooden formers and stringers over the truss to give the angular form a smooth streamlined shape.

One of the major breakthroughs in aircraft structure was made in the latter years of World War I, when welded thin-wall steel tubing was first used for the fuselage truss instead of the wood that required wire bracing.

The next logical step in structural development came with the discovery of a form of construction without the truss that provided its strength, but which had the streamlined form that had previously been furnished by the superstructure. This is known generally as a **stressed-skin structure**, because all of the structural loads are carried by the skin itself.

The Lockheed Aircraft Company pioneered stressed-skin construction with their popular Vega series of airplanes in the 1920's and 1930's. Thin sheets of wood veneer were held under heat and pressure in a large concrete mold and formed into a plywood eggshell-like structure. Laminated wood rings were built into this shell for support at its critical spots to provide attachment points for the engine, the wing, the tail and the landing gear.

Thin aluminum-alloy sheets were next used for the skin of stressed-skin aircraft structure. These sheets are formed with their compound curves either in hydropresses or by drop hammers. The

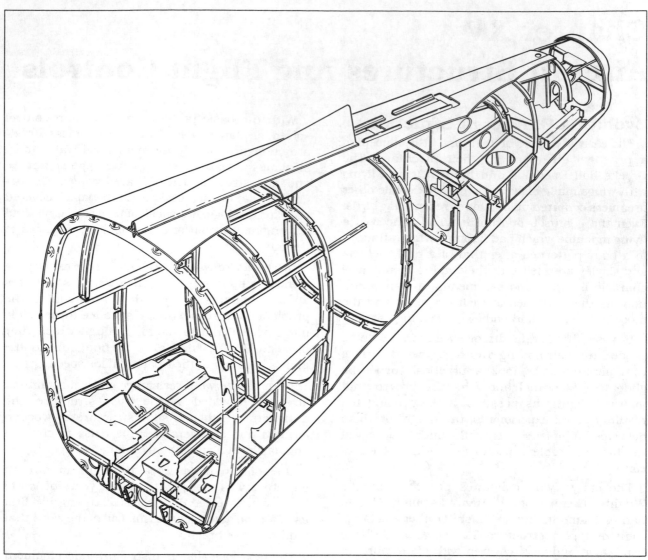

Figure 15-2. Monocoque structure uses a minimum of internal structure to support the stress-carrying skin.

formed skins are then riveted onto thin sheet metal formers and ribs.

Some of the smallest airplanes today use a **monocoque** structure in which there is virtually no internal framework, but the vast majority of modern all-metal aircraft use a **semi-monocoque** form of structure in which an internal arrangement of formers and stringers is used to provide additional rigidity and strength to the skin. See figures 15-2 and 15-3.

Pressurized airplanes provided a new set of structural challenges that had to be solved before high-altitude flight could become the commonplace means of transportation with which we are familiar today.

Soon after the first jet transport aircraft started flying in the early 1950's, three of them broke apart in the air under mysterious circumstances, two of them in relatively nonturbulent air. An extremely thorough investigation disclosed that the cause of the breakups was **metal fatigue** brought about by the flexing of the structure during the pressurization and depressurization cycles.

As a result of this investigation, a system of fail-safe construction has been devised in which rip-stop doublers are installed at strategic locations throughout, especially around the windows and doors. If a crack starts, rather than causing a major failure, it will stop at a doubler which will support the load.

The airframe of a fixed-wing aircraft is generally considered to consist of five principal units: the fuselage, wings, stabilizers, flight control surfaces and landing gear. See figure 15-4.

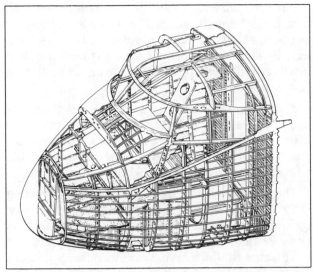

Figure 15-3. Additional strength is provided in a semi-monocoque structure by a substructure that reinforces the skin.

The airframe components are constructed from a wide variety of materials and are joined by rivets, bolts, screws, welding and adhesives. The aircraft components are composed of various parts called structural members (i.e., stringers, longerons, ribs, bulkheads, etc). Aircraft **structural members** are *designed to carry a* **load** *or to resist* **stress**. A single member of the structure may be subjected to a combination of stresses. In most cases, the structural members are designed to carry end loads rather than side loads; that is, to be subjected to tension or compression rather than bending.

Strength may be the principle requirement in certain structures, while others need entirely different qualities. For example, cowling, fairing and similar parts usually are not required to carry the stresses imposed by flight or the landing loads. However, these parts must have such properties as neat appearance and streamlined shapes.

Stresses And Structure

Aircraft structure must be strong, lightweight and streamlined, but at times these requirements seem incompatible. Truss-type structure can be made both lightweight and strong, but its angular

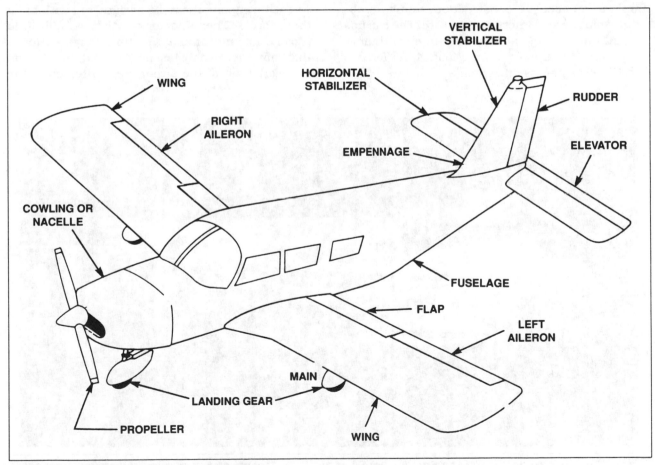

Figure 15-4. Aircraft structural components.

305

form requires a superstructure to make it streamlined. Wood monocoque structure did provide this feature, as it had the needed streamlined form and high strength, but it had the double disadvantage of a relatively limited life of the wood and the high cost of labor.

The high volume of aircraft production we have today has caused riveted or bonded sheet metal construction to become the standard, as this type of fabrication adapts to automation and standardization and gives a lightweight, streamlined and extremely strong structure. Sheet metal construction has the added advantage of the ease with which it may be repaired.

Types Of Sheet Metal Structure

There are two basic types of sheet metal structure used for aircraft: **monocoque** and **semi-monocoque**. Both of these are forms of stressed-skin, meaning that the greatest part of the structural loads are carried in the external skin. But they differ in the amount of internal structure they use.

A thin metal beverage can is an excellent example of monocoque construction, using an absolute minimum of internal structure. The can has two thin metal ends attached to an even thinner body, but will support a large load if the force is applied evenly across its ends.

The problem with a monocoque structure is that just a very small dent will destroy its ability to support a load. An empty aluminum beverage can may be able to support you if you stand on it, but it will crumple easily if there is even a small dent in the side.

Semi-monocoque structure minimizes the problem of dents by supporting the external skin on a framework of formers and stringers. The internal structure stiffens the skin so it is less susceptible to strength-destroying dents and deformation.

The skin may be made more rigid by riveting stiffeners across any large unsupported panels or by using laminated honeycomb material for the skin. Some of the modern high-speed jet aircraft have skins that are milled for stiffness. Stiffeners are machined on the inner surface of the skin by either conventional machining on a tape-controlled milling machine, or by chemical or electrochemical milling. See figure 15-23.

Structural Loads

An airplane manufacturer must consider all of the loads to which the structure will be subjected so each component can be designed to withstand these forces. Then a factor of safety is built in to provide for any unusual or unanticipated loads that may conceivably be encountered. In addition to meeting all of the strength requirements, the

Figure 15-5. Modern sheet metal construction produces airplanes that are strong, streamlined and economical.

structure must be as light as it is possible to build it.

As pilots, it is our responsibility to be sure that any repair made restores both the original strength and stiffness to the structure and maintains the original shape of the part. After all, it is we who will fly the repaired aircraft!

Before studying the structures themselves, let's take a look at the stresses these structures must withstand and the materials from which they are made.

Stresses

As pilots, it is important that we understand the stresses that act on the structure. There are only five types of stress with which we are concerned. Two of these are primary, and the other three can, for all practical purposes, be expressed in terms of the first two.

Tension is a primary stress that tries to pull a body apart. When a weight is supported by a chain, the chain is subjected to tension or, as we normally express it, to a **tensile stress**. The weight is attempting to pull the chain apart. See figure 15-6(A).

Compression, the other primary stress, tries to squeeze the part together. A weight supported on a post exerts a force that tries to squeeze the ends of the post together, or to collapse it. This is called a compressive stress.

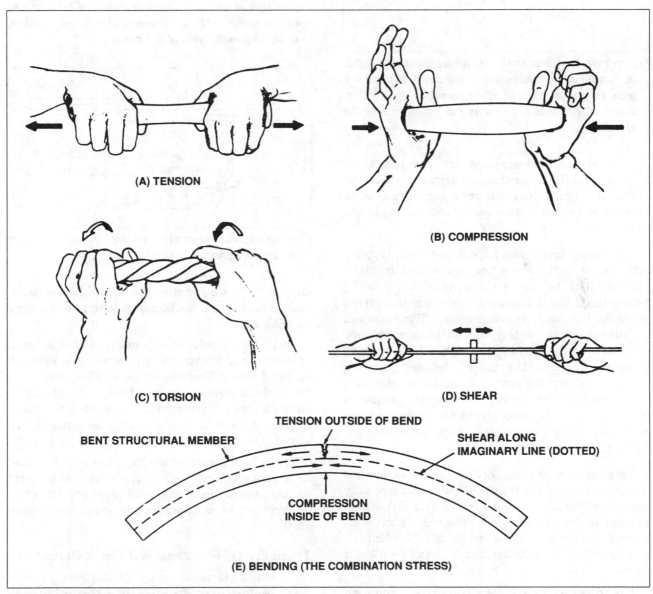

Figure 15-6. Five stresses acting on an aircraft.

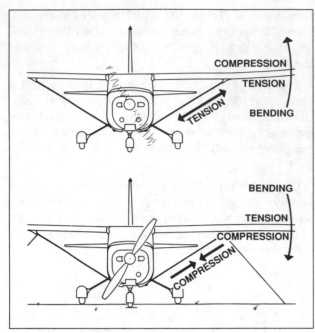

Figure 15-7. The wing struts are under tension in flight, but under compression when the airplane is on the ground and/or experiencing negative "g's." Outboard of the strut, the wing is subjected to bending stresses.

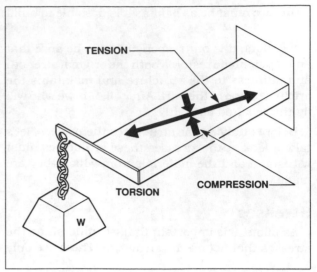

Figure 15-8. A torsional stress consists of tension and compression acting perpendicular to each other, each diagonally across the body.

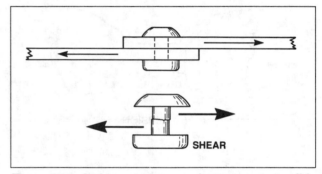

Figure 15-9. A shear stress on a rivet attempts to slide the shank apart.

Tension and compression are the two basic stresses and the other three, bending, torsion and shear, are really just different arrangements of tension and compression working on a body at the same time.

A **bending force** tries to pull one side of a body apart while at the same time squeezing the other side together. When a person stands on a diving board, the top of the board is under a tensile stress while the bottom feels compression. Wing spars of cantilever wings or the section of a wing spar outboard of the struts is subjected to bending stresses. In flight, the top of the spar is being compressed and the bottom is under tension, but on the ground, the top is pulled and the bottom is compressed. The wings struts are under tension in positive g flight but under compression in negative g flight or on the ground.

Torsion is a twisting force. When a structural member is twisted, a tensile stress acts diagonally across the member and a compressive stress acts at right angles to the tension. The crankshaft of an aircraft engine is under a torsional load when the engine spins the propeller. See figures 15-6(C) and 15-8.

Shear forces try to slide a body apart, and if we examine a rivet or bolt that has failed because of

shear forces, we see that the shank has actually been pulled apart, not *along* its length, but *across* its shank.

Shear forces exist in any material that is bent. This is easily illustrated by bending a 1" thick stack of paper. The sheets must slide over each other (shearing) in order to adjust to the changing circumference dimension of the bend. If the stack of papers were a solid, strong shearing forces would exist in the material. See figure 15-6(E).

Rivets hold pieces of aircraft skin together, and in a properly designed riveted joint, the rivets support *shear loads only*. The joint should never be required to support tensile loads. See figure 15-9.

Transfer Of Stresses Within A Structure

An aircraft structure must be designed in such a way that it will accept all of the stresses imposed upon it by the flight and ground loads without any

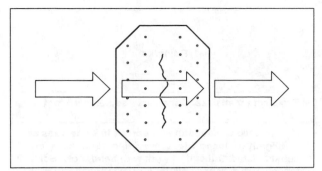

Figure 15-10. Any repair to an aircraft structure must accept all of the loads, support the load, and then transfer it back into the structure.

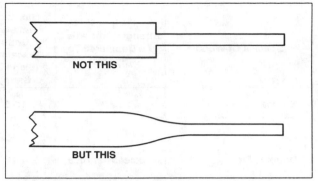

Figure 15-11. Abrupt changes in the cross-sectional area of a part must be avoided. An abrupt area change will concentrate the stresses and cause the part to fail.

permanent deformation. Any repair that is made must accept the stresses, carry them across the repair, and then transfer them back into the original structure. See figure 15-10.

Some **deformation** takes place whenever a load is placed on a structure. If the deformation disappears when the load is removed it is considered to be normal or **nonpermanent deformation**. Pilots of low-wing aircraft will see some wrinkling of the skin on the top of the wing when executing a steep bank (high g) turn. If those wrinkles go away when the aircraft returns to one g level flight, no permanent deformation took place. Wrinkles observed during preflight inspection in the top of the wing or the bottom of the horizontal stabilizer and/or stretch marks on the bottom of the wing or the top of the stabilizer that weren't put there during the manufacturing process are good evidence of **permanent deformation** due to positive g's, indicating that the aircraft has been overstressed and is no longer airworthy. Negative g stress indications would be found on the opposite surfaces.

We can think of these stresses as flowing through the structure, so there must be a complete path for them with no abrupt changes in cross-sectional area along the way. Abrupt changes in area will cause the stresses to concentrate, and it is at such a point that failures occur. A scratch or gouge in the surface of a highly stressed piece of metal will obstruct the flow of stresses and concentrate them so the metal will fail. See figure 15-11.

The thin metal of which most aircraft structure is made is subject to cracks, and when a crack starts in the edge of a sheet that is subjected to a tensile stress, the stresses will concentrate at the extremely fine end of the crack, and the metal is not strong enough to prevent the crack extending until the sheet tears.

For example, if a small crack starts in the edge of a piece of 0.032-inch sheet aluminum alloy which has a tensile strength of 64,000 pounds per square inch, calculations would show that it will take a stress of just over two pounds to extend the crack, and normal vibration will build up stresses far greater than this.

Materials For Aircraft Construction

Wood

Several forms of wood are commonly used in aircraft. Solid wood or the adjective "solid" used with such nouns as beam or spar refers to a member consisting of one piece of wood.

Laminated wood is an assembly of two or more layers of wood which have been glued together with the grain of all layers or laminations approximately parallel. Plywood is an assembled product of wood and glue that is usually made of an odd number of thin plies (veneers) with the grain of each layer at an angle of 45° to 90° with the adjacent ply or plies. High-density material includes compreg, impreg, or similar commercial products, heat stabilized wood, or any of the hardwood plywoods commonly used as bearing or reinforcement plates. The woods listed in figure 15-12 are those used for structural purposes. For interior trim, any of the decorative woods such as maple or walnut can be used since strength is of little consideration in this situation.

Aluminum Alloys

At one time aluminum was considered a novelty, priced somewhere between silver and gold, and it had no practical use as a structural material. But when an economical way of extracting it from its

Species Of Wood	Strength Properties As Compared To Spruce	Maximum Permissible Grain Deviation (Slope Of Grain)	Remarks
Spruce	100%	1:15	Excellent for all uses. Considered as standard for this table.
Douglas Fir	Exceeds spruce	1:15	May be used as substitute for spruce in same sizes or in slightly reduced sizes providing reductions are substantiated. Difficult to work with hand tools. Some tendency to split and splinter during fabrication. Large solid pieces should be avoided due to inspection difficulties. Gluing satisfactory.
Noble Fir	Slightly exceeds spruce except 8 percent deficient in shear	1:15	Satisfactory characteristics with respect to workability, warping, and splitting. May be used as direct substitute for spruce in same sizes providing shear does not become critical. Hardness somewhat less than spruce. Gluing satisfactory.
Western Hemlock	Slightly exceeds spruce	1:15	Less uniform in texture than spruce. May be used as direct substitute for spruce. Gluing satisfactory.
Pine, Northern White	Properties between 85 percent and 96 percent those of spruce	1:15	Excellent working qualities and uniform in properties but somewhat low in hardness and shock-resisting capacity. Cannot be used as substitute for spruce without increase in sizes to compensate for lesser strength. Gluing satisfactory.
White Cedar, Port Orford	Exceeds spruce	1:15	May be used as substitute for spruce in same sizes or in slightly reduced sizes providing reductions are substantiated. Easy to work with hand tools. Gluing difficult but satisfactory joints can be obtained if suitable precautions are taken.
Poplar, Yellow	Slightly less than spruce except in compression (crushing) and shear	1:15	Excellent working qualities. Should not be used as a direct substitute for spruce without carefully accounting for slightly reduced strength properties. Somewhat low in shock-resisting capacity. Gluing satisfactory.

Figure 15-12. Woods for aircraft use.

abundant ore was discovered to bring its cost down, and when it was alloyed with other elements to increase its strength without appreciably increasing its weight, aluminum took on new importance.

Aluminum alloys have one inherent drawback as a structural material; they are susceptible to corrosion—much more so than pure aluminum. But their good strength-to-weight ratio has caused metallurgists to devise ways of overcoming this problem, and today aluminum alloys are the most important metal used for aircraft construction.

The first major breakthrough in the use of aluminum as a structural material came when it was alloyed with copper, with the addition of a small amount of magnesium and manganese. This was done by the Germans who called the alloy Duralumin, which is the origin of the contraction we often use, "dural."

Rather than using names for all the various alloys, we find that it is more convenient to use a numbering system that describes the alloy. For aluminum alloys, a four-digit system is used in which the first digit denotes the primary alloying element and the other digits identify the specific alloy. See figures 15-13 and 15-14.

Commercially pure aluminum is identified as 1100 and is seldom used alone in aircraft construction; but alloy 3003, which is almost the same but has just a small amount of manganese

ALLOY NO. SERIES	CHIEF ALLOYING ELEMENT
1xxx	99% ALUMINUM, MINIMUM
2xxx	COPPER
3xxx	MANGANESE
4xxx	SILICON
5xxx	MAGNESIUM
6xxx	MAGNESIUM AND SILICON
7xxx	ZINC
8xxx	OTHER ELEMENTS

Figure 15-13. Alloys of aluminum.

| PERCENT OF ALLOYING ELEMENTS | | | | | | | |
| ALUMINUM AND NORMAL IMPURITIES CONSTITUTE REMAINDER | | | | | | | |
ALLOY	COPPER	SILICON	MANGANESE	MAGNESIUM	ZINC	CHROMIUM	
1100							Seldom used in aircraft
3003			1.2				Cowling and nonstructural parts
2017	4.0		0.5	0.5			Obsolete—superseded by 2024
2117	2.5			0.3			Most aircraft rivets
2024	4.5		0.6	1.5			Majority of structure
5052				2.5		0.25	Gas tanks and fuel lines
7075	1.6			2.5	5.6	0.3	High strength requirements

Figure 15-14. Nominal composition of wrought aluminum alloys.

in it, is used for cowling and for other non-load-bearing application. Alloy 2017 was at one time a popular alloy, and in some of the older maintenance manuals, you will still find it mentioned. It contains copper, manganese and magnesium and has been superseded by alloy 2024, which has the same alloying elements but in slightly different proportions. Alloy 2024 is used for most of the modern all-metal aircraft. Zinc increases the strength of aluminum, and alloy 7075 is used for many of our high-strength requirements.

Welded aluminum parts such as fuel and oil tanks and fluid lines that carry low pressures are usually made of aluminum alloyed with magnesium and chromium, known as alloy 5052. It cannot be hardened by heat-treating, but it may be softened by annealing it with heat.

Corrosion Prevention

Susceptibility of aluminum alloys to corrosion is one of their limiting factors as structural material. But this problem has been minimized by three methods of protection: cladding the alloy with pure aluminum (**"alclad"**), covering the surface with an impenetrable oxide film, and covering the surface with coating such as primer and paint.

Before we look at each of these methods, let's review corrosion to see what causes it and what can be done to prevent it.

Corrosion is an electrochemical action in which an element in the metal is changed into a porous salt of the metal. This salt is the white or gray powder seen on a piece of corroded aluminum, or the dark "shadow" indicating that aluminum has been worn off, forming a powder of corroded aluminum.

There are three requirements that must be met for corrosion to form on aluminum alloy. There must be an area of electrode potential difference within the metal; there must be a conductive path within the metal between these areas; and some

form of electrolyte must cover the surface between these areas to complete the electrical circuit.

Let's look at an example of dissimilar-metal corrosion to see the way it works. When a steel bolt holds two pieces of aluminum alloy together, there is the set-up for **galvanic, or dissimilar-metal, corrosion** to form. Aluminum is more anodic, or more active, than steel and will furnish electrons for the electrical action that occurs when the surface is covered with an electrolyte such as water. When electrons flow from the aluminum to the steel, they leave positive aluminum ions that attract negative hydroxide ions from the water. This results in the formation of aluminum hydroxide, or corrosion, and the aluminum metal is eaten away.

There are several things that can be done to prevent corrosion. Almost all steel aircraft hardware is plated with a thin coating of cadmium which has an electrical potential almost the same as aluminum. As long as the cadmium is not scratched through, there will be no contact between the steel and the aluminum and almost no electrode potential difference. To further protect the aluminum from damage, the bolt can be dipped in primer before it is installed. This will exclude water and air from the joint, and corrosion cannot form.

Pure aluminum will not corrode since there is no electrode potential difference within the metal, but it is too weak for use as an aircraft structural material. However, if a thin coating of pure aluminum is rolled onto the surface of the strong aluminum alloy, we have the good features of both the pure metal and the alloy, a strong and corrosion-resistant material. For this process, called **cladding**, we pay the penalty of a loss of only about five percent of the strength of the alloy.

While pure aluminum does not corrode, it does oxidize; that is, it readily unites with oxygen in the air to form a dull-looking film on its surface. This film is extremely tight and prevents any more oxygen reaching the metal, so the oxidizing action stops as soon as the film is completely formed on the surface. Clad aluminum may be used as the outside skin for airplanes, and it gives them a nice silvery appearance without being painted. Care must be taken, however, to prevent any scratches through the thin cladding, as the alloy would then be exposed and would corrode. It is normal for corrosion to form along the edges of the sheets where the alloy is exposed.

An **oxide film** can protect aluminum alloy in the same way it protects pure aluminum, by excluding air and moisture from the metal. And since an aluminum alloy cannot corrode unless an electrolyte is in contact with the metal, the oxide film insulates the surface from the electrolyte, and corrosion cannot form. This protecting oxide film may be formed either electrolytically or chemically, and both methods produce a film that not only excludes air from the surface, but roughens it enough for paint to bond tightly.

When all-metal airplanes first became popular, they were seldom painted. Their skin was usually of clad aluminum alloy which had a shiny silver appearance. But, in order to keep them shiny, the oxide film had to be continually rubbed off, and since this type of surface requires so much care, the modern trend is to paint all of the aircraft.

The majority of the modern high-volume production aircraft are primed with a two-part wash primer that etches the surface of the metal so paint will adhere. Then, when the primer is completely cured, the entire airplane is sprayed with acrylic lacquer.

When the cost of the finish allows it and when there is sufficient time in the production schedule, the surface may be primed with epoxy primer and the aircraft finished with polyurethane enamel. Polyurethane enamel is far more durable than acrylic lacquer and gives a much more attractive finish. It is interesting to note that there is 500 pounds of paint on the typical Boeing 747, in the stripe and tail paint.

Honeycomb

Aircraft structure must not only be strong but rigid as well, and very thin aluminum skins, while providing adequate strength, often lack the rigidity needed. A structural material that provides both the strength and rigidity with light weight is the honeycomb structure. In honeycomb material, a core of metal, paper or fiberglass in a cellular structure has face sheets of fiberglass or aluminum alloy bonded to either side.

Complex aerodynamic shapes may be constructed of aluminum alloy honeycomb core, faced with thin sheet aluminum alloy. Many control surfaces are made by this method. Flat sheets are also used for floorboards and for compartment bulkheads.

Some of the popular Grumman-American light aircraft use honeycomb material for fuselage panels and for many of the smaller structural components. Helicopter rotors are often made of aluminum alloy sheets bonded to a honeycomb core to produce a rotor that has both the required strength and rigidity.

The greatest amount of research that has been done in bonded structures has been in conjunction with our space program and our development of high-speed military aircraft. Some aircraft that fly at speeds far greater than the speed of sound have skins made of stainless steel, furnace-brazed to a stainless steel honeycomb core. Titanium is another material that is used for some of these skins because it will keep its strength at the high temperatures caused by the friction of the air passing over them.

Magnesium

Magnesium weighs only about 65% as much as aluminum and finds a great many applications in aircraft structure. In its cast form it is used as housings for many of the components in engine and airframe systems, and for wheels. As sheet metal it is often used for control surfaces where light weight is so important.

Magnesium has two major drawbacks as a structural material. It is quite brittle and is therefore subject to cracking when it is exposed to vibration, and it is highly susceptible to corrosion. If the structure is properly designed and built and has been properly protected against corrosion, magnesium makes a highly useful structural material.

Stainless Steel

Stainless steel is used for many components in modern aircraft, but in this text we are concerned only with its application to the structure itself.

There have been at least two commercially built airplanes whose structures were made entirely of stainless steel. One of these was a seaplane. But its weight and the complexities of construction make stainless steel a poor choice as a structural material. This is not true, however, for some of the extremely high-speed military aircraft which use stainless steel honeycomb panels for their outer skins. A cellular structure of thin stainless steel is used as the core, and thin face sheets of stainless steel are furnace-brazed to each side of the core. These panels maintain their strength at the temperatures reached by these skins at the high Mach number airspeed they fly.

Structures

Wing Construction

Truss-type Wing Construction

Fabric-covered airplane wings have a truss-type structure that has changed very little throughout the evolution of the airplane.

The main lengthwise members in a wing truss are the **spars**. In the past, these were all made of wood, but the more modern construction uses spars of extruded aluminum alloy.

Wood spars are usually made of Sitka spruce and may be either solid or laminated. Because of the difficulty in getting a single piece of near-perfect wood of the size needed for wing spars, many manufacturers use laminated spars, in which strips of wood are glued together with their grain all running in the same direction. A properly laminated spar has essentially the same strength as that of a solid spar, yet it is considerably less expensive.

The spars are separated by compression members, or **compression struts**, that may be either steel tubing or heavy-wall aluminum alloy tubing. **Compression ribs** are sometimes used, ribs which have been especially strengthened to take compressive loads.

The **truss** is held together with high-strength solid steel wires that cross the bays formed by the compression struts. The wires that extend from the front spar inboard to the rear spar outboard oppose the forces that tend to drag against the wing and pull it backward; these are called **drag wires**. The wires that attach to the rear spar inboard and go to the front spar outboard are called **anti-drag wires**, since they oppose any force that tends to move the tip of the wing forward. A **wing truss** consisting of spars, compression members, and drag and anti-drag wires, when properly assembled and rigged, provides the lightweight and strong foundation needed for a wing. See figures 15-15 through 15-17.

An adaptation of the truss-type wing is one using a box spar. This was pioneered in World War I on some of the all-metal Junkers airplanes and the wood and fabric Fokker triplane. A box structure built between the spars stiffens the spars so they can carry all of the bending and torsional loads to which the wing is subjected in flight. See figure 15-18.

The **former ribs** in the wing attach to the spars to give the wing the aerodynamic form or shape it

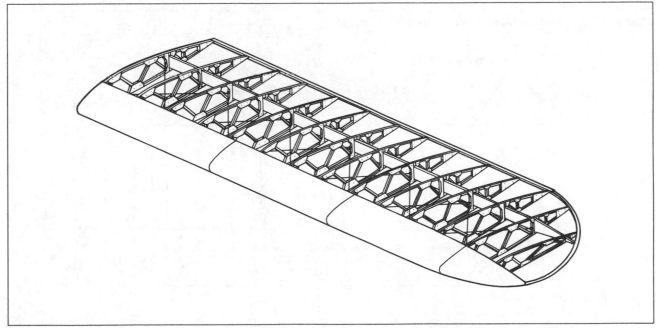

Figure 15-15. The fabric-covered truss-type wing has changed very little in its construction over the past three decades.

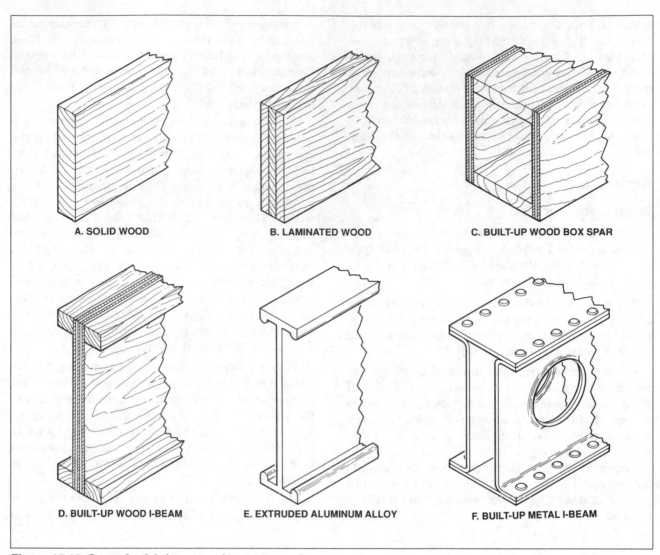

A. SOLID WOOD   B. LAMINATED WOOD   C. BUILT-UP WOOD BOX SPAR

D. BUILT-UP WOOD I-BEAM   E. EXTRUDED ALUMINUM ALLOY   F. BUILT-UP METAL I-BEAM

Figure 15-16. Spars for fabric-covered truss-type wings.

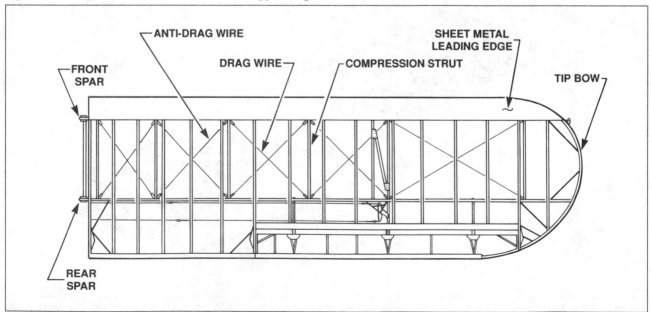

Figure 15-17. Truss-type wing.

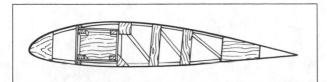

Figure 15-18. A built-up box spar accepts torsional as well as bending loads.

needs to produce lift when air flows over its fabric covering.

Before the cost of labor became so high, some wing ribs were built up of strips of Sitka spruce. The strips that form the top and bottom of the rib are called cap strips, and those between the cap strips are called cross members. Since end grain glue joints have very little strength, each intersection of a cap strip and a cross member has a gusset of thin mahogany plywood glued to the strips of wood to carry the stresses from one strip to the other. See figure 15-19.

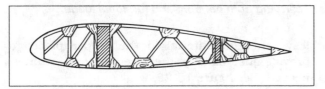

Figure 15-19. A built-up wing rib made of wood.

Metal wing ribs may be either built up by riveting together cap strips and cross members made of formed, thin sheets of aluminum alloy, or may be pressed from aluminum alloy sheets in a hydropress.

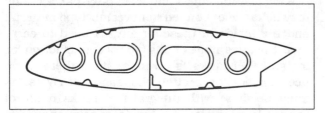

Figure 15-20. Wing rib made of pressed sheet metal.

The most critical part of a wing, as far as the production of lift is concerned, is the front end, or the leading edge. To prevent air loads distorting the leading edge, most wings have **nose ribs**, or **false ribs**, that extend from the front spar forward and are placed between each of the full-length former ribs. A sheet of thin aluminum alloy is wrapped around the leading edge so the fabric will conform to the shape of the ribs.

The trailing edge is normally formed of aluminum alloy and ties the back end of the ribs together to give the wing its shape. See figure 15-21.

Cloth reinforcing tape is laced diagonally between the ribs from the top of one rib to the bottom of the adjacent rib near the point of their greatest thickness to hold the ribs upright until the fabric is stitched to them.

The fabric covering is placed over the wing and is laced to each of the ribs with strong rib-lacing cord to hold the fabric to the shape of the rib and to transmit the air loads from the surface to the wing spars.

Stressed-skin Wing Construction

In the same manner as the fuselage, wings have generally evolved from the truss form of construction to one in which the outer skin carries the greatest amount of the stresses. Semi-monocoque construction is generally used for the main portion of the wing, while the simple monocoque form is often used for the control surfaces.

Wing ribs may be pressed from sheet aluminum alloy in a hydropress, or they may be built up of sheet metal channels and hat sections riveted to the skin to give it both the shape and rigidity it needs. One of the advantages of an all-metal wing is the ease with which it can be built to carry all of the flight loads within the structure so it does not need any external struts or braces. Such an internally braced wing is called a **cantilever wing**.

The Douglas DC-2 was one of the first highly successful airplanes to use the configuration that has become standard for transport aircraft; cantilever low wing, with retractable landing gear. The airfoil section of a cantilever wing is normally quite thick, and the wing has a strong center section built into the fuselage. The engines and landing gear attach to this center section. Rather than using the familiar two-spar construction, most of these wings are of the multi-spar construction in which several spars carry the flight loads, and spanwise stiffeners run between the spars to provide even greater strength.

As airspeeds increased with their higher flight loads, it became apparent that not only was more strength needed for the skins of all-metal wings, but more stiffness was also needed. And to gain the strength and stiffness needed and yet keep the weight down, the manufacturers of some of the high-speed military aircraft begin the construction of wing skins with thick slabs of aluminum alloy. Then they machine away some of the thickness but

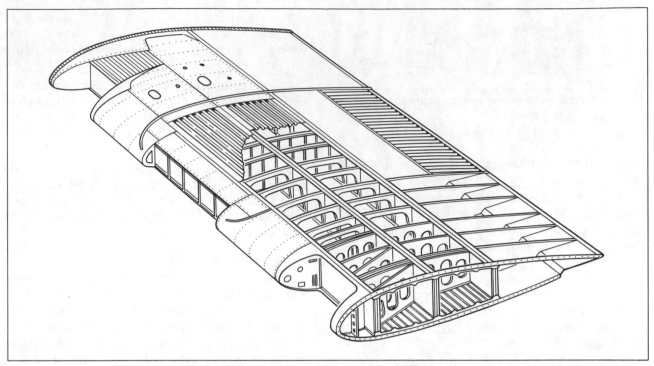

Figure 15-21. A multi-spar metal wing structure is capable of carrying both bending and torsional loads.

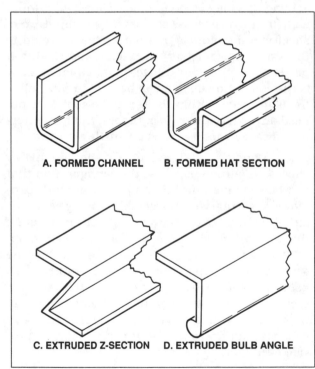

A. FORMED CHANNEL   B. FORMED HAT SECTION

C. EXTRUDED Z-SECTION   D. EXTRUDED BULB ANGLE

Figure 15-22. Structural forms used in all-metal aircraft structure.

leave enough material in the proper places to provide just exactly the strength and stiffness needed. See figure 15-23.

To gain the maximum amount of stiffness for the weight, some aircraft have wing skins made of laminated structure in which thin sheets of metal are bonded to a core of fiberglass, paper or metal honeycomb material. And some airplanes that travel at supersonic speeds have outer skins made of stainless steel brazed to cores of stainless steel honeycomb.

Wing leading edges and even box spar sections may be made of bonded honeycomb-type material and the inside of these structures used to carry fuel. The advantage of this type of construction for integral fuel tanks is obvious, since there is no need for sealing around thousands of rivets, as must be done with integral fuel tanks made of conventional riveted sheet metal construction. See figure 15-24.

Some of the extremely light wing structures, such as are used for high-performance sailplanes and for some home-built airplanes, are built using

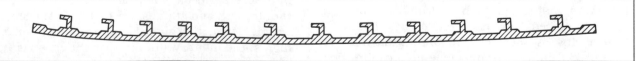

Figure 15-23. Milled wing skins give maximum strength and rigidity with minimum weight.

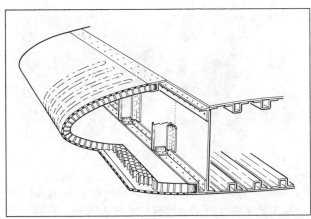

Figure 15-24. A leading edge made of laminated honeycomb material can be used as an integral fuel tank.

an inner structure of polystyrene foam covered with sheets of fiberglass cloth bonded to the foam and filled with either an epoxy or polyester resin.

Wing Alignment

Cantilever wings have very little adjustment potential, as this is all taken care of when the airplane is built; but some airplanes do have either a cam arrangement or a serrated washer at the rear spar attachment bolt, and a few degrees of **wash-in** may be set in the wing (increased angle of incidence) to correct for a wing-heavy flight condition.

Strut-braced wings using V-struts normally have provisions for adjusting both the dihedral angle and the incidence angle of the wings. The wing is installed and the fuselage checked to be sure that it is level both longitudinally and laterally, then the fittings in the end of the front struts are adjusted to get the correct dihedral. This is determined by using a dihedral board that has a specific taper. It is held against the main spar on the bottom of the wing at the location specified by the manufacturer, and the fitting in the end of the strut is screwed either in or out until the bottom of the dihedral board is level. On some aircraft, rather than measuring the dihedral with a dihedral board, a string is stretched between the wing tips at the front spar. When the dihedral is correctly adjusted, there will be a specific distance between the wing root fitting and the string.

When the dihedral is correctly adjusted, the **wash-in** or **wash-out** may be set. This is normally done by adjusting the length of the rear strut. An incidence board similar to a dihedral board is held under a specified wing rib, and the strut length is adjusted until the bottom of the board is level. On airplanes having this adjustment, the initial set-

ting will likely have to be changed after the first flight to trim the airplane for straight and level hands-off flight. Increasing the angle of incidence, that angle between the chord line of the wing and the longitudinal axis of the airplane, is called "washing the wing in" and it increases the lift. **Washing out** a wing is done by rigging it with a lower angle of incidence to decrease its lift.

Control Surface Construction

Fabric-covered Control Surfaces

Most of the simpler truss-type fabric-covered airplanes have all of their tail surfaces made of welded thin-wall steel tubing. The vertical fin of this type airplane is built as an integral part of the fuselage, and the rudder attaches to the fin with hinge pins through steel tubes welded to both the fin and the rudder. See figure 15-25(A).

The horizontal stabilizer bolts to the fuselage and is held rigid with high-strength steel wires. The elevators hinge to its trailing edge in the same way the rudder hinges to the vertical fin.

The tail surfaces on almost all modern airplanes are of the cantilever type and are bolted to fittings in the fuselage. Special care must be exercised by the pilot during the preflight inspection to see that there is no excessive play or movement up and down as well as fore and aft of the stabilizers which would indicate wear of failure of the stabilizer attach mechanisms.

The ailerons of fabric-covered airplanes are built up in much the same way as the wings. The aileron ribs conform to the shape of the rear end of the wing former ribs, and the aileron trailing edge is made of the same material as the trailing edge of the wing. The aileron leading edge is normally covered with thin sheet aluminum alloy so it will retain its shape under all flight loads. The hinge line of the aileron is usually well back behind its leading edge, so some of the aileron nose will protrude below the wing when the aileron is fully up. This will decrease **adverse yaw**.

Construction Of Control Surfaces For All-metal Airplanes

Control surface flutter is one of the more serious problems high-speed airplanes have had in their design evolution. To eliminate flutter, it is extremely important that the control surfaces be **mass balanced** so that their center of gravity does not fall behind their hinge line (figure 15-25(B)). For this reason, many surfaces have extensions ahead of the hinge line in which lead weights are installed. A number of the higher speed airplanes

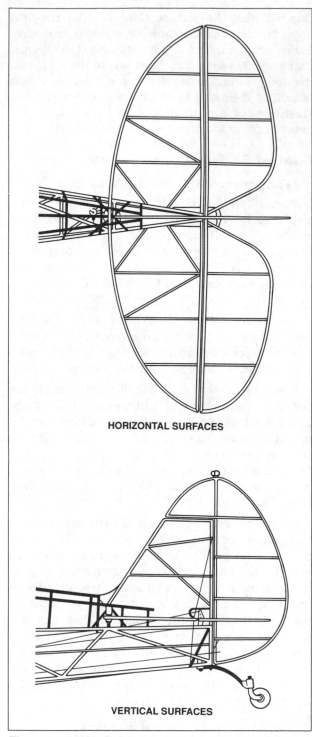

HORIZONTAL SURFACES

VERTICAL SURFACES

Figure 15-25(A). Tail surfaces of welded steel tubing are covered with cloth fabric.

of World War II vintage were of all-metal construction except for the control surfaces, and to keep their weight to a minimum, these surfaces were covered with cotton or linen fabric. Today, almost all of the new all-metal airplanes have their control surfaces covered with either thin aluminum alloy or magnesium alloy sheets.

Figure 15-25(B). The overhang on this rudder houses a lead weight to move the center of gravity of the surface ahead of its hinge line to prevent flutter.

Figure 15-26. This rudder is corrugated to give the surface stiffness, while requiring an absolute minimum of internal structure.

Many of the lighter aircraft gain rigidity in their control surfaces by corrugating the skin. The stiffness provided by the corrugation minimizes the amount of substructure needed (figure 15-26).

Fuselage Construction

The fuselage is the body of the aircraft, to which the wings, tail, engine and landing gear attach. Because of the tremendous loads that are imposed upon the fuselage structure, it must have maximum strength and, as with all of the parts of an aircraft, it must also have minimum weight.

There are two types of construction used in modern aircraft fuselages: the truss and the stressed-skin type.

Truss Fuselage Construction

By definition, a **truss** is a form of construction in which a number of members are joined to form a rigid structure. Many early aircraft used the Pratt truss, in which wooded longerons served as the main lengthwise structural members and were held the proper distance apart by wood struts. Each bay, or space between the struts, was

Figure 15-27. A Pratt-truss fuselage.

crossed by two piano wire stays whose tension was adjusted by brass turnbuckles. The basic characteristic of a Pratt truss is that its struts carry only compressive loads, while the strays carry only the tensile loads. See figure 15-27.

When technology progressed to the extent that fuselages could be built of welded steel tubing, the Warren truss became popular. In this type of truss, the longerons are separated by diagonal members that can carry both compressive and tensile loads. See figure 15-28.

The smooth aerodynamic shape required by an airplane fuselage is provided for those using both Pratt and Warren trusses by the addition of a non-load-carrying superstructure, and the entire fuselage is covered with cloth fabric.

Stressed-skin Structure

The necessity of having to build a non-load-carrying superstructure over the structural truss led designers to develop the stressed-skin form of construction, in which all of the loads are carried in the outside skin. This skin does not require the angular shape that is necessary for a truss, but

Figure 15-28. A Warren-truss fuselage.

can be built with a very clean, smooth and aerodynamically efficient shape.

One of the best examples of a natural stressed-skin structure is the common hen egg. The fragile shell of an egg can support an almost unbelievable load, when it is applied in the proper direction, as long as the shell is not cracked.

The main limitation of a stressed-skin structure is that it cannot tolerate any dents or deformation in its surface. We have all seen this characteristic demonstrated with a thin aluminum beverage can. When the can is free of dents, it will withstand a great amount of force applied to its ends, but if we put only a slight dent in its side, it can be crushed very easily from top or bottom.

That portion of the fuselage behind the cabin of some of the smaller training airplanes is built with **monocoque-type construction**. The upper and lower skins are made of thin sheet aluminum alloy that have been formed into compound curved shapes with a drop hammer or a hydropress. The edges of both of these skins are bent to form a lip which gives the skin rigidity. These skins are riveted to former rings that have been pressed from thin sheet aluminum in a hydropress. The sides of the fuselage between the top and bottom skins are made of flat sheet aluminum, riveted to the skins and to the former rings. See figure 15-29.

This type of construction is economical and has sufficient strength for these relatively low-stress areas. It is extremely important that all repairs to monocoque structure restore the original shape, rigidity and strength to any area that has been damaged. Pilots should be alert for dent-damage during preflights as this weakens the compressive load capabilities of curved surfaces.

Most aircraft structure requires more strength than that provided by pure monocoque construction, and to provide this strength, a substructure

Figure 15-29. A monocoque structure carries all of the stresses in its skin.

of formers and stringers is built and the skin is riveted to it, creating a semi-monocoque structure. The former rings and bulkheads, which are formers that also serve as compartment walls, are made of relatively thin sheet metal that have been formed in hydropresses, and the stringers are made of extruded aluminum alloy. The stringers usually have a bulb on one of their sides to provide added strength needed to oppose bending loads. The longerons are also made of extruded aluminum alloy, but are heavier than the stringers and carry a good amount of the structural loads in the fuselage. See figure 15-30.

Pressurized Structure

High-altitude flight places the occupants in a hostile environment in which life cannot be sustained unless supplemental oxygen is supplied. Since wearing oxygen masks is both uncomfortable and inefficient, in order for flying to appeal to the mass of air travelers, other provision had to be made. Increasing the air pressure in the cabin of the airplane provides sufficient oxygen in the air for the passengers to breathe normally without the need of supplemental oxygen.

The first airliners to be pressurized were powered by piston engines and were unable to cruise at the extremely high altitudes that are common for our jet transports; their cabins were pressurized to a pressure differential of only about two psi. This low pressurization created no big problems, but when the first jet transports, the British Comets, were put in service with their pressurization of 8¼ psid, real problems did arise. The continued flexing of the structure caused by the pressurization and depressurization cycles fatigued the metal to such an extent that a crack developed at a square corner of a cutout in the structure, and the large amount of pressure differential caused the structure to virtually explode. When the cause of the structural failure was determined, new emphasis was placed on fail-safe design of aircraft structures.

Stress risers, or portions of the structure where the cross section changes abruptly, have been eliminated. Joints and connections are carefully pre-stressed to minimize the cyclic stresses from the flight loads, and most important, the structure is designed with more than one load path for the stresses. If a crack does develop and weaken the structure in one place, there is another path through which the stresses can be supported, and no serious failure will occur.

Figure 15-30. A semi-monocoque fuselage has a substructure to stiffen the external skln.

Figure 15-31. *The conventional elevators and rudder rotate this airplane about its lateral and vertical axes.*

Figure 15-32. *The all-moving horizontal tail surface (stabilator) rotates this airplane about its lateral axis.*

Flight Controls

The flight controls of an airplane do no more than modify the aerodynamic shape of the surface to which they are attached. This change in shape changes the lift and drag produced by the surface, with the immediate result of rotating the airplane about one of its three axes. It is this rotation that produces the change in flight path that gives the desired control.

Pitch Control

The movable horizontal tail surface changes the tail load and causes the airplane to pitch about its lateral axis. Some airplanes use a conventional fixed horizontal stabilizer with a movable elevator hinged to its trailing edge, figure 15-31.

Some airplanes have the entire horizontal surface pivoted about its main spar, utilizing a full-span tab to stabilize it, figure 15-32.

The **stabilator**, or all-movable tail, is one form of horizontal tail surface that is finding a good deal of popularity. This type of horizontal tail surface has no fixed stabilizer, but rather there is an almost full-length anti-servo tab on its trailing edge. This tab cannot only be adjusted from the cockpit to change the longitudinal trim for hands-off flight at various airspeeds, but the adjustment jack screw is attached to the fixed structure in such a way that when the leading edge of the stabilator moves down to increase the downward tail load, the anti-servo tab moves up. This tab action tends to restore the stabilator to a normal streamlined position. In the same way, when the nose of the stabilator moves up, the tab moves down and the resultant air load tends to streamline the stabilator.

A stabilator requires a rather heavy weight on a long arm inside the fuselage to give it static balance.

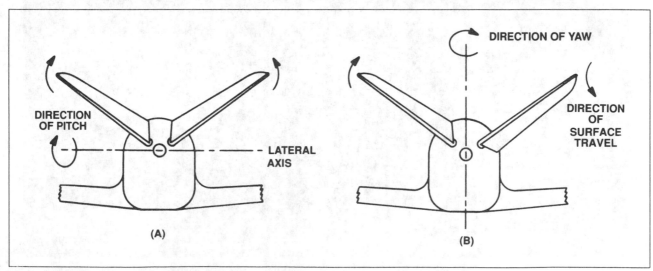

Figure 15-33. *Ruddervator action. (A)—Functioning as an elevator; (B)—Functioning as a rudder.*

Figure 15-34. The two surfaces of this V-tail serve the same functions as the three surfaces of the conventional tail.

Another very popular configuration for providing longitudinal control is the **ruddervators** used on the V-tail Beech Bonanza. The tail of this airplane has two fixed and two movable surfaces that provide both longitudinal and directional stabilization and control. The movable surfaces move together to provide pitch control and move differentially for yaw control. See figures 15-33 and 15-34. Since one less tail is being dragged through the air, the V-tail is very efficient. A control box of rather complex geometry is located in the tail cone to properly mix the elevator and rudder inputs from the pilot. Control use is completely conventional.

Even though the movable horizontal surfaces are usually called **elevators**, they in no way elevate an airplane. Their sole function is to change the angle of attack of the airplane, which alters its speed, lift and drag.

Consider a condition of flight where the aircraft is trimmed for level flight at 100 KTS.

Now, without changing anything else, the pilot can pull back on the control column, and the trailing edge of the elevator will move up. This change in shape of the horizontal tail increases the tail-down force and rotates the airplane nose-up about its CG and lateral axis. Disregarding the *momentary* rise when the wheel is first pulled back, we will see that the airplane will fly along with its nose high, but at an airspeed lower than it was before the pilot pulled back the wheel. If the pilot adds power from the engine, the aircraft will climb at this speed, and if the pilot decreases throttle a little bit, the aircraft will descend at this speed.

Therefore, pilots will find aircraft operation much simpler by remembering that the **elevator controls airspeed** and the **throttle controls ver-**

tical speed while the aircraft is airborne. If you run across a pilot who doesn't believe this, bet him whatever you think "the traffic will bear," go flying with him, letting him control the throttle while you control the elevator, and see who controls the airspeed!

Lateral Or Roll Control

The basic controls for rolling an airplane are its **ailerons**. These are hinged control surfaces on the trailing edge of the wing, usually as near the tip as they can be mounted. They are connected to the control wheel or stick by a cable and pulley system or by **pushrods** and **bellcranks** so that rotating the wheel to the left or moving the stick to the left will lower the aileron on the right wing and raise the one on the left wing. See figure 15-35. These deflections will change the effective shape of the airfoils and increase the lift on the right wing, while decreasing the lift of the left wing. This rolls, or banks, the airplane.

The rolling action produced by the ailerons is the primary method of lateral control on most aircraft, but directional and lateral control are so closely interrelated that we cannot consider this to be the only type of lateral control, and we must look at both of these types of control together.

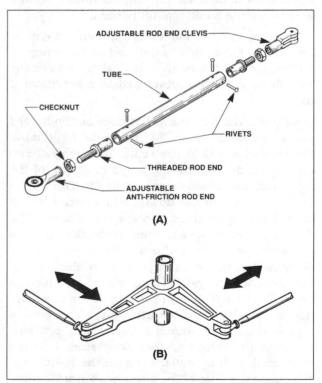

Figure 15-35(A)—Push-pull control rod. (B)—Control system bellcrank with control rods attached.

Directional Control

It is easy to think of the **rudder**, the movable portion of the vertical tail surface of an airplane, as an airplane's directional control, but this is not true, because the rudder does not turn an airplane. The rudder rotates the airplane about its vertical axis, an action called **yawing**. The rudder also provides a form of roll control because application of rudder causes yaw which will induce a roll.

Normally, an airplane is turned by banking it with the ailerons. Since lift always acts perpendicular to the lateral axis, the horizontal component of the lift vector will move the airplane in the direction it is banked. The wind pushing on the tail surface (application of rudder) will align the nose of the airplane with its relative wind, and the airplane will turn in a smooth circular flight path, with the inclinometer ball centered. See figure 15-36(A).

The vertical component of the lift vector must be equal to the weight of the airplane, and while the airplane is turning, the centrifugal force adds an apparent weight, so the total lift must be equal to both the weight of the airplane and the centrifugal force. In order to prevent the nose of the airplane dropping in a turn, the pilot must pull the control wheel back to increase the angle of attack enough to produce the additional lift needed.

To appreciate the amount of apparent weight the centrifugal force adds to the load on the wings in a turn, look at figure 15-36(B), where we see the load factors that are developed in a coordinated turn.

Actually, the ailerons and rudder are both used in turning the airplane. To turn to the right, move the control wheel to the right. The right aileron moves up, decreasing the angle of attack of the right wing tip, which decreases its lift. The aileron on the left wing moves down and increases the lift of the left wing, and the airplane starts to roll. The deflected ailerons which cause the bank also cause the airplane to yaw (*adverse aileron yaw*—to be discussed in detail shortly). The problem is that the direction of yaw is *opposite* that which is desired. The left aileron moving down increases not only the lift of the left wing, but it also creates a good deal of additional induced drag centered out near the wing tip. This drag, since it is not countered with a similar drag on the right wing, will cause the nose of the airplane to start to move to the *left*. A smooth turn entry can be made only by the coordinated use of ailerons and rudder. When the pilot moves the wheel to the right, just

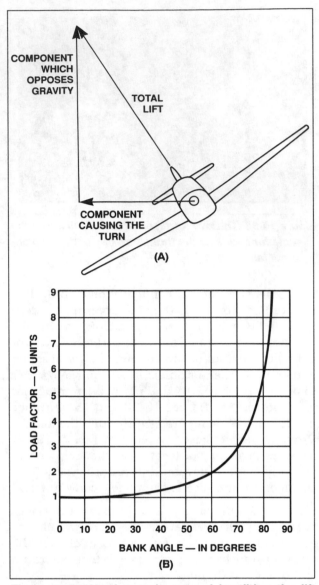

Figure 15-36(A)—A turn is caused by tilting the lift produced by the wing. The vertical tail surfaces keep the airplane turned into its relative wind. (B)—Load factor curve.

enough right rudder pressure is applied to prevent the nose moving to the left. Once the bank is established and ailerons neutralized, the rudder is no longer needed and it can be streamlined.

Turning an airplane requires rotation about both the longitudinal and vertical axes, and so the pilot uses both the rudder and ailerons. In most modern airplanes there is some form of mechanical interconnection between these two systems, usually not a positive one, but one that can be overridden if it is necessary to slip the airplane.

Figure 15-37(A) shows the basic aileron control system for a high-wing airplane. Rotation of the

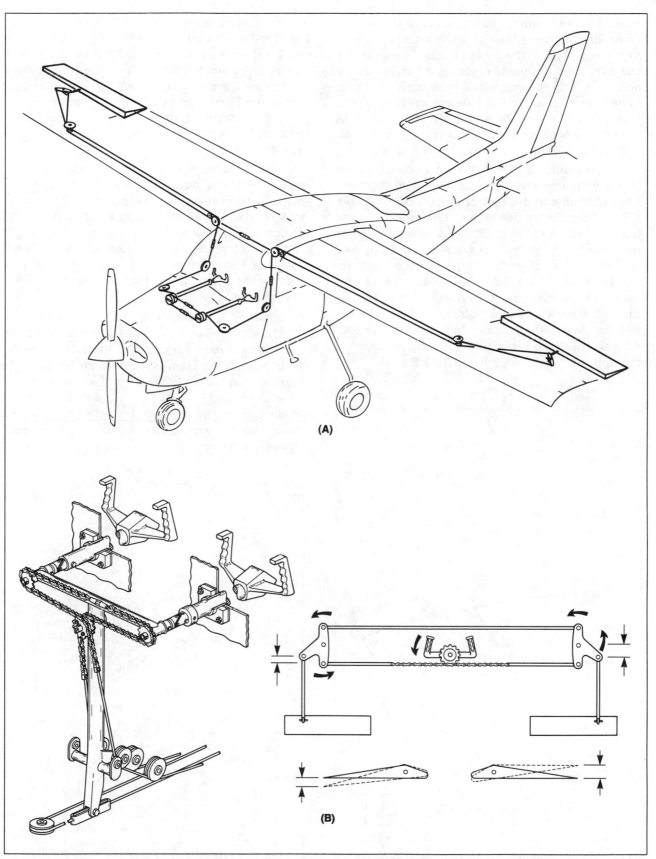

Figure 15-37(A)—A typical aileron system. (B)—Turning the control wheel moves the ailerons differentially to rotate the airplane about its longitudinal axis.

control wheel turns the drum to which the aileron control cables are attached. If the wheel is rotated to the right, the right cable is pulled and the left one is relaxed. The cable rotates the right aileron bellcrank, and the push-pull tube connected to it raises the right aileron. A **balance cable** connects both aileron bellcranks, and as the right aileron is raised, the balance cable pulls the left bellcrank and its push-pull tube lowers the left aileron.

Adverse aileron yaw is the big problem caused by the displacement of the ailerons, and there are two expedients in use to decrease this undesirable effect. The aileron that moves downward creates both lift and induced drag, and this drag way out near the wing tip pulls the nose of the airplane around in the direction opposite the way the airplane should turn. The geometry of the bellcranks is made such that the aileron moving upward travels a greater distance than the one moving down, thus producing enough parasite drag to counteract much of the induced drag on the opposite wing. See figure 15-37(B).

The **Frise aileron** is the type most commonly used today. It minimizes adverse yaw because of the location of its hinge point. These ailerons have their hinge point back a way from the leading edge, and when the aileron is raised, its nose sticks out below the lower surface of the wing and produces enough parasite drag to counter much of the induced drag from the down aileron. See figure 15-38.

Since adverse yaw is produced each time the control wheel deflects the ailerons, many manufacturers connect the control wheel to the rudder control system through an interconnecting spring. When the wheel is moved to produce a right roll, the interconnect cable and spring pulls forward on the right rudder pedal just enough to prevent the nose of the airplane starting to the left. See figure 15-39.

Airplanes whose rudder pedals are connected rigidly to the nosewheel for steering have the interconnect cables attached to the rudder cables with connector clamps in the aft end of the fuselage. The effect is the same for connection at either location. A small amount of rudder force is applied when the ailerons are deflected, but this force can be overridden because it is applied through a spring.

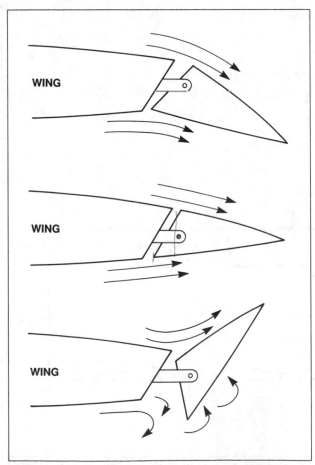

Figure 15-38. The Frise aileron creates parasite drag in the up position.

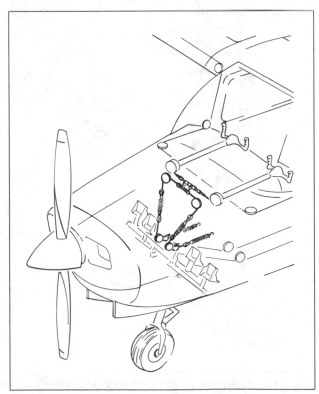

Figure 15-39. Rudder-aileron interconnecting springs.

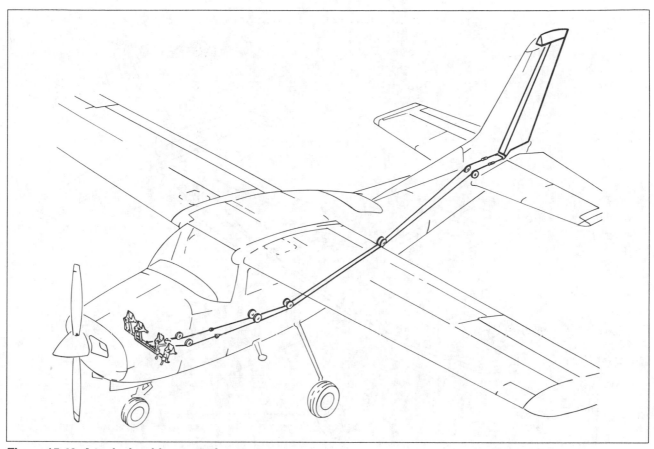

Figure 15-40. A typical rudder control system.

The rudder pedals are connected to the rudder horn with steel control cables and, on an airplane with a nosewheel, also to the nosewheel steering mechanism. Forward movement of the right rudder pedal will deflect the rudder to the right. See figure 15-40.

Pitch Control

The typical pitch control system consists of a fixed horizontal stabilizer on the rear end of the fuselage. Hinged to its trailing edge are movable elevators. On the trailing edge of the elevator is a trim tab to adjust the down load of the tail for hands-off flying at any desired airspeed.

The elevator is connected to the control yoke or stick in the cockpit typically with steel control cable and moves up or down as the wheel is moved in or out. Figure 15-41 shows a typical elevator system. Pulling back on the wheel pulls the elevator-up cable and rotates the top of the elevator bellcrank forward. The control horn on the bottom of the elevator torque tube is attached to the bellcrank with a push-pull rod, and as the bottom of the bellcrank moves back, it pushes the elevator up. Pushing in on the control wheel has

the opposite results. The elevator-down cable is pulled and the bottom of the bellcrank moves forward, causing the push-pull rod to pull the elevator down.

Many modern airplanes that have several rows of seats or otherwise have a possibility of inadvertently being loaded with their center of gravity far back, have provisions for automatically lowering the nose to prevent an approach-to-landing stall caused by the center of gravity being far back. This provision is the **elevator down spring**. If an airplane with the center of gravity far back is slowed down for landing, the trim tab will be unable to hold the nose down, causing the nose to pitch up, slowing the aircraft even more. If the airplane in this unstable condition encounters turbulence and slows down further, the elevator will streamline, and at this slow speed the trim tab cannot force it back down. The nose of the airplane will pitch up, aggravating the situation and possibly causing a stall at this critical altitude.

The elevator down spring holds a mechanical load on the elevator, forcing it down. This mechanical force is balanced by the aerodynamic force of the trim tab in normal flight, and the airplane may

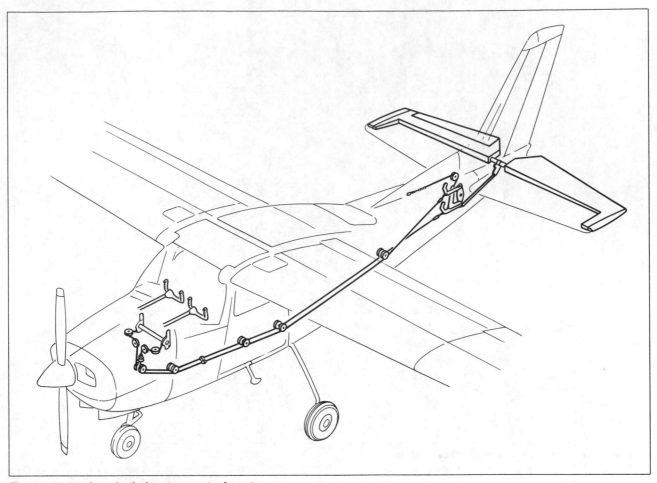

Figure 15-41. A typical elevator control system.

be trimmed for its approach speed in the normal way. Now, if the airplane encounters turbulence that slows it down, the trim tab will lose its effectiveness but the down spring will pull the elevator down and drop the nose so the airspeed will not decrease to stall. Thus, the down spring provides an artificial stability for an aircraft with an aerodynamic pitch trim system. See figure 15-42.

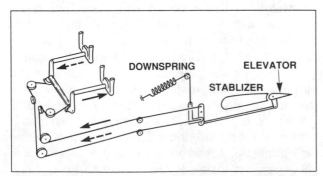

Figure 15-42. The elevator down spring supplies a mechanical load to the controls to lower the nose when the aerodynamic tail load becomes ineffective in an aft center of gravity flight condition.

Airplanes with stabilators have essentially the same type of control system. The control yoke pivots and pulls on the stabilator-up cable when it is pulled back. This pulls down on the balance arm of the stabilator and raises its trailing edge, rotating the airplane nose up. Pushing the yoke in lifts the stabilator balance arm and lowers the trailing edge of the stabilator. This system uses stabilator down springs in the same way as those just described for the elevator, to improve the longitudinal stability of the airplane during conditions of low airspeed with a far aft center of gravity.

Auxiliary, Or Trim, Controls

Trim tabs are small movable portions of the trailing edge of the control surface. These tabs are controlled from the cockpit to alter the camber of the surface and create an aerodynamic force that will hold the control surface deflected.

Trim tabs may be installed on any of the primary control surfaces and if only one tab is used, it is normally on the elevator, to adjust the tail load so the airplane can be flown hands-off at any given

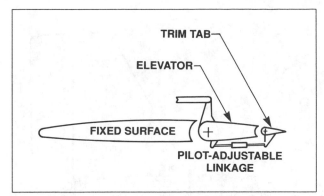

Figure 15-43. Trim tab.

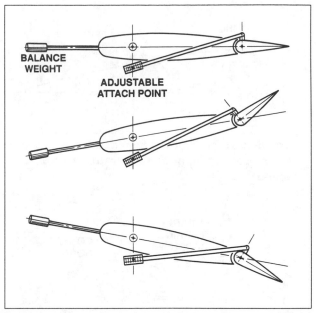

Figure 15-45. Anti-servo tab on an all-movable tail surface.

airspeed and power setting. The airplane speed is set with the control wheel, and then the trim tab is adjusted until the airspeed can be maintained with no force needed on the wheel.

Control forces may be excessively high in some airplanes, and in order to decrease them, the manufacturer may use a **balance tab**. This tab is located in the same place as a trim tab and in many installations one tab serves the function of both. The basic difference is that the control rod for the balance tab is connected to the fixed surface on the same side as the horn on the tab. In figure 15-44 we see the way a balance tab works. If the control surface is deflected upward, the connecting linkage will pull the tab down, and when the tab moves in the direction opposite that of the control surface, it will create an aerodynamic force that aids the movement of the surface, making control forces lighter.

If the linkage between the tab and the fixed surface is adjustable from the cockpit, the tab will act as a combination trim and balance tab. It can be adjusted to any desired deflection to trim the airplane for a steady flight condition, and any time the control surface is deflected, the tab will move in the opposite direction and ease the load on the elevator controls.

All-movable horizontal tail surfaces do not have a fixed stabilizer in front of them, and the location

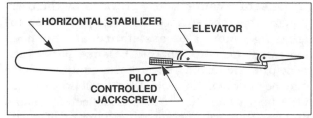

Figure 15-44. A balance tab aids the pilot in moving the control surface.

of their pivot point makes them extremely sensitive and unstable. To decrease this instability, a full length **anti-servo tab** may be installed on the trailing edge. This tab works in the same manner as the balance tab except that it moves in the opposite direction. The fixed end of the linkage is on the *opposite* side of the surface from the horn on the tab, and when the trailing edge of the stabilator moves up, the linkage forces the trailing edge of the tab up. When the stabilator moves down, the tab also moves down.

The fixed end of the linkage may be attached to a jackscrew so the tab may be used as a trim tab as well as an anti-servo tab. See figure 15-45.

Large aircraft are usually equipped with power-operated irreversible flight control systems. In these systems, the control surfaces are operated by hydraulic actuators controlled by valves moved by the control yoke and rudder pedals. An artificial feel system gives the pilot resistance that is proportional to the flight loads on the surfaces.

The control forces are too great for the pilot to manually move the surfaces, so, in the event of a hydraulic system failure, they are controlled with **servo tabs**. In the manual mode of operation, the flight control column moves the tab on the control surface, and the aerodynamic forces caused by the deflected tab move the main control surface.

Another device for aiding the pilot of high-speed aircraft is the **spring tab**. The control horn is free to pivot on the hinge axis of the surface, but it is

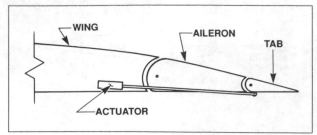

Figure 15-46. A servo tab creates an aerodynamic force that moves the control surface to which it is attached.

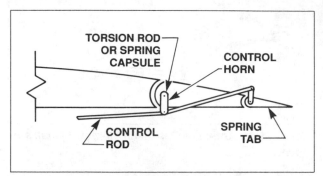

Figure 15-47. A spring tab deflects only when the control forces are great enough to distort the torsion rod or collapse the spring capsule.

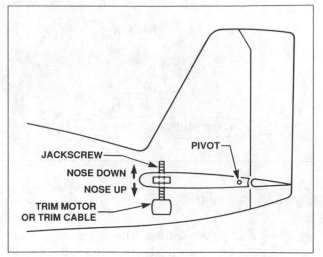

Figure 15-48. Adjustable stabilizer.

restrained by a spring. For normal operation when the control forces are light, the spring is not compressed and the horn acts as though it were rigidly attached to the surface. At high airspeeds when the control forces are too high for the pilot to properly operate, the spring collapses and the control horn deflects the tab in the direction to produce an aerodynamic force that aids the pilot in moving the surface.

The **adjustable stabilizer** trim system pivots the horizontal stabilizer about its rear spar and mounts its leading edge on a jackscrew that is controllable from the cockpit. On the smaller airplanes the jackscrew is cable-operated from a trim crank, and on the larger airplanes it is motor-driven. The trimming effect of the adjustable stabilizer is the same as that obtained from a trim tab, except this system is more effective (more tail down force produced) at low speeds and more efficient (less drag) at high speeds (figure 15-48).

It is very seldom that a lightweight simple airplane can be assembled and rigged so that it will fly at cruise speed without there being a need to hold some pressure on the controls to keep the wings level. This out-of-trim condition is usually corrected by the use of a small sheet metal **fixed-trim tab** on the trailing edge of the control. This tab is bent so it will deflect the control just enough

to counteract the unbalanced condition at the desired cruising speed. See figure 15-49(A).

A tab on the rudder corrects a slip or skid flight condition. It should not be used to force the airplane to fly in a yaw to create the lift difference between the wings that is needed to correct a wing-heavy condition.

Airplanes with wing flaps can have a wing-heavy flight condition corrected by rigging the flaps so the flap on the side with the heavy wing extends down far enough in flaps-up flight to create the extra lift needed to correct the wing-heavy condition. The amount that the flap must be extended is seldom enough even to notice, except possibly in a full-flap, power-on stall, where the aircraft may break sharply. Care and skill is needed when rigging and trimming aircraft.

Even on small, light airplanes aerodynamic assistance in the movement of the controls has been used. The simplest form of this assistance is the **aerodynamically balanced control surface**. In the case of the rudder, the balance portion, or overhang, deflects to the opposite side of the fuselage from the main rudder surface to produce an aerodynamic force that aids the pilot in moving the surface (figure 15-49(B)).

The **aerodynamic boost** provided by the overhanging balance surface does not provide an increasing amount of assistance as the need increases. An **aerodynamic balance panel** connected to the leading edge of the control surface may be built in such a way that it will provide very little help when the surface is deflected a small amount, but will increase the amount of assistance that it gives as the surface deflection is increased.

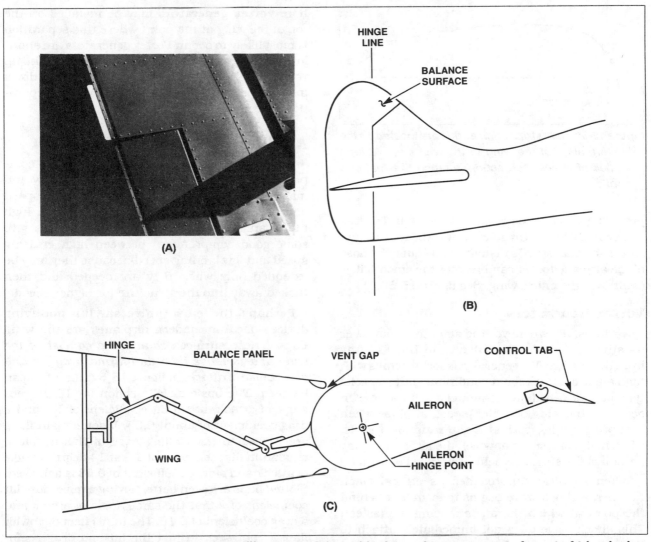

Figure 15-49(A)—Fixed trim tab on the trailing edge of the rudder is used to compensate for out-of-trim rigging. (B)—The overhang on the control surface ahead of the hinge line aids the pilot in the movement of the surface. (C)—An aerodynamic balance panel aids the pilot in the movement of the control surface. The amount of aid increases as the deflection of the surface increases.

As seen in figure 15-49(C), the compartment ahead of the aileron is divided by a hinged lightweight, rigid balance panel which has a relatively large area. The panel divides this space into two smaller compartments, with one connected through a vent to the gap in the upper surface between the wing and the aileron, and the other through a vent to the same gap in the lower surface.

When the aileron is deflected upward, the high-velocity air over the lower vent gap decreases the air pressure under the balance panel and pulls it down. This downward force on the leading edge of the aileron causes the trailing edge to move up. The greater the deflection, the lower the pressure, and the more assistance will be provided by the balance panel.

When the aileron is moved downward, the high-velocity air over the top of the aileron will produce a pressure drop that will cause the balance panel to assist in moving the aileron down.

It is extremely important that an airplane wing stall progressively from the root out to the tip, so the ailerons will be effective throughout the stall. If a wing does not naturally have this stall progression characteristic, it is possible for the manufacturer to place a small triangular strip of metal on the leading edge of the wing in the root area. When a high angle of attack is reached, this triangular **stall strip** will break the airflow over the section where the stall strip is installed, causing the portion of the wing behind the stall strip to stall while the airflow is still smooth over the aileron, thus

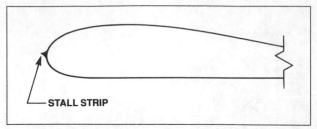

Figure 15-50. Stall strips such as this are located on the leading edge of the wing in the root area. At high angles of attack their action is much like that of a spoiler.

providing effective roll control in the stall. The loss of lift caused by the turbulence over the root section where the stall strip is installed will cause the nose of the airplane to drop and restore the smooth flow of air over the entire wing. See figure 15-50.

Vortex Generators

A wing stalls whenever the smooth airflow over its surface becomes turbulent and breaks away from the surface. We generally associate stalls with high angle of attack flight conditions, but a special type of stall called shock-induced separation can occur on the wing of a high-speed airplane when it approaches its **critical Mach number**; that is, when it flies at a speed at which the airflow over any portion of the surface reaches the speed of sound.

When an airfoil approaches its critical Mach number, a shock wave begins to form just behind the point at which the air is moving the fastest. This shock wave does not immediately attach to the trailing edge of the wing, but it moves back and forth and causes the air to separate from the upper surface of the wing. This separation causes a buffeting of the controls. To prevent this separa-

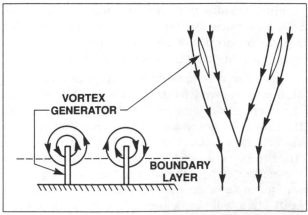

Figure 15-51. Vortex generators bring high-energy air to the surface of the wing to prevent shock-induced separation.

tion, **vortex generators** may be installed on the top of the wing at the point where this separation is most likely to occur. Vortex generators are short, low-aspect-ratio airfoils arranged in pairs. The tip vortices of these airfoils pull high-energy air down into the boundary layer and prevent the separation. See figure 15-51.

Auxiliary Lift Devices

An airplane is a series of engineering compromises. We must choose between stability and maneuverability and between high cruising speed and low landing speed, as well as between high utility and low cost. Lift-modifying devices give some good compromises between high cruising speed and low landing speed because they may be extended only when they are needed and then tucked away into the structure for higher speed.

Perhaps the most universal lift-modifying devices used on modern airplanes are the **wing flaps**. These surfaces *change the camber of the wing and increase both its lift and drag for any given angle of attack*. In figure 15-52 this effect can be seen. The basic airfoil section, at 15 degrees angle of attack, has a lift coefficient of 1.5 and a drag coefficient of about 0.05. If some plain flaps are hinged to the trailing edge of this airfoil, a maximum lift coefficient of 2.0 at 14 degrees angle of attack and a drag coefficient of 0.08 is achieved. Slotted flaps are even better, giving a maximum lift coefficient of 2.6 at the same angle of attack and a drag coefficient of 0.10. The total effect of Fowler flaps is not seen in just the lift and drag coefficients, because they not only give an excellent increase in the lift coefficient, but they also increase the wing area which also has an important effect on both lift and drag.

Plain flaps are simple devices that are merely sections of the trailing edge of the wing, inboard of the ailerons. They are typically about the same size as the aileron and are hinged so they can be deflected, usually in increments of 10, 25, and 40 degrees. Generally speaking, the effect of these flaps is minimal, and they are seldom found on modern airplanes.

Split flaps is another design of flap that was used with a great deal of success in the past, but it is seldom used today. On the extremely popular Douglas DC-3, a portion of the lower surface of the trailing edge of the wing from one aileron to the other, across the bottom of the fuselage, could be hinged down into the airstream. As we see in figure 15-52, the lift change was similar to that produced by a plain flap, but it produced much more drag

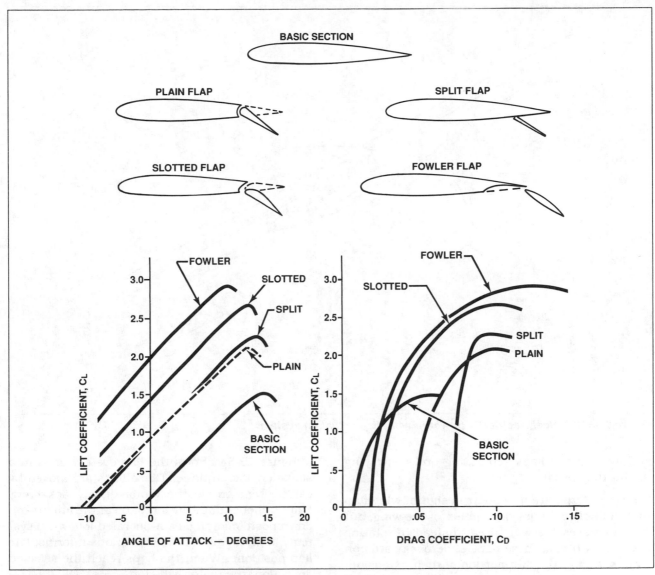

Figure 15-52. The effect of different types of flaps.

at low lift coefficients, and this drag coefficient changed very little with the angle of attack.

The most popular flap on airplanes today is the **slotted flap**, with variants of this design used for small airplanes as well as for the largest. You will notice that the slotted flap increases the lift coefficient a good deal more than the simple flap. On small airplanes, the hinge is located below the lower surface of the flap, and when it is lowered, it forms a duct between the flap well in the wing and the leading edge of the flap.

When the flap is lowered all of the way and there is a tendency for the airflow to break away from its surface, air from the high pressure area below the wing flows up through the slot and blows back over the top of the flap. This high energy flow on the surface pulls air down and prevents the flap stall-

ing. It is not uncommon on large airplanes to have double- and even triple-slotted flaps (figure 15-53) to allow the maximum increase in drag without the

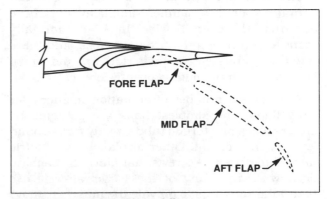

Figure 15-53. Triple-slotted flaps.

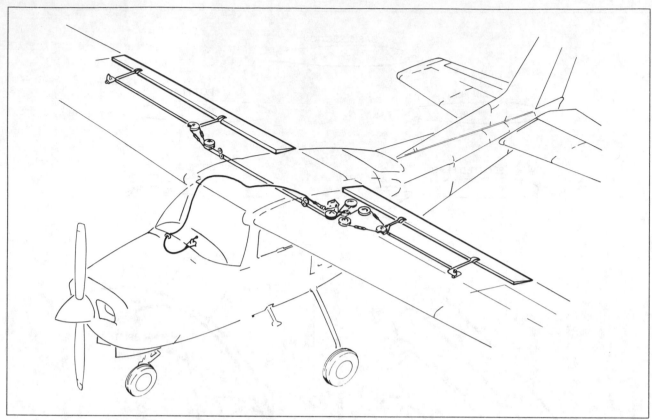

Figure 15-54. Typical electrically powered wing flap actuation system.

airflow over the flaps separating and destroying the lift they produce.

Fowler flaps are a popular design of wing flap that not only changes the camber of the wing, but also increases the wing area. Instead of rotating down on a hinge, these flaps slide backwards on **tracks**, and in the first portion of their extension, drag is very little but lift increases a great deal as both the area and camber are increased. As the extension continues, the flap deflects downward, and during the last portion of its travel, it increases the drag with little additional increase in lift.

The flaps are connected to their hinges and actuator rods in a manner similar to that of the ailerons. However, Fowler flaps are normally mounted on rollers that ride in tracks, and these must be adjusted so they ride up and down smoothly with no binding or interference.

There are a number of actuation methods for wing flaps. The simplest flaps are actuated by either cables or a torque tube directly from a hand lever in the cockpit. Other airplanes use electric motors to drive jackscrews that move the flaps up or down, and many of the larger aircraft use hydraulic actuators to provide the muscle to move the flaps against the air loads.

Figure 15-54 shows the flap system used in a single-engine airplane. These flaps are moved by cables from an electric motor-driven jackscrew. Limit switches shut the motor off at the full up and down position, and a cam-operated follow-up system allows the pilot to select various intermediate flap positions. When the flaps reach the selected deflection, the motor will stop.

Because the flaps on most airplanes are not physically connected to each other except by a control cable/pulley or push-pull rod/bellcrank system, it is conceivable that the system controlling one flap could fail, allowing one flap to quickly streamline causing an asymmetrical lift condition that would roll the aircraft unexpectedly. For this albeit remote reason, some pilots (including the author) have developed a personal policy of planning flap extensions and retractions to occur only when wings are level. Such planning is easy to do and is only a small part of each experienced aviator's risk management system.

Leading Edge Devices

We have seen that a stall occurs when the angle of attack becomes so great that the energy in the air flowing over the wing can no longer pull air

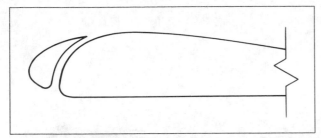

Figure 15-55. Wing section showing a fixed slot which ducts air over the top of the wing at high angles of attack.

down to the surface. The boundary layer thickens and becomes turbulent and the airflow separates from the surface.

This separation can be delayed to a higher angle of attack by any expedient that increases the energy of the air flowing over the surface. One method used is a **slot** in the leading edge of the wing. This slot is simply a duct for air to flow from below the wing to the top where it is directed over the surface in a high-velocity stream. Slots are usually placed ahead of the aileron to keep the outer portion of the wing flying after the root has stalled. This keeps the aileron effective and provides lateral control during most of the stall. See figure 15-55.

Many high-performance airplanes have a portion of the wing leading edge called a **slat** mounted on tracks so it can extend outward and create a duct to direct high-energy air down over the surface and delay separation to a very high angle of attack.

In some airplanes these slats are actuated by aerodynamic forces and are entirely automatic in their operation. As the angle of attack increases, the low pressure just behind the leading edge on top of the wing increases and pulls the slat out of

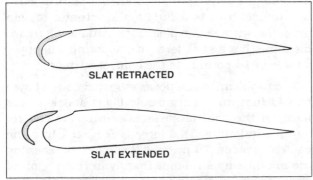

Figure 15-56. A movable slat moves out of the leading edge of the wing at high angles of attack to form a duct for high-energy air.

the wing. When the slat moves out, it ducts the air from the high-pressure area below the wing to the upper surface and increases the velocity of the air in the boundary layer. When the angle of attack is lowered, air pressure on the slat moves it back into the wing where it has no effect on the airflow. See figures 15-56 and 15-57.

Some airplanes have slats operated by either hydraulic or electric actuators, and they are lowered when the trailing edge flaps are lowered. These slats prevent the airflow breaking away from the upper surface when the flaps increase the camber of the wing. Flaps that are used with slats are usually slotted and they duct high-energy air over the deflected flap sections so the air will not break away over their surface.

To augment the action of trailing edge flaps, many of the large jet transport airplanes also have flaps on the leading edge as well.

The **drooped leading edge flap** uses a jackscrew arrangement to push the leading edge of the wing against a hinge on its lower surface. This causes the leading edge to droop and increase the camber of the wing to increase C_L at high angles of attack. See figure 15-58.

The Krueger flap is similar to the drooped leading edge flap, except that a hinged linkage extends out ahead of the leading edge and produces a special high-lift leading edge shape (figure 15-59).

Special wing tips have found a place in aircraft design because air flowing over the top of a wing creates a low pressure, while the air passing below the wing has been slowed down somewhat and its pressure is higher than ambient. This differential of pressure causes air to spill over the wing tip and create **vortices** that effectively kill some of the lift, create drag and cause instability at high angles of attack and low airspeed.

There are a number of expedients that have been used to prevent this loss of lift and stability. Some manufacturers install fuel tanks on the wing tips that serve the triple function of increasing the range of the airplane, distributing the weight over a greater portion of the wing and preventing the air spilling over the wing tip. Smaller airplanes that do not use tip tanks may have tip plates installed on the tip. These plates have the same shape as the airfoil but are larger and prevent the air spilling over the tips. See figure 15-60.

Far less drastic than tip tanks or tip plates are specially shaped wing tips. Some wing tips have a special droop and a square trailing edge designed to decrease vortex development (figure 15-61).

Figure 15-57. This full-span slat is opened by aerodynamic forces at high angles of attack.

More recent designs feature **tipsails**, sometimes referred to as **winglets**. These vertical airfoils at the wing tip take many forms, shapes and sizes but their principal function is to inhibit wing tip vortex development, thus reducing drag. They offer the advantage of being able to develop a forward component of lift (thrust) sufficient to offset their own drag. Secondarily, tipsails are pleasing to the eye and make the aircraft look "high-tech"—a good sales tool.

Control Systems For Large Aircraft

As aircraft increase in size and speed, airloads cause their controls to become more difficult to operate and systems must be used to aid the pilot. The power-boosted control system is similar in principle to power steering in an automobile. A hydraulic actuator is in parallel with the mechanical operation of the controls, and in addition to moving the control surface, the normal control movement by the pilot also moves a control valve that directs hydraulic fluid to the actuator to help move the surface. A typical boost ratio is about 14, meaning that a stick force of one pound will apply a force of 14 pounds to the control surface.

The problem with a power-boosted control system is that during transonic flight shock waves form on the control surfaces and cause control surface buffeting. This force is fed back into the control system. To prevent these forces reaching the pilot, many airplanes that fly in this region of airspeed use a power-operated **irreversible control system**. The flight controls in the cockpit

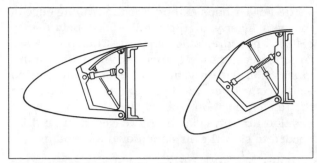

Figure 15-58. Drooped leading edge flap.

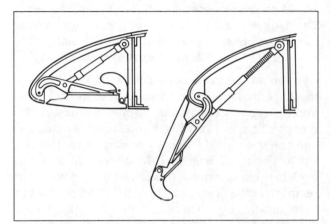

Figure 15-59. Krueger-type leading edge flap.

Figure 15-60. Tip tanks minimize the amount of high-pressure air from below the wing spilling over into the low-pressure area above the wing, decreasing the wing tip vortex and, therefore, decreasing induced drag.

Figure 15-61. Drooped wing tips minimize loss of lift and decrease drag from wing-tip vortices.

actuate control valves which direct hydraulic fluid to control surface actuators. Since the pilot has no actual feel of the flight loads, some form of **artificial feel** must be built into the system that will make the control stick force proportional to the flight loads on the control surfaces.

The flight control system for the Boeing 727 jet transport airplane is typical for large aircraft. Rotation about the three axes is controlled by conventional ailerons, rudders and elevators, and these primary controls are assisted by **spoilers** which also double as speed brakes, by an adjusting horizontal stabilizer, and by both leading and trailing edge flaps.

The primary control surfaces are moved by a power-operated irreversible control system powered by the two hydraulic systems in the airplane, and in the event of the failure of both systems, manual flight control is possible by the use of servo tabs. In normal flight, these tabs serve as balance tabs.

This airplane has two sets of ailerons and fourteen spoilers. For high speed flight the inboard ailerons and the flight spoilers provide **roll control**, and the outboard ailerons are locked in their trim position. When the trailing edge flaps are lowered, the outboard ailerons are unlocked and both sets of ailerons and the flight spoilers operate together. The ailerons use aerodynamic balance panels to minimize the control force needed to move them.

Five spoilers on each wing operate in flight. They move proportional to the aileron displacement and decrease the lift on the wing the pilot wants to move down. When the speed brake control is moved back in flight, all ten flight spoilers are lifted, and they produce a great deal of drag. If the control wheel is rotated while the spoilers are deployed they will move differentially, decreasing the lift on the wing that should go down, and decreasing some of the drag on the wing that should rise. When the airplane is on the ground, the four ground spoilers, as well as the ten flight spoilers, deploy up from the top of the wing. The ground spoilers have only two positions, fully raised and fully closed, and they can be raised only

337

when the weight of the airplane is on the landing gear. They are used to destroy lift and slow the airplane in its landing roll.

Pitch control is conventional, by the use of elevators that are hydraulically powered. If both hydraulic systems should fail, the elevators can be operated by their servo tabs. Electric motors operate jackscrews to raise or lower the leading edge of the horizontal stabilizer for longitudinal trim. The pilot, copilot or the autopilot can operate the trim motors. In the event of electrical failure, the stabilizer can be adjusted by turning the trim wheel on the control pedestal.

The Boeing 727 has two independent rudders to provide **yaw control**, and each of these rudders is equipped with an anti-balance (anti-servo) tab. Both rudders are powered by hydraulic pressure, but from different hydraulic systems.

In addition to control from the pilot's and copilot's pedals, the rudder receives signals from the **yaw damper** to counteract **Dutch roll**. When the yaw damper receives a signal from the turn and slip indicator, it produces a rudder deflection that is proportional to the rate of yaw but is in the opposite direction.

Each wing of the Boeing 727 is equipped with two **triple-slotted Fowler flaps** on the trailing edge, and on the outboard portion of the leading edge there are four **retractable slats**. Three **Krueger flaps** are on the inboard portion of the leading edge. The trailing edge flaps are actuated by hydraulic motors and have, for emergency backup, electric motors that will raise or lower them.

The Fowler flaps extend out the trailing edge of the wing nearly to their full chord before they deflect an appreciable amount. This provides a large increase in wing area and lift, with a minimum amount of drag, and this is the position used for takeoff. When the flaps are extended beyond the takeoff position, they deflect to their full 40° position and provide the increased drag and lift needed to slow the airplane down for landing.

Both the Krueger flaps and the slats alter the shape of the wing's leading edge to provide the added camber needed for high lift at slow speed. The flaps will cause the inboard portion of the wing to stall before that portion behind the slats. This gives better lateral control during stall.

In the two-degree position of the flap control, the trailing edge flaps deflect two degrees, and the two middle slats on each wing extend. In the five-degree position, all of the slats and flaps are extended and the trailing edge flaps are down five degrees. Up to about 25 degrees, the Fowler flaps move back to their maximum travel, but they move down relatively little. Beyond this, they deflect rather than moving out. There are detents in the flap control at 2°, 5°, 15°, 25°, 30°, and 40°. A gate at 25° allows the pilot to position the flap here for go-around without having to look at the control. A gate at the two-degree position requires the flap handle to stop here during flap retraction, to allow the airspeed to build up sufficiently high before all of the leading edge slats are retracted.

For a greatly expanded discussion of large aircraft flight controls, see Wild (Additional Reading, item 3).

Additional Reading

1. Airframe and Powerplant Mechanic's Airframe Handbook; EA-AC65-15A; IAP, Inc., Publ.; 1976; Chapters 1 and 2.
2. A&P Technician Airframe Textbook; EA-ITP-A2; IAP, Inc., Publ.; 1985; Chapters 1 and 4.
3. Wild, T.; Transport Category Aircraft Systems; EA-363; IAP, Inc, Publ.; 1990.

Study Questions And Problems

1. What is the difference between monocoque and semi-monocoque construction?

2. What is a truss?

3. What is meant by the term "stressed-skin structure?"

4. What are the two primary and three secondary stresses an airframe undergoes?

5. Why isn't pure aluminum used by itself in aircraft structures? How is it used and why?

6. From the pilot's point of view, what is the difference between permanent and temporary deformation?

7. How does the pilot recognize permanent deformation during a preflight?

8. What type of construction is used for most horizontal stabilizers?

9. What is cantilever construction?

10. Why should a low-wing trim condition not be corrected with rudder trim?

11. How is a stabilator balanced?

12. Why do control surfaces need to be balanced?

13. Explain the difference between leading edge flaps, slots and slats.

14. Explain the function of a stall strip and a vortex generator.

15. What is a balanced control surface? (Hint: this is different than a "balanced control.")

16. How is an aerodynamic balance panel an improvement over a balanced control surface?

17. What surfaces on a B-727 are capable of providing roll control?

18. What control in the cockpit controls airspeed when the airplane is on the ground during takeoff? When the aircraft is airborne?

19. Using figure 15-37(A), analyze the effect on flight of a broken balance cable between the left and right aileron bellcranks.

20. Analyze the effect on flight of a broken cable controlling rudder or elevator.

21. Rank, in order of seriousness, the breaking of each cable in the airplane that is connected to flight controls.

22. List four ways to make control forces lighter in the pitch mode and explain each.

Chapter XVI
Weight And Balance, Inspections And Pilot Maintenance

Weight And Balance

Importance Of Weight And Balance

An aircraft is a dynamic device that requires a careful balance between all of its forces to maintain safe and efficient flight. The **lift** produced by the wing is concentrated at a point approximately 20-35% of the way back on the wing from the leading edge, and to provide stability, the **center of gravity**, or that point at which all of the aircraft weight can be considered to be concentrated, is located ahead of this center of lift (figure 16-1). This location results in a nose down pitching moment that is balanced by a **tail down force** (TDF).

The amount of TDF is determined by the airspeed, trim and elevator position. It decreases as the airplane slows down. The weight remains constant and ahead of the center of lift, pulling the nose down so the airplane will automatically regain the speed it has lost, providing **positive pitch stability** (figure 16-2).

If the center of gravity falls outside of the rather narrow limits allowed by the aircraft designer, serious control problems can result.

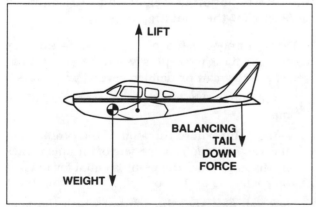

Figure 16-1. *The center of gravity of an airplane is ahead of its center of lift. The resulting nose-down pitch is balanced by a tail down force that varies with the airspeed, trim and elevator position.*

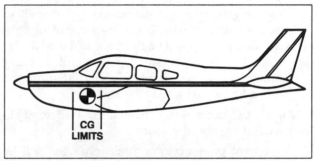

Figure 16-2. *The center of gravity of an airplane must fall within rather narrow center-of-gravity limits.*

When an aircraft is designed, limits are put on its maximum weight, and restrictions are set up regarding the range within which the center of this weight is allowed to vary. A part of the certification procedure for an airplane is to determine that its weight and balance are within the allowable limits, and this information is furnished with the aircraft as part of its operations manual.

Therefore, if the aircraft is over its gross weight or loaded outside the balance (CG) limits, it is not within certification limits and is not airworthy.

It is the responsibility of the pilot to know before each flight that his aircraft is properly loaded, that it does not exceed the allowable gross weight, and that the center of gravity is within the allowable range.

There is another reason for knowing the location of the center of gravity, and too few pilots seem to be aware of this one:

CG location determines *how* the aircraft will fly, so calculating CG location allows the pilot to predict, before flight, how the aircraft will react in flight.

For aircraft with aft mounted stabilizers, a **forward CG** increases tail down force (TDF) which increases **effective weight** (which equals real weight + TDF), therefore:

—all parameters of aircraft performance will decrease except maximum power off glide distance. Takeoff and landing distance and stall speed will increase while cruising speed and rate of climb will decrease.

—control pressures needed to accomplish a given pitch change will be greater.

—pitch stability will be very positive (more effort needed to cause pitch changes).

—pitch trim status indicator will be aft (nose up).

—since ground effect decreases elevator effectiveness, the most dangerous part of a flight with forward CG will be the flare-to-land. It may be advisable to use some power and/or carry some extra airspeed and/or use a flat approach with a known forward CG.

An **aft CG** decreases the need for TDF so effective weight is less, therefore:

—control pressures for pitch changes will be lighter.

—pitch stability will be poor (nose of the aircraft tends to wander up and down more—the aircraft has to be flown with more attention and the aircraft becomes less affected by pitch trim).

—stall speed decreases.

—all performance parameters improve, except maximum power off glide distance (the indicated airspeed for best maximum distance glide decreases).

—the most dangerous part of flight with aft CG loading occurs in slow flight. With a very aft CG a stall and spin may lead to a flat spin which is usually unrecoverable.

Note: If the aircraft is loaded within CG limits, the dangerous conditions mentioned above should be controllable. The manufacturer has probably demonstrated the aircraft's ability to flare for landing at the forward CG limit and, if the aircraft is certified for spins, recovery from a spin has probably been demonstrated (with the aircraft wearing a spin chute and the test pilot wearing a parachute).

The weight of an aircraft changes during its operational life as equipment is added or removed and as repairs are made. All of these changes must be monitored and the weight and balance information used by the pilot must be kept up to date. This is done by the aviation maintenance technician. However, the final responsibility for accurate, up-to-date aircraft empty weight and balance lies with the pilot.

Any time equipment is added or removed, or any repair or alteration is made that could affect either the empty weight of the aircraft or its empty-weight

Figure 16-3. *The great distance between the rows of seats and the baggage compartments of this type of aircraft makes careful loading important to keep the center of gravity within the allowable limits.*

center of gravity, this change must be recorded in the weight and balance information, and all of the old computations must be plainly marked so the pilot can easily see that they have been superseded.

Very close track must be kept of the weight and balance of aircraft used to carry passengers or cargo for hire, so they must be reweighed periodically and have their center of gravity recomputed.

The fuel tanks of some of the smaller training aircraft are located in the wing, and the seats and baggage compartment are located directly below the wing, so it is less likely that these aircraft will be loaded in such a way that the center of gravity will fall outside of the allowable center-of-gravity range. Most of the larger aircraft, however, have several rows of seats, some of which are ahead of the center-of-gravity range and some behind it, and there are often both forward and aft baggage compartments. This wide range of loading possibilities makes the determination of weight and CG location a necessity for the pilot (figure 16-3).

Determining the CG of any mass is an easy calculation if the concept is well understood. **Center of gravity** can be defined several ways. It is:

—the center of balance of all components of the mass.

—the point where the sum of the products of the weight of each component of the mass times its distance from the point is equal for masses on each side of the point.

—for an airplane, the point at which the nose down and tail down moments are of equal magnitude.

What other definitions of CG can you list that you feel are worthwhile? This might be a good time to consult some other books on the subject.

The CG of any mass can be determined if only two relationships are remembered, which are:

1. weight × arm = moment

2. $\dfrac{\text{sum of all moments}}{\text{total weight}}$ = CG (an arm)

Perhaps now, before proceeding with some calculations, is a good time to review some weight and balance terminology that is important, so here is a short glossary:

all up weight Weight when the aircraft is loaded and ready to go (large aircraft).

arm The horizontal distance of any object from the **datum**. It is expressed in inches and may be negative or positive depending on its relationship to the datum.

ballast Weight used to obtain a favorable center-of-gravity location. It is often made of lead. It may be moveable or permanent and must be marked as such.

basic empty weight See basic weight.

basic weight Weight of the aircraft to include unusable fuel and full oil, excluding all other useful load—sometimes referred to as basic empty weight.

center of gravity The point at which the nose and tail moments are of equal magnitude. It is abbreviated CG and is sometimes shown symbolically as ⊕ .

datum An imaginary vertical line from which all horizontal measurements are made or indicated. The datum may be located at any convenient location by the manufacturer.

empty center of gravity range The most forward and rearward limits of the empty aircraft as determined by the manufacturer. When these limits are met, no loading limits are necessary for standard loads. It may be abbreviated ECGR. Most aircraft do not have an ECGR.

empty weight The weight of the aircraft with unusable fuel and no useful load. Oil may or may not be included in the empty weight of the aircraft. For many years, oil was not part of the empty weight except for residual oil or undrainable oil. Today, due to a change in FAR 23, aircraft are being manufactured which include full oil as a part of the empty weight. To determine whether a specific aircraft's empty weight includes engine oil, the Aircraft Data Sheet must be checked. Other operating fluids, such as the hydraulic fluid, are included in the empty weight. In most cases, the manufacturer only weighs every tenth aircraft in order to establish the **licensed empty weight** of a particular type of aircraft. This is done prior to adding the optional equipment. Then, mathematically, the optional equipment with a fixed location is added to the empty weight. This weight, without optional equipment, is referred to as the *basic empty weight* and should not be confused with the empty weight. A review of aircraft POH's from the past 30 years shows some variation of the use of the terms *empty weight, basic empty weight, standard empty weight* and *licensed empty weight*. The pilot must be careful to determine how the term is used and, therefore, what its meaning is, before proceeding.

equipment list A comprehensive list of equipment installed on a particular aircraft. This includes required and optional equipment.

fleet weight The average weight of several aircraft of the same model and with the same equipment. This weight may be used by 121 and 135 operators.

gross weight Maximum weight.

landing weight The weight of the aircraft at touchdown. This is often limited to less than the takeoff weight by the manufacturer for structural reasons.

LEMAC Abbreviation for leading edge of mean aerodynamic chord.

load manifest An itemized list of weights and moments of a particular load taken on a specific flight. Used by 121 and 135 operators.

MAC Abbreviation for mean aerodynamic chord.

mean aerodynamic chord The chord of a rectangular wing, or the average chord of an irregularly shaped wing. Often the CG is expressed in a percentage of MAC.

minimum fuel An amount of fuel used when computing adverse loading for most forward and rearward conditions.

moment A rotational force caused by a weight acting on an arm. The product of the weight multiplied by the arm. The moment may be positive or negative depending on the sign of the weight and arm.

moment index Moment reduced by 100,000 or 10,000 or 1,000 for ease in balance calculations.

maximum takeoff weight The greatest design weight at the time of liftoff.

maximum weight Maximum weight allowed for a specific aircraft in a category. This is sometimes referred to as gross weight.

maximum zero fuel weight Maximum weight of the aircraft including useful load excluding the weight of the fuel. Some manufacturers designate this weight for structural reasons.

operating weight Term used by aircraft operators to include the empty weight of the aircraft and items always carried in the aircraft, such as crew, water, food, etc.

payload Maximum design, zero fuel weight minus basic empty weight. Or see useful load.

ramp weight See taxi weight.

required equipment Items necessary on a type of aircraft to be airworthy. See FAR 91.205.

station An imaginary vertical line denoting a horizontal distance from the datum.

takeoff weight Weight of the aircraft at liftoff. Often used as maximum takeoff weight.

tare Any material used during the weighing process between the scale and the aircraft. Example: chocks.

taxi-weight Maximum weight allowed for ground maneuvering. Sometimes referred to as ramp weight.

TEMAC Abbreviation for trailing edge of the mean aerodynamic chord.

undrainable fuel Amount of fuel that remains in the system after draining. This is considered a part of the empty weight of the aircraft.

undrainable oil Oil remaining after draining the oil from an engine. This oil is considered a part of the empty weight of the aircraft.

unusable fuel Fuel that cannot be consumed by the engine. This fuel is considered equivalent to undrainable fuel for weight and balance purposes.

unusable oil Oil that cannot be used by the engine. Unusable oil is usually slightly more than undrainable oil. It is not considered in weight and balance computations.

usable fuel Portion of the total fuel available for consumption by the aircraft in flight.

useful load Difference between the empty weight of the aircraft and the maximum weight of the aircraft. Sometimes referred to as the "payload."

zero fuel weight The weight of the aircraft to include all useful load except fuel.

Principles Of Weight And Balance

The actual principle involved in finding the center of gravity of an aircraft is quite simple, and it is easy to visualize when considering the playground teeter-totter, or seesaw, which is a practical example of weight and balance.

Let's work through some seesaw problems of increasing complexity, then apply them to airplanes.

When a large child and a small child get on a seesaw, the large child must slide up close to the support, or the fulcrum, to balance the small child who is farther away from the fulcrum.

The distance from the fulcrum to the center of gravity of the weight is called the **arm** of the weight, and it may be measured in such units as feet, inches or meters. The amount of force—in this case, the weight of the child—is measured in pounds, grams or kilograms, and is considered to be concentrated at a point called its center of gravity.

The product of the weight and the arm is the **moment** of the force and is expressed in pound-feet, pound-inches, or in gram- or kilogram-meters. And since a moment is a force that causes rotation, one must specify the direction, either clockwise or counterclockwise, in which the force causes the weight to rotate.

To best understand the principles of weight and balance, let's consider that the board used does not have any weight of its own, and that all of the weight is concentrated at the center of gravity of the weights themselves.

In the illustration of figure 16-5, there is a board on which two weights are located. The weight on

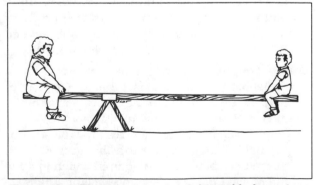

Figure 16-4. The principle of weight and balance is as simple as the balance of these children on the seesaw.

344

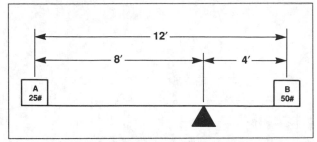

Figure 16-5. A simple lever problem, using weight A as the datum.

the left is 25 pounds, and the one on the right is 50 pounds, and there are 12 feet between the centers of gravity of the two weights. To find the location of the fulcrum about which the two weights will balance, choose the location of a datum, or a reference line, from which all measurements will be made. This line can be anywhere, but for this initial explanation, let's assume it to be located at the center of gravity of one of the weights. In this case we will choose the weight A, the one on the left side.

To visualize the computations more clearly, let's make a chart such as the one in figure 16-6. Since weight A is on the datum, its arm is zero and when we multiply any number by zero, the product, or the moment, is also zero. The arm of weight B is 12 feet, and its moment is 12 × 50, or 600 pound-feet, and its direction of rotation is clockwise.

To find the balance point, we must divide the total moment by the total weight. The total moment is 600 pound-feet, and the total weight is 75 pounds; this places the balance point (CG) eight feet to the right of the datum. In this example, there are no counterclockwise moments and so the total moment is clockwise.

To check the above work and prove that the board is really balanced about the point we have just discovered, make a chart similar to that in figure 16-7. Here the datum has been moved from the center of gravity of weight A to the fulcrum,

ITEM	WEIGHT	ARM	MOMENT
A	25	0	0
B	50	12′	600
TOTAL	75	8′	600

Figure 16-6. Chart for the lever problem of figure 16-5.

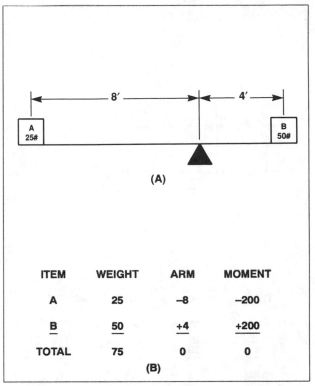

(A)

ITEM	WEIGHT	ARM	MOMENT
A	25	–8	–200
B	50	+4	+200
TOTAL	75	0	0

(B)

Figure 16-7(A)—Simple lever balance problem, using the fulcrum as the datum. (B)—Chart for the problem of figure 16-7(A).

and all of the moments from this new location are calculated. Any counterclockwise moment is considered to be negative, and a moment that causes a clockwise rotation is positive.

Weight A has an arm of negative eight feet, and its moment is –200 pound-feet. The arm of weight B is positive four feet, and when this is multiplied by its weight of 50 pounds, it gives us a moment of +200 pound-feet. The sum of the moments is zero, which means that the board does actually balance about the fulcrum.

It is apparent that the datum can be placed anywhere by working this same problem, using two different locations for the datum. In figure 16-8, the datum is placed between the two weights, three feet to the right of weight A. The arm of A is now negative three feet, and its moment is –75 pound-feet. The arm of weight B is positive nine feet and its moment is +450 pound-feet.

The total moment is +375 and the weight is 75 pounds, so the balance point is five feet to the right of the datum, which places it in exactly the same location previously found, eight feet to the right of A.

Some aircraft manufacturers place the datum a given distance ahead of the aircraft so all of the moments will be positive, and we can see by the

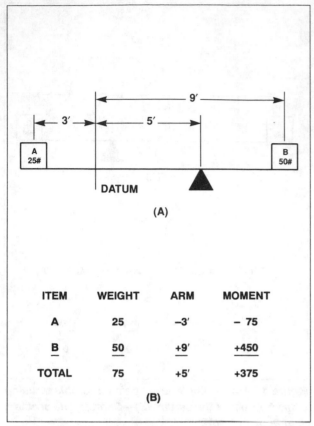

ITEM	WEIGHT	ARM	MOMENT
A	25	–3'	– 75
B	50	+9'	+450
TOTAL	75	+5'	+375

(B)

Figure 16-8(A)—Simple lever problem, placing the datum between weight A and the fulcrum. (B)—Chart for the problem of figure 16-8.

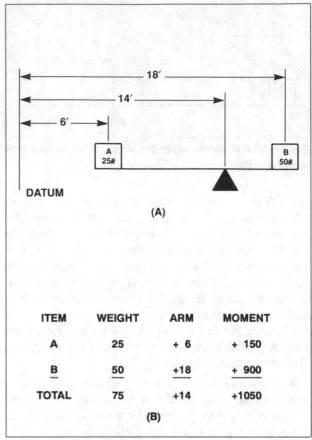

ITEM	WEIGHT	ARM	MOMENT
A	25	+ 6	+ 150
B	50	+18	+ 900
TOTAL	75	+14	+1050

(B)

Figure 16-9(A)—Simple lever problem, placing the datum to the left of the lever. (B)—Chart for the problem of figure 16-9(A).

example of figure 16-9 that this does not change the answer. The datum in this example is located six feet to the left of weight A and the moment of A is +150 pound-feet. Weight B is 18 feet from the datum, and its moment is +900 pound-feet. The total moment is +1,050 pound-feet and when this is divided by the total weight of 75 pounds, the balance is found to be 14 feet to the right of the datum. This again is the same location as was found in the previous two computations, eight feet to the right of weight A.

We can continue with our teeter-totter explanation to find where we would place a third weight to balance the board. In figure 16-10, since the moment of weight B is greater than that of A, there will be a net force tending to rotate the board in a clockwise direction. The moment of this force is +550 pound-inches.

In order to balance the board by placing 50-pound weight C the proper distance from the fulcrum, weight C must have a moment of –550 pound-inches, because for a board to balance about a point, the sum of the moments about that

point must equal zero. The moment of C is –550 pound-inches and its weight is 50 pounds, so its arm must be –11 inches, or the center of weight C must be 11 inches to the left of the fulcrum.

In figure 16-11, we see this balance proven. The sum of the two negative moments is –1,800 pound-inches, and the positive moment is +1,800 pound-inches, so the board balances.

It is necessary for an aircraft center of gravity to fall within a given range, and we sometimes need to add ballast to the aircraft to move the empty-weight center of gravity into the allowable range. We can again use the seesaw to see the way.

Let's assume that the seesaw in figure 16-12 balances at a point 37.5 inches from item A, but we want to balance at a point 42.5 inches away. Weight B is 180 inches from weight A, and there is a location 170 inches from A at which we can place our ballast weight. Our problem is to find the amount of weight we will have to add 170 inches from weight A in order to move the point of balance five inches to the right.

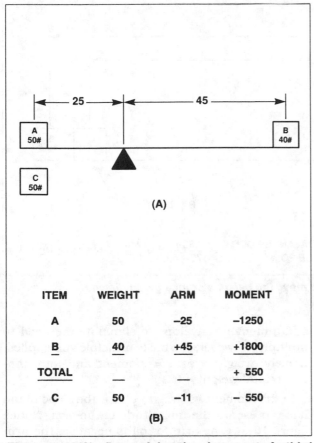

ITEM	WEIGHT	ARM	MOMENT
A	50	−25	−1250
B	40	+45	+1800
TOTAL	—	—	+ 550
C	50	−11	− 550

(B)

Figure 16-10(A)—Determining the placement of a third weight to balance lever. (B)—Chart for the problem of figure 16-10(A).

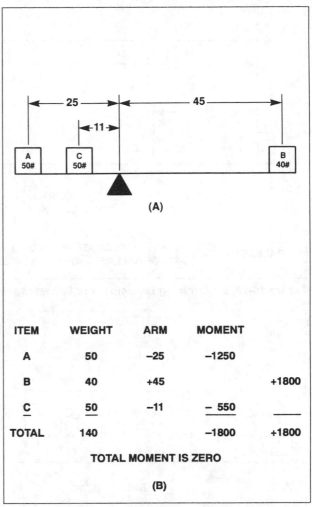

ITEM	WEIGHT	ARM	MOMENT	
A	50	−25	−1250	
B	40	+45		+1800
C	50	−11	− 550	
TOTAL	140		−1800	+1800

TOTAL MOMENT IS ZERO

(B)

Figure 16-11(A)—Proof of the balance found in the problem of figure 16-10. (B)—Chart for the problem of figure 16-11(A).

The formula to use is:

Ballast =

$$\frac{\text{Total weight} \times \text{distance needed to shift balance point}}{\text{Arm of ballast} - \text{arm of desired balance point}}$$

The total weight is 480 pounds, and we need to shift the balance point five inches. The arm of the ballast is 170 inches, and the arm of the new balance point is 42.5 inches. When we work the problem, it is found that we must add 18.82 pounds of ballast at 170 inches to move the balance point five inches to the right.

We can check our computations by posing the problem of figure 16-13. If the sum of all of the moments about the new balance point is equal to zero, we have added the correct amount of ballast.

Item *A* has a weight of 380 pounds, and it is located at an arm of −42.5 inches. This gives it a moment of −16,150 pound-inches. Item *B* has a weight of 100 pounds and an arm of +137.5 inches, giving it a moment of +13,750 pound-inches. the

ballast weighs 18.82 pounds and is located at +127.5, and so it has a moment of +2,400 pound-inches. The total positive moment is +16,150 pound-inches and the total negative moment is −16,150 pound-inches. The sum of the moments about the new balance point is zero, so the ballast was correct.

A typical loading problem would be to determine whether or not an airplane is within its legally loaded envelope when it carries four standard-weight occupants (170 pounds each), full fuel, full oil and 80 pounds of baggage.

You will notice that in figure 16-14, rather than using moments, **moment index** is used. This is simply the moment divided by an appropriate reduction factor to get a number that is convenient to use (usually 1,000 for small aircraft and 10,000 or 100,000 for large and heavy aircraft).

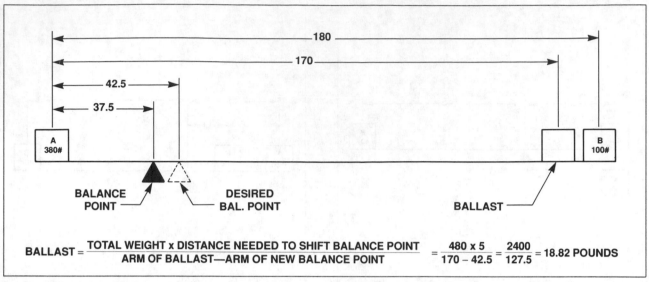

BALLAST = $\dfrac{\text{TOTAL WEIGHT x DISTANCE NEEDED TO SHIFT BALANCE POINT}}{\text{ARM OF BALLAST—ARM OF NEW BALANCE POINT}}$ = $\dfrac{480 \times 5}{170 - 42.5}$ = $\dfrac{2400}{127.5}$ = 18.82 POUNDS

Figure 16-12. Determining the amount of ballast required to move the balance point a given distance.

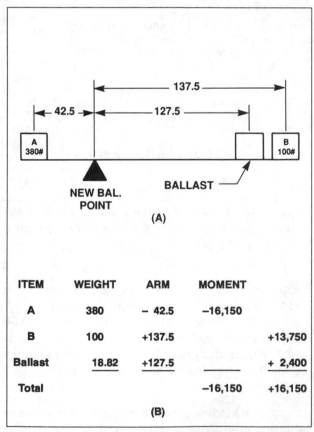

ITEM	WEIGHT	ARM	MOMENT	
A	380	– 42.5	–16,150	
B	100	+137.5		+13,750
Ballast	18.82	+127.5		+ 2,400
Total			–16,150	+16,150

(B)

Figure 16-13(A)—Proof of the accuracy of the ballast determined in the problem of figure 16-12. (B)—Chart for the problem in figure 16-13(A).

Aircraft manufacturers have developed several different schemes to assist the pilot with weight and balance computations. Cessna is one that provides the pilot with a weight vs. moment chart and a moment envelope to eliminate the need to multiply. However, the basic principle still applies, namely: weight × arm = moment and total moment/total weight = CG.

In the moment vs. weight chart, the slope of the line represents the magnitude of the arm. Notice figure 16-14: the arm for oil is negative, the arm for the pilot is small positive and the arm for baggage is large positive. The arm for any item can be found by taking from the chart a moment and weight from the line for any item. Choose values near the highest point of the line to decrease any reading error. Then, moment/weight = arm. For example, the arm for the baggage compartment of the aircraft represented by figure 16-14 is 12,000 pound-inches/126 pounds = 95.2 inches.

The empty weight of the airplane and its moment index is recorded in the chart of figure 16-15. We know that eight quarts, or two gallons, of oil is on board, and since oil weighs 7.5 pounds per gallon, locate 15 pounds on the vertical scale on figure 16-14 and follow it over to the oil curve. Then follow a vertical line from the intersection down to –0.3 on the moment index scale. There are to be two occupants, or a weight of 340 pounds, in the front seats and the moment index for this is +12. The rear seat also carries 340 pounds, and its moment index is +24.5. The 80 pounds of baggage has a moment index of +7.3, and the 40 gallons of fuel weighs 240 pounds and has a moment index of +11.5. The total weight is 2,355 pounds, and the total moment index is +106.6

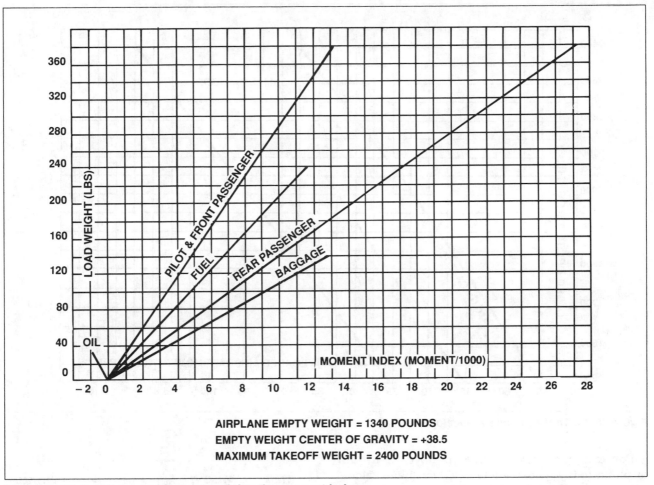

AIRPLANE EMPTY WEIGHT = 1340 POUNDS
EMPTY WEIGHT CENTER OF GRAVITY = +38.5
MAXIMUM TAKEOFF WEIGHT = 2400 POUNDS

Figure 16-14. Loading graph for determining the moment index.

ITEM	WEIGHT	MOMENT INDEX
AIRPLANE	1340	51.6
OIL (8 QTS)	15	−0.3
FRONT SEAT	340	12.0
REAR SEAT	340	24.5
BAGGAGE	80	7.3
FUEL (40 GAL)	240	11.5
TOTAL	2355	106.6
LIMITS	2400	SEE CHART

Figure 16-15. Finding the total moment index of a loaded airplane.

To determine whether or not the aircraft is loaded so that its center of gravity falls within the allowable range, transfer the information just computed to the graph of figure 16-16. Draw a horizontal line through the 2,355-pound weight line and a vertical line through the 106.87 pound-inch/1000 moment index line. These two lines must intersect within the envelope. The actual intersection is within the envelope, near the rearward center of gravity limit.

Notice in figure 16-16 that this airplane is approved for both the normal and the utility categories. In the normal category the allowable gross weight is 2,400 pounds and the most rearward center of gravity is 45.83 inches aft of the datum (moment divided by weight). In the utility category, the maximum gross weight is only 2,050 pounds and the most rearward center of gravity is 40.85 inches aft of the datum.

Although working directly with moments (or moment index) offers the advantage of no multiplication, there is a price to pay. The pilot is unable to glance at the data to gain a sense of the effect of loading on CG that is available when working directly with arm and CG.

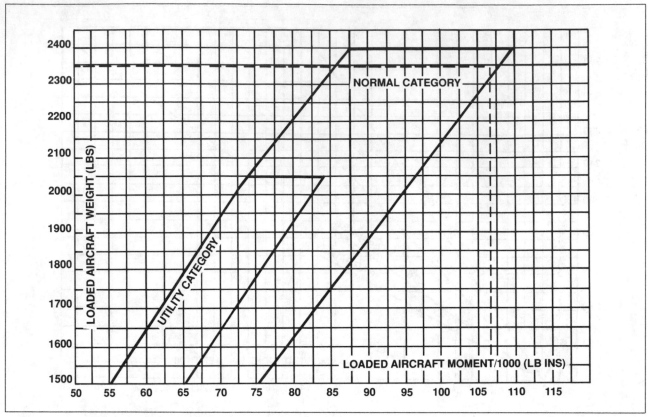

Figure 16-16. Loaded moment envelope.

For example, if you were to have to fly an airplane today whose loaded CG you have calculated to be 102.2″ and the forward CG limit is 102.0″. If you were working directly with arms, at a glance you could see that the arm of the auxiliary fuel tank is 138″. It would be apparent immediately that using fuel from that tank would move the CG forward—perhaps out of limits. It is obvious that a weight and balance computation should be done to explore the consequences of burning fuel from that tank, to determine if the forward CG limit would be breached while enroute.

Getting a feel for CG movements just isn't possible when using moments directly unless problems are worked for all contingencies. Figure 16-17 shows the same problem of figure 16-15 with arms given and moments calculated. To find CG, divide the moment by the total weight:

$$CG = \frac{106,600 \text{ pound–inches}}{2355 \text{ pounds}} = 45.27 \text{ inches}$$

A quick glance at the calculated CG and the fuel arm tells the pilot that the fuel is located only 2.5 inches aft of the aircraft's CG, so it is expected that the CG will move forward somewhat during flight.

ITEM	WEIGHT	x	ARM	=	MOMENT/1000
Airplane	1340	x	38.5	=	51.6
Oil (8 qts)	15	x	−20.0	=	−.3
Front Seat	340	x	35.3	=	12.0
Rear Seat	340	x	72.1	=	24.5
Baggage	80	x	91.3	=	7.3
Fuel (40 gal)	240	x	47.9	=	11.5
Total	2355				106.6
Limits	2400				

Figure 16-17. Calculations with arms given to find moments (from data of figure 16-15).

If the calculated CG is near the forward limit, a destination CG should be calculated to be sure the aircraft is still within limits upon arrival.

Air Taxi Loading

Airplanes that are operated under Part 135 of the Federal Aviation Regulations, that governs Air Taxi and Commercial Operators, must compile a

STANDARD WEIGHTS	
GASOLINE:	6 POUNDS PER GALLON
TURBINE FUEL:	6.7 POUNDS PER GALLON
LUBRICATING OIL:	7.5 POUNDS PER GALLON
CREW & PASSENGERS:	170 POUNDS PER PERSON
WATER:	8.35 POUNDS PER GALLON

NOTE: SINCE A GALLON IS A VOLUME MEASURE, THE ABOVE WEIGHTS VARY SLIGHTLY WITH TEMPERATURE.

FAR 135 OPERATIONS STANDARD WEIGHTS	
ADULTS:	160 POUNDS (SUMMER) FROM MAY 1 TO OCT. 31 165 POUNDS (WINTER) FROM NOV. 1 TO APRIL 30
CHILDREN:	2-12 YEARS: 80 POUNDS
CREW:	FEMALE ATTENDANTS: 130 POUNDS MALE ATTENDANTS: 150 POUNDS OTHER CREW: 170 POUNDS
BAGGAGE:	CHECK IN: 23.5 POUNDS CARRY ON: 5 POUNDS

Figure 16-18. Standard weights for weight and balance purposes.

passenger and cargo manifest for *each* flight. This allows the pilot to know the exact takeoff weight and whether or not the airplane is loaded within the allowable center-of-gravity limits.

The operator is given the choice of using either the actual weight of the passengers and crew, or of using standard weights for the occupants and the baggage. Standard weights are shown in figure 16-18.

A typical passenger and cargo manifest is shown in figure 16-19, in which the actual weight of the passengers and the moment indexes found in figure 16-20 are used.

The fuel weight and moment indexes found in figure 16-21 are entered in the appropriate blanks as well as the weight and moment indexes of the crew, the oil and the empty airplane. The total weight and moment index is the sum of these individual values. The maximum allowable weight for this flight is 10,000 pounds.

By referring to figure 16-24 we find the minimum and maximum moment indexes that give the allowable forward and aft center-of-gravity limits for each weight. In the example, the loaded weight is 9,435 pounds, and in order to find the forward and aft limits we must interpolate between the figures for 9,400 and 9,450 pounds. Figure 16-24 shows the difference in the minimum index is 89.8 index numbers for each 100 pounds, and we are 35 pounds, or 35%, above 9,400 pounds; so we must add 35% of 89.8 index numbers to get the minimum index number of 8,467.9. The maximum index interpolation requires 102.8 index numbers for each 100 pounds. Thirty-five percent of this is

36.0, so the maximum index for 9,435 pounds is 9,694.5. Both the loaded weight and loaded moment index are within the allowable limits.

Shifting CG

Temporary Ballast—Moving The CG Back Within Limits

For certain flight conditions, it may be necessary to carry temporary ballast to keep the aircraft within the allowable center-of-gravity limits. Some tandem-seat trainers must be flown solo from the rear seat because, with one occupant in the front seat and a full tank of fuel ahead of the front seat, the center of gravity will be ahead of the forward limit.

If a pilot wants to fly solo from the front seat, enough ballast must be carried in the baggage compartment behind the rear seat to bring the loaded center of gravity into range. In this instance, we want to move the center of gravity aft two inches. In the example of figure 16-25, the loaded weight of the aircraft is 1,045 pounds and the loaded CG is +10 inches. The center of the baggage compartment is 36 inches behind the forward center-of-gravity limit. A weight of 58 pounds must be carried in the baggage compartment for solo flight from the front seat. This ballast must be clearly marked, stating that it is to be carried in the aircraft only when the aircraft is being flown solo from the front seat. As a practical consideration, it must be ensured that the baggage compartment is structurally strong enough to carry this amount of weight.

PASSENGER-CARGO LIST	WEIGHT	SEAT COMPT.	INDEX
	180	1	126.0
	170	2	119.0
	140	3	126.0
Passenger Names	150	4	135.0
entered here	200	5	220.0
	160	6	176.0
	210	7	273.0
	130	8	169.0
	120	9	180.0
	140	10	210.0
	130	11	221.0
PASSENGER — CARGO — MAIN CABIN TOTAL			

FUEL						
WEIGHT	TANK	INDEX	BAGGAGE COMPT	100	NOSE	20.0
300	R—MAIN	270.0	BAGGAGE COMPT	200	REAR	380.0
300	L—MAIN	270.0	PILOT—COPILOT	330		165.0
120	R—AUX	144.0	OIL	85		7.7
120	L—AUX	144.0	FUEL	840	TOTAL	828.0
	NOSE		EMPTY WEIGHT	6,150		5842.5
840	TOTAL	828.0	TAKEOFF WEIGHT	9,435		9198.2
			TAKEOFF LIMITS	10,000		8467.9 9694.5

AIRCRAFT N 123QC

FLIGHT 45

ROUTE OKC – MKC

DATE _____

I HEREBY CERTIFY THAT THE ABOVE TAKEOFF WEIGHT AND INDEX IS WITHIN LIMITS AS SPECIFIED IN THE FLIGHT OPERATIONS MANUAL.

_____
PILOT IN COMAND

Figure 16-19. Passenger and cargo manifest.

WEIGHT	PILOT & COPILOT	SEATS 1 & 2	SEATS 3 & 4	SEATS 5 & 6	SEATS 7 & 8	SEATS 9 & 10	SEAT 11
100	50.0	70.0	90.0	110.0	130.0	150.0	170.0
110	55.0	77.0	99.0	121.0	143.0	165.0	187.0
120	60.0	84.0	108.0	132.0	156.0	180.0	204.0
130	65.0	91.0	117.0	143.0	169.0	195.0	221.0
140	70.0	98.0	126.0	154.0	182.0	210.0	238.0
150	75.0	105.0	135.0	165.0	195.0	225.0	255.0
160	80.0	112.0	144.0	176.0	208.0	240.0	272.0
170	85.0	119.0	153.0	187.0	221.0	255.0	289.0
180	90.0	126.0	162.0	198.0	234.0	270.0	306.0
190	95.0	133.0	171.0	209.0	247.0	285.0	323.0
200	100.0	140.0	180.0	220.0	260.0	300.0	340.0
OCCUPANTS							MOMENT/100

Figure 16-20. Moment indexes for occupants.

WEIGHT	NOSE	MAIN	AUX.
30	6.0	27.0	36.0
60	12.0	54.0	72.0
90	18.0	81.0	108.0
120	24.0	108.0	144.0
150	30.0	135.0	180.0
180	36.0	162.0	216.0
210	42.0	189.0	252.0
240	48.0	216.0	288.0
270	54.0	243.0	324.0
300	60.0	270.0	360.0
330	66.0	297.0	396.0
360	72.0	324.0	432.0
390	78.0	351.0	468.0
420	84.0	378.0	504.0
450	90.0	405.0	540.0
480	96.0	432.0	576.0
510	102.0	459.0	612.0
540	108.0	486.0	648.0
570	114.0	513.0	684.0
600	120.0	540.0	720.0
FUEL MOMENT/100			

Figure 16-21. Moment indexes for fuel.

GALLON	WEIGHT	MOMENT
2	17	1.5
4	34	3.1
6	51	4.6
8	68	6.1
10	85	7.7

Figure 16-23. Moment indexes for oil.

Shifting Weight

Large aircraft having several rows of seats and more than one baggage compartment may be kept in balance without adding ballast, by shifting some of the weight that is carried. For example, figure 16-26 shows a large aircraft with a baggage compartment at station 26 and one at station 246. To find the amount of weight that must be shifted to bring the center of gravity back 1.5 inches, the ratio of amount of weight shifted to the total weight of the aircraft is proportional to the ratio of the change in center of gravity required to the distance the

WEIGHT	NOSE BAGGAGE	COMPT. A	COMPT. B	COMPT. C	COMPT. D	COMPT. E	REAR BAG.
20	4.0	16.0	20.0	24.0	28.0	32.0	38.0
40	8.0	32.0	40.0	48.0	56.0	64.0	76.0
60	12.0	48.0	60.0	72.0	84.0	96.0	114.0
80	16.0	64.0	80.0	96.0	112.0	128.0	152.0
100	20.0	80.0	100.0	120.0	140.0	160.0	190.0
120	24.0	96.0	120.0	144.0	168.0	192.0	228.0
140		112.0	140.0	168.0	196.0	224.0	266.0
160		128.0	160.0	192.0	224.0	256.0	304.0
180		144.0	180.0	216.0	252.0	288.0	342.0
200		160.0	200.0	240.0	280.0	320.0	380.0
300		240.0	300.0	360.0	420.0	480.0	
400		320.0	400.0	480.0	560.0	640.0	
500		400.0	500.0	600.0	700.0	800.0	
		BAGGAGE	MOMENT/100				

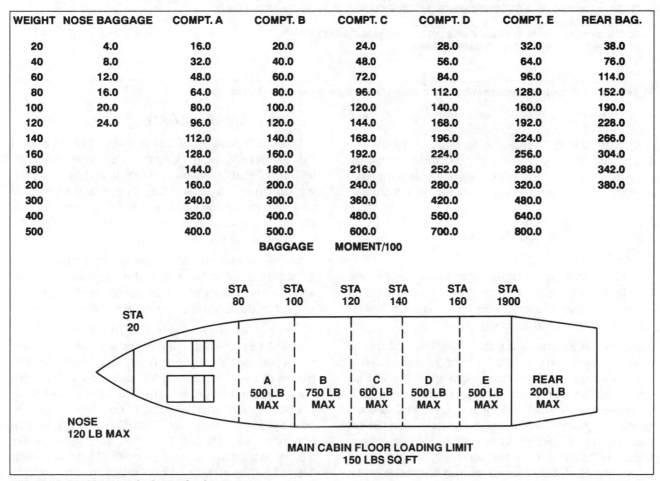

Figure 16-22. Moment indexes for baggage.

WEIGHT	MIN. INDEX	MAX. INDEX	WEIGHT	MIN. INDEX	MAX. INDEX	WEIGHT	MIN. INDEX	MAX.INDEX
7000	6282.5	7192.5	8000	7180.0	8220.0	9000	8077.5	9247.5
7050	6327.4	7243.9	8050	7224.9	8271.4	9050	8122.4	9298.9
7100	6372.4	7295.3	8100	7296.8	8322.8	9100	8167.3	9350.3
7150	6417.1	7346.6	8150	7314.6	8374.1	9150	8212.1	9401.6
7200	6462.0	7398.0	8200	7359.5	8425.5	9200	8257.0	9453.0
7250	6506.9	7449.4	8250	7404.4	8476.9	9250	8301.9	9504.4
7300	6551.8	7500.8	8300	7449.3	8528.3	9300	8346.8	9555.8
7350	6596.6	7552.1	8350	7494.1	8579.6	9350	8391.6	9607.1
7400	6641.5	7603.5	8400	7539.0	8631.0	9400	8436.5	9658.5
7450	6686.4	7654.9	8450	7583.9	8682.4	9450	8481.4	9709.9
7500	6731.3	7706.3	8500	7628.8	8733.8	9500	8526.3	9761.3
7550	6776.1	7757.6	8550	7673.6	8785.1	9550	8571.1	9812.6
7600	6821.0	7809.0	8600	7718.5	8836.5	9600	8616.0	9864.0
7650	6865.9	7860.4	8650	7763.4	8887.9	9650	8660.9	9915.4
7700	6910.8	7911.8	8700	7808.3	8939.3	9700	8705.8	9966.8
7750	6955.6	7963.1	8750	7853.1	8990.6	9750	8750.6	10018.1
7800	7000.5	8014.5	8800	7898.0	9042.0	9800	8795.5	10069.5
7850	7045.4	8065.9	8850	7942.9	9093.4	9850	8840.4	10120.9
7900	7090.3	8117.3	8900	7987.8	9144.8	9900	8885.3	10172.3
7950	7135.1	8168.6	8950	8032.6	9196.1	9950	8930.1	10223.6
						10000	8975.0	10275.0

THE DIFFERENCE IN MIN. INDEX FOR EACH 100 POUNDS IS 89.8 INDEX NUMBERS
THE DIFFERENCE IN MAX. INDEX FOR EACH 100 POUNDS IS 102.8 INDEX NUMBERS

TO FIND LIMITS FOR A SPECIFIC WEIGHT: FOR EXAMPLE 9435 FIND LIMITS FOR 9400
8436.5 — 9658.5
FOR 9435 POUNDS, ADD 35% OF 89.8 OR 31.4 TO 8436.5 = 8467.9
ADD 35% OF 102.8 OR 36.0 TO 9658.3 = 9694.5

Figure 16-24. Chart for determining the moment envelope using moment index.

weight is shifted. Using the formula in figure 16-26, we find that by shifting 55.9 pounds of baggage from the front to the rear baggage compartment, we will shift the center of gravity aft by 1.5 inches.

Adverse-loaded Center Of Gravity

It is possible to determine whether or not one can load an aircraft in such a way that it will fall outside of the allowable center-of-gravity limits. On some aircraft, it is easily done!

Figure 16-27 shows a typical four-place airplane powered by an engine of 230 METO horsepower. It carries 12 quarts of oil, 50 gallons of fuel, and four occupants and has a rear baggage compartment that is limited to 120 pounds of baggage. The forward center-of-gravity limit is at +53 inches, and the aft limit is at +58 inches. The airplane has an empty weight of 2,024 pounds and an empty-weight center of gravity of +52.5. The maximum allowable gross weight for this airplane is 3,150 pounds.

Forward Adverse-loading Check

First, let's find out if it is possible to legally load this airplane in such a way that its center of gravity will fall ahead of the forward limit without exceeding maximum gross weight. To check this, we will *use the maximum weight of everything that is ahead of the forward limit and minimum weight for all items aft of the forward limit.*

We begin our check by compiling a chart such as figure 16-28. We enter the airplane with its weight, its arm and its moment. Enter the full weight of the oil and its arm and moment.

The center of gravity of the fuel is behind the forward center-of-gravity limit, so use the minimum fuel for this computation. The minimum fuel for weight and balance computations is no more than the amount of fuel required for one half-hour of operation at rated maximum continuous power, or maximum except takeoff (**METO power**), which, according to FAR (14CFR) Part 43.13A, Chapter 13, is $1/12$ gallon of fuel per METO horsepower. Therefore the weight of minimum fuel in pounds is equal to METO horsepower divided by two.

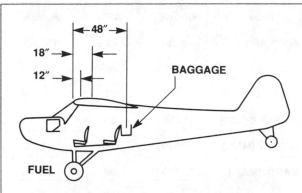

AIRPLANE WITH PILOT & FULL FUEL = 1045 POUNDS
WITH A CENTER OF GRAVITY OF + 10
FIND BALLAST NEEDED IN BAGGAGE COMPARTMENT
AT + 48 TO MOVE CG BACK TO + 12

BALLAST NEEDED =
$$\frac{\text{AIRCRAFT WEIGHT} \times \text{DISTANCE TO MOVE CG}}{\text{DISTANCE FROM BALLAST TO DESIRED CG}}$$

$$\text{BALLAST NEEDED} = \frac{1045 \times 2}{36} = 58 \text{ POUNDS}$$

PROOF

ITEM	WEIGHT	ARM	MOMENT
AIRPLANE	1045	10	10,450
BALLAST	58	48	2,784
	1103		13,234

$$CG = \frac{13234}{1103} = 11.99 \text{ INCHES}$$

Figure 16-25. Use of temporary ballast to correct for an unusual flight condition.

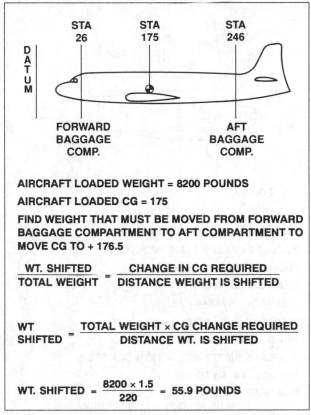

AIRCRAFT LOADED WEIGHT = 8200 POUNDS

AIRCRAFT LOADED CG = 175

FIND WEIGHT THAT MUST BE MOVED FROM FORWARD BAGGAGE COMPARTMENT TO AFT COMPARTMENT TO MOVE CG TO + 176.5

$$\frac{\text{WT. SHIFTED}}{\text{TOTAL WEIGHT}} = \frac{\text{CHANGE IN CG REQUIRED}}{\text{DISTANCE WEIGHT IS SHIFTED}}$$

$$\text{WT. SHIFTED} = \frac{\text{TOTAL WEIGHT} \times \text{CG CHANGE REQUIRED}}{\text{DISTANCE WT. IS SHIFTED}}$$

$$\text{WT. SHIFTED} = \frac{8200 \times 1.5}{220} = 55.9 \text{ POUNDS}$$

Figure 16-26. Shifting weight to move the operational center of gravity.

We must have a pilot, and since the center of gravity of the front seat is behind the forward center-of-gravity limit, only one occupant should be calculated to be in the front seat. The standard weight of a person is 170 pounds.

The rear seats and the baggage compartment are both behind the forward center-of-gravity limit, so we do not want any occupants in the rear seats nor do we want any baggage in the baggage compartment.

When computing the weight and moment of the airplane, full oil, minimum fuel and one front-seat occupant, a total weight for the maximum forward center-of-gravity condition is found to be 2,331.5 pounds, and a most forward center of gravity is +52.54 inches. This loading is out of limits!

Rearward Adverse-loading Check

Check the rearward loading in the same way. Since the center of gravity of the fuel is ahead of the rear limit, use minimum fuel. The front seat is ahead of the rear limit, so use only one occupant in the front seat. Both the rear seat and the baggage compartment are behind the rear center-of-gravity limit, so use two occupants in the rear

Part 43 may be the most authoritative in this case but operationally, FAR Part 91.151 applies, indicating that minimum fuel required is 30 minutes at "normal" (whatever that is) power. METO power for most of today's piston engines is 75% power. I would interpret "normal power" to be 65%. The two regulations aren't far apart, with Part 43 being the most conservative.

So, if the aircraft manufacturer has not given this figure for computing the weight and balance, it is permissible to use $\frac{1}{12}$ gallon of gasoline for each maximum-except-takeoff (METO) horsepower. Aviation gasoline has a nominal weight of six pounds per gallon, so we can find the weight of the minimum fuel by dividing the METO horsepower by two. In our example we have an engine with 230 METO horsepower, so our minimum fuel weight is 115 pounds.

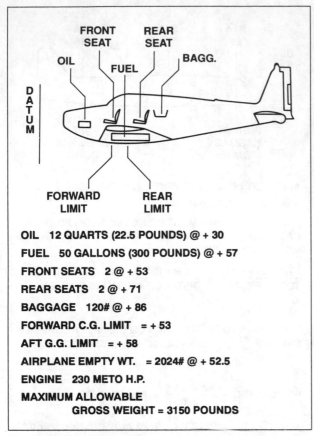

OIL   12 QUARTS (22.5 POUNDS) @ + 30
FUEL   50 GALLONS (300 POUNDS) @ + 57
FRONT SEATS   2 @ + 53
REAR SEATS   2 @ + 71
BAGGAGE   120# @ + 86
FORWARD C.G. LIMIT   = + 53
AFT G.G. LIMIT   = + 58
AIRPLANE EMPTY WT.   = 2024# @ + 52.5
ENGINE   230 METO H.P.
MAXIMUM ALLOWABLE
        GROSS WEIGHT = 3150 POUNDS

Figure 16-27. Airplane for use in determining adverse-loaded center of gravity.

ITEM	WEIGHT	ARM	MOMENT
AIRPLANE	2,024	52.5	106,260
OIL	22.5	30	675
FUEL (MIN.)	115	57	6,555
F. SEAT (PILOT)	170	53	9,010
	2,331.5	52.54	122,500

Figure 16-28. Forward adverse-loaded center of gravity check.

seat and the maximum allowable amount of baggage, 120 pounds (figure 16-29).

Find the total moment and divide this by the total weight, producing a weight for maximum rearward center-of-gravity conditions of 2,791.5 pounds and the most rearward center of gravity of +56.23 inches. This is within the allowable limits of both weight and balance.

ITEM	WEIGHT	ARM	MOMENT
AIRPLANE	2024	52.5	106,260
OIL	22.5	30	675
FUEL (MIN.)	115	57	6,555
F. SEAT (PILOT)	170	53	9,010
R. SEAT (MAX.)	340	71	24,140
BAGG. (MAX.)	120	86	10,320
	2,791.5	56.23	156,960

Figure 16-29. Aft adverse-loaded center of gravity check.

ITEM	WEIGHT	ARM	MOMENT
AIRPLANE	2024	52.5	106,260
OIL	22.5	30	675
FUEL (FULL)	300	57	17,100
F. SEAT (MAX.)	340	53	18,020
R. SEAT (MAX.)	340	71	24,140
BAGG. (MAX.)	120	86	10,320
	3,146.5	56.09	176,515

Figure 16-30. Maximum gross weight center of gravity check.

Maximum Gross Weight Condition Check

We know now that the airplane cannot be loaded in such a way that it will fall outside of the aft center-of-gravity limits, let's see if we can exceed the maximum gross weight.

Construct a chart like figure 16-30. Add the total weight and moment of the airplane, full oil, full fuel, maximum baggage and four occupants. The maximum weight is 3,146.5 pounds, which is less than the maximum allowable gross weight. The fully loaded center of gravity is +56.09 inches, which is well within the allowable center-of-gravity range. However, the pilot would do well to remember that the average weight of four occupants may not exceed 171 pounds with full fuel and baggage!

The example aircraft discussed above probably isn't typical. Very few aircraft can be operated with full fuel and baggage and all seats occupied by adults without being over gross weight.

Although Part 91 implies that weight and balance be determined *before every flight* for takeoff, minimum fuel (and probably at fuel exhaustion), and modern personal computers make this job very easy, operational reality makes this difficult, especially in corporate operations. However, pilots with a professional attitude will compute numerous "worst-case scenarios" covering all possible loadings of the aircraft they fly and keep these computations in the aircraft flight manual, referring to them as needed to see if an out-of-limits loading situation might exist.

Weight And Balance Changes After An Alteration

The A&P maintenance technician will have to find the new empty weight and empty-weight center of gravity after he/she has altered an aircraft.

Normally, if the weights of the items installed or removed from the aircraft are known and the location of the alteration is known, a new empty weight and empty-weight center of gravity can be computed without having to reweigh the aircraft. The new empty weight and balance can be determined either way.

In the example given in figure 16-31, an alteration was done that removed an old radio from the instrument panel, along with its power supply that was mounted in the baggage compartment. In its place a smaller radio was installed in the instrument panel. This new radio did not require a separate power supply. At the same time, an anti-collision beacon was installed on the tail.

Figure 16-31 shows a chart that allows the computation of the change in the empty weight and the empty-weight center of gravity. The empty weight of the airplane is 2,024 pounds and its empty-weight center of gravity is +52.5 inches. The radio that is removed from the instrument panel has a negative weight of 16 pounds at an arm of +49 inches. The power supply weight is also negative, –12, at an arm of +86. The radio that is to be installed has a positive weight of +9 pounds at a positive arm of +49 inches. The anti-collision beacon has a weight of +4 pounds at an arm of +127 inches. The sum of the positive moments is +107,209 pound-inches and the negative moments are –1,816 pound-inches, for a total

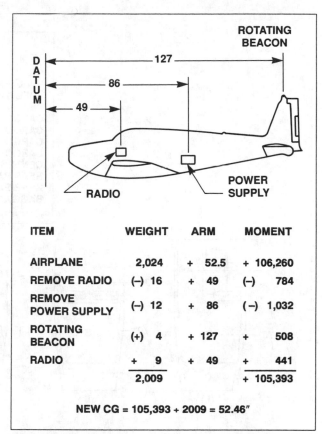

ITEM	WEIGHT	ARM	MOMENT
AIRPLANE	2,024	+ 52.5	+ 106,260
REMOVE RADIO	(–) 16	+ 49	(–) 784
REMOVE POWER SUPPLY	(–) 12	+ 86	(–) 1,032
ROTATING BEACON	(+) 4	+ 127	+ 508
RADIO	+ 9	+ 49	+ 441
	2,009		+ 105,393

NEW CG = 105,393 ÷ 2009 = 52.46″

Figure 16-31. Weight and balance changes resulting from an alteration.

moment of +105,393 pound-inches. The new total weight is 2,009 pounds. When the total moment is divided by the total weight, the new empty-weight center of gravity is 52.46 inches aft of the datum.

The pilot will find a computation of this sort, done by a technician, in the aircraft's weight and balance documentation by looking for the most recently dated Form 337 or other form of weight and balance document.

Cargo Aircraft

Most aircraft used for or convertible to cargo operations provide the pilot with loading diagrams like figure 16-22 or like the one for the Cessna 180 in figure 16-32(A) through (C). From this information, the pilot can compute loading for any cargo before loading efforts begin. Recommended cargo tie-down (a must) is shown in figure 16-32(C).

In areas where airplanes are the only means of transportation, many types of cargo are carried. Figure 16-33 shows the loading of a barrel of fuel aboard a Maule on floats, to be flown to a fuel cache somewhere in the wilderness.

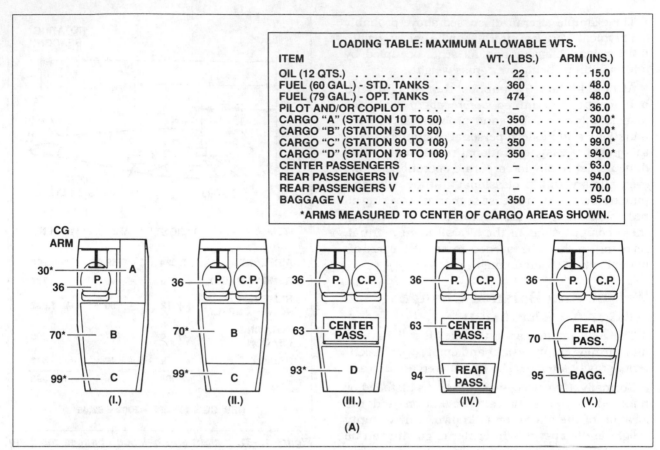

LOADING TABLE: MAXIMUM ALLOWABLE WTS.		
ITEM	WT. (LBS.)	ARM (INS.)
OIL (12 QTS.)	22	15.0
FUEL (60 GAL.) - STD. TANKS	360	48.0
FUEL (79 GAL.) - OPT. TANKS	474	48.0
PILOT AND/OR COPILOT	–	36.0
CARGO "A" (STATION 10 TO 50)	350	30.0*
CARGO "B" (STATION 50 TO 90)	1000	70.0*
CARGO "C" (STATION 90 TO 108)	350	99.0*
CARGO "D" (STATION 78 TO 108)	350	94.0*
CENTER PASSENGERS	–	63.0
REAR PASSENGERS IV	–	94.0
REAR PASSENGERS V	–	70.0
BAGGAGE V	350	95.0
*ARMS MEASURED TO CENTER OF CARGO AREAS SHOWN.		

Figure 16-32(A). Seating-cargo arrangements.

Aircraft Inspections

Purpose Of Inspection Programs

When an aircraft is certified by the Federal Aviation Administration, it is issued a **Type Certificate**, a legal document that states that the FAA has thoroughly examined the design of this aircraft and has determined that it meets all of the requirements for airworthiness as specified in the appropriate part of the Federal Aviation Regulations.

The **Type Certificate** is an approval for the *design* of the aircraft, but the aircraft is actually built by the manufacturer under a **Production Certificate**, a legal document that states that the FAA has examined and continually monitors the production facilities of the manufacturer. The FAA certifies that the facilities, personnel and methods used in the production of the particular aircraft are such that will ensure that when completed, the aircraft will be equivalent to that specified by the Type Certificate. Because of this, any aircraft of this particular design, fabricated under the production certificate, is considered to meet all of the airworthiness requirements, and the aircraft is legal to fly, and is issued a **Certificate of Airworthiness**.

If any condition arises after the aircraft is certificated that indicates that it has some defect rendering it incapable of meeting its certification requirements, an **Airworthiness Directive (AD)** is issued by the FAA that becomes a part of the certification, and its provisions must be met within a specified time period for the aircraft to remain airworthy. These Airworthiness Directives are sent to the registered owner of every affected aircraft, and all maintenance technicians and repair stations have access to them.

It is possible that damage or even normal wear could cause an aircraft to fail to meet its conditions of certification, and when this happens the aircraft cannot be legally flown until the debilitating condition is corrected. To ensure that its condition is continually monitored, the FAA has established a series of required inspections for all aircraft.

Required Inspections

Preflight

The pilots of all aircraft are charged with the responsibility of ensuring that the aircraft is in condition for the flight that is proposed. This means that the aircraft must have had the

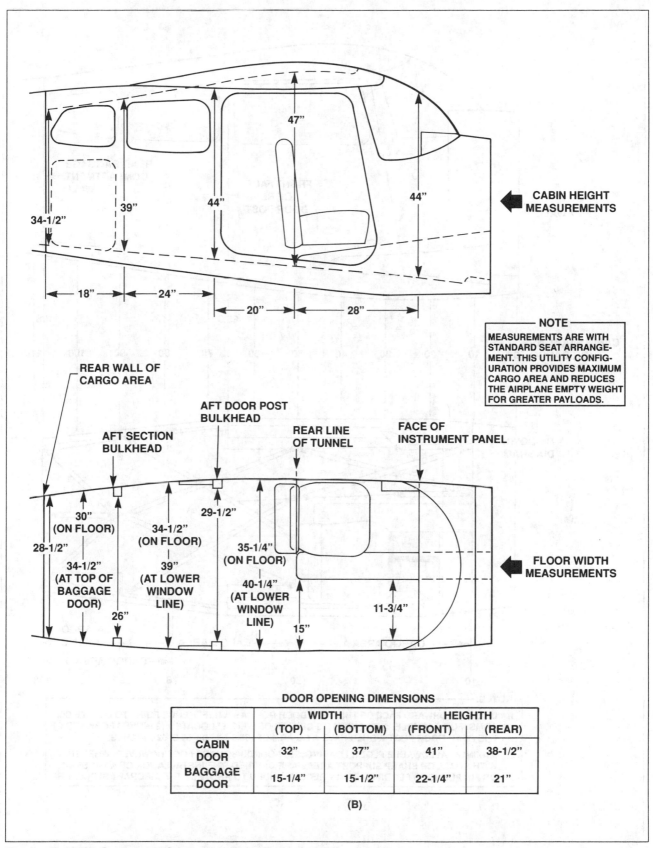

DOOR OPENING DIMENSIONS

	WIDTH		HEIGHT	
	(TOP)	(BOTTOM)	(FRONT)	(REAR)
CABIN DOOR	32"	37"	41"	38-1/2"
BAGGAGE DOOR	15-1/4"	15-1/2"	22-1/4"	21"

(B)

Figure 16-32(B). Cessna 180G internal cabin dimensions.

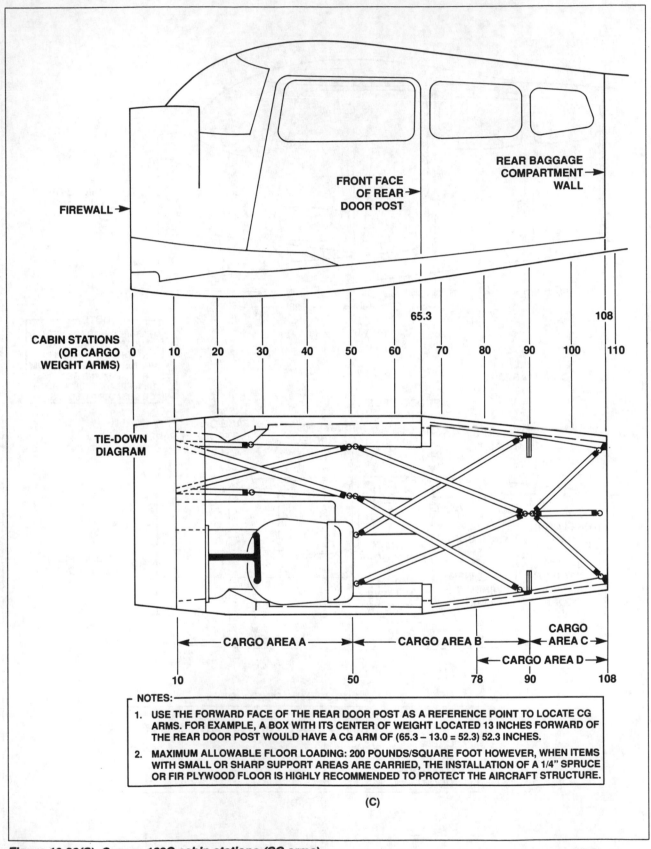

NOTES:

1. USE THE FORWARD FACE OF THE REAR DOOR POST AS A REFERENCE POINT TO LOCATE CG ARMS. FOR EXAMPLE, A BOX WITH ITS CENTER OF WEIGHT LOCATED 13 INCHES FORWARD OF THE REAR DOOR POST WOULD HAVE A CG ARM OF (65.3 – 13.0 = 52.3) 52.3 INCHES.

2. MAXIMUM ALLOWABLE FLOOR LOADING: 200 POUNDS/SQUARE FOOT HOWEVER, WHEN ITEMS WITH SMALL OR SHARP SUPPORT AREAS ARE CARRIED, THE INSTALLATION OF A 1/4" SPRUCE OR FIR PLYWOOD FLOOR IS HIGHLY RECOMMENDED TO PROTECT THE AIRCRAFT STRUCTURE.

(C)

Figure 16-32(C). Cessna 180G cabin stations (CG arms).

Figure 16-33. Loading a barrel of fuel for transportation into the wilderness.

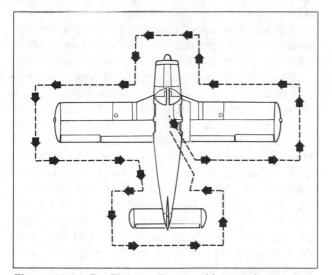

Figure 16-34. Preflight walkaround inspection.

required major inspection within the proper time limit. There must be on board all of the items of required equipment and a current weight and balance report. The pilot must have determined that under the flight load condition proposed, the aircraft will be within the approved weight and balance limitations at the beginning, during and at the end of the flight. The aircraft must have on board sufficient fuel and oil for the flight, and all of the required systems must be functioning properly.

Mr. Jeff Boerboon, who was the 1989 regional and national champion in the preflight inspection National Intercollegiate Flying Association competition, had these words of advice for pilots about doing preflight inspections:

"Have you ever been ramp checked? If an FAA examiner came up while you were performing a preflight, you might be expected to be able to

answer lots of questions about the process. For example, if the examiner asked why there was a superseded page missing out of the weight and balance, what would you say?

You should be able to tell him that not all the superseded pages need to be in the airplane as long as the most current page is there. A knowledge of the wording of the regulations pertaining to inspection and licensure records is important, so a periodic review of 14CFR Parts 91 and 43 on this subject is worthwhile. Part 91.3 tells us that the pilot in command of an aircraft is directly responsible for, and is the final authority as to, the operation of that aircraft. No person may operate a civil aircraft unless it is in airworthy condition. The pilot in command of a civil aircraft is responsible for determining whether that aircraft is in condition for safe flight, according to Part 91.7. This last regulation really makes necessary a complete and accurate preflight inspection.

It is most important for every pilot to create their own repeatable system to use while doing each preflight. A generic system is discussed below that could be used as an outline when making up your own system to match the particular aircraft you fly. An outline appears below to make the process easier to follow.

I. Cockpit
    a. paperwork
    b. seats and floor
    c. instrument panel
    d. placards
    e. electrical

II. Empennage
    a. skin
    b. inspection panels
    c. control surfaces
    d. fuselage attachment
    e. counterweights

III. Left/right wing
    a. flaps
    b. ailerons
    c. skin
    d. inspection panels
    e. tips and lights
    f. landing gear

IV. Nose section
    a. landing gear
    b. cowling
    c. propeller/spinner
    d. engine (including oil level)

V. Fuel

Now let's look at each section of this outline and discuss some things to look for that are not covered on the standard checklist. First, **the cockpit**. The acronym ARROW is often used to help with the task of paperwork inspection. The A stands for Airworthiness certificate. This certificate does not have an expiration date so it is good as long as the aircraft is kept in an airworthy condition. It must have the correct N number of the aircraft on it, and must have a blue seal stamped on the front. If the certificate is a replacement, it must say so.

The R stands for Registration certificate. This certificate comes first as a temporary and is good for 90 days or until the permanent certificate is issued which is good until the aircraft changes owners or registration numbers. The temporary certificate is not good for international flight.

The second R stands for Radio station license. It must be aboard the aircraft, and not expired, in order to legally transmit (use the communications radio, transponder, DME, radar or anything that transmits a signal).

O stands for operating limitations, placards and markings. This includes the pilot's operating handbook or flight manual and the necessary placards and markings required by Part 91.

Finally, the W refers to the weight and balance data. The original W&B issued with the aircraft should be present as it contains the required equipment list. Usually, there is a paper trail from that first data sheet showing each addition and deletion, marked superseded until the current form. This trail may be broken and started anew by weighing the aircraft. When you are finished with the paperwork, remember to place the Airworthiness Certificate where it can be seen (FAR 91.203b).

When looking at the **seats and floor**, be sure that the seat stops are all in place and correctly located with locking pins fastened; seat belts properly attached with FAA approval tag in place on each belt. Floor panels, flap handle, brake and rudder pedals and anything else that appears as though it should be tied down should be inspected for anything out-of-the-ordinary. Be sure to ask a qualified maintenance person if you're not sure about the way things look.

Moving on to the **instruments**, some important things to look for here include the general condition of the instruments. The DG should be controllable, as should the adjustable aircraft symbol on the attitude indicator. The indicated altitude should change 100 feet with each 0.1" change in the Kollsman window reading and the altimeter, when set to the current altimeter reading, should indicate the field elevation within 75 feet (an IFR requirement but a good standard for VFR flight, too). Check the markings on the airspeed indicator for Vne, Vno and flap extension speed against the published values in the POH, as many airspeed indicators aren't correctly marked on older aircraft. The magnetic compass should point generally in the right direction and the deviation card must be readable. To continue with **placards**, the list of placards in the flight manual should agree with what is seen on the aircraft. Unreadable or missing placards make the aircraft unairworthy.

Electrical equipment is something else that is not covered well on most preflight checklists. When the master switch is turned on, there is more to check than the fuel quantity indicators and flap extension. Developing a system sequence is really important here. Remember the fuel quantity gauge readings to compare later with the visual inspection of the tanks. Pick a pattern that flows smoothly when checking the electrical system of your airplane. If the fuel pump switch is next to the master, turn it on and listen for its operation. If the pitot heat is next, turn it on momentarily and note the ammeter reading to check its operation. Using the appropriate rheostats and switches, check the interior lights, then the exterior lights (per FAR 91.213, they are required for day flight unless deactivated and placarded). Return to the cockpit, shut off the lights, master and be sure the flaps are down. Avionics may be checked, also.

Outside, look for signs of fatigue in the empennage, such as wrinkled or cracked skin and popped rivets. All inspection panels must be in place and all screws intact. Elevator and rudder hinge bushings should be smooth and without play and travel stops must be secure and working. Control surface counterweights must be in place and secure as well as hingebolt cotterpins or locknuts (with threads showing). It is not uncommon to find a bird's nest or other foreign objects tucked away in the tail cone or stabilizers, so check for that. The stabilizer-to-fuselage attach points should be checked by trying to move the tip of the stabilizer fore-and-aft and up-and-down with thumb and forefinger. *Any* detectable play here should make you not want to ride in that airplane until checked out by a qualified person.

Moving to the wing, look again for signs of overstress, both top and bottom. Check the flaps to see that they both come down the same distance, that they are both secure with a minimum of play and that the hingepin retaining fasteners are properly in place. Do the same check on the aileron but include a check to see that, when the aileron is deflected, the yoke is deflected in the correct direction and that the counterweights are present and secure. Check the wing for missing screws, fairings, lights and inspection covers as well as dents and general condition.

The landing gear should be checked for proper oleo strut inflation, hydraulic leaks, brake pad and shoe condition and wear, tire inflation and condition. Take a good, long look for anything unusual such as debris, lumps on tires that would bind in the wheel well, missing cotterpins, worn hydraulic lines, etc.

The cowling is an area of high vibration from engine and propeller slip stream so watch for missing screws, and especially loose rivets in the cowl flap hinges and inspection openings. Check inside the cowl for foreign objects (bird nests, animals, trash). Look for dings in the propeller leading edge and face and evidence of oil leakage from the propeller seals as well as missing screws, dents and general security of the spinner. An attempt to move the exhaust stack by hand should prove that it is secure. A loose exhaust system is a good way to start an engine compartment fire in flight or cause CO poisening of the occupants. Fuel tanks should be sumped in a definite order (see Chapter 6).

The preflight is one of the most important parts of a safe and fun flight. The more airplanes that you fly and the more FBOs that you rent from, the more you will become convinced of the importance of going through the airplane very carefully. A good preflight should take about 15 minutes, which gets hard sometimes in the winter (when it is 40 below outside) but once you get a system that works for you the enjoyment with doing it and your confidence will increase. Here are some samples of things you might find during a thorough preflight:

—missing weight and balance and/or equipment list

—landing gear panel lights switched (the nose gear light in the main gear position)

—seat stops missing or improperly located

—fire extinguisher missing, discharged or out-of-date

—outside air temperature gauge loose

—missing cotter pin on wheel axle (they have been known to fall off in flight)

—safety wire on gear door missing

—plugged oil breather tube

—position lights switched (red where the green should be)

—plugged pitot tube

—obstructed static port

—inspection panel loose

—fuel placard missing (placards are required at filler port AND at fuel selector valve)

—wing tiedown missing

—static wick missing

—fuel injector nut loose

—wrong dipstick for oil

—tools in the engine compartment

—airspeed indicator speeds don't agree with the book

—compass deviation card missing"

Hopefully, Jeff's comments will help all of us improve our preflight inspections. Let's move on, now, to some other types of inspections.

Periodic Maintenance Inspections

Federal Aviation Regulation Part 91 contains the General Operating and Flight Rules, and it specifies the inspections that are required to determine the condition of airworthiness of most general aviation aircraft. There are two subparts of this regulation that deal with inspection. Subpart E deals with Maintenance, Preventive Maintenance, and Alterations, and Subpart F is involved with Large and Turbine-Powered Multi-Engine Airplanes. Both of these subparts describe inspection systems that may be used.

14CFR Part 91, Subpart E is the portion of the General Operating Rules that deals with all aircraft except large and turbine-powered multi-engine airplanes, and as such it covers the vast majority of the airplanes in the general aviation fleet. Aircraft whose maintenance is governed by this portion of the FAR must have a complete inspection every 12 calendar months, and if the aircraft is operated for compensation or hire, including flight instruction, it must have an inspection of the same scope each one hundred hours of operation.

The only difference between the annual and 100-hour inspection is *who* does it. The 100-hour can be done by an A&P licensed technician but the annual must be done by the IA licensed technician.

Annual Inspection

The most generally used type of inspection for Part 91, Subpart E (91.409), aircraft is the **annual inspection**. Within every 12 calendar months the aircraft must have an inspection that is complete enough to determine that the aircraft meets all of the requirements for its certification. A calendar month is one that ends at midnight of the last day of the month. For example, if the inspection was completed on January 14 of this year, it will remain valid until midnight of January 31, next year.

The 100-hour limitation may be exceeded by not more than 10 hours if necessary to reach a place at which the inspection can be done. However, the time in excess of 100 hours must be deducted from the 100-hour interval in computing the time in service until the next 100-hour inspection.

The FAA does not specify the details of this inspection, but Appendix D of FAR 43 includes a list of items entitled "Scope and Detail of Items to be Included in Annual and 100-hour Inspections." This list is not at all comprehensive, but it is typical of the scope of the inspection the FAA recommends. The manufacturer of the aircraft is the final authority on what should be inspected, and a detailed inspection list is included in the service manual for each aircraft.

Although the annual inspection is only one of the approved inspection programs, it is the one most commonly used, as it is applicable to small and simple aircraft as well as to those that are more complex. And the procedures used are similar to those that apply to both progressive inspections and the continuous airworthiness inspection programs.

Reproduced in figure 16-35 is a typical inspection schedule recommended by the manufacturer of a popular single-engine retractable landing gear airplane. In this schedule there are four columns indicating the times specified for the various items. Some items should be inspected every 50 hours of operation, and even though this is not an FAA requirement, one must remember that the FAA-*required* inspections are the absolute minimum allowed; but if the manufacturer recommends inspection of some items more frequently, one should definitely follow that program. If the aircraft is operated for hire, it must be inspected *every* one hundred hours, but there are some items that the manufacturer does not feel needs to be inspected every one hundred hours, and these are listed in the 200-hour column. Regardless of the number of hours an aircraft accumulates in a 12-month period, it must be given an annual inspection, and this includes *all* of the items listed in the 50-hour, 100-hour and 200-hour columns, as well as any of the special instructions that apply.

It is well to mention at this point that an annual inspection and a **100-hour inspection** are identical as far as the inspections themselves are concerned. The only differences are the person authorized to perform them and the fact that a list of discrepancies must be provided in the event the aircraft fails an *annual* inspection. If an aircraft is operated for hire and has several 100-hour inspections during a 12-month period, and if they are all performed by an A&P maintenance technician holding an Inspection Authorization, any or all of them may be considered to be an annual inspection and may be written up as such in the aircraft records.

Many pilots fly airplanes yet never develop any knowledge of how the machine they are flying is maintained. These pilots become the "weak link" in the team effort needed to keep the complex machine they fly in good operating condition. It is the pilot's responsibility to notice and report to the maintenance facility any discrepancy in normal operation of the machine, to test-fly the aircraft after any maintenance that may affect the aircraft's operation in flight (Part 91.407) and to make the determination that the aircraft may return to service and enter that fact in the aircraft's logbooks (also Part 91.407). Please take a moment to read 91.407 and to look over figure 16-35. For a comprehensive description of the inspection process and maintenance records, the reader is referred to item 1 of the additional reading list at the end of this chapter.

Preparation For An Annual Inspection

The Paperwork
Perhaps the most important part of an annual inspection takes place before the aircraft is physically examined: the paperwork is started.

The Work Order
The work order must be started and signed by the owner of the aircraft stating that he/she understands exactly what is to be done. The owner must be given an idea of what can be expected in the way of the cost of the operation. This is an area that must be carefully handled, because the cost of the *inspection* is not necessarily the final cost of the job. And yet by the careless use of the word "inspection," the owner may think that is all that he/she is getting. When an aircraft comes into the shop for an annual inspection, it normally gets, in addition to the inspection itself, a thorough cleaning, an oil change, and numerous items of minor maintenance that can be done most economically while the aircraft is open for inspection and which will extend the life of the aircraft. The inspection may reveal parts that need to be replaced, parts such as bolts and nuts, hingepins, spark plugs, and often seals and gaskets.

	EACH 50 HOURS	EACH 100 HOURS	EACH 200 HOURS	SPECIAL INSPECTION ITEM
PROPELLER				
1. Spinner	•			
2. Spinner bulkhead			•	
3. Blades	•			
4. Bolts and nuts			•	
5. Hub			•	
6. Governor and control			•	11
7. Anti-ice electrical wiring	•			
8. Anti-ice brushes, slip ring and boots	•			
ENGINE COMPARTMENT				
Check for evidence of oil and fuel leaks, then entire engine and compartment, if needed, prior to inspection.				
1. Engine oil screen filler cap, dipstick, drain plug and external filter element	•			1
2. Oil cooler		•		
3. Induction air filter	•			2
4. Induction airbox, air valves, doors and controls		•		
5. Cold and hot air hoses			•	
6. Engine baffles	•			
7. Cylinders, rocker box covers and push rod housings		•		
8. Crankcase, oil sump, accessory section and front crankshaft seal		•		
9. Hoses, metal lines and fittings	•			3
10. Intake and exhaust systems	•			4
11. Ignition harness		•		
12. Spark plugs		•		
13. Compression check			•	
14. Crankcase and vacuum system breather lines			•	
15. Electrical wiring		•		
16. Vacuum pump		•		

Figure 16-35. (1 of 7) Typical inspection schedule for single-engine retractable landing gear airplane.

	EACH 50 HOURS	EACH 100 HOURS	EACH 200 HOURS	SPECIAL INSPECTION ITEM
17. Vacuum relief valve filter			•	5
18. Engine controls and linkage		•		6
19. Engine controls and linkage			•	
20. Cabin heat valves, doors and controls			•	
21. Starter, solenoid and electrical connections		•		
22. Starter brushes, brush leads and commutator			•	
23. Alternator and electrical connections		•		21
24. Alternator brushes, brush leads, commutator or slip ring				7
25. Voltage regulator mounting and electrical leads		•		
26. Magnetos (externally) and electrical connections		•		
27. Magneto timing				8
28. Fuel/air (metering) control unit		•		
29. Firewall			•	
30. Fuel injection system	•			
31. Engine cowl flaps and controls	•			
32. Engine cowling		•		
33. Turbocharger			•	9
34. All oil lines to turbocharger, waste gate and controller		•		20
35. Waste gate, actuator and controller		•		
36. Turbocharger pressurized vent lines to fuel pump, discharge nozzles and fuel flow gauge	•			
37. Turbocharger	•			
38. Alternator support bracket for security	•			
FUEL SYSTEM				
1. Fuel strainer, drain valve and control, bay vents, cap and placards	•			
2. Fuel strainer screen and bowl		•		
3. Fuel injector screen	•			
4. Fuel reservoirs			•	

Figure 16-35. (2 of 7) Typical inspection schedule for single-engine retractable landing gear airplane.

	EACH 50 HOURS	EACH 100 HOURS	EACH 200 HOURS	SPECIAL INSPECTION ITEM
5. Drain fuel and check bay interior, attachment and outlet screens				5
6. Fuel bays and sump drains			•	
7. Fuel selector valve and placards	•			
8. Auxiliary fuel pump		•		
9. Engine-driven fuel pump		•		
10. Fuel quantity indicators and sensing units	•			
11. Fuel lines, check valve and vapor return line		•		
12. Turbocharger vent system		•		
13. Engine primer		•		
LANDING GEAR				
1. Brake fluid, lines and hose, linings, disc, brake assemblies and master cylinders			•	19
2. Main gear wheels	•			
3. Wheel bearings				10
4. Main gear springs			•	
5. Tires	•			
6. Torque link lubrication	•			
7. Parking brake system			•	
8. Nose gear strut and shimmy dampener (service as required)	•			
9. Nose gear wheel	•			
10. Nose gear fork			•	
11. Nose gear steering system			•	
12. Park brake and toe brakes operational test	•			

LANDING GEAR RETRACTION SYSTEM

NOTE:

When performing an inspection of the landing gear retraction system, the aircraft must be placed on jacks and an external electrical power source of at least 60A should be used to prevent drain on the aircraft battery while operating the system.

Figure 16-35. (3 of 7) Typical inspection schedule for single-engine retractable landing gear airplane.

	EACH 50 HOURS	EACH 100 HOURS	EACH 200 HOURS	SPECIAL INSPECTION ITEM
1. Operate the landing gear through five fault-free cycles			•	
2. Check landing gear doors for positive clearance with any part of the landing gear during operation, and for proper fit when closed	•			
3. Check all hydraulic system components for security, hydraulic leaks and any apparent damage to components or mounting structure			•	19
4. Check doors, hinges, hinge pins and linkage for evidence of wear, other damage and security of attachment			•	
5. Inspect internal wheel well structure for cracks, dents, loose rivets, bolts and nuts, corrosion or other damage			•	
6. Check electrical wiring and switches for security of connections, switch operation and check gear position indicator lights for proper operation. Check wiring for proper routing and support			•	
7. Perform operational check and ensure proper rigging of all systems and components including downlocks, uplocks, doors, switches, actuators and power pack (observing cycle time)			•	
8. Check main gear strut-to-pivot attachment		•		
9. Check condition of all springs		•		
10. Hydraulic fluid contamination check				12
11. Power pack check valve screen cleaned		•		
12. Landing gear and door manifold solenoids (mounted on power pack)				

AIRFRAME

	EACH 50 HOURS	EACH 100 HOURS	EACH 200 HOURS	SPECIAL INSPECTION ITEM
1. Aircraft exterior	•			
2. Aircraft structure			•	
3. Windows, windshield, doors and seals	•			
4. Seat stops, seat rails, upholstery, structure and mounting			•	
5. Seat belts and shoulder harnesses	•			
6. Control column bearings, sprockets, pulleys, cables, chains and turnbuckles			•	
7. Control lock, control wheel and control column mechanism			•	
8. Instruments and markings	•			
9. Gyros central air filter			•	13
10. Magnetic compass compensation				5

Figure 16-35. (4 of 7) Typical inspection schedule for single-engine retractable landing gear airplane.

	EACH 50 HOURS	EACH 100 HOURS	EACH 200 HOURS	SPECIAL INSPECTION ITEM
11. Instrument wiring and plumbing			•	
12. Instrument panel, shock mounts, ground straps, cover, decals and labeling			•	
13. Defrosting, heating and ventilating systems and controls		•		
14. Cabin upholstery, trim, sun visors and ash trays			•	
15. Area beneath floor, lines, hose, wires and control cables			•	
16. Lights, switches, circuit breakers, fuses and spare fuses		•		
17. Exterior lights		•		
18. Pitot and static systems			•	
19. Stall warning unit and pitot heater			•	
20. Radios, radio controls, avionics and flight instruments		•		
21. Antennas and cables			•	
22. Battery, battery box and battery cables		•		
23. Battery electrolyte				14
24. Emergency locator transmitter		•		15
25. Oxygen system			•	
26. Oxygen supply, masks and hose		•		16
27. Deice system plumbing			•	
28. Deice system components			•	
29. Deice system boots			•	

CONTROL SYSTEMS

In addition to the items listed below, always check for correct direction of movement, correct travel and correct cable tension.

	EACH 50 HOURS	EACH 100 HOURS	EACH 200 HOURS	SPECIAL INSPECTION ITEM
1. Cables, terminals, pulleys, pulley brackets, cable guards, turnbuckles and fairleads . .			•	
2. Chains, terminals, sprockets and chain guards			•	
3. Trim control wheels, indicators, actuator and bungee		•		
4. Travel stops			•	
5. Decals and labeling			•	
6. Flap control switch, flap rollers and flap position indicator		•		

Figure 16-35. (5 of 7) Typical inspection schedule for single-engine retractable landing gear airplane.

	EACH 50 HOURS	EACH 100 HOURS	EACH 200 HOURS	SPECIAL INSPECTION ITEM
7. Flap motor, transmission, limit switches, structure, linkage, bell cranks, etc.			•	
8. Flap actuator jackscrew threads				17
9. Elevators, trim tab, hinges and push-pull tube	•			
10. Elevator trim tab actuator lubrication and tab free-play inspection				18
11. Rudder pedal assemblies and linkage			•	
12. External skins of control surfaces and tabs	•			
13. Ailerons, hinges, and control rods	•			
14. Internal structure of control surfaces			•	
15. Balance weight attachment			•	

SPECIAL INSPECTION ITEMS

1. First 25 hours, refill with straight mineral oil (MIL-L-6082) and use until a total of 50 hours have accumulated or oil consumption has stabilized; then change to ashless dispersant oil. Change filter element each 50 hours and oil at each 100 hours, or every six months.

2. Clean filter, replace as required.

3. Replace hoses at engine overhaul or after 5 years, whichever comes first.

4. General inspection every 50 hours.

5. Each 1000 hours, or to coincide with engine overhaul.

6. Each 100 hours for general condition, lubrication and freedom of movement. These controls are not repairable. Replace every 1500 hours or sooner if required.

7. Each 500 hours.

8. Internal timing and magneto-to-engine timing limits are described in the engine service manual.

9. Remove insulation blanket or heat shields and inspect for burned area, bulges or cracks. Remove tailpipe and ducting; inspect turbine for coking, carbonization, oil deposits and impeller for damage.

10. First 100 hours and each 500 hours thereafter. More often if operated under prevailing wet or dusty conditions.

11. If leakage is evident, refer to Governor Service Manual.

12. At first 50 hours, first 100 hours, and thereafter each 500 hours or one year, which ever comes first.

13. Replace each 500 hours.

Figure 16-35. (6 of 7) Typical inspection schedule for single-engine retractable landing gear airplane.

14. Check electrolyte level and clean battery compartment each 50 hours or each 30 days.

15. Refer to manufacturer's manual.

16. Inspect masks, hose and fittings for condition, routing and support.

17. Refer to maintenance manual.

18. Lubrication of the actuator is required each 1000 hours or three years.

19. Each five years replace all rubber packings, back-ups and hydraulic hoses in both the retraction and brake systems. Overhaul all retracton and brake system components.

20. Replace check valves in turbocharger oil lines each 1000 hours.

21. Check alternator belt tension.

Figure 16-35. (7 of 7) Typical inspection schedule for single-engine retractable landing gear airplane.

Minor maintenance and parts replacement are both a part of the overall job, but it must be understood by the owner that they will be *in addition* to the charge for the inspection. Most shops charge a flat rate for the inspection plus an hourly charge for all of the time the aircraft is being worked on. This is, naturally, in addition to the cost of the parts.

Before the inspection can be accomplished, the owner must supply to the inspector the maintenance records of the aircraft itself. These records may be in either a bound or looseleaf form and must include at least:

The total time in service of the airframe.

The current status of any life-limited parts.

The time since last overhaul of any items the FAA requires to be overhauled at a given time basis.

The identification of the type of inspection program, including the time since the last required inspection.

The current status of applicable Airworthiness Directives and the method used for their compliance.

A list of current major alterations that have been made to the airframe, engine, propeller, rotor or any appliance.

An A&P maintenance technician holding an Inspection Authorization is required to supervise the conduct of and certify an annual inspection.

Others who may complete and certify an annual inspection include:

1. An authorized representative of the manufacturer of the aircraft, and

2. Personnel of a properly certificated repair station, approved by the FAA to perform that function, usually the chief inspector.

If the aircraft passes the inspection, the mechanic with Inspection Authorization (IA) can write up the results of the inspection in the maintenance records and approve the aircraft for return to service. If for any reason the aircraft does not meet all of the airworthiness requirements, the inspector must write this up in the maintenance records: "I certify that this aircraft has been inspected in accordance with an annual inspection, and a list of discrepancies and unairworthy items dated (date) has been provided for the aircraft owner or lessee."

Then, any certificated A&P maintenance technician may correct the discrepancies the inspector has listed, and when he/she is satisfied that the work has been done properly, may approve the aircraft for return to service. The date the next annual inspection is due is based on the date of the original inspection and not on the date the discrepancies were corrected and the aircraft was approved for return to service.

One-Hundred Hour Inspection

If the aircraft is operated for compensation or hire, it must be given an inspection of the same scope and detail as the annual inspection every 100 hours of operation. The only difference between a 100-hour inspection and an annual is that the A&P maintenance technician conducting the 100-hour inspection need not have an Inspection Authorization, and there is no requirement that an aircraft failing the inspection have a list of the discrepancies furnished the owner and sent to the FAA.

Progressive Inspections

If the operator of an aircraft does not feel that it is wise to keep the airplane out of commission long enough to perform the complete annual inspection at one time, an option is to use a progressive inspection instead. A **progressive inspection** is exactly the same in scope and detail as the annual inspection, but it is conducted according to a time schedule that must be approved by the FAA. For example, the engine may be inspected at one time, the airframe inspection may be conducted at another time, and the landing gear at another. The basic stipulations for this type of inspection are that the aircraft has had a complete annual inspection before the progressive is started, and the schedule must be approved in advance by the FAA. An A&P maintenance technician holding an Inspection Authorization is required to conduct each phase of the inspection, and all items in the inspection schedule must be covered within the 12 calendar months allowed for an annual inspection.

Altimeter And Static System Inspection

According to Part 91.411, every aircraft that is operated under Instrument Flight Rules must have its altimeters and static systems inspected every 24 calendar months according to the details specified in FAR Part 43 Appendix E. The altimeter must be checked for operation and accuracy up to the highest altitude that it will be used, and a record must be made of this inspection in the aircraft maintenance records. For IFR flight, the pilot should not operate unless the altimeter, when correctly set, reads within 75 feet of the actual altitude.

The altimeter inspection may be conducted by either the manufacturer of the aircraft, or by a certificated repair station holding the rating that authorizes this particular inspection. A technician holding an Airframe rating may conduct the static system test.

ATC Transponder Inspection

Per 91.413, the radar beacon transponder that is required for aircraft operating in most areas of controlled airspace must be inspected each 24 calendar months by a certificated repair station approved for this inspection. This test is described in FAR Part 43, Appendix F.

FAR Part 91, Subpart F Inspections

This subpart deals with large aircraft and multi-engine turbine aircraft and requires an inspection program that is tailored to the specific aircraft and its unique operating condition. The operator of these aircraft must have a person in charge of maintenance and must submit a maintenance and inspection schedule to the FAA describing in detail the inspections that will be conducted on the aircraft. These inspections do not necessarily follow the same format as an annual or 100-hour inspection, but they are adapted to the specific aircraft and its unique operating conditions. Some may, for example, require landing gear inspection more often than every 100 hours if the airplane is operated at high gross weight on rough fields, and may allow a longer period between inspections for some items that have a history of giving very little trouble.

The FAA has made provision for five different sources of the continuous airworthiness inspection program specified in FAR 91.409.d.

It may be the same program that is approved for the same aircraft for an air carrier or commercial operator operating under FAR Part 121.

It may be the same program that is approved for the same aircraft for an air taxi or commercial operator operating under FAR Part 135.

It may be the same program that is approved for the same aircraft for an air travel club operating under FAR Part 123.

It may be a special program recommended by the manufacturer of the aircraft.

It may be a program developed by the operator specifically for his operation. This must be individually approved by the FAA.

Pilot Accomplished Maintenance

Probably the three most important maintenance activities the pilot can accomplish are:

1. Fly the aircraft in such a way that the need for additional maintenance isn't created. Make power changes slowly and smoothly, lean the mixture properly on the ground so spark plugs aren't fouled, keep RPM as low as possible while taxiing (especially when accelerating from zero groundspeed), do runups only on a clean, hard surface and taxi slowly over rough ground, etc. to minimize propeller blade damage.

2. Communicate clearly with the mechanic or technician when a maintenance need occurs—knowledge of the system on the aircraft is important here.

3. Be sure the maintenance paperwork is correct and up to date. Keep track of life-limited parts and schedule them for overhaul rather than waiting until they quit. Example: magnetos should usually be overhauled every 800-1200 hours, depending on the make and model.

Probably, *no one should be expected to care more about the well-being of the aircraft than the person who rides in it all the time—the pilot.* Therefore, pilots should make it their business to find out about the "care and feeding" of the aircraft they fly from experienced maintenance personnel in order to be able to take an active role in maintaining the condition of their aircraft.

One way to do this is for the pilot to participate, to whatever extent possible, in the maintenance of the aircraft. The result will be a more intimate knowledge of the machine and more opportunities to communicate with experienced technicians. When they observe a pilot's interest in learning more about the airplane, most technicians will take an interest in helping to educate, providing the pilot approaches the technician in a courteous way, considerate of the technician's time and respectful of the technician's tools.

Fortunately for the pilot with mechanical aptitude (mostly a sincere interest in learning) who has found a willing maintenance technician, the FAR's approve of a pilot's participation in any type of maintenance provided it is properly supervised (the maintenance technician must be on-site, but not necessarily looking over the pilot's shoulder continuously) to the extent necessary to ensure that the work is done correctly.

There are many maintenance tasks the pilot may do without any supervision. Pilots may accomplish **preventative maintenance** which, according to 14CFR Part 1, is defined as: "simple or minor preservation operations and the replacement of small standard parts not involving complex assembly operations." This definition has been interpreted many different ways. A more detailed list of approved procedures is found in figure 16-36 which is a synopsis of the section of 14CFR Part 43 dealing with pilot-accomplished maintenance.

Whenever maintenance work is accomplished, the pilot should, upon completion of the work, make a logbook entry stating date, aircraft and engine total time, type of work completed, pilot's name and certificate number.

Maintenance Forms And Records

Permanent Records

The accepted method of keeping aircraft maintenance records has been in a bound airframe and powerplant logbook, in which the flight time is recorded and a record made of all maintenance and inspections.

The requirements for a bound logbook have been changed so the records can now be kept in a loose-leaf format, and the records that are needed have been clearly defined and divided into two categories: those that must be kept with the aircraft as long as it is operating and those that may be destroyed after a given period of time, after there is a separate logbook for the airframe, each engine and propeller.

FAR 91.417 describes the records that must be kept. Permanent records that must stay with the aircraft and be transferred with it when it is sold are:

a. The total time in service of the airframe.

b. The current status of any life-limited parts.

c. The time since last overhaul of all items which are required by the FAA to be overhauled on a given time basis.

d. The identification of the current inspection status, including the time since the last required inspection.

e. The current status of applicable Airworthiness Directives. This must include the method of their compliance.

f. A list of the current major alterations that have been made on the airframe, engine, propeller, rotor or appliance.

The allowable operational life of life-limited parts is specified by the FAA, and at each inspection the status of these parts must be determined.

Temporary Records

In addition to the records that must be permanently maintained, the owner or operator of the aircraft must keep the following records for a period of one year, or until the operation is repeated or superseded:

a. Records of the maintenance that has been done to the airframe, engine, propeller, rotor or appliance.

b. Records of 100-hour, annual or progressive inspections.

These records must include:

a. Description of the work performed.

b. Date of completion of the work performed.

c. Signature and certificate number of the person approving the aircraft for return to service.

FAR Part 43 Maintenance, Preventive Maintenance, Rebuilding, and Alteration (Synopsis)

Part 43 of the Federal Aviation Regulations permits the holder of a pilot certificate to perform specific types of preventive maintenance on any aircraft owned or operated by the pilot. Some examples of work considered as preventive maintenance are:

1. Removal, installation, and repair of landing gear tires.
2. Replacing elastic shock absorber cords on landing gear.
3. Servicing landing-gear shock struts by adding oil, air, or both.
4. Servicing landing-gear wheel bearings, such as cleaning and greasing.
5. Replacing defective safety wiring or cotter keys.
6. Lubrication not requiring disassembly other than removal of nonstructural items such as cover plates, cowlings, and fairings.
7. Making simple fabric patches not requiring rib stitching or the removal of structural parts or control surfaces. In the case of balloons, the making of small fabric repairs to envelopes (as defined in, and in accordance with, the balloon manufacturer's instructions) not requiring load tape repair or replacement.
8. Replenishing hydraulic fluid in the hydraulic reservoir.
9. Refinishing decorative coating of fuselage, balloon baskets, wings tail group surfaces (excluding balanced control surfaces), fairings, cowlings, landing gear, cabin, or cockpit interior when removal or disassembly of any primary structure or operating system is not required.
10. Applying preservative or protective material to components where no disassembly of any primary structure or operating systems is involved and where such coating is not prohibited or is not contrary to good practices.
11. Repairing upholstery and decorative furnishings of the cabin, cockpit, or balloon basket interior when the repairing does not require disassembly of any primary structure or operating system or interfere with an operating system or affect the primary structure of the aircraft.
12. Making small, simple repairs to fairings, nonstructural cover plates, cowlings, and small patches and reinforcements, but not changing the contour so as to interfere with proper air flow.
13. Replacing side windows where that work does not interfere with the structure or any operating system such as controls, electrical equipment, etc.
14. Replacing safety belts.
15. Replacing seats or seat parts with replacement parts approved for the aircraft, not involving disassembly of any primary structure or operating system.
16. Troubleshooting and repairing broken circuits in landing-light wiring circuits.
17. Replacing bulbs, reflectors, and lenses of position and landing lights.
18. Replacing wheels and skis where no weight and balance computation is involved.
19. Replacing any cowling not requiring removal of the propeller or disconnection of flight controls.
20. Replacing or cleaning spark plugs and setting spark-plug gap clearance.
21. Replacing any hose connection except hydraulic connections.
22. Replacing prefabricated fuel lines.
23. Cleaning or replacing fuel and oil strainers or filter elements.
24. Replacing and servicing batteries.
25. Cleaning of balloon burner pilot and main nozzles in accordance with the balloon manufacturer's instructions.
26. Replacement or adjustment of nonstructural standard fasteners incidental to operations.
27. The interchange of balloon baskets and burners on envelopes when the basket or burner is designated as interchangeable in the balloon type certificate data and the baskets and burners are specifically designed for quick removal and installation.
28. The installation of anti-misfueling devices to reduce the diameter of fuel tank filler openings provided the specific device has been made a part of the aircraft type certificate data by the aircraft manufacturer, the aircraft manufacturer has provided FAA-approved instructions for installation of the specific device, and installation does not involve the disassembly of the existing tank filler opening.
29. Removing, checking, and replacing magnetic chip detectors.

In addition to preventive maintenance, a pilot can perform other maintenance if he works under the direct supervision of a properly certificated mechanic.

Figure 16-36. FAR Part 43 —Synopsis of maintenance the pilot is authorized to perform.

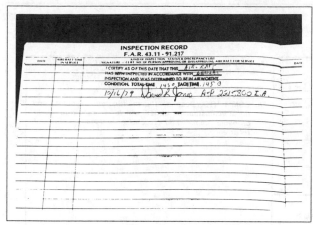

Figure 16-37. An airframe logbook can be used as part of the required maintenance records.

Figure 16-38. It is extremely important that records be kept of all maintenance done to an aircraft.

Maintenance Forms

Malfunction Or Defect Report Form 8330-2

If a malfunction or defect shows up on an aircraft, powerplant, propeller or appliance that is not caused by poor maintenance practices, a malfunction or defect report should be filled out and mailed to the FAA GADO. In figure 16-39 is an example of this form with spaces for the make, model and serial number of the unit that had the difficulty and a place for the identification of the specific part that failed or caused the problem. There is space on the report for recommendations for a fix that would prevent any recurrence of the problem.

These forms are forwarded by the GADO to the FAA Aeronautical Center in Oklahoma City, Oklahoma, where all of the reports that have been received are analyzed and any trends that would indicate a dangerous situation are noted. The information is made available to A&P's in the form of General Aviation Airworthiness Alerts, and if the problem has occurred frequently enough and is of a serious enough nature, an Airworthiness Directive is issued to require the fix to be made before a similar failure can cause an accident. This is a service that is very valuable to pilots as it acts as an "early warning system" for weak design features on all aircraft.

Major Repair And Alteration Form 337

When a **major repair** or **major alteration** is made on an airframe, powerplant, propeller or appliance, a Form 337 must be filled out to record the work that has been done.

Major repair and major alteration are defined in detail in Appendix A of 14CFR Part 43. Essentially, a major repair is a repair involving the strengthen-ing, reinforcing, splicing, manufacturing and replacement of primary structural members of the airframe, engine or propeller. A **major alteration** is an alteration or change of a primary structural member of an airframe, engine or propeller.

The original form stays with the aircraft records and becomes part of the permanent records. The duplicate copy must be sent to the FAA GADO within 48 hours after the aircraft is approved for return to service. See figures 16-40(A) and (B).

The information required on the front side of the form is quite straightforward, but the back side is just a big blank space in which to describe the work that has been done. Some basic rules that are followed when filling out this space are:

A statement will be found to the effect that changes in the weight and balance have been noted in the weight and balance statement of the aircraft records, or that no weight and balance change was effected.

When the description is finished, a line is drawn across the bottom of the form and the word END written in the center of the line.

Certificate Of Registration And Bill Of Sale, Form 8050

When an aircraft is sold, the seller must return the Aircraft Registration form to the FAA in Oklahoma City, Oklahoma, and the purchaser must file his application for a new Registration.

The form 8050-1 is filled out in triplicate, and the original, which is the white copy, and the green copy are sent to the FAA in Oklahoma City, Oklahoma, along with the registration fee and proof of ownership. The pink copy of the form is kept in the aircraft until the registration form is returned by the FAA.

1. REGISTRATION NO.	DEPARTMENT OF TRANSPORTATION FEDERAL AVIATION ADMINISTRATION MALFUNCTION OR DEFECT REPORT			Form Approved Budget Bureau No. 04–R0003	8. DATE SUB.	FOR FAA USE ONLY CONTROL NO.

	A. MAKE	B. MODEL	C. SERIAL NO.
2. AIRCRAFT			
3. POWERPLANT			
4. PROPELLER			

7A. COMMENTS (Describe the malfunction or defect and the circumstances under which it occurred. State probable cause and recommendations to prevent recurrence.)

5. APPLIANCE/COMPONENT (assy. that includes part)

A. NAME	B. MAKE	C. MODEL	D. SERIAL NO.

6. SPECIFIC PART (of component) CAUSING TROUBLE

A. NAME	B. NUMBER	C. PART/DEFECT LOCATION

Continue on reverse

SUBMITTED BY

FAA USE	E. PART TT	F. PART TSO	G. PART CONDITION	B. REP. STA.	C. OPER.	D. MECH.	E. AIR TAXI	F. MFG.	G. FAA	H. OTHER	I.
D. ATA CODE											

FAA FORM 8330-2 (9-70) SUPERSEDES PREVIOUS EDITIONS

USE THIS SPACE FOR ADDITIONAL COMMENTS IF NEEDED

DEPARTMENT OF TRANSPORTATION
FEDERAL AVIATION ADMINISTRATION
WASHINGTON, D.C. 20590

OFFICIAL BUSINESS
PENALTY FOR PRIVATE USE, $300

POSTAGE AND FEES PAID
FEDERAL AVIATION ADMINISTRATION

Figure 16-39. Malfunction or Defect report form.

DEPARTMENT OF TRANSPORTATION
FEDERAL AVIATION ADMINISTRATION

Form Approved
Budget Bureau No. 04-R060.1

MAJOR REPAIR AND ALTERATION
(Airframe, Powerplant, Propeller, or Appliance)

FOR FAA USE ONLY GAD
OFFICE IDENTIFICATION 2-2-11

INSTRUCTIONS: Print or type all entries. See FAR 43.9, FAR 43 Appendix B, and AC 43.9-1 (or subsequent revision thereof) for instructions and disposition of this form.

1. AIRCRAFT	MAKE	MODEL
	SERIAL NO.	NATIONALITY AND REGISTRATION MARK
2. OWNER	NAME (As shown on registration certificate)	ADDRESS (As shown on registration certificate)

3. FOR FAA USE ONLY

4. UNIT IDENTIFICATION

UNIT	MAKE	MODEL	SERIAL NO.	5. TYPE REPAIR	ALTER-ATION
AIRFRAME	◆◆◆◆◆◆◆ (As described in item 1 above) ◆◆◆◆◆◆◆				
POWERPLANT					
PROPELLER					
APPLIANCE	TYPE				
	MANUFACTURER				

6. CONFORMITY STATEMENT

A. AGENCY'S NAME AND ADDRESS	B. KIND OF AGENCY	C. CERTIFICATE NO.
	U.S. CERTIFICATED MECHANIC	
	FOREIGN CERTIFICATED MECHANIC	
	CERTIFICATED REPAIR STATION	
	MANUFACTURER	

D. I certify that the repair and/or alteration made to the unit(s) identified in item 4 above and described on the reverse or attachments hereto have been made in accordance with the requirements of Part 43 of the U.S. Federal Aviation Regulations and that the information furnished herein is true and correct to the best of my knowledge.

DATE	SIGNATURE OF AUTHORIZED INDIVIDUAL

7. APPROVAL FOR RETURN TO SERVICE

Pursuant to the authority given persons specified below, the unit identified in item 4 was inspected in the manner prescribed by the Administrator of the Federal Aviation Administration and is ☐ APPROVED ☐ REJECTED

BY	FAA FLT. STANDARDS INSPECTOR	MANUFACTURER	INSPECTION AUTHORIZATION	OTHER (Specify)
	FAA DESIGNEE	REPAIR STATION	CANADIAN DEPARTMENT OF TRANSPORT INSPECTOR OF AIRCRAFT	

DATE OF APPROVAL OR REJECTION	CERTIFICATE OR DESIGNATION NO.	SIGNATURE OF AUTHORIZED INDIVIDUAL

FAA Form 337 (7-67)

(8320)

Figure 16-40(A). Major Repair and Alteration Form.

NOTICE

Weight and balance or operating limitation changes shall be entered in the appropriate aircraft record. An alteration must be compatible with all previous alterations to assure continued conformity with the applicable airworthiness requirements.

8. DESCRIPTION OF WORK ACCOMPLISHED *(If more space is required, attach additional sheets.   Identify with aircraft nationality and registration mark and date work completed.)*

☐ ADDITIONAL SHEETS ARE ATTACHED

Figure 16–40(B). Major Repair and Alteration Form (continued).

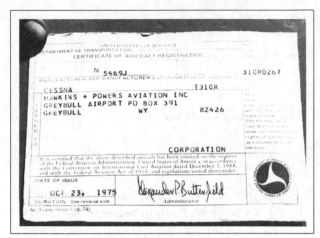

Figure 16-41. Certificate of Registration.

Figure 16-42. Airworthiness Certificate.

Proof of ownership may be substantiated by the Bill of Sale Form 8050-2 or by any other form of proof that the purchaser may want to use. Before buying an aircraft, the purchaser should have the title of the aircraft verified by a title search that can be conducted by any of a number of companies in Oklahoma City, Oklahoma, that provide this service. If a search is not made, it is possible that the person that sells the aircraft may not be the same as the person listed on the FAA records as the legal owner of the aircraft, and the ownership can be disputed at some later date.

Standard Aircraft Airworthiness Certificate Form 8100-2

When an aircraft is built and is found to meet all of the requirements for its certification, it is issued an Airworthiness Certificate, Form 8100-2. This certificate must be displayed in the aircraft and is transferred with the aircraft when it is sold, unless it is sold to a foreign purchaser. A standard Airwor-

thiness Certificate is issued to aircraft licensed under the normal, utility, acrobatic or transport category, or manned free balloons. These standard certificates remain in effect as long as the aircraft receives the required maintenance and is registered in the United States. See figure 16-42.

Special Airworthiness Certificate Form 8130-7

Aircraft that are certificated under any category other than those of the standard classifications are issued a Special Airworthiness Certificate. This includes aircraft that are operating under the Experimental, Restricted, Limited and Provisional categories.

Weight And Balance Statement

This document is not standardized so it may take many forms but it will show how the empty weight and balance was determined (by calculation from a previous value or by weighing the aircraft).

Additional Reading

1. Enga, J.; Aircraft Inspection and Maintenance Records; EA-IAR; IAP, Inc., Publ.; 1979.

Study Questions And Problems

1. Should CG be considered an arm or a moment? Why?

2. Define CG as many different ways as you can.

3. Compute the change of moment about the CG that the Part 135 pilot must trim for when a standard passenger moves from their seat at station +70.0" to the aft toilet at station +160", if the aircraft weighs 15,000 pounds. Use winter values. Would the Part 91 pilot have to trim more, less or the same amount? Up or down?

4. A Cessna 180G landplane's center of gravity moment envelope is enclosed by a line from 60,000 pound-inches at 1,800 pounds to 70,000 pound-inches at 2,100 pounds to 108,000 pound-inches at 2,800 pounds for the forward limit and a line from 131,500 pound-inches at 2,800 pounds to 84,500 pound-inches at 1,800 pounds for the aft limit. Draw this aircraft's weight vs. moment and weight vs. center of gravity envelopes on two graphs side-by-side on one sheet of proper graph paper.

5. The C-180's loading data appears in figure 16-32(A). The airplane empty weight is 1,704 pounds and its CG is located at station 36.65". Accomplish a forward and aft adverse loading check for this aircraft. See Question #4 for CG limits. The 0-470 engine in the C-180 is rated at 230 HP. Use configuration V.

6. Develop a weight vs. moment chart similar to figure 16-14 for the C-180G and use it to prove your answers to Question #5.

7. Do a Part 135 passenger and cargo manifest for N123QC for the flight of figure 16-19, using Part 135 standard weights.

8. Do the task assigned in Question #5 for the aircraft you fly.

9. If the aircraft of Question #3 has a CG of 92.3" with passengers in their seats, what is the new CG when the passenger of Question #3 reaches the toilet? Prove your answer using another method.

10. Prepare moment vs. weight and CG vs. weight graphs for the airplane depicted by the data of figure 16-24.

Chapter XVII
Aircraft Instrument Systems

Introduction

A very important aspect of the growth of aviation into an effective transportation system has been the development of efficient flight instruments.

Prior to World War II, only a few pilots could fly by instruments, and very few airplanes were equipped for flight without reference to the ground. Navigation was done almost exclusively by **pilotage**—that is, by flying from one recognizable landmark to another. A low cloud cover, therefore, required flying at a dangerously low altitude or not flying at all.

On September 24, 1929, the famous engineering pilot Jimmy Doolittle made a flight in which he had absolutely no outside visual reference. He had, in the Consolidated NY-2 airplane he flew, an artificial horizon to give him an indication of the longitudinal and lateral attitude of the airplane relative to the earth's surface, and a sensitive altimeter that showed his altitude above the ground to within a few feet. A radio direction finder allowed him to determine his position relative to the landing area.

With this equipment, Doolittle proved that "blind" flight was indeed possible, but it was not until World War II came along with the absolute necessity for flight under all kinds of weather conditions that flight by instrument became widely practiced.

Thousands of military-trained pilots came home after World War II, and the aircraft factories turned out many thousands of new airplanes, most of which were equipped with instruments salvaged from the military bombers and fighters that were being melted down into aluminum ingots.

But the availability of trained pilots and instrument equipped airplanes is not all that is needed for an effective transportation system. There was no problem of keeping the airplane straight and level with no outside reference, but this is of little use without some method of guiding the airplane once it is off the ground. This guidance became a reality with the VOR, or the Very high frequency Omnidirectional Radio range, the most common system of radio navigation in use today.

Figure 17-1. Flight instruments were no problem in the early Curtiss airplanes such as this one flown by the pioneer Glenn Curtiss. There were no instruments.

Figure 17-2. The instrument panel of this 1929 Cessna had only an airspeed indicator and an altimeter in addition to the basic engine instruments.

Figure 17-3. The instrument panel of a modern twin-engine business airplane provides the pilot with all of the information (s)he needs to conduct a safe flight.

Today our modern **avionics**, as these aviation oriented electronic systems are called, has become so sophisticated and so elaborate that it is not uncommon for the price of the avionics package to rival that of the machine which carries it.

Avionics and aircraft instrumentation are advancing more rapidly than any other phase of aviation technology. Vacuum tube electronics first made radio communications and some of the primitive control systems possible; then, solid-state electronics hastened the proliferation of electronic navigation and automatic flight control systems. Today, integrated circuits and microprocessors have allowed digital electronics to revolutionize all forms of flight instrumentation and control systems. Now, another generation of flight instruments is here; in this one, the display is a color cathode ray tube with the information given in numbers or letters as well as with an analog display when this makes the information easier for us to interpret.

Classification Of Instruments

The instruments carried in an aircraft are classified into three groups: flight instruments, engine instruments, and auxiliary instruments.

Flight instruments allow the pilot to visualize the attitude, location and vertical and horizontal speeds of the aircraft. **Engine instruments** allow monitoring of the performance and condition of the powerplants, and the **auxiliary instruments** provide information on the status of the hydraulic, pneumatic and fuel systems, as well as on the positions of the various components such as the landing gear and flaps.

Flight Instruments

By our definition, flight instruments are those that help visualize the attitude, location and speeds of the aircraft. Most of those which allow visualization of location are electronic navigation instruments. This chapter is concerned with two very important types of flight instruments; those which indicate the relationship to the air through which we are flying and those which relate to our position in space without considering the air. Instruments in the first group are actuated by air pressure and tell how high we are in the atmosphere, and how fast we are moving through it. The instruments in the second group are actuated by gyroscopes which remain rigid in space and give reference from which we can measure attitude or direction.

Pitot-static Systems

In order for the pitot-static instruments to function properly, they must be connected into a system that senses **dynamic air pressure** and **ambient static air pressure**. Dynamic air pressure, or P_q, is pressure caused by moving air. It can be calculated by dividing air density by 2, then multiplying that value times the true airspeed.

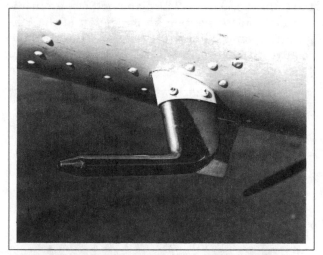

Figure 17-4. A combination pitot and static pickup.

Figure 17-5. A flush static pressure pickup point.

382

Ambient static air pressure, or P_s, is the pressure of non-moving air just outside the aircraft. The aircraft's static system attempts to sample P_s with as little error as possible (which is hard to do,

considering the changing speeds at which the static port is moving through the air).

Pressure in the static lines is assumed to be equal to P_s. Pressure in the pitot line is equal to $P_s + P_q$. The pressure sampled by the pitot tube may have errors caused by (1) the pitot is rarely located in undisturbed air, (2) the pitot is oriented to the relative wind at different angles, depending on airspeed (angle of attack) and, (3) the pitot may be blocked by insects or dust (or mud or ice after a rain).

Figures 17-6 and 17-7 show how both simple and complex pitot-static systems are plumbed.

Provisions must be incorporated in the system to prevent water causing an inaccurate reading and to prevent ice blocking the pickup points and lines.

Pitot System

The various types of airspeed indicators or Machmeters are connected into the pitot system. The pitot tube is usually installed in the wing of a single-engine aircraft, outside of the propeller slipstream, or on the fuselage of a multi-engine

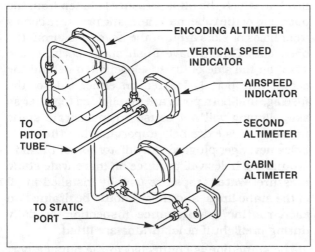

Figure 17-6. Pitot-static system for a small pressurized aircraft.

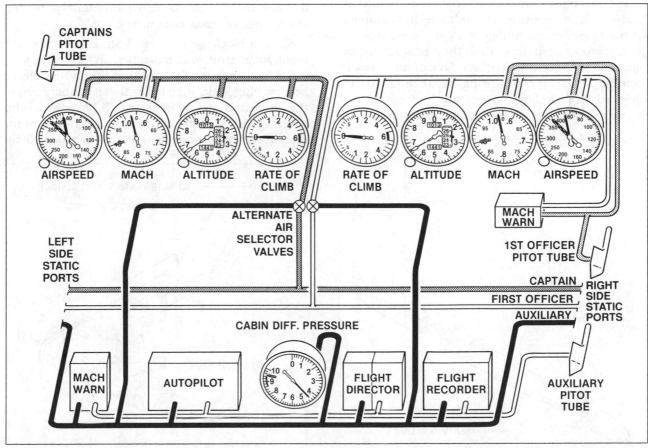

Figure 17-7. Pitot-static system for a jet transport airplane. (Instruments are not located as they would be in the aircraft.)

aircraft. The pitot head (figure 17-4) is mounted in such a way that it points directly into the airflow so the air that enters it will be stopped and the pressure that builds up inside will be proportional to the velocity of the air entering the tube.

The plumbing that connects the pitot tube to the airspeed indicator is run as directly as possible, and there is usually a T-fitting or some form of sump at a low point in the line that will collect any moisture that should get into the line. If water should get into the pitot line, it can cause the airspeed indicator to oscillate as the water moves back and forth in the line. To remove the water, the pitot line is disconnected from the instrument and the water blown out with low-pressure air, blowing from the instrument end of the line, or, if a drain valve is fitted to the low point in the system, it may be opened to clean the system.

If ice should form inside the pitot head or in the pitot line, the airspeed indication will no longer be accurate because the pressure inside the system will be trapped. As the aircraft descends, the static pressure will increase and cause the instrument to show a decrease in airspeed. To prevent ice forming inside the pitot head, those installed on aircraft that will be flown into icing conditions are equipped with an electric heater (figure 17-8). These heaters produce so much heat that they should not be operated without an adequate flow of air over the tube to prevent it overheating. The heater may be checked on a preflight inspection by momentarily turning it on. During the walk-around, cautiously

feel to be sure the head has heated up. In flight, the operation of the heater may be verified by watching the ammeter or loadmeter.

Static System

Some aircraft have a combination pitot-static head, in which the pitot pressure is taken from the open end of the tube, and the static pressure comes from holes or slots around the head, back from the open end. Others have flush static ports, figure 17-5, on the side of the fuselage, often behind the cabin. One port is located on either side of the fuselage and both ports are manifolded together so side slipping will not cause an inaccurate static pressure. It is extremely important that the static holes never get plugged. Any distortion of the skin around the holes will produce an inaccurate static pressure. **Water traps** are usually installed in all of the static lines, and these should be drained on each routine maintenance inspection or daily during preflight if quick drains are fitted.

The static line is connected to the outside port on the back of the airspeed indicator, to the altimeter, the vertical speed indicator, the altitude encoder and, in more sophisticated aircraft, other devices that need to know ambient pressure (autopilots, air data computers, etc.).

Since a blockage or loss of static pressure will result in an erroneous indication on all of the pitot-static instruments, it is vital that this system never become plugged in such a way that it cannot sample the ambient static air pressure. If blockage of the static system should occur, many aircraft have an **alternate static source** that can be selected by

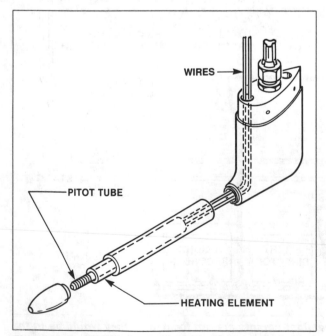

Figure 17-8. Electric heater installed in a pitot head.

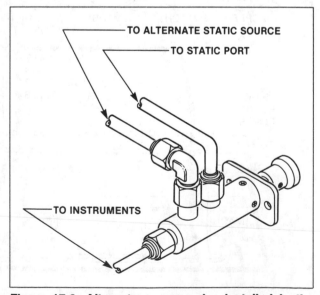

Figure 17-9. Alternate source valve installed in the static system.

the pilot. The **alternate source valve**, figure 17-9, is located on or under the instrument panel, and for unpressurized aircraft, the alternate air is picked up from behind the instrument panel. On pressurized aircraft, the alternate air is taken from some portion of the fuselage outside of the pressure vessel.

FAR Part 91.411 requires that all aircraft that are operated on instrument flight rules must have the static system checked every 24 calendar months by a licensed airframe maintenance technician and the results of this inspection recorded in the aircraft records. This inspection is described in Federal Aviation Regulation 43, Appendix E, and it requires that the static system be checked to be sure it is free from entrapped moisture and leakage. This is done by applying a pressure of one inch of mercury for unpressurized aircraft, and for pressurized aircraft, a pressure equal to the maximum cabin pressure differential. The system is then sealed off and the altimeter watched for one minute. On an unpressurized aircraft, the system should not leak more than 100 feet of altitude indication in one minute. On a pressurized aircraft, the leakage should not be more than 2% of the altitude that is equivalent to the maximum cabin differential pressure or 100 feet, whichever is the greater.

If a static port heater is installed, it must be checked for operation during preflight inspection, and it must be determined that there is no deformation in the surface surrounding the static port that would cause a disturbance of the air.

Airspeed Indicators

An **airspeed indicator** is a differential pressure gauge that measures the difference between $P_s + P_q$ and P_s. It consists of an airtight case in which a thin metal bellows is mounted. Pitot pressure is taken into the bellows and the inside of the case is connected to the static pressure source. The bellows expand in proportion to the difference between the pitot and the static pressure, and this expansion is measured by a mechanical linkage and is displayed as a pointer moves over the dial which is graduated in miles per hour, knots, or kilometers per hour. See figure 17-10(A).

The uncorrected reading of an airspeed indicator is called **indicated airspeed**, and while it relates to the stalling speed of the aircraft, it is of limited use to the pilot for navigational purposes. For navigation, indicated airspeed must be converted into **true airspeed**, but there is an intermediate step: **calibrated airspeed**.

It is almost impossible to find a location for the static port that is entirely free from airflow distor-

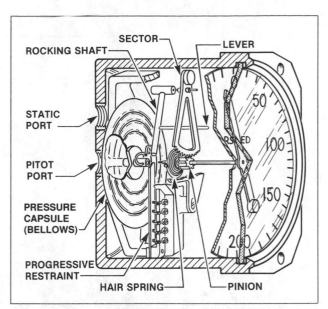

Figure 17-10(A). Internal mechanism of an airspeed indicator.

tion, and any distortion produces an error. To compensate for this, a flight test must be performed and a calibration made between an airspeed indicator connected to the aircraft static system and one connected to a test static pickup that has no airflow disturbance. A calibration card is made and is included with the aircraft flight manual. For most modern production aircraft, this error is so small that for practical purposes it is often ignored, and true airspeed is computed directly from indicated airspeed.

Calibrated and true airspeed are the same under standard sea level atmospheric conditions. To find the true airspeed under non-standard conditions a correction must be applied. This is normally done with a flight computer, or with one of the hand-held electronic calculators. Many airspeed indicators have a movable dial that may be rotated to align a set of temperature and altitude scales so the pointer will indicate the computed true airspeed.

True Airspeed Indicator

There are true airspeed indicators that incorporate not only the airspeed capsule, but a temperature sensor and an altitude bellows that modifies the indication of the airspeed indicator and produces a true airspeed indication. See figure 17-11.

Machmeter

When airplanes fly at or near the speed of sound, a measurement is needed that compares the true airspeed of the airplane with the speed of sound. This measurement is called the **Mach number**.

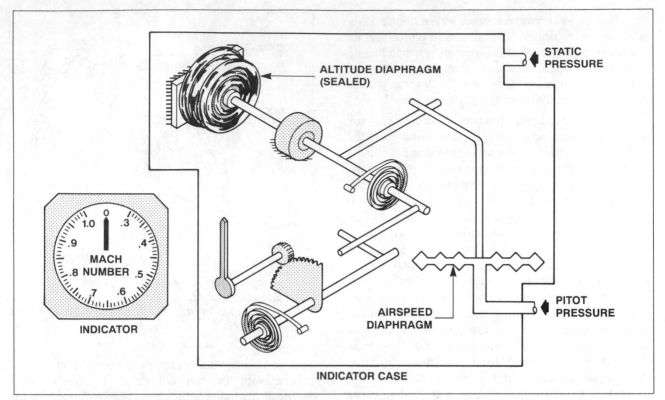

Figure 17-10(B). Altitude diaphragm modifies indicated airspeed information to determine Mach number.

The speed of sound varies in air only with temperature, but at higher altitudes (above 25,000 ft.) temperature varies little from standard atmosphere values, so using air density as an indicator of temperature in determining the speed of sound is justified. Since TAS is required to develop Mach number, an altimeter mechanism approximates temperature and accomplishes the conversion to TAS as well as determination of the speed of sound by measuring ambient air density.

A Machmeter is similar to an airspeed indicator in that the pointer is moved by an expanding bellows that compares pitot pressure with static pressure. But the mechanical advantage of the measuring system is varied by the action of an altimeter mechanism. See figure 17-10(B).

An indication of Mach one occurs when the airplane is flying at the speed of sound. Below the speed of sound, the indication is given as a decimal fraction, and above Mach one, the indication is an integer with a decimal. For example, flight at Mach 1.25 is flight at an airspeed of 1.25 times the speed of sound at that altitude. Mach 0.75 is flight at an airspeed of 75% of the speed of sound (figure 17-12). See figure 17-13 for speed of sound values.

Maximum Allowable Airspeed Indicator

Airplanes that are not designed to fly at sonic airspeed must never be allowed to reach their **critical Mach number**. That is, they must never be flown at a speed that will allow the airflow over any part of the aircraft to reach sonic velocity. When this happens, shock waves form and serious aerodynamic problems can result.

Airplanes whose maximum speed is limited by structural considerations have their never-exceed speed marked by a fixed red line on the dial of the airspeed indicator. But if the maximum speed is limited by the critical Mach number, the fixed red line is replaced by a red pointer that is driven by an altimeter bellows. If, for example, the airplane is never to be flown at a speed in excess of Mach 0.75, the pointer is set for standard sea level conditions at 497 knots. This is 75% of 661.7 knots, the speed of sound under standard sea level conditions. At 10,000 feet, the pointer will move back, limiting the indicated airspeed to 479 knots. You can visualize this decrease in allowable airspeed by referring to the chart of figure 17-13 that shows the way the speed of sound decreases with temperature.

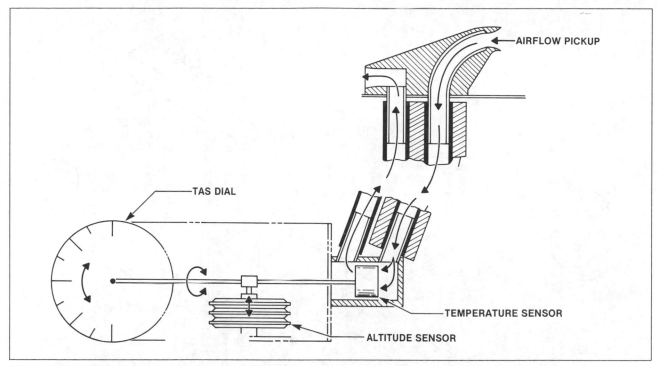

Figure 17-11. Outside air temperature and altitude correction devices for a true airspeed indicator.

Combination Airspeed Indicator

The increased value of instrument panel space aboard modern aircraft, and the need to integrate as much of this information as possible has brought out one instrument that combines the airspeed indicator with the Machmeter and also shows the maximum allowable operating airspeed. This instrument also includes **"bugs"** that are small indicators around the periphery of the dial that may be manually set to indicate the correct speed for certain flight conditions, such as that needed during takeoff or an approach to landing (figure 17-14).

Altimeter

An altimeter is simply a barometer that measures the absolute pressure of the air. This pressure is caused by the weight of the air above the instrument and, naturally, as the ocean of air above the earth's surface moves, this pressure constantly changes.

The altimeter is one of the oldest flight instruments, and some of the early balloon flights carried some form of primitive barometer which served to indicate the height. The standard altimeter used in many of the early airplanes has a simple evacuated bellows whose expansion and contractions are measured by an arrangement of gears and levers that transmit the changes in dimensions into movement of the pointer around the dial. The dial is calibrated in feet, and since a change in the barometric pressure changes the pointer position, the dial can be rotated so the instrument will read zero when the aircraft is on the ground. This form of operation is adequate for aircraft that seldom fly cross-country and for flights that have little need for accurate altitude information.

For serious flight, however, it is extremely important that the altitude indication be accurate, and that the pilot be able to quickly read the altitude within a few feet. These requirements are complicated by the fact that the pressure lapse rate, the decrease in pressure with altitude, is not linear; that is, the pressure for each thousand feet is greater in the lower altitudes than it is in the higher levels. The bellows are designed with corrugations that allow the expansion to be linear with a change in altitude, rather than with a change in pressure, and this allows the use of a uniform altitude scale and multiple pointers.

For many years, all of the best altimeters had three pointers, the long one making a complete round each 1,000 feet, a short, fat pointer making a complete round for each 10,000 feet, and a third pointer geared so that it would have made one trip around the dial for 100,000 feet if the instrument were to go that high. But the range of these altimeters is usually 20,000, 35,000, 50,000 or 80,000 feet. See figure 17-15.

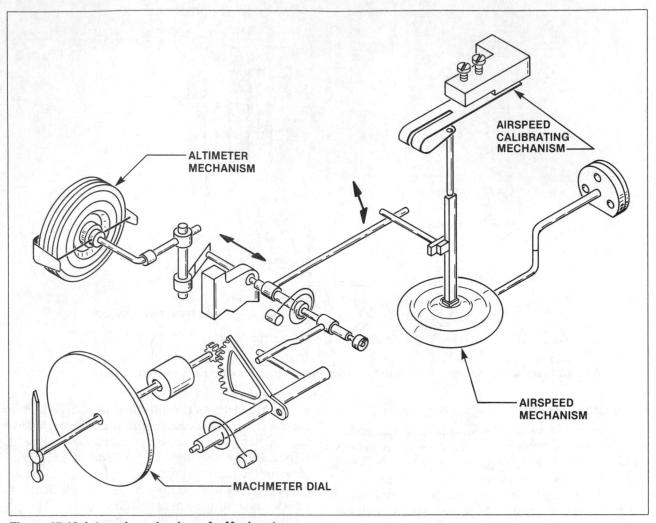

ALTIMETER
MECHANISM

AIRSPEED
CALIBRATING
MECHANISM

AIRSPEED
MECHANISM

MACHMETER DIAL

Figure 17-12. Internal mechanism of a Machmeter.

Because of the ease of misreading these altimeters at higher altitudes, the more modern instruments combine a drum scale with a single pointer. The drum gives the thousands of feet in digital form, and the pointer indicates the hundreds of feet as it makes one trip around the dial for one thousand feet. Each mark on the dial is only twenty feet of altitude. See figure 17-16.

Encoding altimeters are pneumatic altimeters such as we have just been discussing, except they have an encoding device in them that sends a digital code to the radar beacon transponder. When the transponder replies to the interrogation of the air traffic control radar on the ground, a numerical altitude readout appears on the screen beside the return for the aircraft.

The encoding device contains no provision for adjusting it for changes from standard pressure, so it "reads out" just like an altimeter whose Kollsman window is always set at 29.92" Hg.

Therefore, the encoder portion is *not* pilot adjustable. It is found either in the back of the altimeter case (**encoding altimeter**) where it shares commonality with the altimeter only in that it is in the same case and uses the same static source, or it resides in its own case and can be located anywhere in the aircraft that it can be plumbed into the static system (**blind encoder**).

The encoder transmits only **pressure altitude**. Corrections for non-standard pressure are made in the ground radar computer so the radar-displayed altitude should agree with the **indicated altitude** that the pilot sees. Changes of altimeter adjustment by the pilot will not change the encoder altitude readout.

Some altimeters combine a pneumatic altimeter and a radio, or radar altimeter. In figure 17-17, we see the dial of such a combination altimeter. Drums are used to indicate the tens of thousands and thousands of feet, and a single pointer makes

ALTITUDE FT.	DENSITY RATIO σ	PRESSURE RATIO δ	TEMPER-ATURE °F	SPEED OF SOUND α KNOTS
0	1.0000	1.0000	59.00	661.7
1000	0.9711	0.9644	55.43	659.5
2000	0.9428	0.9298	51.87	657.2
3000	0.9151	.08962	48.30	654.9
4000	0.8881	0.8637	44.74	652.6
5000	0.8617	0.8320	41.17	650.3
6000	0.8359	0.8014	37.60	647.9
7000	0.8106	0.7716	34.04	645.6
8000	.07860	0.7428	30.47	643.3
9000	0.7620	0.7148	26.90	640.9
10000	0.7385	0.6877	23.34	638.6
15000	0.6292	0.5643	5.51	626.7
20000	0.5328	0.4595	−12.32	614.6
25000	0.4481	0.3711	−30.15	602.2
30000	0.3741	0.2970	−47.98	589.5
35000	0.3099	0.2353	−65.82	576.6
*36089	0.2971	0.2234	−69.70	573.8
40000	0.2462	0.1851	−69.70	573.8
45000	0.1936	0.1455	−69.70	573.8
50000	0.1522	0.1145	−69.70	573.8
55000	0.1197	0.0900	−69.70	573.8
60000	0.0941	0.0708	−69.70	573.8
65000	0.0740	0.0557	−69.70	573.8
70000	0.0582	0.0438	−69.70	573.8
75000	0.0458	0.0344	−69.70	573.8
80000	0.0360	0.0271	−69.70	573.8
85000	0.0280	0.0213	−64.80	577.4
90000	0.0217	0.0168	−56.57	583.4
95000	0.0169	0.0134	−48.34	589.3
100000	0.0132	0.0107	−40.11	595.2

* GEOPOTENTIAL OF THE TROPOPAUSE

Figure 17-13. Table of standard atmospheric conditions.

Figure 17-14. Maximum allowable airspeed indicator with digital Mach readout and speed bug.

Figure 17-15. Three-pointer altimeter.

one revolution for each one thousand feet. Radio altitude is displayed in digital form with light emitting diodes (LEDs). In this instrument, a barber-pole stripe is visible on the tens of thousands foot drum when the aircraft is below ten thousand feet. There is a dual barometric scale where pressure either in inches of mercury or in millibars may be set into the instrument.

Position error is inherent with static systems, and is caused by the static port not always being in undisturbed air or the pitot head not always being aligned with the relative wind. This error varies with each aircraft design and it changes with airspeed and altitude. The **servo altimeter** has a built-in compensation system that tailors the instrument to the particular aircraft and minimizes this error for the full range of flight speeds and altitudes.

Types Of Altitude Measurement

An altimeter can measure height above almost any convenient reference point, and for most flying, it measures the altitude above the existing sea level pressure level. This is called **indicated altitude** and is read directly from the indicator

Figure 17-16. Drum-pointer-type encoding altimeter.

Figure 17-17. Drum-pointer-type altimeter with a digital readout of radar (radio) altitude.

when the altimeter setting is placed in the barometric window.

The barometric **(Kollsman) window** is a hole in the dial through which a scale calibrated in either inches of mercury or millibars is visible. A knob on the outside of the instrument case usually located at the seven o'clock position rotates the scale and, through a gear arrangement, the mechanism inside the case. Airport control towers and flight service stations along the flight route give the pilot the **altimeter setting** which is their *local barometric pressure corrected to sea level*. When the pilot puts this pressure indication into the barometric window, the **indicated altitude** (altitude read directly from the altimeter) is a very good approximation of **true altitude** (height above sea level). All elevations on aeronautical charts are measured from mean sea level (MSL), and therefore with a bit of simple arithmetic, the pilot can easily and accurately find height above any charted position **(absolute altitude)**. When the airplane is on the ground with the local altimeter setting in the barometric window, the altimeter should indicate the surveyed elevation of the airport reference point (ARP).

Indicated altitude gives a measure of true altitude at low altitudes, but for vertical separation between aircraft flying at higher altitudes, **pressure altitude** is used. When the barometric pressure scale is adjusted to standard sea level pressure, 29.92 inches of mercury or 1013.2 millibars, the altimeter measures the height above this standard pressure level. This is not an actual

point, but is a constantly changing datum. The reason for using it, however, is that all aircraft in the upper level airspace have their altimeters set to the same reference level, and even though an airplane flying at a constant 30,000 feet pressure altitude, for example, may vary its height above sea level, all of the aircraft flying in this same area will vary the same amount and the separation between the aircraft will remain the same. When an aircraft is flying with the altimeter set to indicate pressure altitude, it is operating at a flight level. Flight level 320 is 32,000 feet, pressure altitude, for example.

The performance of an aircraft and its engines is determined by the density of the air, not just its pressure. And since air density is affected by its temperature as well as its pressure, we must consider **density altitude**, which by definition is the altitude in standard air that corresponds to the existing air density. Density altitude is not a direct measurement, but must be computed by correcting the pressure altitude for non-standard temperature. This may be done by using a chart or with a computer, although many professional pilots can make the computation in their heads.

Required Tests For Altimeters

The altimeter is one of the most important instruments used on an instrument approach. Here, an error of only a few feet can be the difference between a successful approach and one that may lead to an accident. At one time there were no requirements for instruments to be periodically checked for accuracy, but after a number of accidents that were plainly attributable to inaccurate altimeters, the FAA instituted a set of required test and inspections for all altimeters that are used while flying under instrument flight rules. FAR 91.411 requires that "No person may operate an airplane in controlled airspace under IFR unless, within the preceding 24 calendar months, each static pressure system and each altimeter instrument has been tested and inspected and found to comply with Appendix E of Part 43." The regulation goes on to specify the persons or facilities that can conduct this test. The altimeter must be inspected by either the manufacturer of the airplane or a certificated repair station equipped and approved to perform these tests. Many fixed base operators have the equipment and approval for this particular test, since it is a rather simple test and a facility with the ability to perform it can attract new customers.

The pilot should require this test be done and the altimeter be calibrated if the altimeter, when properly set, disagrees with the field elevation by more than 75 feet.

FAR Part 43, Appendix E requires that the altimeter instrument have the following tests:

SCALE ERROR The altimeter must indicate the same altitude shown on the master indicator or manometer within a specified allowable tolerance.

HYSTERESIS The reading taken with the altitude increasing must agree with the readings at the same pressure level when the altitude is decreasing. A specified tolerance is allowed for this test.

AFTER EFFECT The altimeter must return to the same indication, within tolerance, after the test as it had when the test began.

FRICTION Two altitude readings are to be taken at each pressure level, one before and one after the instrument is vibrated. There should be no more than a specified difference between the two readings.

CASE LEAK A low pressure is trapped inside the case and it should not leak down more than a specified amount in a given period of time.

BAROMETRIC SCALE ERROR The correlation between the barometric scale and the indication of the altimeter pointers must be correct within the allowable tolerance.

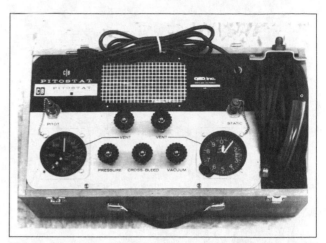

Figure 17-18. Portable tester for conducting the static system test and altimeter tests required by FAR Part 91.411.

The altimeter must be tested to the highest altitude the aircraft will be flown, and after the test is completed, the results of the test must be recorded in the aircraft records.

Vertical Speed Indicator (VSI)

The rate of climb indicator, more properly called the vertical speed indicator, serves only as a backup for the altimeter and airspeed indicator as a pitch indicating instrument because it is slow to indicate a change in pitch. Its main function is that of helping the pilot establish a rate of ascent or descent that will allow a specified altitude to be reached at a given time.

The vertical speed indicator has as its operating mechanism a **bellows**, or pressure capsule, similar to that of an altimeter, except that rather than being evacuated and sealed, it is vented to the inside of the instrument case through a diffuser (orifice) which is an accurately **calibrated leak**. In figure 17-19 is seen the principle of operation of one type of vertical speed indicator.

When the aircraft begins to climb, the pressure inside the bellows begins to decrease to a value below that inside the instrument case, and the bellows compresses, causing the levers and gears to move the pointer so it will indicate a climb. The pressure inside the case now begins to decrease by leaking through the diffuser. This leak is calibrated so that there will always be a difference between the pressure inside the bellows and that inside the case that is proportional to the rate of change of the outside air pressure. As soon as the aircraft levels off, the pressure inside the case and that inside the bellows will equalize, and the indicator will show a zero rate of change.

391

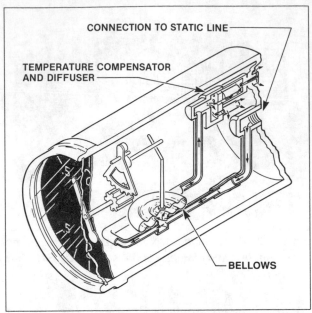

Figure 17-19. Internal mechanism of a vertical speed indicator.

Instantaneous Vertical Speed Indicator (IVSI)

The vertical speed indicator could be used as a pitch indicator if its indication did not lag behind the actual pressure change by 6-8 seconds. To rectify this problem, the instantaneous vertical speed indicator, the IVSI, has been developed. An IVSI uses a mechanism similar to a conventional VSI but it also has an **accelerometer-operated dashpot**, or air pump, across the capsule. When the aircraft noses over to begin a descent, the inertia of the accelerometer piston causes it to move upward, instantaneously increasing the pressure inside the capsule and lowering the pressure inside the case. This change in pressure gives an *immediate* indication of a descent. By this time, the lag of the ordinary VSI has been overcome and it begins to indicate the descent, there is no more inertia from the nose-down rotation, and the accelerometer piston will be centered so the instrument will be ready to indicate the leveling off from the descent. See figure 17-20.

Gyroscopic Instruments

The characteristics possessed by a small spinning gyroscope have not only intrigued us by their seemingly odd behavior, but they have made possible the flight of aircraft without reference to a visible horizon.

In 1851, the French physicist Leon Focault devised a small wheel with a heavy outside rim that, when spun at a high speed, demonstrated the strange characteristic of remaining rigid in the plane in which it was spinning. He deduced that because the wheel remained rigid in space, it could show the rotation of the earth, and because of this he named the device the **gyroscope**, a name that translated from the Greek means "to view the earth's rotation."

A spinning **gyroscope** possesses two characteristics we use in aircraft instrumentation. The first of these is **rigidity in space**, as just discussed. Let's assume that a gyro having no friction in its bearings, but with a power source to keep it spinning, were positioned as we see in figure 17-21. If we could view it from the United States, at noon we would see the tail of the arrow. By the time the earth rotated 90 degrees, at six P.M., we would see the side of the wheel with the arrow pointing to the right. At midnight we would again be in line with the arrow, only this time it would be pointing *at* us. By six A.M., we would again see the side of the wheel. Now, however, the arrow would be pointing to the left. This characteristic makes the gyroscope valuable to us as a stable reference for determining both the attitude and the direction of the aircraft carrying the gyro.

In our description of the attitude gyro, we mention that there must be no friction in the bearings. The reason for this lies in the second characteristic of the gyro—**precession**.

If a force is applied to a spinning gyroscope, its effect will be felt, not at the point of application, but at a point 90 degrees from the point of application in the direction of rotation of the wheel. If a gyro is spinning in the plane shown in figure 17-22, and a force is applied to the top of the wheel, it will not topple over as a static body would; it will rather rotate about its vertical axis. This rotation is called the precession of the gyro. If one of the bearings which supports the gyro shaft has friction, it will produce a force that will cause precession. Precession is a much stronger force than rigidity in space.

Precession is not desired in an attitude gyro, but it may be used in a rate gyro because the amount of precession is related to the amount of force that caused it. We use rate gyros to measure the rate of rotation of the aircraft about one or more of its axes. Most aircraft have either a **turn and slip indicator** or a **turn coordinator**, both of which use precession as the actuating force.

Attitude Gyros

Gyro Horizon

This basic flight instrument shows the relationship between the pitch and roll axes of the aircraft

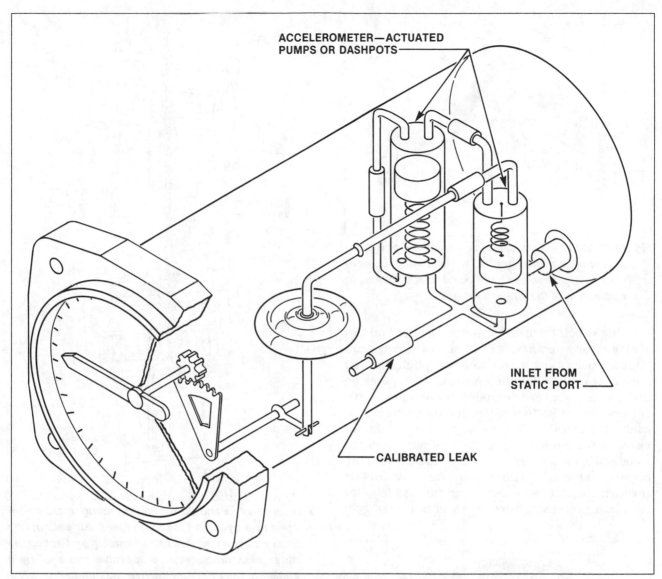

Figure 17-20. Internal mechanism of an instantaneous vertical speed indicator.

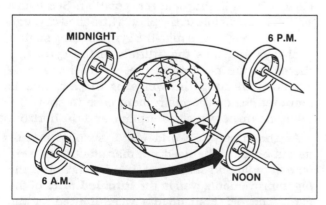

Figure 17-21. Rigidity in space causes a spinning gyro-scope to remain in one position as the earth rotates.

and a vertical line through the center of the earth, and it gives a stable reference so we can keep the wings level. But it tells nothing about the horizontal direction in which the nose of the aircraft is pointing.

The instrument we see in figure 17-23 is the gyro horizon that was used in World War II and appeared in many hundreds of airplanes in the postwar years. These instruments were salvaged from ex-military aircraft, and while they were large and heavy and their sky pointer indication is backward to the actual flight condition, they were available in large quantities at a very low cost. Almost all pilots were familiar with their unusual indication and were skilled in flying with them.

The World War II artificial horizon uses a heavy brass rotor supported in ball bearings with its spin axis vertical. It is spun by a jet of air impinging on buckets cut into the periphery of the wheel. The

393

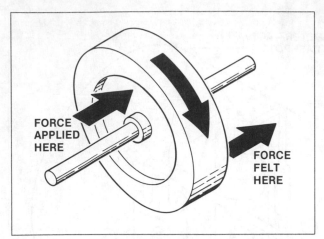

Figure 17-22. The precessive characteristic of a gyroscope causes a force applied to a spinning wheel to be felt ninety degrees to the point of application, in the direction of the wheel's rotation.

housing which holds the rotor is mounted on two **gimbals**, or supports, which act as a universal joint allowing the aircraft to freely pitch and roll about the gyro. When the gyro is erect, air leaving the gyro housing exits equally through four vertical slots in the bottom of the housing. One half of each of these slots is covered with a **pendulum valve**, mounted in such a way that any tilt of the rotor will open one valve and close the valve on the opposite side of the housing. Air now leaving through the slot in one side and not in the other creates a precessive force that will bring the gyro

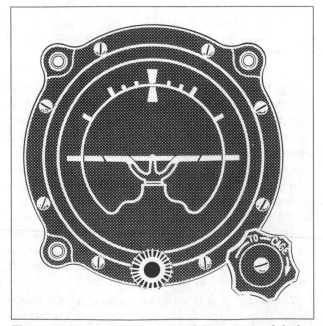

Figure 17-23. Artificial horizon of the type used during World War II and the years following.

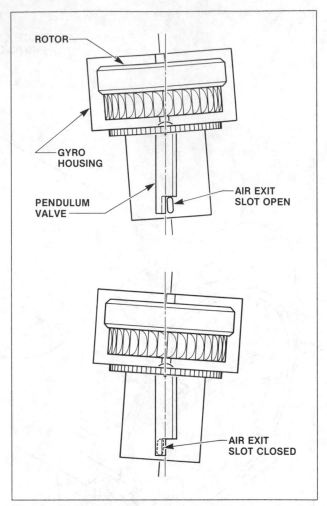

Figure 17-24. Pendulum valve erecting mechanism used in a gyro horizon instrument. Air exhausting from a slot on one side of the housing and not on the other side produces a precessive force at right angles to the direction the gyro has tilted.

back to its upright, or erect, position. See figure 17-24. A bar visible to the pilot is attached through a counterweighted arm to the gyro housing so that it always indicates the attitude of the gyro, and therefore, the position of the earth's horizon. A symbol indicating the wings of the airplane is mounted inside the instrument case to show the relationship between the airplane and the horizon.

As the supply of World War II surplus instruments began to be used up, manufacturers were able to build and sell new instruments. One of the big improvements was in the **tumble limits** of the gyro. The old instruments were limited in the amount of pitch and roll they would tolerate; beyond this limit, the indicating mechanism would apply a force to the gyro housing and create such a precessive force that the gyro would actually tumble over. The indicating system of many of the newer instruments is designed in such a way that the aircraft has complete freedom about the gyro and can actually loop or roll without tumbling the gyro. If the aircraft carrying one of the older instruments was going to be put into a maneuver that would exceed its tumble limits, a caging knob on the front of the instrument could be turned by the pilot to lock the gyro housing rigid in the instrument case so the gyro could not tumble and damage the mechanism. But if the gyro were left caged during landing, the impact could easily damage the bearings and shorten the service life of the instrument. For the gyros that have no tumble limits, there is usually no caging mechanism on the instrument.

Another feature of the newer instruments is the way the information is presented to the pilot. Rather than representing the horizon with a bar, these instruments use a two-color movable **dial**. Above the horizon, the dial is light colored, usually blue, to represent the sky, and below the horizon, it is brown or black representing the ground. The lower half is marked with lines which meet at the center to help the pilot visualize this as the horizon. These lines also provide angular references to help establish the desired bank angle, although this is not their primary purpose. Short horizontal lines both above and below the horizon help the pilot to establish pitch angles, and across the top of the instrument, a **sky pointer** may be aligned with index marks to establish the desired bank angle. These marks are located at 10, 20, 30, 60 and 90 degrees. See figure 17-25.

Many of the newer attitude gyros, as they are properly called, are driven by electric motors. Good installations have either the rate or attitude instruments driven by air and the others by electricity, so in the event of failure of one of the systems, there will still be enough gyro instruments functioning to get safely on the ground.

Attitude Director Indicator (ADI)

A more sophisticated instrument for gyroscopic attitude indication is the **attitude director** indicator of the **flight director system** (FDS). A flight director is the portion of an automatic flight control system that provides cues to the pilot so (s)he can fly the aircraft at the command of the flight control system. In other words, the human pilot is given the same commands as the automatic pilot, and (s)he functions in the same way as the servos do when the flight control system is in the automatic pilot mode.

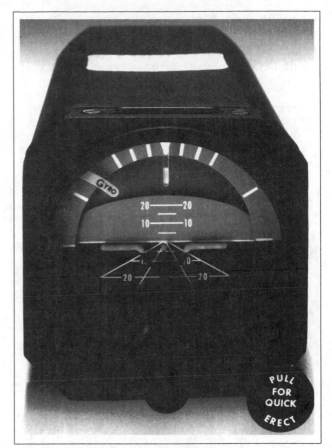

Figure 17-25. Modern attitude indicator.

The attitude director indicator looks much like a normal attitude gyro, except that it has **command bars** driven by the flight computer. When the pilot answers the commands of the flight director, the command bars are in one type of display, at the wingtips of the symbolic aircraft, figure 17-26. But when the flight director commands a climb, the command bars move up and the pilot must raise the nose of the aircraft to place the wingtips on the bars. (The symbolic aircraft actually does not move, but since both the command bars and the horizon card behind the aircraft move, it *appears* to the pilot that the symbolic aircraft has moved to answer the command.) The command bars can move to command a climb, descent or turn, or any combination of turns with climbs and descents.

Directional Gyro

The most commonly used **magnetic compass** consists of one or two small permanent magnets soldered to a metal float and suspended in a bowl of liquid. A graduated card surrounds the float so the pilot can see the direction the nose of the airplane is pointed, with respect to the earth's magnetic field. This primitive type of direction indicator is quite adequate for visual flight when it is only occasionally referred to, but since it oscillates back and forth so much and reads incorrectly during turns, it cannot be used as a heading indicator when flying on instruments.

COMMAND BAR
GYRO HORIZON DIAL
SYMBOLIC AIRCRAFT ALIGNMENT KNOB
SYMBOLIC AIRCRAFT
AIRCRAFT IS FLYING STRAIGHT AND LEVEL AND ALL COMMANDS ARE SATISFIED.
(A)
AIRCRAFT IS FLYING STRAIGHT AND LEVEL AND THE FLIGHT DIRECTOR IS COMMANDING A CLIMB.
(B)
THE AIRCRAFT HAS SATISFIED THE CLIMB COMMAND.
(C)

Figure 17-26. Attitude director indicator for a flight director system.

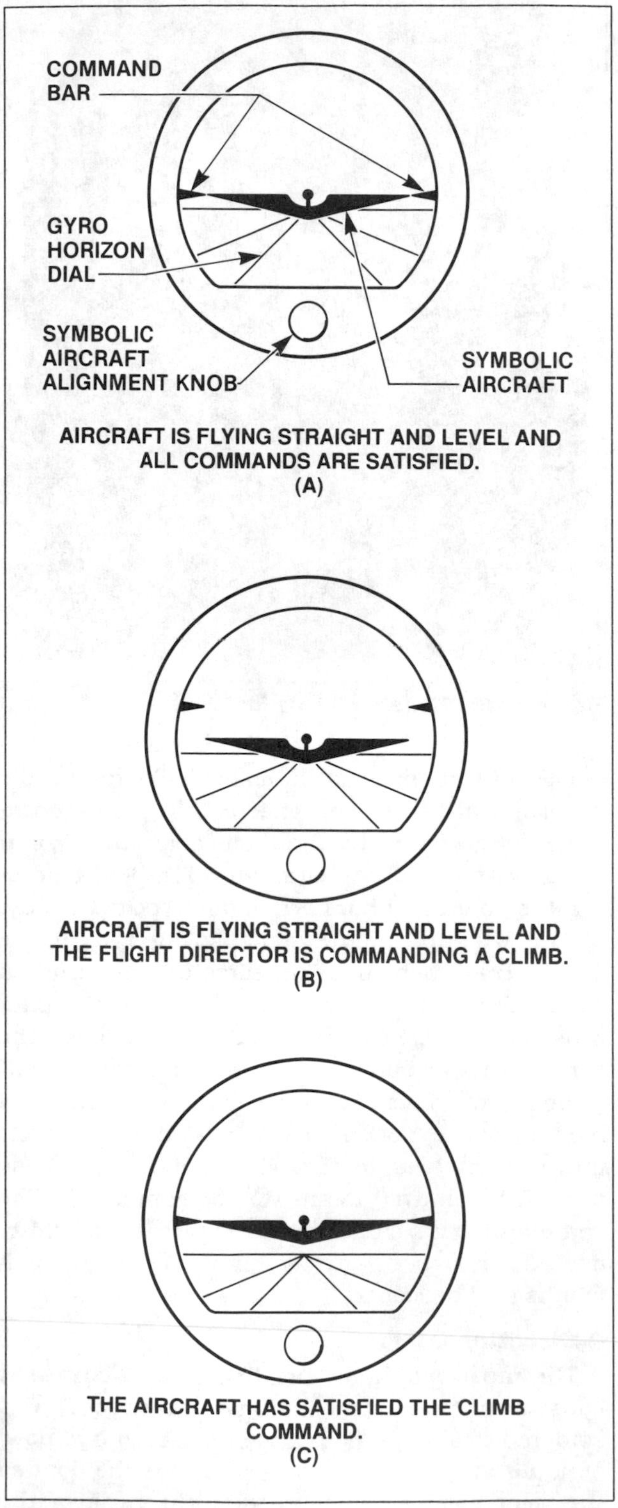

Remember that one of the two primary characteristics of a gyroscope is its ability to remain rigid in space. If we have a freely spinning gyroscope set to align with the earth's magnetic field, we can visualize our heading with respect to it, and since it does not oscillate or hang up in a bank or pitch, it can be used as a heading indicator for instrument flight. The main problem is that this instrument has no north-seeking tendency, and so it must be set to agree with the magnetic compass.

Early directional gyros called **horizontal card gyrocompasses** resembled the magnetic compass with its gyro rotor suspended in a double gimbal with its spin axis in a horizontal plane inside the calibrated scale, figure 17-27. The rotor was spun by a jet of air impinging on buckets cut into its periphery. Pushing in on the caging knob in the front of the instrument leveled the rotor and locked the gimbals. The knob could then be turned to rotate the entire mechanism and bring the desired heading opposite the reference mark, or lubber line. Pulling the knob out unlocked the gimbals so the rotor could remain rigid in space. As the aircraft turned about the gyro, the pilot had a reference between the heading of the aircraft and the earth's magnetic field.

The bearings in these instruments had enough friction to cause the gyro to precess, so it had to be reset to agree with the magnetic compass about every ten or fifteen minutes.

INNER GIMBAL
OUTER GIMBAL
33
30
27
24
21
GYRO ROTOR
CARD
CAGING KNOB

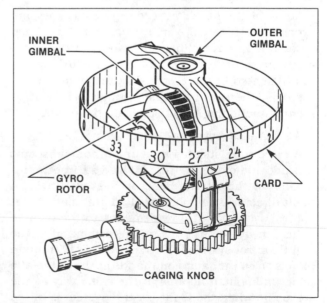

Figure 17-27. Mechanism for a horizontal card directional gyro.

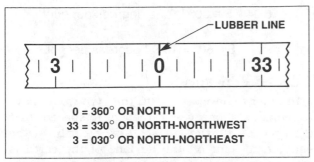

0 = 360° OR NORTH
33 = 330° OR NORTH-NORTHWEST
3 = 030° OR NORTH-NORTHEAST

Figure 17-28. The dial of a drum-type directional gyro is marked opposite to the way flight direction is normally visualized.

Notice the card, as the dial is called, of the directional gyro in figure 17-28. When the pilot is flying on a heading of zero degrees, or north, and wants to turn *left* to a heading of 330°, (s)he sees the number 33 (for 330 degrees) to the *right* of the zero mark, but (s)he must turn left to bring the 33 under the lubber line.

This is no major problem to the thousands of pilots who learned to fly instruments using the floating magnet compass and the drum-type directional gyro, but to the new instrument pilots it did create a problem—one that was easily solved, however, by the **vertical card directional gyro** as is pictured in figure 17-29. Instead of a simple lubber line in front of the card, this instrument has a symbol of an airplane on its face, in front of the dial, with its nose pointing straight up, representing straight ahead. The circular dial is con-

Figure 17-29. A vertical card directional gyro.

nected to the gyro mechanism, so it remains rigid in space and, as the airplane turns about it, the dial rotates. The knob in the lower left-hand corner of the instrument may be pushed in to cage the instrument. This locks the gimbals so the pilot can turn the mechanism to get the indication on the dial under the nose of the symbolic airplane that corresponds to the heading shown on the magnetic compass. When the knob is released, a spring pushes it back out and disengages it. The rotor support of these instruments is such that the gyro will not reach its stops during any flight maneuver, and therefore there is no need to cage the gyro. The bearings are of such high quality that friction is minimized and precession is not as big a problem as it is on older instruments. This type of directional gyro, like the older types, must be set to agree with the magnetic compass, and it too must be periodically checked to be sure that it has not drifted out of agreement with the compass. If it has, the knob may be pushed in and the dial reset.

The modern directional gyro, like the gyro horizon, has been combined with other instruments to make it the versatile flight instrument found in more sophisticated aircraft today. One of the most useful combinations has been that of **slaving** the gyro to a magnetic compass. A flux gate, or flux valve, picks up an induced voltage from the earth's magnetic field and after processing it, directs it to a slaving torque motor in the instrument that rotates the dial until the airplane's magnetic heading is under the nose of the symbolic airplane on the face of the instrument. This slaving gives the directional gyro all of the advantages of a magnetic compass without its most disturbing faults. We will discuss these faults further in the section on the magnetic compass.

In the more exotic direction-indicating instruments, the slaved directional gyro is combined with radio navigation systems so it will display information from the VOR, ADF or RNAV system, as well as from the glide slope. These instruments are called Horizontal Situation Indicator (HSI) and Radio Magnetic Indicator (RMI).

Gyrocompass Errors

There are three errors associated with the unslaved gyrocompass, two of which are totally predictable and one which is not predictable (**precession**). Often, the sum of the three errors (total error) is erroneously referred to as precession.

Drift, sometimes called apparent precession, is caused by the fact that the earth is turning but the gyro wheel remains rigid in space. This error increases as one goes from the equator (zero drift error) to the true north or south pole where the error is slightly more than 15 degrees per hour.

To help understand this, let's park an airplane on the ice at the true north pole, facing an igloo we built there. The engine is running, so we have suction to drive the gyros. The brakes are set and the airplane is tied down. We would be correct in setting the gyrocompass to 180 degrees because no matter what direction we are headed, we are headed south, right?

As every hour passes, the gyrocompass will show that we have turned left about 15 degrees, even though we are still tied down and pointed at the igloo. The gyrocompass is correct. The earth has turned under us. Had we been airborne over the north pole, flying a south heading, we would be pointed at the same point in space that we were an hour ago but we would be flying a path over the ground in a southwesterly direction. If we flew for 24 hours by the gyrocompass' south heading, we would cross every line of longitude!

If you are having a little problem with this discussion, it is time to find a globe of the earth. Start the globe turning (counterclockwise as viewed from the north pole, right?). Now launch your flight from the north pole, headed south (in any direction!). Note that the path over the ground is a spiral which crosses every meridian of longitude as the globe completes one revolution.

Please keep in mind that this is a theoretical discussion, aimed only at understanding *gyro drift*, and that I said "flying over the north pole." Keep the airplane at or very near the north pole for the 24-hour flight mentioned above.

Now let's get practical for awhile. It is obvious that flying in the *extreme* north requires some special techniques and applications, which won't be dealt with any more here because so few readers will have the opportunity to use them. The following discussion of gyrocompass errors has application for all pilots.

Gyro drift *is* a predictable error which the pilot can compensate for. Here is how its value is calculated.

Gyro drift <H> of a *stationary* airplane can be found with the use of the formula:

15.04 × Sine of the latitude = <H>, degrees per hour

For example, at 30 and 60 degrees north latitude, the drift of a stationary gyro would be:

15.04 × Sin 30° = 7.52 degrees per hour

15.04 × Sin 60° = 13.02 degrees per hour

Gyro Drift In Flight

In order to have zero drift, the gyro would have to remain stationary with respect to space rather than move with the surface of the earth. At the equator, a point on the surface of the earth is moving at a speed of 900 knots (as viewed from space), so the speed of a point on the earth at any latitude can be found by:

900 × Cosine of the latitude = knots

For example, at 30 degrees north, the surface of the earth is moving at a speed of:

900 × Cos 30° = 780 knots

If we departed a point on the earth at 30 degrees north latitude and flew west at 780 knots, from space it would appear that we were remaining stationary and the earth was turning under us at a speed of 780 knots. Since the aircraft would be stationary with respect to space, the gyrocompass would have zero drift error.

At 60 degrees north, the aircraft would only have to fly west at a speed of:

900 × Cos 60° = 450 knots

in order to remain stationary with respect to space, and have zero gyro drift error. Now we know that a flight to the west will reduce gyro drift error and a flight to the east will increase it, over the stationary value. We can now compute the amount of drift of our gyrocompass while moving in flight in any direction.

On a flight to the north or south, drift <H> will be:

15.04 × Sin of the average latitude = <H>.

For example, if we fly from 49 degrees north to 51 degrees north, the average latitude is 50 degrees north so drift will be:

15.04 × Sin 50° = 11.52 degrees (decrease)/HR

To correct the gyro drift on this trip, the pilot will need to *add* about 11.52/4 = 3 degrees every 15 minutes.

If the flight is to the east or west, gyro drift will be <H> (the stationary drift rate) minus (westerly) or plus (easterly) the drift rate due to aircraft speed.

The drift value for aircraft speed easterly or westerly can be determined by reasoning that, since the earth's speed <J> at 50 degrees north is 578.5 knots, and the stationary (zero speed) drift rate <H> is 11.52 degrees per hour, the drift rate will be zero if the aircraft flew west at <J>, it would be <H> if the aircraft was parked and twice <H> if the aircraft flew east at <J>.

At 100 knots, drift due to east or west movement would be:

$$[\text{east or west movement per hour (knots)}/<J>] \times <H> = <K>$$

and total drift <D> for flight in any direction is:

$$\text{westerly: } <H> \text{ minus } <K> = <D>, \text{ and}$$

$$\text{easterly: } <H> \text{ plus } <K> = <D>$$

Our gyrocompass should be corrected by adding the amount of the drift <D> each hour or, better yet, add one-fourth of <D> every 15 minutes.

To help the reader-pilot get a feel for the values of drift, figure 17-30 shows values of drift for various latitudes for flights to the north, south, east and west, at 100 knots.

Flights to other points of the compass are easy to estimate. Just use the average latitude (actually, at these speeds, latitude of any point on the flight is close enough). For the east-west speed component, use the number of nautical miles

LATITUDE, DEGREES	FLIGHT DIRECTON, AT 100 KNOTS		
	N. or S.	WEST	EAST
30	7.52	6.5	8.5
40	9.67	8.3	11.1
50	11.52	9.5	13.5
60	13.02	10.1	15.9
70	14.1	9.5	18.7
80	14.8	5.3	24.3

Figure 17-30. Gyrocompass drift values, degrees per hour.

you will go to the east or west in one hour as the entry value.

Remember, as we fly north, gyro drift error increases to a maximum of 15 degrees per hour at the pole. Flying east increases and flying west decreases the error. Drift is a predictable gyro error and can be corrected by adding (in the northern hemisphere) the proper correction to the gyrocompass every 15 minutes.

If you didn't understand all of this discussion of drift correction the first time through, pat yourself on the back — you are normal and of above average intelligence! If you really want to understand it (and you can!), here is how:

1. Read and understand Kershner, pages 52-67.
2. Re-read the gyro drift discussion in this chapter.
3. Calculate drift for six trips you have taken or will take (check the results with figure 17-30).
4. Prepare and give a lecture on the subject of drift to your local pilot's group.

If you do the above, congratulations! You have graduated. For a graduate course, you can come try to explain it to my students!

The Other Predictable Gyrocompass Error

When flying by gyrocompass, the pilot should understand all of the errors inherent in the instrument that lead him or her to the destination (or ash-tray, as the case may be). The three major errors of the gyrocompass are *drift*, or apparent precession, changes in *magnetic variation* and *precession*. The first two are predictable. Precession is not.

Changes in magnetic variation encountered enroute will affect the magnetic compass reading but not the gyrocompass. Therefore, the magnetic compass and the gyrocompass will disagree by the amount of *change* in magnetic variation. If the gyrocompass is not corrected, it will show an error with respect to the magnetic compass.

In the north, large changes in magnetic variation occur in relatively short distances. This is one reason why, in the far north, all navigation is done with respect to true north. Magnetic north and the magnetic compass are not used. Even the VOR's up there have their 360 degree radials aligned with true north instead of the customary magnetic north.

If the pilot is using magnetic headings, (s)he should remember that the gyrocompass and magnetic compass will disagree by the amount of any change in magnetic variation. For example, on a trip from Grand Forks, ND (7° E.) to Duluth, MN

(2° E.), magnetic variation changes by 5 degrees. For this example flight, let's assume the gyrocompass has no other errors. Departing GFK, we establish a heading of 112° magnetic. Both magnetic and gyro compasses are reading 112°. The gyro is not reset during the trip. If we steer by the magnetic compass, the gyrocompass will be reading 107° when we get near Duluth. So, it can be said that the gyrocompass has developed, over whatever period of time the flight took, an error of –5 degrees. To be correct, the gyrocompass must have 5 degrees added. This is another "apparent" error (the gyro wasn't really wrong, it was reading correctly with respect to space, but we wanted direction with respect to magnetic north).

The Unpredictable Gyrocompass Error

Precession, it is called. It is caused by forces applied to the axis of the gyro wheel. These forces are due to small amounts of friction in the gimbal bearings that support the gyro wheel. While the gyro wheel attempts to remain rigid in its orientation in space, the aircraft is rolling, pitching and yawing. This causes the gimbal bearings to move with every aircraft movement. Since the bearings are not perfectly frictionless, the aircraft's movements apply small forces to the axis of the gyro wheel, causing it to precess from its original orientation. This translates to changes in the aircraft's indicated heading—an error.

Precession error increases if:

1. There is an increase in the aircraft's movements (turbulence, pilot induced attitude changes).

2. Gimbal bearings are worn or contaminated with oil, tars, nicotine, etc. (including body and machine oils and other residues from smoking).

Precession error (not total error) should be less than 3 degrees in 15 minutes. Pilots should be alert for indications of poor gyro health, such as increasing precession error, short spin-down time, vibration and noise from the gyro after engine shutdown.

Using The Gyrocompass In The North

In the far north, where there are not section lines and the magnetic compass is unreliable, other means are used to periodically reset the gyrocompass.

Occasionally, terrain features of known alignment (from plotting on a sectional chart) can be used. Another old standby for this purpose is the

Figure 17-31. The astrocompass, when aligned with the sun, indicates true heading.

astrocompass (figure 17-31). With inputs of time (Local Hour Angle), approximate latitude and declination, and alignment of the instrument so that sunlight casts a shadow in the sighting tray, true heading can be read from the instrument and the gyrocompass corrected.

The astrocompass is somewhat cumbersome to use in a small airplane but is ideal for use in larger aircraft, especially those fitted with a sighting dome, where the astrocompass can be properly mounted.

With the advent of the small, programmable microcomputer, the **shadow-pin pelorus** has become a useful instrument to keep the gyrocompass set properly. It is far simpler than the astrocompass to use, and takes up far less space in the aircraft. It consists of a compass rose of 3-5 inches in diameter with an upright pin in the center of the rose. It is mounted level on the sun shield on top of the instrument panel in front of the pilot. Sunlight, falling on the pin, casts a shadow on the compass rose. If the sun's azimuth from the aircraft is known, the reciprocal of that number is the true direction from the sun to the aircraft. In the example of figure 17-32, the sun's azimuth is 261 degrees, so the direction from the sun (the direction the sun's shadow falls) is 081 degrees true. Knowing this, the pilot rotates the compass rose of the pelorus until the shadow falls on 81 degrees. The compass rose now indicates true direction. The pilot can read the true heading of the aircraft at the **lubber-line** (line indicating alignment with the longitudinal axis of the aircraft), and reset the gyrocompass accordingly.

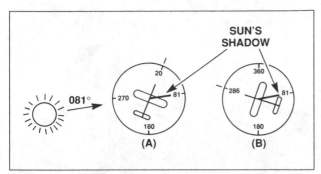

Figure 17-32. Use of the pelorus. In this example, sun's azimuth is 261 degrees true, so the sun's shadow will be 081 degrees. The pilot rotates the compass rose until the shadow falls on 81 degrees and reads the aircraft's true heading of 020 degrees for airplane A and 286 degrees for B.

The difficult task of computing the sun's azimuth is made easy by the use of a small microcomputer such as the Sharp PC1500 series or its equivalent Radio Shack model. A software program will provide the sun's azimuth given date, time, latitude and longitude. As long as the sun is shining, gyrocompass error is a thing of the past!

Rate Gyros

The two basic characteristics of a gyroscope are rigidity in space and precession. Rigidity is the characteristic used for attitude and heading instruments with precession used only incidentally, or effectively eliminated. Precession, however, is the primary characteristic used for rate gyros. There are two basic rate gyros used for flight instrumentation, the **turn and slip indicator** and the **turn coordinator**, and these operate in much the same way. Rate gyros are also incorporated into a number of autopilot systems.

The basic difference between a rate gyro and an attitude gyro is in the mounting of the gyro itself, or in the number of degrees of freedom the gyro is given. An attitude gyro is mounted in a double gimbal and has freedom about two axes, while a rate gyro is mounted in a single gimbal and has freedom about only one axis (neglecting the spin axis).

Turn And Slip Indicator

This is one of the first "blind flight" instruments developed, and it has been called by number of names, the most common having been the **needle and ball** and the **turn and bank** indicator. But in the past decade or so, the name "turn and slip indicator" has become almost universally accepted for this instrument.

There are actually two instruments in one housing, and the simpler instrument is an **inclinometer** set into the dial. This is a curved glass tube filled with a damping liquid, and riding in it is a black glass ball. When the aircraft is perfectly level and there are no other forces acting on it, the ball will rest in the bottom center of the tube between two wires which are used to hold the tube in the case and act as reference lines. In flight, the ball indicates the relationship between the pull of gravity and centrifugal force caused by a turn. The pull of gravity is affected by the bank angle; the steeper the bank, the more the ball wants to roll toward the inside of the turn—toward the low wing. Centrifugal force, on the other hand, pulls the ball toward the outside of the turn. The greater the rate of turn, the greater the centrifugal force. A **coordinated turn** is one in which the bank angle is correct for the rate of turn, and the ball remains centered.

The gyroscopic part of the turn and slip indicator is a brass rotor, spun either by air or by an electric motor. This rotor has its spin axis parallel to the lateral axis of the aircraft, and the axis of the single gimbal is parallel to the longitudinal axis of the aircraft. A **centering spring** holds the gimbal erect when there is no outside force acting on it. When the rotor is spinning and the aircraft rotates about its vertical, or yaw, axis, a force is carried into the rotor shaft by the gimbal in such a way that one side of the shaft is moved forward while the other side is moved back. Precession causes the rotor to tilt, as the force is felt, at 90 degrees to the point of application in the direction of rotor rotation. This tilt is opposed by both a dashpot which smooths out the force and by a centering spring which restricts the amount the gimbal can tilt. A wide paddle-shaped pointer is driven by the gimbal in such a way that it indicates not only the direction of yaw, but the amount of its deflection is proportional to the rate of yaw. (You will notice in figure 17-33 that the pointer actually moves in the direction opposite to the direction the rotor precesses. This is a detail of the mechanics of the instrument and is not disturbing to the pilot as (s)he sees only the needle.)

The dial of a turn and slip indicator is not graduated with numbers, but the amount of turn is measured in needle widths, and there are two standard calibrations. Some instruments are called two-minute turn indicators, and a standard rate turn of three degrees per second (360 degrees in 120 seconds) is indicated by the pointer leaning over one needle width. In a standard rate turn to

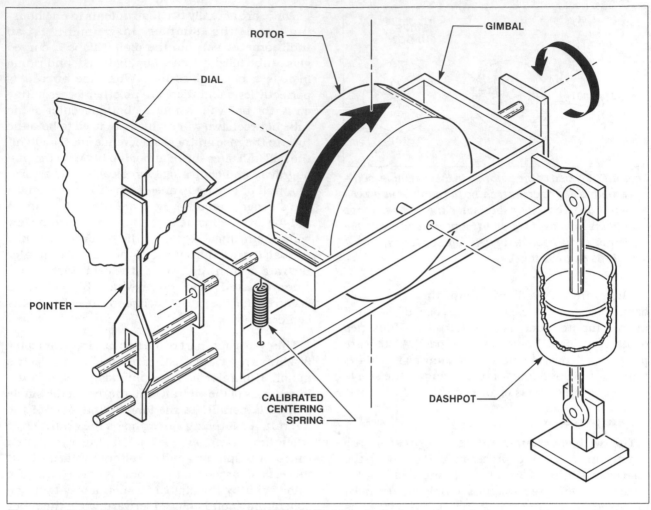

Figure 17-33. Internal mechanism of a turn and slip indicator.

the right, the left edge of the pointer aligns with the right edge of the index mark, figure 17-34. Most of the newer turn and slip indicators are calibrated as four-minute turn indicators. With this calibration, the needle deflects one needle-width for a turn of one and a half degree per second (half-standard rate). Instruments calibrated for four-minute turns have two small doghouse-shaped marks on top of the dial, one needle-width away from either side of the center index mark. These instruments may also be marked "FOUR MINUTE TURN." When the aircraft is rotating about its vertical axis at three degrees per second, the needle of the four-minute turn indicator aligns with the appropriate doghouse.

Turn Coordinator

A turn and slip indicator can show rotation about *only* the vertical axis of the aircraft—yaw. But since a turn is started by banking the aircraft—that is, by rotating it about its lon-

gitudinal axis—a turn indicator would be of more value if it sensed this rotation also. The mechanism of a turn coordinator is similar to that used in a turn and slip indicator, except that its gimbal axis is tilted, usually about 30-35 degrees, so the gyro will precess when the aircraft rolls as well as when it yaws. This is especially handy since a turn and slip indicator is affected by adverse yaw at the beginning of a turn, but a turn coordinator senses enough roll to cancel any deflection caused by adverse yaw. See figure 17-35.

Rather than using a needle for its indicator, the turn coordinator uses a small symbolic airplane with marks on the dial opposite its wing tips. When the aircraft is turned at a standard rate to the left, the left wing of the symbolic airplane aligns with the mark on the left side of the instrument dial, the one marked "L." When the rate of yaw is correct for the bank angle, the ball will be centered between the two lines across the inclinometer. See figure 17-36.

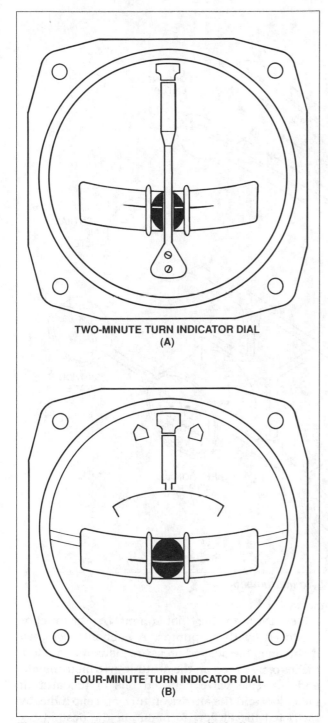

TWO-MINUTE TURN INDICATOR DIAL
(A)

FOUR-MINUTE TURN INDICATOR DIAL
(B)

Figure 17-34. Dials of a turn and slip indicator.

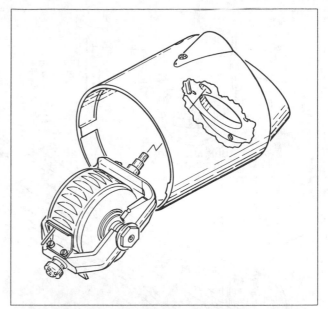

Figure 17-35. The canted rotor of a turn coordinator senses rotation about both the roll and yaw axes of the aircraft.

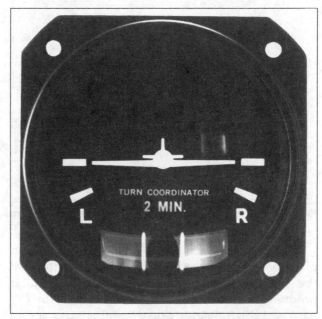

Figure 17-36. Dial of an electric turn coordinator. The small window above the right wing provides a warning flag if power is lost.

Power For Gyros

Pneumatic Gyros

Vacuum or air driven gyro instruments are all operated by air flowing out of a jet over buckets cut into the periphery of the gyro rotor. For early aircraft, a **venturi** was mounted on the outside of the aircraft to produce a low pressure, or vacuum, which evacuated the instrument case, and air flowed into the instrument through a paper filter and then through a nozzle onto the rotor.

Venturi systems have the advantage of being extremely simple and requiring no power from the engine, nor from any of the other aircraft systems; but they do have the disadvantage of being susceptible to ice, and when they are most needed, they may become unusable. See figure 17-37.

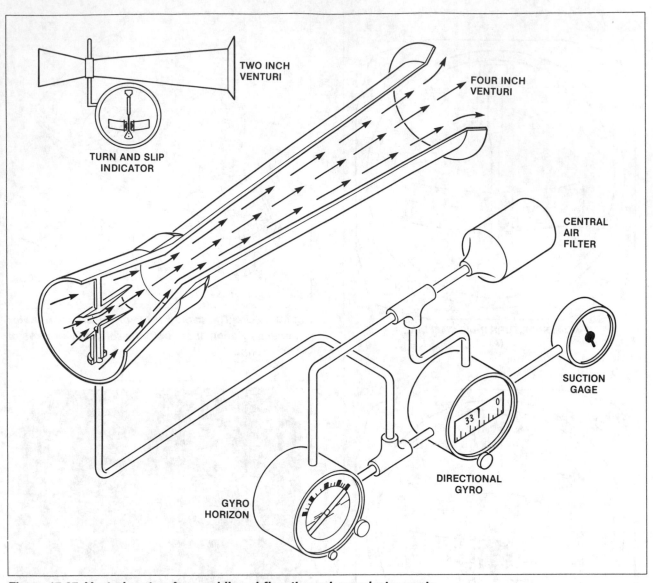

Figure 17-37. Venturi system for providing airflow through gyro instruments.

There are two sizes of venturi tubes: those which produce four inches of suction are used to drive the attitude gyros, and smaller tubes, which produce two inches of suction, are used for the turn and slip indicator. Some installations use two of the larger venturi tubes connected in parallel to the two attitude gyros, and the turn and slip indicator is connected to one of these instruments with a needle valve between them. A suction gauge is connected to the turn and slip indicator, and the aircraft is flown so the needle valve can be adjusted to the required suction at the instrument when the aircraft is operated at its cruise speed.

In order to overcome the major drawback of the venturi tube—that is, its susceptibility to ice—aircraft are equipped with engine-driven **vacuum pumps**. The gyro instruments are driven by air

pulled through the instrument by the suction produced by these pumps. A suction relief valve maintained the desired pressure (usually about four inches of mercury) on the attitude gyro instruments, and a needle valve between one of the attitude indicators and the air-driven turn and slip indicator restricted the airflow to maintain the desired two inches of suction in its case. See figure 17-38. Most of the early instruments used only paper filters in each of the instrument cases, but in some installations a central air filter was used to remove contaminants from the cabin air before it entered the instrument case. See Chapter 14 for more details.

Electric Gyros
Because of the importance of gyro instruments in flight conditions with no outside visibility, the manufacturers of most aircraft use air for operating

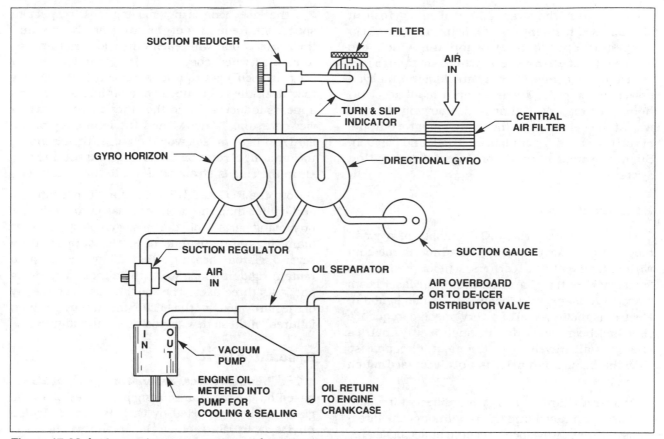

Figure 17-38. Instrument vacuum system using a wet-type vacuum pump.

the two attitude instruments and electricity for the rate gyro. When this is done, failure of either system will still leave the pilot with sufficient gyroscopic instruments to get safely back on the ground.

Electrically operated instruments may use either AC or DC, and some rate gyros have AC motors inside their rotor. A solid-state inverter inside the instrument case converts DC from the aircraft system into AC to run the instrument. All electrical instruments must have some type of failure indicator to inform the pilot of a loss of electrical power to the instrument. Usually a flag is held out of sight by an electromagnet, and if power is lost, a spring pulls the flag into view.

Some gyro instruments are operated by both electricity and air. An electric motor is built inside the rotor and buckets are cut around the outside of the rotor. The instrument may be supplied with a source of air or of DC electricity or, as a safety feature, by both. In this way, a failure of neither system by itself will deprive the pilot of the indication of the gyro. See figure 17-39(E).

New Types Of Rate Gyros

Rate gyros are gyroscopes that are designed in such a way that they measure rate of rotation in one plane. Turn and slip and turn coordinators are rate gyros.

If three rate gyros are installed in an aircraft and oriented so that one unit sensed rotation about each of the three axes (pitch, roll and yaw) and this information was fed into a computer, the computer could keep account of the orientation of the aircraft in space. This is how the AHS or AHRS systems (page 421) function to provide spatial orientation information to modern glass cockpit instruments. These rate gyros can be of the spinning-mass type, or the newer ring-laser (RLG) resonator gyros described below, or slow-spinning piezoelectric sensors that turn at a few thousand RPM, relying on minute gyroscopic forces acting on piezoelectric protrusions from the spinning mass to create meaningful electric signals to the computer. These latter three types are considered **strapdown systems** as they are attached directly to the frame of the aircraft in order to detect rates of yaw, pitch or roll.

Output from the strapped-down rate gyros can also be used to maintain a table (in the aircraft) in a specific spatial orientation on which are mounted **accelerometers** which measure acceleration vertically, laterally and longitudinally. A powerful computer keeps account of all accelerations with respect to time and, knowing where it was at the beginning of the flight, can give outputs of vertical speed, groundspeed, track and position. Such a system is called an **inertial navigation system**.

Ring-laser Gyro

The ring-laser gyro (RLG) consists of a small block of glass or other light-conducting medium with mirrors at the corners which produces a "racetrack" or "ring" around which the laser beam travels. A **laser** is a beam of visible light (an electromagnetic wave) of a very specific frequency that has been generated in such a way that it's energy is all travelling in the same direction, so that the beam's intensity is not decreased much over distance.

Two laser beams of somewhat different frequencies are introduced into the ring in opposite directions. As these waveforms meet in space inside the ring, they amplify or cancel each other according to the principle of superposition (Chapter One). If one could look in a window in the ring (as the pickoff sensor does), pulses of light and no-light would be seen as the waves of two different frequencies, travelling in opposite directions but occupying the same space, interacted with each other.

This pulsed output is ideal for computers as it is already digital in form so it can be counted directly by the computer.

Now let's take a look at how the ring laser is able to act as a rate gyro by sensing the aircraft's speed of rotation about one of the three axes (three RLG's are needed to sense movement about the three axes).

The reader may want to read up on **Doppler effect** in a high school or college physics book or an encyclopedia. Imagine looking at sinusoidal waveform "A" that is emitted from a stationary source and is passing from left to right, and counting the number of peaks and valleys of the wave that passed each second, then imagine counting wave form "B" that was leaving a source at the same frequency but the source was travelling from left to right at 30 MPH. To a stationary observer, wave form "B" would appear to be of higher frequency because more of the peaks and valleys of the wave would pass the observer each second.

If the laser ring is not moving with respect to space, the frequency of the two laser beams and their interaction with each other does not appear to change when observed at the "window" but if the ring itself has a rotational velocity (with respect to space), the frequency of one of the laser beams appears to increase and the other decrease, when viewed at the "window" and the frequency of the light/no-light pulses would change, thus sensing the amount of rotation of the ring (and the aircraft, since the ring is strapped down to the airframe).

Both the RLG and HRG are many times smaller and lighter than the "iron" or mass gyro, and have no moving parts while the mass gyro is spinning at many tens-of-thousands of RPMs, supported by several gimbal bearings that allow the gyro to remain rigid in space while the aircraft is moving about its three axes. The weight, cost, complexity of the hardware and the **MTBF** (mean time between failures) of these new "gyros" is greatly improved.

Resonator Gyro

The **hemispherical-resonator gyro (HRG)** is based on the rotation sensing properties of a wine glass, as first reported by G.H. Bryan, a British physicist, in 1890. When the crystal wineglass is "rung" by striking it, the rim vibrates, forming a wave that precesses (moves around the rim) at a given rate. If the glass is rotated, the wave appears (to a stationary observer) to precess faster or slower, providing a means of sensing rate of rotation of any axis of the aircraft.

The Delco Electronics Corporation unit, part of the Carousel 400 system, consists of three fused quartz parts — a wine-glass-shaped resonator, a pickoff housing and an external-forcer housing. See figure 17-39(D). The resonator and pickoff units employ metalized thin-film electrodes and conductors. When an AC resonating voltage is applied to the resonator, it exhibits the **piezoelectric** characteristics of a quartz radio crystal, physically flexing about 10^{-4} inches in response to the voltage, creating a low amplitude wave that can be sensed by pickoffs. The precession of this wave correlates to rotation rate. The forcer housing excludes air which would impede vibration. It also contains the electrodes which "force" the vibrations to continue indefinitely by injecting energy at the unit's resonant frequency. The location of the precessing standing wave is sensed by capacitance pickoffs found on the pickoff housing/base. These systems are in flight-testing and are expected to be commercially available in 1992-93.

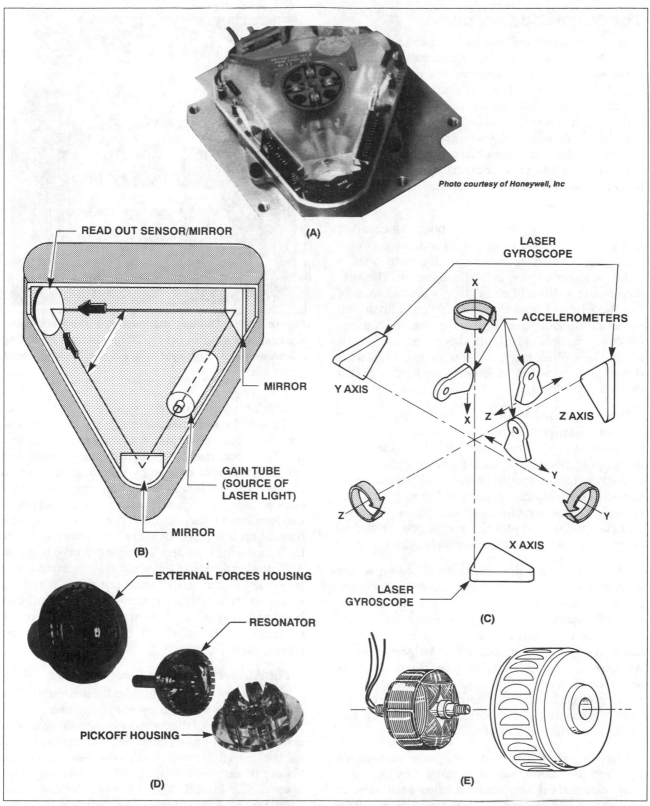

Photo courtesy of Honeywell, Inc

Figure 17-39(A). A ring-laser gyro. (B) Schematic drawing of a ring laser gyroscope. The readout sensor detects differences in frequency, the gain tube provides the light beam, and mirrors turn the laser beam. A rectangular "ring" could also be used. (C) A complete laser navigation system would use three laser gyros and three accelerometers to measure rates and velocities in all six degree-of-freedom for the aircraft. (D) Working parts of a hemispherical-resonator gyro. (E) Combined electrical and air-driven rotor for a turn coordinator.

The Magnetic Compass

In order to fly from one location to another, we must have some form of instrument that will maintain a constant directional relationship as we fly. We have been using just such a device since the 12th century: it is the **magnetic compass**. But within the last couple of decades such great strides have been taken in the development of navigational systems that the magnetic compass is no longer the most used navigational instrument, but has been relegated to a standby position in most of our aircraft.

Not only is the earth a great sphere spinning in space, but it is also a huge permanent magnet with a magnetic north and a magnetic south pole. A freely suspended permanent magnet on the surface of the earth will align itself with the lines of flux linking the two magnetic poles, and it will maintain this alignment anywhere on the surface of the earth. Because of this alignment, navigation should be simple, but there are two problems with this alignment we must understand before we can use a magnetic compass for navigation.

The geographic and the magnetic poles are not located together. The magnetic poles are located somewhere around 1,300 miles from the geographic poles and, to further complicate the situation, the magnetic poles move around continually, not enough to cause a big problem, but enough that aeronautical charts must be periodically updated to give the correction needed to compensate for this difference in location.

Since all charts are laid out according to the geographic poles, and our magnetic compass points to the magnetic poles, we have an error called **variation**. To simplify the correction for this error, our aeronautical charts are marked with lines of equal variation, called **isogonic lines**. Anywhere along an isogonic line, there is a constant angle between the magnetic and geographic north poles. The variation error is the same on any heading we fly, and is determined only by our position on the surface of the earth.

The other error inherent in magnetic compasses is called **deviation**, and it is caused by the magnetic fields in the aircraft interfering with those of the earth. A magnetic field surrounds any wire carrying electricity, and almost all of the steel parts of an aircraft and the engine have some magnetism in them. Magnetos and both alternators and generators have strong magnets in them, and these are all so close to the compass that they influence it.

Figure 17-40. Floating-magnet type magnetic compass, showing compensating magnet adjusting screws below and light bulb keeper housing above the window.

Aircraft compasses are equipped with two or more small **compensator magnets** in the housing. They may be adjusted to cancel the effect of all of the local magnetic fields in the aircraft. Any uncorrected error caused by this local magnetism is called the **deviation** error, and it is different for each heading we fly, but it does not change with the location of the aircraft. The pilot or an A&P technician must compensate for deviation errors by a procedure called **swinging the compass**. After the error has been minimized, a dated chart is made of the error that remains, which is mounted in a holder mounted on the compass bracket or on the instrument panel adjacent to the compass so the pilot will be able to apply the correction in flight.

The Aircraft Magnetic Compass

The compass is a magnificent instrument. It is simple, with few moving parts. Only one moving part, in the case of the "whisky compass" or standard aircraft compass. Older models utilized alcohol as the internal fluid. Now, odorless (don't you believe it) kerosene with additives to keep it clear is used. The fluid dampens the oscillations of the compass card, lubricates the bearing the card rotates on and decreases the loading on this bearing by floating the compass card somewhat. Therefore, it is important that sufficient fluid is in the compass. If you can see a bubble of air in the compass window, the compass should be serviced

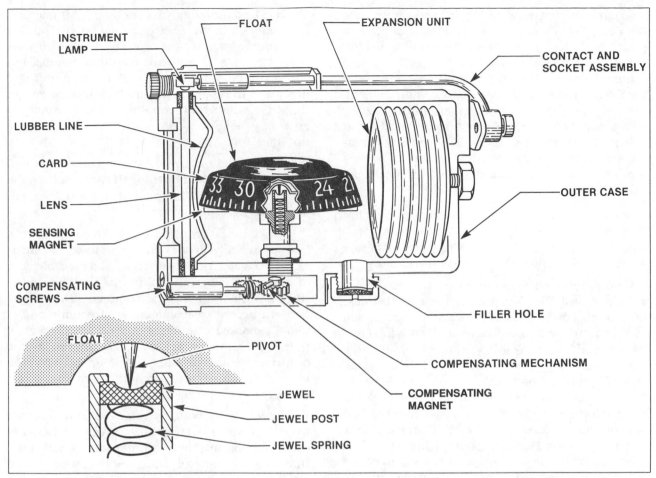

INSTRUMENT LAMP — FLOAT — EXPANSION UNIT

CONTACT AND SOCKET ASSEMBLY

LUBBER LINE

CARD

33 30 24 21

LENS

SENSING MAGNET

OUTER CASE

COMPENSATING SCREWS

FILLER HOLE

FLOAT — PIVOT

COMPENSATING MECHANISM

COMPENSATING MAGNET

JEWEL

JEWEL POST

JEWEL SPRING

Figure 17-41(A). Internal mechanism of a floating magnet-type magnetic compass.

and the rubber diaphragm (which allows for expansion and contraction due to temperature and pressure) replaced.

Its main body is a cast aluminum housing, and one end is covered with a glass lens. Across this is a vertical reference mark called a lubber line. Inside the housing and riding on a steel pivot in a jewel post is a small brass float surrounded by a graduated dial which is part of a cone. Around the full 360 degrees of the dial are 36 marks, representing the tens of degrees. Above every third mark is either a one- or two-digit number representing the number of degrees with the last zero left off. Zero is the same as 360 degrees and is north. Nine is east, or 90 degrees, 18 is south (180 degrees), and 27 is 270 degrees, or west. Two small bar-type permanent magnets are soldered to the bottom of the float, aligned with the zero and 18 marks, north and south. An expansion diaphragm or bellows is mounted inside the housing. A set of compensator magnets is located in a slot in the housing outside of the compass bowl. And a small instrument lamp screws into the front of the housing and shines

inside the bowl to illuminate the lubber line and the numbers on the card. See figures 17-40 and 17-41(A).

Compass Compensation

The magnetic compass should be swung and a new deviation card made up when first installed, any time equipment (such as a new radio in the panel) is added or removed and any time the pilot suspects compass accuracy. I try to do this once a year. A check of deviation cards in aircraft on any given airport usually shows that most cards are very old. There is no reason for this, when it is so easy to swing an aircraft compass. Here is how I do it.

An aircraft compass should not be swung on the ground, except as a first approximation, as many errors will exist there. The greatest error is produced by the fact that the magnetic field produced by the generator or alternator is not present unless the engine is running up to speed. Try this next time you do a runup: with the aircraft on a constant heading and with engine idling, note the compass reading. Then increase RPM until the

409

alternator or generator is showing an output. Note the compass reading. The compass is actually a pretty good alternator/generator output indicator, isn't it? Pilots should know how much effect on the compass there will be if the alternator quits (and in which direction), in case such a failure should occur on a critical leg, going cross-country. Other errors may be present with a "ground swing" but you won't be aware of them. They include variations in the earth's magnetic field due to ferrous (iron) metals in the ground such as rebar in the concrete, metal hangars nearby, etc.

So, swing your compass while airborne. Besides, its a great excuse to go flying. Getting bored on a long, smooth cross-country? Swing your compass—it will only add a few minutes to your trip time.

If you are in an agricultural area, section lines work very well. First you must determine the magnetic direction the section lines are running. If you have forgotten how to apply the old "East is least and West is best" rule, look at a VOR rose on your chart. The chart is laid out on the basis of true direction. The VOR north arrow is pointing to magnetic north. If you are in an area of easterly variation, the VOR north indicator will be pointing east of true north. A straight edge, placed over the VOR station on the chart and aligned North-South (parallel with a meridian) will tell the magnetic direction of the North-South section line and make it clear to you how the magnetic variation must be applied.

Line the centerline of the aircraft up with the section line and set your directional gyro to the MAGNETIC direction of the section line. Example: if the magnetic variation is 7 degrees East, the north section line is pointing 353 degrees magnetic, if variation is 7 degrees West, the section line is pointing 007 degrees magnetic.

If there are not section lines in your area, use the runway centerline of the nearest paved airport. The tower will be happy to give you the runway magnetic heading to the nearest degree. If the field is not controlled, ask the airport manager for the precise runway heading before you go flying. Then, fly down the runway in a low pass, align the aircraft centerline with the runway centerline and set your gyrocompass.

With the gyrocompass set, it's readings become the FOR numbers in your deviation card and the magnetic compass readings are the FLY values. While holding a 360-degree heading on the gyro, read the magnetic compass. If it reads 003 degrees, the first entry on your deviation card should read: FOR 360 (degrees), FLY 003

(degrees). Repeat this process every 30 degrees, then check the gyrocompass against the section line. Your deviation card can now be made up. Be sure to include the date and conditions (radios on, etc.) on the card. It might be well to repeat the flight process in the opposite direction to be sure you get the same numbers. If you don't, then you had an error in your process or the compass is "hanging up" a little bit.

There should not be deviations greater than 10 degrees on any noted heading. If there is, then the compass compensating magnets should be adjusted, as described below.

Most of the larger airports have a compass rose laid out, usually on one of the least used taxi strips as far from electrical interference as possible, and at a location where a technician may be undisturbed while swinging the compass. The rose is laid out according to magnetic directions and is usually marked with a line every thirty degrees. If there is no compass rose available, you can lay one out, using an accurate compass or, better yet, a surveyor's transit, the Sun and a computer or set of Sun's True Bearing tables.

Prepare the aircraft by removing any material from the instrument panel area and glove box that could possibly interfere with the compass. Be sure that all of the normally installed instruments and radio equipment are in place and are properly functioning. Adjust the compensator magnets until the dot on the screw head is aligned with the dot on the instrument case. Align the aircraft headed magnetic north.

Adjust the N-S compensator screw with a non-magnetic screwdriver until the compass reads north (0). Now, turn the aircraft until it is aligned with magnetic east. Adjust the E-W screw until the compass reads east (9). Continue by turning the aircraft south and adjust the N-S screw to remove *one-half* of the south heading error. This will throw the north heading off, but the total north-south error should be divided equally between the two headings. Complete the adjustment by turning the aircraft west (27), and adjust the E-W screw to remove *one-half* of the west error.

Flux-gate Compass

Pilots that operate aircraft equipped with a **slaved gyrocompass** system or a panel-mounted indicator instrument giving magnetic heading information, should be aware that, somewhere in the aircraft is a magnetic-field sensing device that provides electronic output of direction information to the panel indicator or **slaved gyrocompass**. The

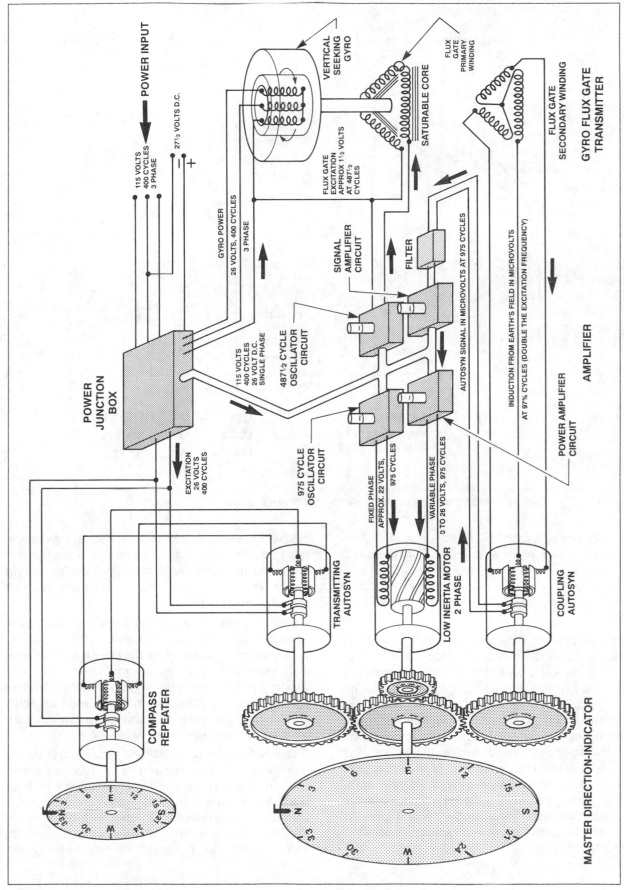

POWER INPUT

115 VOLTS 400 CYCLES 3 PHASE

27½ VOLTS D.C.

VERTICAL SEEKING GYRO

SATURABLE CORE

FLUX GATE PRIMARY WINDING

FLUX GATE EXCITATION APPROX 1½ VOLTS AT 487½ CYCLES

FLUX GATE SECONDARY WINDING

GYRO FLUX GATE TRANSMITTER

GYRO POWER 26 VOLTS, 400 CYCLES 3 PHASE

POWER JUNCTION BOX

SIGNAL AMPLIFIER CIRCUIT

FILTER

115 VOLTS 400 CYCLES 26 VOLT D.C. SINGLE PHASE

487½ CYCLE OSCILLATOR CIRCUIT

AUTOSYN SIGNAL IN MICROVOLTS AT 975 CYCLES

INDUCTION FROM EARTH'S FIELD IN MICROVOLTS AT 97½ CYCLES (DOUBLE THE EXCITATION FREQUENCY)

POWER AMPLIFIER CIRCUIT

AMPLIFIER

EXCITATION 26 VOLTS 400 CYCLES

975 CYCLE OSCILLATOR CIRCUIT

FIXED PHASE APPROX. 22 VOLTS, 975 CYCLES

VARIABLE PHASE 0 TO 26 VOLTS, 975 CYCLES

TRANSMITTING AUTOSYN

LOW INERTIA MOTOR 2 PHASE

COUPLING AUTOSYN

COMPASS REPEATER

MASTER DIRECTION-INDICATOR

Figure 17-41(B). Schematic diagram of Gyro Flux Gate system.

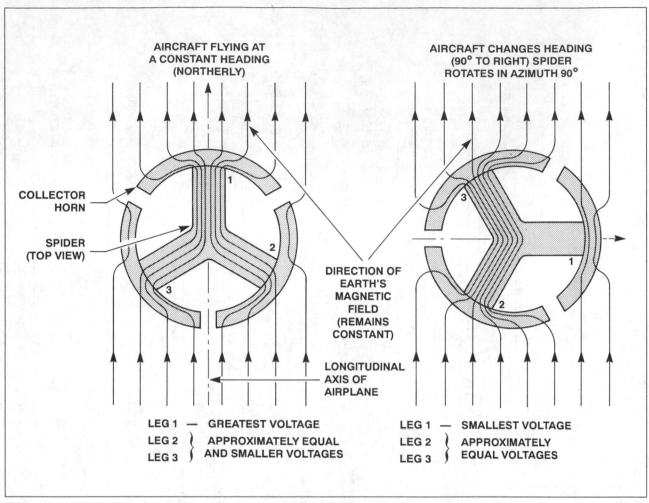

AIRCRAFT FLYING AT
A CONSTANT HEADING
(NORTHERLY)

AIRCRAFT CHANGES HEADING
(90° TO RIGHT) SPIDER
ROTATES IN AZIMUTH 90°

COLLECTOR
HORN

SPIDER
(TOP VIEW)

DIRECTION OF
EARTH'S
MAGNETIC
FIELD
(REMAINS
CONSTANT)

LONGITUDINAL
AXIS OF
AIRPLANE

LEG 1 — GREATEST VOLTAGE
LEG 2 } APPROXIMATELY EQUAL
LEG 3 } AND SMALLER VOLTAGES

LEG 1 — SMALLEST VOLTAGE
LEG 2 } APPROXIMATELY
LEG 3 } EQUAL VOLTAGES

Figure 17-41(C). Effect of aircraft heading on magnitude of induced current.

sensing device is located in a wing or the tail cone in order to locate it as far as possible from magnetic fields generated within the aircraft itself. The system requires a source of electric power and may have to be switched on by the pilot in order to function.

The **flux gate** or **flux valve** is the heart of the system, sensing the direction of the earth's magnetic field (lines of flux). If this unit is not kept level, the lines of flux enter at odd angles (depending on the angle of bank or pitch). This weakens the directional input, decreasing compass accuracy. The more sophisticated of these units incorporate a vertical-seeking gyro to maintain the sensing unit in a level attitude regardless of the attitude of the aircraft, and are called **gyro flux-gate compasses**. If the system is without a gyro (either a vertical-seeking gyro or a slaved gyrocompass), the pilot may not expect accurate readings unless the aircraft is level.

Like all magnetic compasses, the flux-gate system loses its ability to sense direction when operating in far northerly or southerly latitudes due to the angle that the magnetic lines of flux take as they dip into the earth.

The **flux-gate** is a special three-section transformer which develops a signal whose characteristics are determined by the position of the unit with respect to the earth's magnetic field. The flux-gate element consists of three highly permeable cores arranged in the form of an equilateral triangle with a primary and secondary winding on each core. See figure 17-41(B).

Operation of a **flux-valve** is very similar to the flux-gate but construction is a little different. Instead of the three equilaterally spaced coils of the flux-gate, the flux-valve **spider** is constructed with three legs spaced 120° apart with the primary winding around a core in the center. Figure 17-41(C) shows the effect of aircraft heading (as related to the earth's magnetic flux lines) on the three legs of the flux valve.

For a more detailed discussion of flux-gate and flux-valve compasses, see chapter 20 of additional reading item 3.

Vertical Card Compasses

Vertical card compasses have come on the scene recently and are enjoying quite a bit of success. They have a stronger sensing magnet and therefore work quite well in the north country. I have used mine to 65 degrees North latitude, and it was going strong while the whisky compasses I have flown in the North are useless in light turbulence north of 50 degrees North and become totally useless "North of sixty." They just turn slowly around and around.

Vertical card compasses don't exhibit "northerly turning error" and acceleration error like a fluid compass but are roll, pitch and yaw sensitive.

With respect to vertical card compasses: Do not panel mount—they are sensitive to vibration, so they need to be mounted quite loosely. If you have to mount one near the panel support bars that are found on many seaplanes (I did), you will need the external magnets as the internal adjusting magnets will not be sufficient to produce good results. Some people have not had good luck adjusting out large deviations in their vertical card compasses. It is absolutely imperative that the manufacturer's instructions be followed exactly during the adjustment process, as it is easy to get the adjusting magnets aligned so they are fighting each other, producing more errors than are being corrected.

A quick check of the compass before a critical leg is a good idea. There are good directional indicators everywhere. If there are no section lines or paved runways, use the direction of a line drawn between two prominent points on your chart, such

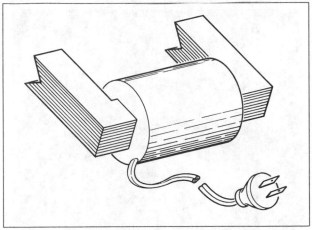

Figure 17-43. A demagnetizer such as this may be used to demagnetize that part of an aircraft structure that is causing a large deviation error on the compass.

as two islands 10-15 miles apart in a lake. While flying over one island, align the aircraft's centerline with the distant island and set your gyrocompass. Now check the compass' indication against the gyro reading when headed in the general direction of the critical leg that is yet to be flown.

As mentioned before, the standard aircraft compass isn't much good in the far north or south. There, the lines of magnetic force point down into the earth, so the horizontal component of that force is small, making for poor direction sensing capability of the parallel magnet compass.

In the far north or south, pilots don't rely on the magnetic compass at all. They use their gyrocompass set to the runway heading at the beginning of the flight and compensate for its known errors enroute, by utilizing an astrocompass (see figure 17-31), or a shadow-pin pelorus and computer, or simply by applying corrections for the gyrocompass' known errors and hoping that precession errors, which are not totally predictable, remain small.

You have now made all of the adjustments you should have to make, so check the compass on all headings and make a calibration card (figure 17-42). The aircraft is now headed west. Start your calibration card here and record the magnetic heading of 270 degrees and the compass reading with the radios off and then with them on. When radios are off, you should try switching off the alternator/generator as well, to see if it makes a difference. Turn the aircraft to align with each of the lines on the compass rose and record the compass reading every thirty degrees. There should not be more than about a ten-degree

FOR	N 0	3	6	E 9	12	15	S 18
STEER RADIO							
STEER NO RADIO							
FOR		21	24	W 27	30	33	
STEER RADIO							
STEER NO RADIO							

Figure 17-42. Compass correction card allows the pilot to apply correction for deviation error. Be sure to date it.

Figure 17-44. Tachometer for a piston-engine airplane with synchronizer.

difference between any of the compass headings and the magnetic heading of the aircraft. But if there is, it may require quite a bit of detective work to find the cause of the error. Steel screws in the vicinity of the compass or magnetized control yokes or structural tubing can cause unreasonable compass errors, as can improperly routed electrical wiring. A degaussing tool (figure 17-43) might help—most TV repair shops have one and can tell you how to use it.

When the compass is swung to your satisfaction, fill out the calibration card and date it.

Engine Instruments For Reciprocating Engines

Tachometer

Reciprocating engines use tachometers to indicate the engine speed. They are calibrated in hundreds of RPM with the most generally used range up to about 3,500 RPM. The number 20 on the dial indicates an engine speed of 2,000 RPM.

The tachometers of small single-engine aircraft are normally mechanically driven and are of the magnetic drag type, very similar in operation and construction to an automobile speedometer. Multi-engine aircraft normally use electrical or electronic tachometers.

Some multi-engine aircraft have synchroscopes in conjunction with the tachometers. These are small indicators that show the difference in speed of two engines.

Manifold Pressure Gauge

The absolute pressure inside the induction system of an engine is an important indicator for the power developed by the engine. It is not a direct measurement, but it does relate. The aircraft flight manual includes charts showing the various combinations of RPM and manifold pressure that produce the desired engine power.

Figure 17-45. Manifold pressure gauge. This instrument shows the absolute pressure of the air entering the cylinders of the engine.

Figure 17-46. Exhaust-gas temperature gauge. This shows the temperature of the exhaust gases as they leave the cylinder.

Figure 17-47. Fuel-flow indicator. This is actually a pressure gauge that measures the pressure drop across the injector nozzles of a fuel injected engine.

Figure 17-48. The engine instruments on most small aircraft are clustered together and use small indicators such as these to save panel space.

For normally aspirated engines, the manifold pressure gauge usually has a range from 10 inches to 40 inches of mercury absolute, and for turbocharged engines, the range goes high enough to adequately cover the highest manifold pressure the engine is allowed to have.

Cylinder Head Temperature Gauge

Thermocouple-type temperature indicators are usually installed on the cylinder shown by flight test to be the one that runs the hottest.

Exhaust-gas Temperature Indicator

This is a thermocouple-type instrument that measures the temperature of the exhaust gas as it leaves the cylinder. The temperature of the exhaust gas is a good indicator of the combustion efficiency of the engine, and modern procedures of mixture control are based on EGT indications.

Some EGT installations use only one probe placed in the exhaust of the cylinder recommended by the engine manufacturer, while others use a probe in the exhaust of each cylinder. Multi-probe installations have either a manual or an automatic scanning switch, to allow the pilot to see the condition of all of the cylinders. With this information, he can detect abnormalities in cylinder operation.

Fuel flow

For large carbureted engines, a fuel-flow transmitter is installed in the fuel line between the engine-driven fuel pump and the carburetor to measure the amount of fuel entering the carburetor.

But fuel injected engines, primarily those of the horizontally opposed type, have as their fuel-flow indicator a pressure gauge that measures the pressure drop across the injection nozzles. This system operates on the basis that the pressure drop across the nozzles is directly proportional to the flow through them. Normally aspirated engines use gauge pressure for this indication, and turbocharged engines require a measurement of differential pressure—the difference between the fuel pressure delivered to the nozzles and the MAP, or the pressure of the air as it enters the cylinders.

Oil Pressure Gauge

This instrument measures the gauge pressure of the oil as it enters the oil passages inside the engine after it has passed through the oil pump.

Oil Temperature Gauge

This measures the temperature of the oil as it enters the engine, and shows the operating effectiveness of the oil cooler.

Turbine Engine Instruments

Tachometer

These are all of the electric or electronic type and measure the speed of the compressor. They are calibrated in percentage, with the upper limit of 110 to 120 percent.

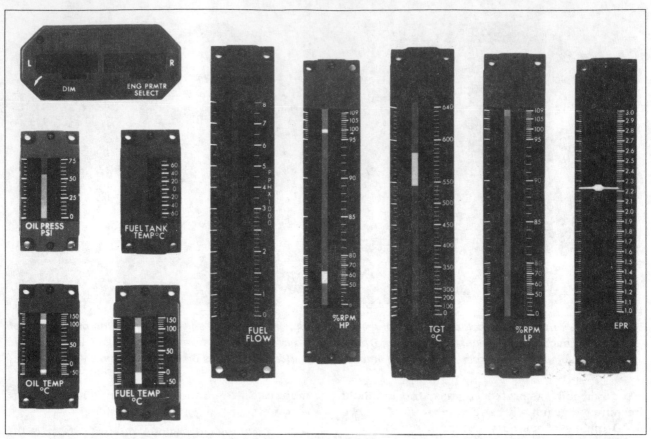

Figure 17-49. Vertical scale instruments are becoming popular because of the minimum amount of panel space they require.

Engines having a dual, or split, compressor use two tachometers. The N_1 tachometer measures the speed of the low-pressure compressor, and N_2, the speed of the high-pressure rotor.

Engine Pressure Ratio Indicator (EPR)

Thrust produced by a turbojet engine using a centrifugal compressor is usually calculated by using the indications of the tachometer and the exhaust gas temperature gauges, but the best indication of thrust produced by a dual-rotor axial-flow turbine engine is computed by using the engine pressure ratio. This is measured with a differential pressure gauge that senses the pressure difference between the tailpipe total pressure, usually abbreviated as P_{T1}, and the compressor inlet total pressure, or P_{T2}. These gauges usually have a range from 1.1 to 2.5 or 3.0

Exhaust Gas Temperature Gauge (EGT)

These instruments receive their input from an averaging system of thermocouples arranged around the discharge of the turbine section to measure the temperature of the gases as they leave the turbine wheel. The range of these instruments

is normally up to somewhere above 800 degrees Celsius.

Fuel-flow Indicator

This is a mass flow indicator that measures the number of pounds of fuel flowing into the engine and is calibrated in pounds per hour.

Oil Pressure Gauge

Like the gauge used on a reciprocating engine, this instrument measures the gauge pressure of the oil after it leaves the engine oil pump.

Oil Temperature Indicator

This measures the temperature of the oil as it enters the engine, after it has passed through the fuel-oil heat exchanger (oil cooler).

Torquemeter

Some of the large reciprocating engines have been equipped with torque meters, to measure their power output, but most use a combination of RPM and manifold pressure. Modern turboprop engines, however, routinely use torque meters to measure the amount of power being delivered to the propeller, or to the output shaft in the case of

Figure 17-50. Outside air temperature gauge.

a turboshaft engine. A hydraulic cylinder measures the pressure generated by the reaction forces in the reduction gearing. And this reading, along with that of the tachometer, is computed with a constant for the engine to measure the power being produced.

Auxiliary Instruments

Clock

This is one of the most fundamental of instruments, as it is used for timing flight maneuvers as well as for navigation and for such engine functions as determining the rate of fuel consumption.

Until the rise in popularity of digital display instruments, aircraft clocks have been of the analog type, all equipped with a sweep second hand. But now, clocks with digital display are gaining in popularity. These not only display the local time, but also Greenwich mean time (Zulu time), as well as the day and date. Most of these clocks are equipped with circuits that allow them to be used as a stop watch and a lapsed time indicator.

All of the electrically operated clocks are installed in the aircraft with their power lead ahead of the master switch, so they will always have power regardless of the position of the master switch. They are protected with a low current fuse.

Outside Air Temperature Indicator (OAT)

Both engine and aircraft performance depend on density altitude which requires temperature to calculate. Measurement of FAT (**free air temperature**) (ambient) is done with a probe in the free airstream, but several factors may create errors in the value indicated by the bulb. Radiation energy from the sun or a hot runway can cause an error. The largest errors occur in fast aircraft due to compressibility

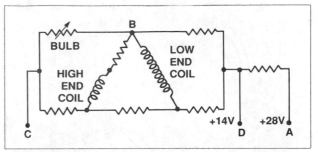

Figure 17-51. Circuit for a ratiometer-type temperature indicator.

and skin friction. A **true outside air temperature indicator** mechanically corrects for compressibility in accordance with the **mach compressibility function**. In modern aircraft equipped with an airdata computer, outputs of TAT and RAT may be available to the pilot. RAT (**ram air temperature**) is FAT plus the temperature rise due to compressibility (important because it is the best indication of the temperature a jet engine "sees"). Rat is typically about 80% of TAT (**total air temperature**) which is FAT plus rise due to compressibility plus rise due to skin friction.

At airspeeds below 150 KTS, there is little temperature rise due to speed, but at Mach. 7 at crusing altitude, it is not unusual to see TAT being 100% greater than FAT.

An extremely simple instrument is used on most of the small single-engine aircraft. It consists of a bimetallic-type thermometer in which strips of two dissimilar metals are welded together into a single strip and twisted into a helix. One end is anchored into a protective tube, and the other end is affixed to the pointer which reads against the calibration on a circular dial. The bimetallic strip in its tube sticks through a hole in the upper portion of the windshield, with the dial in easy view of the pilot.

A more accurate outside air temperature indicator is electrically operated and measures the change in resistance of a coil of nickel wire with changes in temperature. These instruments are usually of the ratiometer type, which measures the resistance of the sensing but by comparing the current flow through the bulb with the flow through a set of resistors inside the instrument case. By using a ratio of current, the effects of variations in system-voltage is minimized.

Figure 17-51 shows the principle of operation of a ratiometer-type temperature indicator. This particular instrument may be used with either a 14- or 28-volt electrical system, the difference being in the pin to which the power lead is connected. The power from a 28-volt system is connected through pin A so

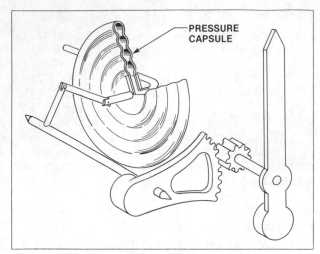

Figure 17-52. Diaphragm-type pressure indicating mechanism.

the current will have to flow through a dropping resistor that lowers the voltage to that required by the instrument.

When the temperature (and thus the bulb resistance) is low, most of the current flows through the low-end coil and the bulb, rather than through resistors B and C and the high-end coil. When the temperature is high, the bulb resistance will be high, and more current will flow through the high-end coil than through the low-end coil. The pointer of these indicators is mounted on a small permanent magnet that aligns with the magnetic fields produced by the low- and high-end coils.

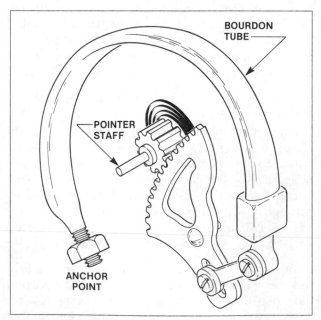

Figure 17-53. Bourdon tube-type pressure indicator mechanism.

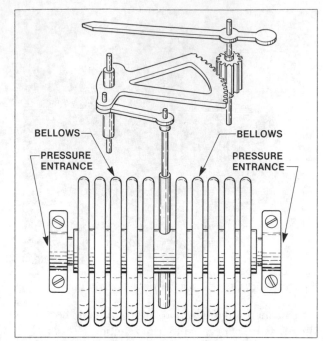

Figure 17-54. Differential bellows-type pressure indicator mechanism.

Pressure Indicators

Oil pressure, fuel pressure, hydraulic pressure and suction, or vacuum, oxygen pressure are all measurements which the pilot needs to determine the operating conditions of the various systems or components.

There are several methods of measuring these various pressures, and the choice of method depends primarily on the type and range of the pressure. Low pressures such as suction or instrument air pressure are measured by a pressure capsule or **bellows**. The pressure to be measured is directed into the capsule, which is expanded by a positive pressure or compressed by a negative pressure. The change in dimensions of the capsule caused by the pressure is measured by the linkage that transmits this movement into rotation of the pointer over the dial. See figure 17-52.

Higher gauge pressures are usually measured by a **Bourdon tube** mechanism. The sensitive portion of this instrument consists of a bronze tube with an elliptical cross section formed into a curved shape, with one end anchored into the case and the other end connected to the sector gear. When pressure enters the instrument, it tries to round out the elliptical cross section, and as the tube rounds out it also straightens out and pulls on the tail of the sector gear. This rotates the pinion gear and moves the pointer across the dial.

Bourdon tube instruments can be made in an almost unlimited range by varying the design of

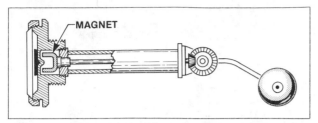

Figure 17-55. Mechanical fuel quantity gauge.

the tube. Some low-pressure instruments have tubes formed into a helix of several turns, while those that measure high pressure use stiff tubes formed into less than a full circle.

We occasionally have a need to measure either absolute or differential pressure in an auxiliary system, and a dual bellows is one the handiest mechanisms for measuring either type of pressure. Two bellows are mounted together and supported inside the instrument case. A linkage joins the rocking shaft to the sector gear which drives the pinion to which the pointer is attached. If the instrument is used to measure differential pressure, one of the bellows is connected to each of the pressures to be measured, and the pointer will move an amount proportional to the difference between the two pressures.

If the instrument is used to measure absolute pressure, one of the bellows is evacuated and the other is connected to the pressure to be measured. The pressure is then referenced from what is effectively a vacuum.

Pressure sensors for electronic systems utilize **strain gauges** or **load cells** which are devices that alter the resistance or capacitance of a circuit depending on the amount of pressure applied to the device. The change in R or C is converted to pressure indications by a microprocessor. Electronic bathroom scales and altitude encoders are examples.

Fuel Quantity Indicator

One of the instruments required for all aircraft by FAR Parts 43 and 91 is a "Fuel gauge indicating the quantity of fuel in each tank." These gauges,

or systems, may be as simple as a wire attached to a cork float sticking out of the fuel tank cap. The amount of wire protruding from the cap indicates the amount of fuel in the tank.

Other direct-reading fuel quantity indicators move a pointer across the dial by magnetic coupling. A float rides on the top of the fuel and drives a bevel gear, which rotates a horseshoe-shaped permanent magnet inside the indicator housing. A pointer attached to a small permanent magnet and mounted on a pivot is separated from the horseshoe magnet by an aluminum alloy diaphragm, but is coupled to it by the magnetic fields. The pointer will move around the dial to indicate the level of fuel in the tank. See figure 17-55.

Fuel quantity indicating systems similar to those used in automobiles are found in many aircraft. These are electrically operated and consist of a variable resistor as a tank unit, or sender, as it is sometimes called, and a current measuring instrument as the indicator.

The tank unit consists of either a wire-wound resistor or a segment of composition resistance material, and a wiper arm driven by a float that moves across this resistance material to change the circuit resistance as a function of the fuel level in the tank. Some units signal a full tank with maximum resistance, and others a full tank by a minimum resistance.

The indicator for most of these systems is a ratiometer-type gauge that minimizes the error that would be caused by variations in system-voltage. Current flows through both coils, figure 17-56, and both the fixed resistor and the tank unit. As the resistance of the tank unit varies, so does the current through coil B, and its magnetic strength will vary. The pointer is mounted on a small permanent magnet and moves across the dial in such a way that it indicates the level of fuel in the tank. See figure 17-56.

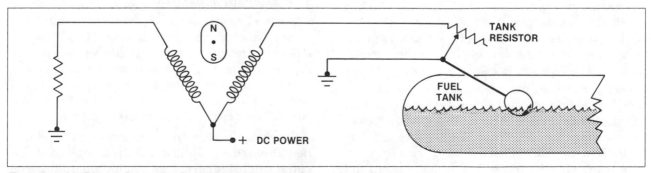

Figure 17-56. Circuit for a simple ratiometer-type fuel quantity indicator.

419

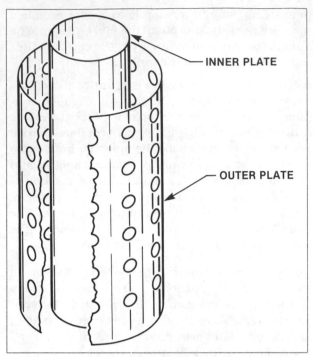

Figure 17-57. Fuel tank probe for a capacitance-type fuel quantity system.

Many modern larger aircraft use a **capacitance-type fuel quantity measuring system** that measures the mass of the fuel, rather than just its level in the tank. This is an electronic system that measures the capacitance of the probe or probes which serve as the tank sender units. Let's review capacitors a bit, so we can better understand this simple but effective system.

A capacitor, as we remember from our study of basic electricity, is a device that can store electrical charges, and it consists of two conductors called plates separated by some form of dielectric or insulator. The capacity of a capacitor depends upon three variables: the area of the plates, the separation between the plates, which is the thickness of the dielectric, and the dielectric constant of the material between the plates.

The probes in a capacitance fuel quantity indicating system are made of two concentric metal tubes which serve as the plates of the capacitor. The area of the plates is fixed, as well as the separation between them, so the only variable is the material which separates them. See figures 17-57 and 17-58.

These probes are installed so they cross the tank from top to bottom, and when the tank is empty, the plates are separated by air which has a dielectric constant of one. When the tank is full, the dielectric is fuel which has a constant of approximately two. In any condition between full

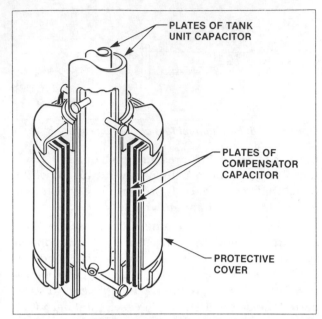

Figure 17-58. Compensating capacitor for a capacitance-type fuel quantity system.

and empty, part of the dielectric is air and part is fuel, and so the capacity of the probe varies according to the level of fuel in the tank.

One of the big advantages of this type of measuring system is that the probes can be tailored for tanks of all sizes and shapes, and all of the probes in the aircraft can be connected so the system integrates their output to show the total amount of fuel on board.

The dielectric constant of the fuel is approximately two, but it varies according to its temperature and so a compensator is built into the bottom of one of the tank units. It is electrically in parallel with the probes and decreases the changes in dielectric constant as the temperature of the fuel changes, thus permitting the system to measure fuel mass rather than fuel quantity.

EFIS And EICAS Systems

The Microprocessor

Most of us are familiar with computers and some of the things they can do. They come in different sizes and capabilities, depending on the capabilities of the computer's **microprocessor** which is a large-scale integration (LSI) digital circuit that contains several of the functional elements required in a computer system.

They vary in size, cost, capability and application through an almost infinite range. A **dedicated microprocessor** is usually manufactured as a self contained system with all of the computer

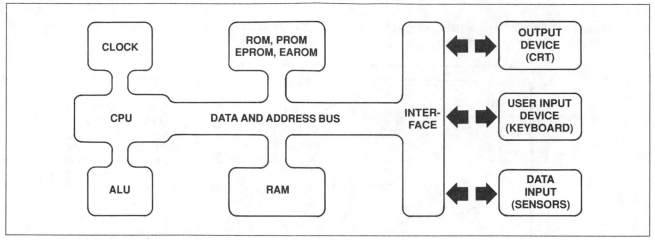

Figure 17-59. A microprocessor system.

(CPU); **arithmetic and logic unit** (ALU); permanent information memory, including **read only memory** (ROM) where the information stored is prewritten and unchangeable, **programmable read only memory** (PROM) which is user programmed but only once, **erasable programmable read only memory** (EPROM) which is erased with ultraviolet light, and **electrically alterable read only memory** (EAROM) which can be electrically erased and reprogrammed; temporary information memory (RAM) which can give or receive information to or from the CPU; system clock (synchronizer); BUS (the information communication and transportation system; and interface, which converts the foreign language of the user and sensor inputs into a language compatible with the CPU and visa-versa.

The microprocessor is designed to do certain tasks by the program in its memory. Task sequencing can be altered by the program depending on information received through the interface (inputs).

The circuitry used to make up the microprocessor is referred to as the system **hardware**. **Software** is the program developed to make the hardware perform the desired functions. When the software is programmed into permanent memory, the combined hardware/software is referred to as **firmware**.

Obviously, given adequate inputs, a microprocessor unit can do tasks with and in an airplane that are limited only by the imagination, including fly the airplane. Development of these systems is well started but still very limited in scope compared to the potential. Two systems that are now in use but still developing very rapidly are EFIS and EICAS.

The EFIS (**Electronic Flight Instrument System**) displays all flight, navigation and weather radar information on two CRTs (sometimes three). The following system description is broken down for you into three parts: (1) input data sources, (2) displays and (3) user controls.

EFIS Inputs

1. **Air data system** (ADS) includes static pressure, pitot pressure, angle of attack data and raw (indicated) air temperature as inputs to the air data computer (ADC).

2. Pitch, roll, heading (and rates thereof) and acceleration data for all axes from the **attitude heading system** (AHS) or the **attitude and heading reference system** (AHRS) which utilize laser, piezoelectric or conventional gyros and accelerometers, or inputs from conventional vertical gyro (attitude gyro), compass and accelerometers. See figure 17-39(C).

3. Flight control system.

4. Weather radar system.

5. Electronic navigation systems

    —VOR/LOC/glideslope/marker beacon

    —MLS

    —DME

    —radio altimeter

    —RNAV (area navigation)

    —ADF

    —long range navigation systems

        —inertial navigation system (INS)

        —VLF/OMEGA

        —LORAN

        —global positioning system (GPS)

        —flight management system (FMS)

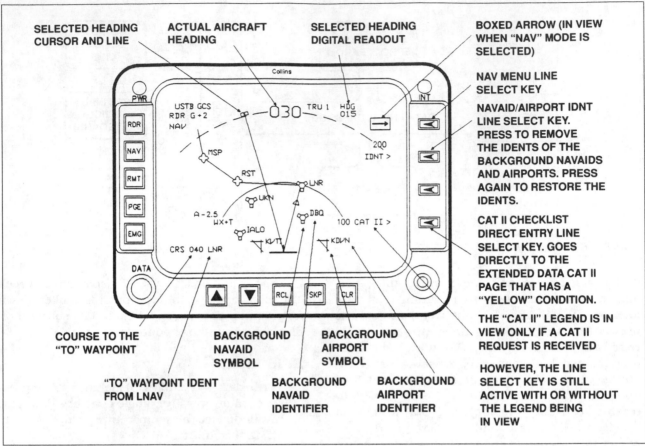

SELECTED HEADING CURSOR AND LINE

ACTUAL AIRCRAFT HEADING

SELECTED HEADING DIGITAL READOUT

BOXED ARROW (IN VIEW WHEN "NAV" MODE IS SELECTED)

NAV MENU LINE SELECT KEY

NAVAID/AIRPORT IDNT LINE SELECT KEY. PRESS TO REMOVE THE IDENTS OF THE BACKGROUND NAVAIDS AND AIRPORTS. PRESS AGAIN TO RESTORE THE IDENTS.

CAT II CHECKLIST DIRECT ENTRY LINE SELECT KEY. GOES DIRECTLY TO THE EXTENDED DATA CAT II PAGE THAT HAS A "YELLOW" CONDITION.

THE "CAT II" LEGEND IS IN VIEW ONLY IF A CAT II REQUEST IS RECEIVED

HOWEVER, THE LINE SELECT KEY IS STILL ACTIVE WITH OR WITHOUT THE LEGEND BEING IN VIEW

COURSE TO THE "TO" WAYPOINT

BACKGROUND NAVAID SYMBOL

BACKGROUND AIRPORT SYMBOL

"TO" WAYPOINT IDENT FROM LNAV

BACKGROUND NAVAID IDENTIFIER

BACKGROUND AIRPORT IDENTIFIER

Figure 17-60. MFD display, showing NAV mode, heading up format.

Courtesy Collins Avionics

EFIS Displays

The data displayed and its format styles are unlimited and presently, are only in the early stages of standardization. The system described is only one example, which is a five-tube display consisting of one **multifunction display** (MFD) and pilot's and co-pilot's **Electronic Attitude Director Indicator** (EADI) and **Electronic Horizontal Situation Indicator** (EHSI).

Multifunction Display

The MFD is a color CRT mounted in the instrument panel where the weather radar indicator would reside. Functions displayed on the MFD include but are not at all limited to:

—weather radar

—navigation maps, displayed heading up, north up-aircraft centered or north up-max view (see figure 17-60).

—page data, including checklists (see figures 17-61 and 17-62).

—EICAS data—anything else can be displayed here that is programmed and FAA

approved, including the sports channel if it could be approved!

The MFD is operated by the MPU (multifunction processor unit).

Electronic Attitude Director Indicator

The EADI is a color CRT that presents a display of aircraft attitude, flight control steering commands (flight director), VOR, LOC or other navigation system deviation, glideslope, marker beacon, radio altitude, decision height, indicated airspeed, reference (bug) airspeed, airspeed trend vector, comparator warnings, altitude alert and excessive deviations of any of the above parameters. Figure 17-63 shows some, but not all, functions of the EADI. Note the one instrument that has survived this high-tech transformation: the "ball" or inclinometer. It was with Lindbergh when he crossed the Atlantic!

Electronic Horizontal Situation Indicator

The EHSI is a color CRT that presents a plan view of the aircraft's horizontal navigation situation. Information displayed includes true or

```
WX + T                                    WX + T
PAGE 1 OF 6              CRUISE           PAGE 4 OF 6              CRUISE
                                         PRES POS        N 42°17.4'
TAS    450 KTS      GS 430                                W117°33.1'
WIND     44°/100
SAT     −15°C                             CID             N 42°12.2'
TAT     −29°C                                             W118°15.1'
ISA      −5°C
TIME TO WPT      00:32 ETE 12:34 ETA      SUX             N 38°29.1'
TIME TO DEST     01:13 ETE 13:13 ETA                      W132°29.5'
TO WPT ID        CIDARPT
   BRG/DIST      79°/147 NM               FSD             N 27°49.0'
CROSS TRACK      12.5 NM LEFT                             W133°36.3'
                              ↓                                        ↕
```

```
WX + T                                    WX + T
PAGE 2 OF 6              CRUISE           PAGE 5 OF 6              CRUISE
                                                 AIRDATA

MAG HDG       123°
TRUE HDG      120°                        ALTITUDES:      12500 FT BARO
MAG VAR        3° WEST                                    12300 FT PRESSURE
                                                          10000 FT PRESELECT
DESIRED TRACK    123° (TRUE)
ACTUAL TRACK     125.5° (TRUE)            SPEEDS:         250 KTS IAS
DRIFT ANGLE       2.5° LEFT                               450 KTS TAS
CROSS TRACK       12.5 NM LEFT                            390 KTS GS
                                                          0.783 MACH
                              ↕                           −650 FPM VS      ↕
```

```
WX + T                                    WX + T
PAGE 3 OF 6              CRUISE           PAGE 6 OF 6              CRUISE
                                                 VNAV

TIME TO WPT      01:32 ETE 13:32 ETA
TIME TO DEST     05:43 ETE 17:43 ETA      ACT VERT SPEED    −800 FPM
                                          SEL OFFSET        −16 NM
TO WPT ID        CIDARPT                  SEL VERT PATH     −3.1°
BRG/DIST         79°/147 NM               ACT VERT PATH     −2.8°
MESSAGE          WPT ALERT DR             PRESSURE ALT      10500 FT
                 XTRK WARN                AIMPOINT ALT      8000 FT
NEXT WPT ID      ALOVOR

                              ↕                                        ↑
```

Figure 17-61. Examples of page data displayable on the MFD.   *Courtesy Collins Avionics*

**PGE DATA FORMAT,
TYPICAL CHECKLIST EXAMPLE**

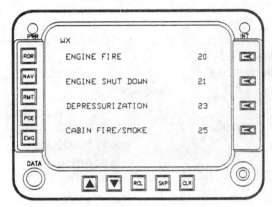

**EMG DATA FORMAT,
TYPICAL CHAPTER TITLES**

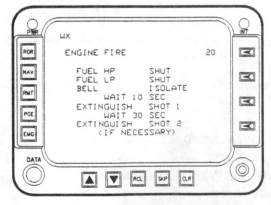

**EMERGENCY DATA FORMAT,
TYPICAL EMERGENCY
CHECKLIST EXAMPLE**

Figure 17-62. Typical checklist page data available on the MFD.

Courtesy Collins Avionics

magnetic heading, DG mode, selected heading, selected nav systems, nav system course and

deviation, TO/FROM/backcourse indications, distance (DME), glideslope, groundspeed, time to go, TAS, wind direction and speed, weather radar target alert, waypoint alert, and VOR driven bearing pointer (RMI). The EHSI can also be operated in map format with or without weather radar overwriting the map. Figure 17-64 shows a typical EHSI format.

EFIS System Controls

Along with the complex displays and myriad of presentable information is a complex set of controls which allow the pilots to select and deselect the information presented, display formats and which CRT to display the presentation. The controls include:

Display Control Panel (DCP) provides EHSI format selection, nav sensor selection, bearing pointer selection and second nav sensor course selection, display dimming, DH selection and radio altimeter test.

Course Heading Panel (CHP) controls course arrows, heading cursors, course function. Also, speeds and times displayed are selected with the CHP.

Display Processor Unit (DPU) provides sensor input processing and switching and power for the EFDs (Electronic Flight Displays).

Weather Radar Panel (WXP) provides the weather radar mode and range selection as well as display format and EFD selection.

EICAS

The **Engine/System Indication and Crew Alerting System** is a sophisticated electronic "glass cockpit" system that accumulates data from sensors all over the aircraft, processes it and displays system condition upon selection by the pilot. It also provides overriding displays of emergencies, classified by level of urgency and color coded to show level of concern, with emergency checklist items.

For example, the current Collins Pro-line EICAS uses two color CRTs that provide complete replacement of conventional engine and systems gauges. The system is smart, meaning that it will bring up displays of gauges that are appropriate to the mode of flight. For example, the "cruise display" for the RJ (Challenger Regional Jet) shows torquemeters, ITT gauges, oil pressure and temperature and advisories. Any other system information is pilot selectable on the second screen.

Other "glass cockpit" terminology:

MND—multi-navigation display

PFD—primary flight display

FMS—flight management system

IAPS—integrated avionics processor system. This is the traffic manager for all displays, sensors, engine and systems data. It records unit failures and intermittences for all systems and makes this information available for display.

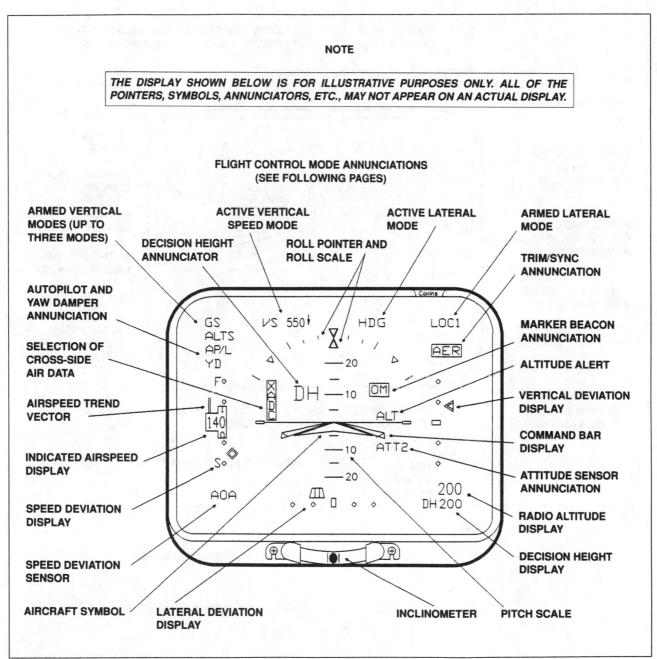

NOTE

THE DISPLAY SHOWN BELOW IS FOR ILLUSTRATIVE PURPOSES ONLY. ALL OF THE POINTERS, SYMBOLS, ANNUNCIATORS, ETC., MAY NOT APPEAR ON AN ACTUAL DISPLAY.

FLIGHT CONTROL MODE ANNUNCIATIONS
(SEE FOLLOWING PAGES)

ARMED VERTICAL MODES (UP TO THREE MODES)

ACTIVE VERTICAL SPEED MODE

ACTIVE LATERAL MODE

ARMED LATERAL MODE

DECISION HEIGHT ANNUNCIATOR

ROLL POINTER AND ROLL SCALE

TRIM/SYNC ANNUNCIATION

AUTOPILOT AND YAW DAMPER ANNUNCIATION

MARKER BEACON ANNUNCIATION

SELECTION OF CROSS-SIDE AIR DATA

ALTITUDE ALERT

AIRSPEED TREND VECTOR

VERTICAL DEVIATION DISPLAY

INDICATED AIRSPEED DISPLAY

COMMAND BAR DISPLAY

ATTITUDE SENSOR ANNUNCIATION

SPEED DEVIATION DISPLAY

RADIO ALTITUDE DISPLAY

SPEED DEVIATION SENSOR

DECISION HEIGHT DISPLAY

AIRCRAFT SYMBOL

LATERAL DEVIATION DISPLAY

INCLINOMETER

PITCH SCALE

Figure 17-63. Some EADI indications from the EFIS system.

Courtesy Collins Avionics

425

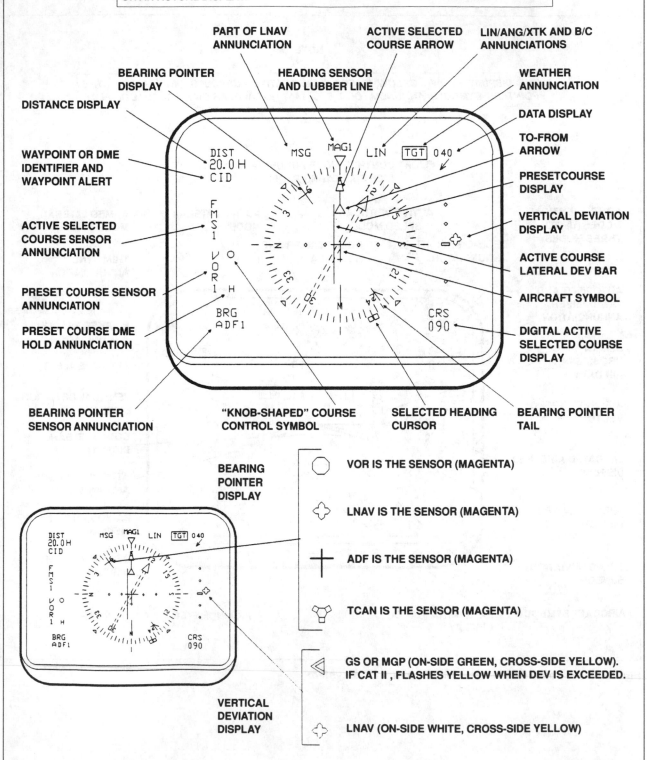

Figure 17-64. A typical EHSI display.

Courtesy Collins Avionics

Additional Reading

1. Kershner, Wm.; The Advanced Pilot's Flight Manual; Iowa State Univ. Press; 1985.

2. Aircraft Instrument Systems; EA-IAS; IAP, Inc., Publ.; 1985.

3. McKinley, J.L. and Bent, R.D.; Electricity and Electronics for Aerospace Vehicles; McGraw-Hill; 1961.

Study Questions And Problems

1. An airspeed indicator measures the difference between _____ and _____ to get _____, a pressure that is related to velocity.

2. What additional factor (information) does a true airspeed indicator need that is not available to the standard airspeed indicator?

3. Diagram the pitot-static system for the aircraft you fly, showing location of the pitot and static ports and ALL devices that are plumbed to this system.

4. What additional data input does a Machmeter need, as compared to the standard airspeed indicator and how does the Machmeter work?

5. What is the barometric pressure at the Flagstaff airport, elevation 7011 feet if the temperature is standard and the altimeter setting is 30.10" Hg? Show calculations and method used for solution.

6. If the temperature at KFLG (Question #5) is 102 degrees F, find pressure altitude and density altitude by calculation and prove your answers by computer. See Kershner, pages 75, 114 if you don't recall how to do this.

7. Why is there a delay (lag) in the indication from a VSI?

8. What would you expect the rate of gyro drift to be for an aircraft on the airport at International Falls, MN?

9. What are the three errors associated with the gyrocompass? Describe each.

10. What would be the gyro drift for the aircraft of Question #8 if it was flying over KINL at 150 knots eastbound? Westbound?

11. What instruments use precession as a driving force? Which ones use rigidity in space?

12. What are the purposes of the calibrated spring in the turn and slip indicator?

13. What is the difference in the turn and slip and turn coordinator instruments with respect to what motion they sense?

14. In the cockpit, how does the pilot tell a turn and slip from a turn coordinator instrument?

15. Explain the basics of how a capacitance-type fuel quantity indicating system works.

16. Diagram a microprocessor and explain the function of its parts.

17. What type of sensing devices are used by pressure sensing instruments?

18. Define and describe the function of these modern acronyms as they apply to present generation avionics: EFIS, EICAS, CPU, MFD, ADS, ADC, EHSI, FMS, EADI, PFD, IAPS.

Solutions To Study Problems

Chapter I

1. $\text{Power} = \dfrac{F \times d}{T} = \dfrac{2500 \text{ lbs.} \times 1000 \text{ ft}}{1 \text{ min}} = 2,500,000 \dfrac{\text{ft-lbs}}{\text{min}}$

$2.5 \times 10^6 \dfrac{\text{ft-lbs}}{\text{min}} \times \dfrac{1 \text{ HP}}{3.3 \times 10^4 \frac{\text{ft-lbs}}{\text{min}}} = 75.7 \text{ HP}$

$75.7 \text{ HP} \times 746 \dfrac{\text{watts}}{\text{HP}} = 5.65 \times 10^4 \text{ watts}$

2. Potential energy due to position:

$\text{P.E.} = 2500 \text{ lbs} \times 1000 \dfrac{\text{ft}}{\text{min}} \times 3 \text{ min} = 7.5 \times 10^6 \text{ ft-lb}$

11. $V = 32 \dfrac{\text{ft}}{\text{sec}^2} \times 5 \text{ sec} = 160 \dfrac{\text{ft}}{\text{sec}}$

$\underset{160 \frac{\text{ft}}{\text{sec}}}{\overset{(1.6 \times 10^2)}{}} \times \underset{3600 \frac{\text{sec}}{\text{hr}}}{\overset{(3.6 \times 10^3)}{}} \times \underset{(5.28 \times 10^3)}{\dfrac{1 \text{ mile}}{5280 \text{ ft}}} = 1.09 \times 10^2 \dfrac{\text{mi}}{\text{hr}}$

$\dfrac{1.6 \times 10^2 \text{ ft}}{1 \text{ sec}} \times \dfrac{3.6 \times 10^3 \text{ sec}}{1 \text{ hr}} \times \dfrac{1 \text{ NM}}{6.08 \times 10^3 \text{ ft}} = 94.7 \text{ KTS}$

$\dfrac{1.09 \times 10^2 \text{ mi}}{1 \text{ hr}} \times \dfrac{1 \text{ KM}}{.62 \text{ mi}} = 1.76 \times 10^2 \text{ KM/hr}$

NOTE: Conversion units are from figure 1-1.

Chapter II

3. See the discussion of power and weight on the second page of Chapter II.

5. $THP = \dfrac{T \times V \text{ mph}}{375 \text{ mi–lbs/hr}}$ Converting, $T = \dfrac{THP \times 375 \text{ mi–lbs/hr}}{V \text{ mph}}$

   Substituting, $T = \dfrac{100 \times 375}{100}$

   $T = 375$ lbs

6. From #3, $T = \dfrac{THP \times 325 \text{ mi–lbs/hr}}{V \text{ KTS}}$

   $= \dfrac{100 \times 325}{90}$

   $T = 361.1$ lbs

   $\therefore$ 375 lbs – 361.1 lbs = 13.9 lbs less thrust

Chapter III

4. Horsepower needed to climb =

$$P = \frac{Fd}{t} = \frac{3000 \text{ lbs} \times 1200 \text{ ft}}{1 \text{ min}}$$

$$= 3.6 \times 10^6 \text{ ft–lbs/min} \times \frac{1 \text{ hp}}{3.3 \times 10^4 \text{ ft–lbs/min}}$$

$$= 109 \text{ HP}$$

Horsepower needed to overcome drag:

Drag = 350 lbs = thrust (steady–state flight)

$$\text{THP} = \frac{T \times V}{325 \text{ NM–lbs/hr–HP}} = \frac{350 \text{ lbs} \times 120 \text{ KTS}}{325}$$

$$= 129 \text{ HP}$$

Total HP = 109 + 129 = 238 HP

Both of these power forms are derived from aircraft performance data. For propeller driven aircraft, thrust horsepower is the power form that causes the aircraft to perform.

5. $$T = \frac{325 \text{ THP}}{V \text{ KTS}} = \frac{325 \times 238}{120} = 644.6 \text{ lbs}$$

6. You can handle it, so go for it! Remember, speed is not involved in calculating power needed to climb.

7. Gauge pressure + ambient pressure = absolute pressure.
    Manifold absolute pressure *is* absolute pressure!

Alt.	Gauge Pressure	+	Ambient Pressure	=	Absolute Pressure
S.L.			29.92″		22″
6000 ft					22″

Chapter V

1. Consider station 36 as a point that scribes a circle as the prop turns, that has a radius of 36″. Each revolution, that point travels a distance equal to the circumference of the circle and in one minute, it makes 2400 circles. C = 2πr. See the index for a discussion of Mach number and assume density altitude is sea level, 59 °F ambient temperature. The Mach number is a little more than .67. Have fun!

2. Using figure 5-4, blade angle is angle B in the diagram. It can be found by adding together angle A, which is 4° and angle C which can be found by:

$$\tan \angle C = \frac{OPP}{ADJ} = \frac{\text{Forward Velocity}}{\text{Linear Velocity}}$$

   where forward velocity is 120 KTS and linear velocity was found in Question #1. The blade angle should be about 19°.

3. It is the same problem as #2, except forward velocity is less. The answer should be about 8°.

8. It is the same problem as #1, but this big prop is making a lot of noise because it's Mach number is getting quite close to the speed of sound. You can see why this prop on a Cessna 185 turning 2850 RPM is called the "Alaska Rooster" — it wakes everyone up during early morning takeoffs!

Chapter VII

2. Before engine start:

   $30.12'' - 7'' = 23.12''$ Hg MAP

   During full throttle takeoff, approximately

   $23'' - 2''$ (friction loss) $= 21''$ Hg MAP

3. Before engine start:

   $30.42'' + .21'' = 30.63''$ Hg MAP

   or, using Press. Alt:

   PA $= 30.42 - 29.92 = .5 = 500$ ft down, therefore $-211' + (-500') = -711'$
   Therefore, ambient pressure $= 30.42 - (-.711) = 31.13''$.

7. The altitude should equal the altitude of an airport where takeoff MAP $= 18''$!
   If the last altimeter setting received was $29.92''$, then

   $29.92'' - 2''$ (friction loss) $- 9.92'' = 18''$ MAP.

   It appears the altitude is 9.92×1000 ft, or about 10,000 ft. With knowledge, the MAP gauge becomes an emergency altimeter!

Chapter IX

1. 8000 ft cabin = (from figure 9.1)   10.91 PSI
   Differential =   −9.00 PSI
   Ambient pressure at altitude   1.91 PSI

   To find altitude of 1.91 PSI (figure 9.1) by linear interpolation:

   46,000 ft = 2.05   Estimate: 1.91 PSI corresponds to an altitude between 46,000 and
   48,000 ft = 1.86   48,000 and is closer to 48,000 ft. Estimate is 47,500.
              .19

   .19x = 100

   $\dfrac{.19}{2000} = \dfrac{.05}{x}$ x = 526   48,000 − 526 = 47,474 ft.

2. Flight profile:

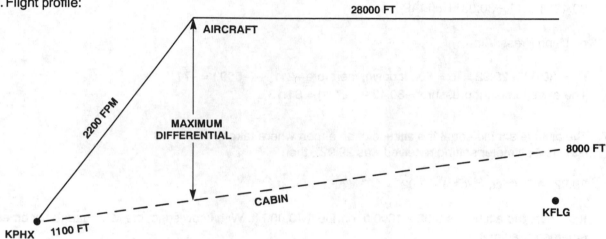

   Rate of ascent: $\dfrac{8000 - 1100}{40 \text{ min}} \cong 175$ FPM

3. Maximum differential occurs just as aircraft reaches cruising altitude.

   Time to climb to cruisng altitude: $\dfrac{28000 - 1100 \text{ ft}}{2200 \text{ FPM}} = 12.2$ min

   In 12.2 minutes, the cabin climbs from 1100 ft to:

   1100 ft + (12.2 mins × 175 FPM) = 3235 ft.

   ∴   PSID =

       Cabin: 3235 ft* = 12.69 PSI + $\left(\dfrac{765}{2000} \times .97\right)$ = 13.06 PSI
      Ambient: 28000 ft =   4.78 PSI
                                            8.28 PSID
                            8.83 if stated from SL

   *HINT: Construct an accurate graph of pressure vs. altitude from which you can extract these pressures
          to avoid lengthy interpolation.

4. The aircraft is climbing at 2,000 FPM and maintaining a 4.2 psid, the rate of pressure change, ambient and cabin is:

Pressure @ 14000' = 8.63 PSI
        16000' = 7.96 PSI
        .67 PSI/min when passing through 15,000 ft.

While the aircraft climbs from 14,000 to 16,000 ft in one minute, the cabin climbs:

14,000 ft** = 8.63 PSI + 4.2 PSI = 12.83 PSI = 3,680 ft*
16,000 ft   = 7.96 PSI + 4.2 PSI = 12.16 PSI = 5,450 ft*
 5,450 ft   − 3,680 ft = 1,770 ft in one minute = 1770 FPM

*From the graph constructed for Problem 3.
**NOTE: A slightly more accurate answer would be expected if altitudes equally on either side of 15,000
        were used.

5. Assume that the maximum desirable cabin rate of descent = 300 FPM.

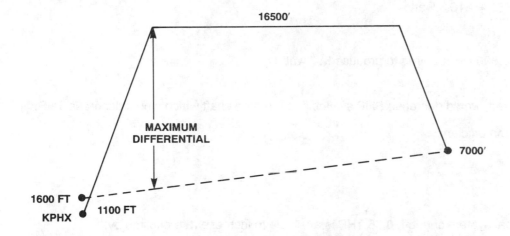

Time to descend aircraft = 16,500 ft − 1,600 ft = $\dfrac{15,400 \text{ ft}}{2,000 \text{ FPM}}$ = 7.5 mins

Cabin descends at: 7,000 − 1,600 ft = $\dfrac{5,400 \text{ ft}}{40 \text{ mins}}$ = 135 FPM

Maximum differential occurs at the time the aircraft descent is begun.

Cabin altitude at time descent begins:

1,600 ft + (135 FPM × 7.5 mins) = 2,612 ft.

PSID = 2,612 ft = 13.46 PSI*
        16,500 ft = 7.80 PSI
                     5.66 PSID

* From the graph constructed for Problem #3.

Chapter X

11. a) $P = IE$

$I = 22.5A + 18A = 40.5$ amps

$E = 14$ volts output from the alternator in a 12–volt system!

$P = 40.5 \times 14$

$= 567$ watts output

b) $EFF = .92 = \dfrac{567 \text{ watts output}}{X \text{ watts input}}$

$x = \dfrac{567}{.92} = 616.3$ watts

c) $HP = \dfrac{616.3}{746} = .826$ HP required to produce 567 watts

This power, if used, would decrease BHP available to the prop shaft which would decrease THP:

$THP = BHP \times$ prop efficiency

$= .826 \times .8$

$= .66$ THP

With the electrical system shut off, 0.66 THP is available to increase rate of climb by:

$ROC = \dfrac{ETHP \times 33,000}{\text{weight}}$

$= \dfrac{.66 \times 33,000}{2,500}$

$= 8.71$ FPM

Chapter XI

5. Total load $= 4a + 1.5a + 1a + 5a = 11.5$ amps

Battery will last:

25 amp hrs $\times$ (50% rule) = 12.5 amps

$\dfrac{12.5 \text{ a–h}}{11.5a} = 1.087$ hours $= 1$ hr 05 min

Estimated failure time: 00:00Z + 1 hr 05 min = 01:05Z

Chapter XII

2. $P = IE = 2.5a \times 14v = 350$ watts

    $\text{\% Efficiency} = \dfrac{\text{Power out}}{\text{Power in}} \times 100$

    $74 = \dfrac{350 \text{ watts}}{X} \times 100$

    $74X = 35{,}000$ watts

    $X = \dfrac{35{,}000 \text{ watts}}{74} = 473$ watts

    $473 \text{ watts} \times \dfrac{1 \text{ HP}}{746 \text{ watts}} = .63 \text{ HP}$

3. THP = BHP × prop. efficiency

    = .63 HP × .82

    = .516 THP (Decrease in available or excess THP)

    $\text{Decrease in ROC} = \dfrac{\text{ETHP} \times 33{,}000}{\text{WT}} = \dfrac{.516 \times 33{,}000}{2{,}500 \text{ lbs}} = 6.82 \text{ FPM}$

    (Not much of a change in performance unless, on a real short field takeoff, your aircraft hits the trees four feet from the top!)

Chapter XIII

3. High-side pressure is 1,500 PSI, low side is 100 PSI, so 1,400 PSI exists across the actuator piston. From the fluid power discussion:

    $HP = GPM \times PSI \times 5.823 \times 10^{-4}$

    Converting from figure 1-1:

    $2{,}000 \text{ cc} \times \dfrac{1 \text{ liter}}{1{,}000 \text{ cc}} \times \dfrac{.26417 \text{ gal}}{1 \text{ liter}} = \dfrac{.52834 \text{ gals}}{15 \text{ sec}} \times \dfrac{60 \text{ sec}}{1 \text{ min}} = 2.11 \text{ GPM}$

    $\text{Power} = 2.11 \text{ GPM} \times 14 \times 10^{2} \text{ PSI} \times 5.83 \times 10^{-4}$

    $= 172 \times 10^{-2} = 1.72 \text{ HP}$

Chapter XVI

3. Standard passenger weight = 165 lbs (from figure 16-18)

Weight × Arm = Moment

165 lbs × (160″ − 70″) = 14,850 in-lbs

5. Forward adverse loading check:

Forward CG Limit = $\dfrac{180,000 \text{ lb–in}}{2,800 \text{ lbs}}$ = 38.57 in

Aft CG Limit = $\dfrac{131,500 \text{ lb–in}}{2,800 \text{ lb}}$ = 46.96 in

*Fuel Required:

METO Power: 75% of 230 HP = 172.5 HP

Fuel = $\dfrac{172.5}{12}$ = 14.4 gals × 6 $\dfrac{\text{lb}}{\text{gal}}$ = 86.25 lbs

Or: $\dfrac{172.5 \text{ HP}}{2}$ = 86.25 lbs

(See METO power discussion)

A. Items ahead of forward limit (maximize):

Oil:	22 lbs	× 15″	=	330 lb-in
Pilot and Pax:	340 lbs	× 36″	=	12,240 lb-in

B. Items aft of forward limit (minimize):

Fuel:	86.25 lbs*	× 48″	=	4140 lb-in
C. Aircraft:	1704 lbs	× 36.65″	=	62,452 lb-in
Totals	2152.25 lbs			79,162 lb-in

CG (ARM) = $\dfrac{\text{Sum of moments}}{\text{Total weight}}$ = $\dfrac{79,162 \text{ lb–in}}{2,152.25 \text{ lbs}}$ = 36.78 in

Read carefully the aft adverse loading check discussion and do it. When I did it, I came up with the CG at 48.77 inches. Be sure not to put any more weight in the front two seats than you have to, and be sure you use the correct baggage arm for loading configuration V. Be sure you write a conclusion statement for this aircraft's adverse loading check, to complete the check professionally.

9. The loaded aircraft moment with all strapped in is:

Weight × Arm = Moment

15,000 × 92.3″ = 1,384,500 in–lbs

The passenger making the potty-call adds 14,850 in-lbs (#3):

1,384,500 in-lbs + 14,850 in-lbs = 1,399,350 in-lbs

With no change in weight, the new CG is:

$$\frac{1,399,350 \text{ lb–in}}{15,000 \text{ lbs}} = 93.29 \text{ in.}$$

NOTE: Use the formula in figure 16-26 to solve for proof.

Chapter XVII

5. The altimeter setting of 30.10″ Hg at KFLG is the barometric pressure corrected to sea level, 7011 feet under the airport, so the decrease in pressure caused by ascending 7011 feet must be considered. Approximately, the pressure decreases one inch per 1000 feet increase in altitude. Therefore:
Barometric pressure at KFLG = 30.10″ Hg – 7.01″ Hg = 23.09″ Hg.

6. Pressure altitude at KFLG:

Kollsman Window	Indicated Altitude
30.10	7011 ft
29.92	7011 – 180 = 6831 ft = Pressure Altitude

Density Altitude = P. Alt. corrected for NST

NST = Ambient temperature – Standard temperature at 6831 ft

$$ST = 59\ °F - (\frac{3.5\ °F}{1000\ ft} \times 6.831\ (1000\ ft))$$

$$ST = 35.1\ °F$$

$$NST = 102\ °F - 35.1\ °F = 66.9\ °F$$

$$Correction\ for\ NST = \frac{66.9\ °F}{15\ °F/1000\ ft} = 4.46\ thousand\ feet\ to\ correct\ P.A.$$

KFLG is warmer so D.A. is higher than P.A.

D.A. = P.A. corrected for NST

D.A. = 6831 ft + 4460 ft = 11,291 ft.

8. Gyrodrift for a stationary aircraft = 15.04 Sin latitude

$$= 15.04\ Sin\ 48°34'$$

$$= 15.04 \times .75$$

$$= 11.28\ degrees/hour$$

Approximate proof: see figure 17-30.

10. Speed (with respect to space) of a stationary aircraft at KINL

= 900 × Cos 48.6°

= 900 × .662

= 595.8 KTS

Therefore, if the aircraft flies west at 595.8 KTS, it will appear (from space) to be standing still and gyro drift error will be zero. Flying east at 595.8 KTS, the aircraft will appear (from space) to be moving twice as fast as when it was stationary on the KINL airport with a gyro error of 11.28 degrees per hour. Eastbound at 595.8 KTS, the gyro error would be 11.28 × 2 = 22.56 degrees/hour. So at latitude 48°34′:

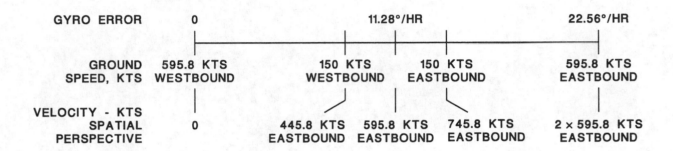

Therefore, at 150 KTS westbound,

$$\text{Gyro drift error} = \frac{445.8 \text{ KTS}}{595.8 \text{ KTS}} \times 11.28°/hr = 8.44°/hr$$

and, at 150 KTS eastbound,

$$\text{Gyro drift error} = \frac{745.8 \text{ KTS}}{595.8 \text{ KTS}} \times 11.28°/hr = 14.12°/hr$$

An approximate proof again comes from figure 17-30.

Index

447

449